SOCIAL
AND
ECONOMIC
NETWORKS

SOCIAL
AND
ECONOMIC
NETWORKS

Matthew O. Jackson

PRINCETON UNIVERSITY PRESS

Princeton and Oxford

ISBN-13: 978-0-691-13440-6

Library of Congress Control Number: 2008926162

British Library Cataloging-in-Publication Data is available

This book was composed in Times Roman and Slimbach using ZzTEX
by Princeton Editorial Associates, Inc., Scottsdale, Arizona.

Printed on acid-free paper. ∞

www.press.princeton.edu

Printed in the United States of America

1 3 5 7 9 10 8 6 4 2

To Sara, Emily, and Lisa

Contents

Preface

This book provides an overview and synthesis of models and techniques for analyzing social and economic networks. It is meant to serve as both a resource for researchers and a text on the subject for graduate students. The focus is primarily on the modeling of and theory behind the structure, formation, and implications of social networks. Statistical and experimental analyses of networks are also discussed throughout, especially when they help set the stage for issues to be investigated. The main emphasis is on providing a foundation for analyzing and understanding social and economic networks.

The book is organized into four parts. The first part introduces network analysis and provides some background on various networks, how they are measured, and useful ways of representing them. The second part presents some of the models that have been used to understand how networks are formed. This part draws on two very different perspectives: random-graph models, for which there is some stochastic process that governs the development of the links in a network, and strategic models of network formation, in which the development of links is based on costs and benefits, and game-theoretic techniques are used. These approaches to modeling provide different insights into networks, how they are formed, and why they exhibit certain characteristics. The third part of the book looks at the implications of network structure. Much of the interest in networks is due to the fact that their structure is an important determinant of how societies and economies function. This part examines how network models are used to predict the spread of disease, the dissemination of information, the behavior of people, and how markets function. The final part of the book covers empirical analyses of networks and methods of identifying social interaction.

Although the table of contents represents a categorization of the chapters by subject, the relationship between the chapters is not entirely linear. I have intermingled some subjects to tie together different approaches, and the chapters are cross referenced. There is also a progression in the book, with some of the more technically demanding chapters coming later, as well as those that draw on concepts from earlier chapters.

The modeling of networks requires some mathematical background, but I have made the book as self-contained as possible. I do not presume any knowledge beyond some familiarity with linear algebra (vectors and matrices), calculus, and probability and statistics. The discussions employing graph theory and game theory are self-contained, and there are appendixes with introductions to some of the topics, including various useful results from graph theory, math, game theory, and probability theory. Those sections and exercises that are more mathematical in nature are marked with an asterisk (*).

I had several reasons for writing this book. First and foremost, networks of relationships play central roles in a wide variety of social, economic, and political interactions. For example, many, if not most, markets do not function as centralized and anonymous institutions, but rather involve a variety of bilateral exchanges or contracts. As a case in point, most jobs are filled by people who were informed about the job through a social contact. This fact has consequences for patterns of employment, inequality in wages across groups, and social mobility. Understanding social network structure and how it influences human interaction is not merely important to science (and the social sciences in particular), it is essential. Second, the topic is timely. Recent technological advances have made information networks much more prominent (e.g., the world wide web), and people are more conscious of the role of networks in their lives. In addition, the formal modeling of networks has now reached a maturity across fields that permits a book-length treatment. A third motivation for writing this work is the inter- and multidisciplinary nature of research on networks: the pertinent knowledge is diffuse, and there is much to be gained by collecting aspects of it from different fields in a unified treatment. Substantial research on networks has been conducted in sociology, economics, physics, mathematics, and computer science, and these disciplines take different approaches and ask varied questions.[1] Thus it is important to bridge the literatures and produce a text that collects and synthesizes different modeling approaches and techniques and makes them all available to researchers from any discipline who are interested in the study of networks.

At the end of each chapter in this book you will find exercises. The exercises are meant to serve several purposes. They serve the usual purpose of problems in a textbook: to give students a chance to work with concepts and more fully familiarize themselves with the ideas presented in the chapter; thus the exercises can be made part of courses. And the interested researcher can work the exercises as well. But the exercises also introduce *new* material. I have used them to introduce new concepts not covered in the text. These are meant to be closely related to material in the text but complementary to it. Often the ideas were important enough to include in the book, but did not fit easily with the main thread of a chapter without making it longer than I desired or taking us off on a tangent. Thus researchers consulting this book as a reference should not ignore the exercises and in many instances may actually find what they are looking for in them.[2]

---

1. Social network analysis is a central and well-developed area of study in sociology, with societies, journals, conferences, and decades of research devoted to it. With occasional overlap, a literature on graph theory has matured in mathematics during the same period. While the literature on networks has been thriving in sociology for more than five decades, it has emerged in economics primarily over the past 10 to 15 years. Its explosion in computer science and statistical physics has been rapid and has taken place mostly during the past decade.

2. This book is obviously not the first to do this, as one sees important results appearing as exercises in many mathematical texts. The usefulness of this technique was made obvious to me through the superb text on axiomatic social choice by Hervé Moulin [492].

As with any such undertaking, there are many acknowledgments due, and they do not adequately represent the scope and depth of the help received. This project would not have been possible without financial support from the Center for Advanced Studies in the Behavioral Sciences, the Guggenheim Foundation, and the Lee Center for Advanced Networking, as well as the National Science Foundation under grants SES-0316493 and SES-0647867. I began this project while I was at the California Institute of Technology and concluded it while at Stanford University, and the support of both institutions is gratefully acknowledged.

As for the content of this monograph, I have been deeply influenced by a number of collaborators. My initial interest in this subject arose through conversations and subsequent research with Asher Wolinsky. I have continued to learn about networks and had enjoyed interaction with a group of coauthors (in chronological order): Alison Watts, Bhaskar Dutta, Anne van den Nouweland, Toni Calvó-Armengol, Francis Bloch, Gary Charness, Alan Kirman, Jernej Copic, Brian Rogers, Dunia Lopez-Pintado, Leeat Yariv, Andrea Galeotti, Sanjeev Goyal, Fernando Vega-Redondo, Ben Golub, Sergio Currarini, and Paolo Pin. Their collaboration and friendship are greatly appreciated.

Although not coauthors on network-related projects, Salvador Barbera, Darrell Duffie, and Hugo Sonnenschein have been great mentors (and friends) and have profoundly shaped my approach and writing. I thank Lada Adamic, Marika Cabral, Toni Calvó-Armengol, Jon Eguia, Marcel Fafchamps, Ben Golub, Carlos Lever, Laurent Mathevet, Tomas Rodriguez Barraquer, Tim Sullivan, and Cyd Westmoreland for extensive comments on earlier drafts.

For all the emotional support and enthusiasm needed to keep such a project afloat, I owe profound thanks to my wife, Sara, my daughters, Emily and Lisa, and my parents. Special thanks are due to Sara, whose encouragement through the persistent asking of the question "Did you get to work on your book today?" always kept me pointed in the right direction, and whose juggling of many tasks allowed me to answer "yes" more often than "no."

PART I

BACKGROUND AND FUNDAMENTALS OF NETWORK ANALYSIS

CHAPTER 1

Introduction

This chapter introduces the analysis of networks by presenting several examples of research. These examples provide some idea not only of why the subject is interesting but also of the range of networks studied, approaches taken, and methods used.

1.1 ▪ Why Model Networks?

Social networks permeate our social and economic lives. They play a central role in the transmission of information about job opportunities and are critical to the trade of many goods and services. They are the basis for the provision of mutual insurance in developing countries. Social networks are also important in determining how diseases spread, which products we buy, which languages we speak, how we vote, as well as whether we become criminals, how much education we obtain, and our likelihood of succeeding professionally. The countless ways in which network structures affect our well-being make it critical to understand (1) how social network structures affect behavior and (2) which network structures are likely to emerge in a society. The purpose of this monograph is to provide a framework for an analysis of social networks, with an eye on these two questions.

As the modeling of networks comes from varied fields and employs a variety of different techniques, before jumping into formal definitions and models, it is useful to start with a few examples that help give some impression of what social networks are and how they have been modeled. The following examples illustrate widely differing perspectives, issues, and approaches, previewing some of the breadth of the range of topics to follow.

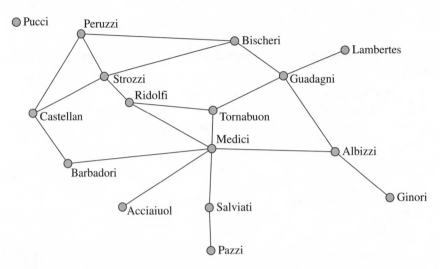

FIGURE 1.1   Network showing fifteenth-century Florentine marriages. Data from Padgett and Ansell [516] (drawn using UCINET).

1.2 ▪ A Set of Examples

1.2.1 Florentine Marriages

The first example is a detailed look at the role of social networks in the rise of the Medici in Florence during the 1400s. The Medici have been called the "godfathers of the Renaissance." Their accumulation of power in the early fifteenth century in Florence was orchestrated by Cosimo de' Medici even though his family started with less wealth and political clout than other families in the oligarchy that ruled Florence at the time. Cosimo consolidated political and economic power by leveraging the central position of the Medici in networks of family inter-marriages, economic relationships, and political patronage. His understanding of and fortuitous position in these social networks enabled him to build and control an early forerunner to a political party, while other important families of the time floundered in response.

Padgett and Ansell [516] provide powerful evidence for this consolidation by documenting the network of marriages between some key families in Florence in the 1430s. Figure 1.1 shows the links between the key families in Florence at that time, where a link represents a marriage between members of two families.[1]

---

1. These data were were originally collected by Kent [387], but were first coded by Padgett and Ansell [516], who discuss the network relationships in more detail. The analysis provided here is just a teaser that offers a glimpse of the importance of the network structure. The interested reader should consult Padgett and Ansell [516] for a much richer analysis.

During this time the Medici (with Cosimo de' Medici playing the key role) rose in power and largely consolidated control of business and politics in Florence. Previously Florence had been ruled by an oligarchy of elite families. If one examines wealth and political clout, however, the Medici did not stand out at this time and so one has to look at the structure of social relationships to understand why the Medici rose in power. For instance, the Strozzi had both greater wealth and more seats in the local legislature, and yet the Medici rose to eclipse them. The key to understanding the family's rise, as Padgett and Ansell [516] detail, can be seen in the network structure.

If we do a rough calculation of importance in the network, simply by counting how many families a given family is linked to through marriages, then the Medici do come out on top. However, they only edge out the next highest families, the Strozzi and the Guadagni, by a ratio of 3 to 2. Although suggestive, it is not so dramatic as to be telling. We need to look a bit closer at the network structure to get a better handle on a key to the success of the Medici. In particular, the following measure of betweenness is illuminating.

Let $P(ij)$ denote the number of shortest paths connecting family i to family j.[2] Let $P_k(ij)$ denote the number of these paths that include family k. For instance, in Figure 1.1 the shortest path between the Barbadori and Guadagni has three links in it. There are two such paths: Barbadori-Medici-Albizzi-Guadagni and Barbadori-Medici-Tornabuon-Guadagni. If we set $i =$ Barbadori and $j =$ Guadagni, then $P(ij) = 2$. As the Medici lie on both paths, $P_k(ij) = 2$ when we set $k =$ Medici, and $i =$ Barbadori and $j =$ Guadagni. In contrast this number is 0 if $k =$ Strozzi, and is 1 if $k =$ Albizzi. Thus, in a sense, the Medici are the key family in connecting the Barbadori to the Guadagni.

To gain intuition about how central a family is, look at an average of this betweenness calculation. We can ask, for each pair of other families, on what fraction of the total number of shortest paths between the two the given family lies. This number is 1 for the fraction of the shortest paths the Medici lie on between the Barbadori and Guadagni, and 1/2 for the corresponding fraction that the Albizzi lie on. Averaging across all pairs of other families gives a betweenness or power measure (due to Freeman [255]) for a given family. In particular, we can calculate

$$\sum_{ij:i \neq j, k \notin \{i,j\}} \frac{P_k(ij)/P(ij)}{(n-1)(n-2)/2} \tag{1.1}$$

for each family k, where $\frac{P_k(ij)}{P(ij)} = 0$ if there are no paths connecting i and j, and the denominator captures that a given family could lie on paths between as many as $(n-1)(n-2)/2$ pairs of other families. This measure of betweenness for the Medici is .522. Thus if we look at all the shortest paths between various families (other than the Medici) in this network, the Medici lie on more than half of them! In contrast, a similar calculation for the Strozzi yields .103, or just over 10 percent. The second-highest family in terms of betweenness after the Medici is

---

2. Formal definitions of path and some other terms used in this chapter appear in Chapter 2. The ideas should generally be clear, but the unsure reader can skip forward if helpful. Paths represent the obvious thing: a series of links connecting one node to another.

the Guadagni with a betweenness of .255. To the extent that marriage relationships were keys to communicating information, brokering business deals, and reaching political decisions, the Medici were much better positioned than other families, at least according to this notion of betweenness.[3] While aided by circumstance (for instance, fiscal problems resulting from wars), it was the Medici and not some other family that ended up consolidating power. As Padgett and Ansell [516, p. 1259] put it, "Medician political control was produced by network disjunctures within the elite, which the Medici alone spanned."

This analysis shows that network structure can provide important insights beyond those found in other political and economic characteristics. The example also illustrates that the network structure is important for more than a simple count of how many social ties each member has and suggests that different measures of betweenness or centrality will capture different aspects of network structure.

This example also suggests other questions that are addressed throughout this book. For instance, was it simply by chance that the Medici came to have such a special position in the network, or was it by choice and careful planning? As Padgett and Ansell [516, footnote 13] state, "The modern reader may need reminding that all of the elite marriages recorded here were arranged by patriarchs (or their equivalents) in the two families. Intra-elite marriages were conceived of partially in political alliance terms." With this perspective in mind we then might ask why other families did not form more ties or try to circumvent the central position of the Medici. We could also ask whether the resulting network was optimal from a variety of perspectives: from the Medici's perspective, from the oligarchs' perspective, and from the perspective of the functioning of local politics and the economy of fifteenth-century Florence. We can begin to answer these types of questions through explicit models of the costs and benefits of networks, as well as models of how networks form.

1.2.2 Friendships and Romances among High School Students

The next example comes from the the the National Longitudinal Adolescent Health Data Set, known as "Add Health."[4] These data provide detailed social network information for more than 90,000 students from U.S. high schools interviewed

---

3. The calculations here are conducted on a subset of key families (a data set from Wasserman and Faust [650]), rather than the entire data set, which consists of hundreds of families. As such, the numbers differ slightly from those reported in footnote 31 of Padgett and Ansell [516]. Padgett and Ansell also find similar results for centrality between the Medici and other families in terms of a network of business ties.

4. Add Health is a program project designed by J. Richard Udry, Peter S. Bearman, and Kathleen Mullan Harris and funded by grant P01-HD31921 from the National Institute of Child Health and Human Development, with cooperative funding from 17 other agencies. Special acknowledgment is due Ronald R. Rindfuss and Barbara Entwisle for assistance in the original design. Persons interested in obtaining data files from Add Health should contact Add Health, Carolina Population Center, 123 West Franklin Street, Chapel Hill, NC 27516-2524 (addhealth@unc.edu). The network data that I present in this example were extracted by James Moody from the Add Health data set.

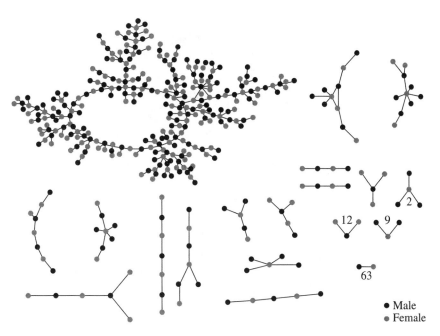

FIGURE 1.2   A network based on the Add Health data set. A link denotes a romantic relationship, and the numbers by some components indicate how many such components appear. Figure from Bearman, Moody, and Stovel [51].

during the mid-1990s, together with various data on the students' socioeconomic background, behaviors, and opinions. The data provide insights and illustrate some features of networks that are discussed in more detail in the coming chapters.

Figure 1.2 shows a network of romantic relationships as found through surveys of students in one of the high schools in the study. The students were asked to list the romantic liaisons that they had during the six months previous to the survey.

The network shown in Figure 1.2. is nearly a *bipartite* network, meaning that the nodes can be divide into two groups, male and female, so that links only lie between groups (with a few exceptions). Despite its nearly bipartite nature, the distribution of the degrees of the nodes (number of links each node has) turns out to closely match a network in which links are formed uniformly at random (for details, see Section 3.2.3), and we see a number of features of large random networks. For example, there is a "giant component," in which more than 100 of the students are connected by sequences of links in the network. The next largest component (the maximal set of students who are each linked to one another by sequences of links) only has 10 students in it. This component structure has important implications for the diffusion of disease, information, and behaviors, as discussed in detail in Chapters 7, 8, and 9, respectively.

In addition, note that the network is quite treelike: there are few loops or cycles in it. There are only a very large cycle in the giant component and a couple of smaller ones. The absence of many cycles means that as one walks along the links of the network until hitting a dead-end, most of the nodes that are met are new ones

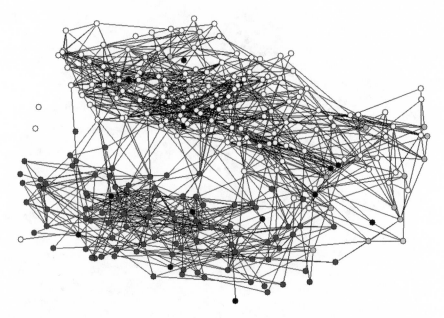

FIGURE 1.3   Add Health data set friendships among high school students coded by race. Hispanic, black nodes; white, white nodes; black, gray nodes; Asian and other, light gray nodes.

that have not been encountered before. This feature is important in the navigation of networks. It is found in many random networks in cases for which there are enough links to form a giant component but so few that the network is not fully connected. This treelike structure contrasts with the denser friendship network pictured in Figure 1.3, in which there are many cycles and a shorter distance between nodes.

The network pictured in Figure 1.3 is also from the Add Health data set and connects a population of high school students.[5] The nodes are coded by their race rather than sex, and the relationships are friendships rather than romantic relationships. This network is much denser than the romance network.

A strong feature present in Figure 1.3 is what is known as *homophily,* a term from Lazarsfeld and Merton [425]. That is, there is a bias in friendships toward similar individuals; in this case the homophily concerns the race of the individuals. This bias is greater than what one would expect from the makeup of the population. In this school, 52 percent of the students are white and yet 85 percent of white students' friendships are with other whites. Similarly, 38 percent of the students are black, and yet 85 percent of these students' friendships are with other blacks. Hispanics are more integrated in this school, comprising 5 percent of the population but having only 2 percent of their friendships with other Hispanics.[6] If friendships

5. A link indicates that at least one of the two students named the other as a friend in the survey. Not all friendships were reported by both students. For more detailed discussion of these particular data see Currarini, Jackson, and Pin [182].

6. The hispanics in this school are exceptional compared to what is generally observed in the larger data set of 84 high schools. Most racial groups (including hispanics) tend to have a greater

were formed without race being a factor, then whites would have roughly 52 percent of their friendships with other whites rather than 85 percent.[7] This bias is referred to as "inbreeding homophily" and has strong consequences. As indicated by the figure, the students are somewhat segregated by race, which affects the spread of information, learning, and the speed with which things propagate through the network—themes that are explored in detail in what follows.

1.2.3 Random Graphs and Networks

The examples of Florentine marriages and high school friendships suggest the need for models of how and why networks form as they do. The last two examples in this chapter illustrate two complementary approaches to modeling network formation.

The next example of network analysis comes from the graph-theoretic branch of mathematics and has recently been extended in various directions by the literature in computer science, statistical physics, and economics (as will be examined in some of the following chapters). This model is perhaps the most basic one of network formation imaginable: it simply supposes that a completely random process is responsible for the formation of links in a network. The properties of such random networks provide some insight into the properties of social and economic networks. Some of the properties that have been extensively studied are how the distribution of links across different nodes, the connectedness of the network in terms of the presence of paths from one node to another, the average and maximal path lengths, and the number of isolated nodes present. Such random networks serve as a useful benchmark against which we can contrast observed networks; comparisons help identify which elements of social structure are not the result of mere randomness but must be traced to other factors.

Erdös and Rényi [227]–[229] provided seminal studies of purely random networks.[8] To describe one of the key models, fix a set of n nodes. Each link is formed with a given probability p, and the formation is independent across links.[9] Let us examine this model in some detail, as it has an intuitive structure and has been a springboard for many recent models.

---

percentage of own-race friendships than the percentage of their race in the population, regardless of their fraction of the population. See Currarini, Jackson, and Pin [182] for details.

7. There are a variety of possible reasons for the patterns observed, as race may correlate with other factors that affect friendship opportunities. For more discussion with respect to these data see Moody [482] and Currarini, Jackson, and Pin [182]. The main point here is that the resulting network has clear patterns and those patterns have consequences.

8. See also Solomonoff and Rapoport [611] and Rapoport [551]–[553] for related predecessors.

9. Two closely related models that Erdös and Rényi explored are as follows. In the first alternative model, a precise number M of links is formed out of the $n(n-1)/2$ possible links. Each different graph with M links has an equal probability of being selected. In the second alternative, the set of all possible networks on the n nodes is considered and one is picked uniformly at random. This choice can also be made according to some other probability distribution. While these models are clearly different, they turn out to have many properties in common. Note that the last model nests the model with random links and the one with a fixed number of links (and any other random graph model on a fixed set of nodes) if one chooses the right probability distributions over all networks.

Consider a set of nodes $N = \{1, \ldots, n\}$, and let a link between any two nodes, i and j, be formed with probability p, where $0 < p < 1$. The formation of links is independent. This is a binomial model of link formation, which gives rise to a manageable set of calculations regarding the resulting network structure.[10] For instance, if $n = 3$, then a complete network forms with probability p^3, any given network with two links (there are three such networks) forms with probability $p^2(1 - p)$, any given network with one link forms with probability $p(1 - p)^2$, and the empty network that has no links forms with probability $(1 - p)^3$. More generally, any given network that has m links on n nodes has a probability of

$$p^m(1 - p)^{\frac{n(n-1)}{2} - m} \tag{1.2}$$

of forming under this process.[11]

We can calculate some statistics that describe the network. For instance, we can find the degree distribution fairly easily. The *degree* of a node is the number of links that the node has. The degree distribution of a random network describes the probability that any given node will have a degree of d.[12] The probability that any given node i has exactly d links is

$$\binom{n-1}{d} p^d(1 - p)^{n-1-d}. \tag{1.3}$$

Note that even though links are formed independently, there is some correlation in the degrees of various nodes, which affects the distribution of nodes that have a given degree. For instance, if $n = 2$, then it must be that both nodes have the same degree: the network either consists of two nodes of degree 0 or two of degree 1. As n becomes large, however, the correlation of degree between any two nodes vanishes, as the possibility of a link between them is only 1 out of the $n - 1$ that each might have. Thus, as n becomes large, the fraction of nodes that have d links will approach (1.3). For large n and small p, this binomial expression is approximated by a Poisson distribution, so that the fraction of nodes that have d links is approximately[13]

---

10. See Section 4.5.4 for more background on the binomial distribution.

11. Note that there is a distinction between the probability of some specific network forming and some network architecture forming. With four nodes the chance that a network forms with a link between nodes 1 and 2 and one between nodes 2 and 3 is $p^2(1 - p)^4$. However, the chance that a network forms that contains two links involving three nodes is $12p^2(1 - p)^4$, as there are 12 different networks we could draw with this shape. The difference between these counts is whether we pay attention to the labels of the nodes in various positions.

12. The degree distribution of a network is often given for an observed network, and thus is a frequency distribution. When dealing with a random network, we can talk about the degree distribution before the network has actually formed, and so we refer to probabilities of nodes having given degrees, rather than observed frequencies of nodes with given degrees.

13. To see this, note that for large n and small p, $(1 - p)^{n-1-d}$ is roughly $(1 - p)^{n-1}$. Write $(1 - p)^{n-1} = (1 - \frac{(n-1)p}{n-1})^{n-1}$, which, if $(n - 1)p$ is either constant or shrinking (if we allow p to vary with n), is approximately $e^{-(n-1)p}$. Then for fixed d, large n, and small p, $\binom{n-1}{d}$ is roughly $\frac{(n-1)^d}{d!}$.

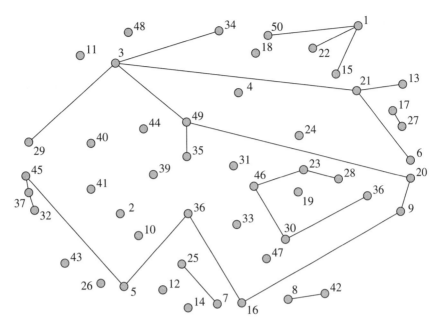

FIGURE 1.4 A randomly generated network with probability .02 on each link.

$$\frac{e^{-(n-1)p}((n-1)p)^d}{d!}.\tag{1.4}$$

Given the approximation of the degree distribution by a Poisson distribution, the class of random graphs for which each link is formed independently with equal probability is often referred to as the class of *Poisson random networks*. I use this terminology in what follows.

To provide a better feel for the structure of such networks, consider a couple of Poisson random networks for different values of p. Set $n = 50$ nodes, as this number produces a network that is easy to visualize. Let us start with an expected degree of 1 for each node, which is equivalent to setting p at roughly .02. Figure 1.4 pictures a network generated with these parameters.[14] This network exhibits a number of features that are common to this range of p and n. First, we should expect some isolated nodes. Based on the approximation of a Poisson distribution (1.4) with $n = 50$ and $p = .02$, we should expect about 37.5 percent of the nodes to be isolated (i.e., have $d = 0$), which is roughly 18 or 19 nodes. There happen to be 19 isolated nodes in the network.

Figure 1.5 compares the realized frequency distribution of degrees with the Poisson approximation. The distributions match fairly closely. The network also has some other features in common with other random networks with p and n in

---

14. The networks in Figures 1.4 and 1.6 were generated and drawn using the random network generator in UCINET (Borgatti, Everett, and Freeman [96]). The nodes are arranged to make the links as easy as possible to distinguish.

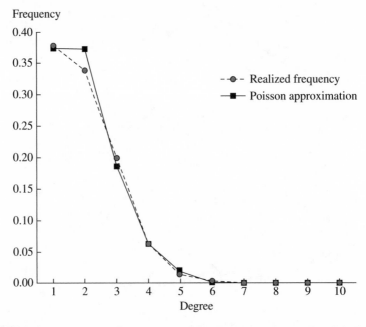

FIGURE 1.5   Frequency distribution of a randomly generated network and the Poisson approximation for a probability of .02 on each link.

this relative range. In graph theoretical terms, the network is a *forest,* or a collection of trees. That is, there are no cycles in the network (where a *cycle* is a sequence of links that lead from one node back to itself, as described in more detail in Section 2.1.3). The chance of there being a cycle is relatively low with such a small link probability. In addition, there are six *components* (maximal subnetworks such that every pair of nodes in the subnetwork is connected by a path or sequence of links) that involve more than one node. And one of the components is much larger than the others: it has 16 nodes, while the next largest only has 5 nodes in it. As we shall discuss shortly, this behavior is to be expected.

Let us start with the same number of nodes but increase the probability of a link forming to $p = \log(50)/50 = .078$, which is roughly the threshold at which isolated nodes should disappear. (This threshold is discussed in more detail in Chapter 4.) Based on the approximation of a Poisson distribution (1.4) with $n = 50$ and $p = .08$, we should expect about 2 percent of the nodes to be isolated (with degree 0), or roughly 1 node out of 50. This is exactly what occurs in the realized network in Figure 1.6 (again, by chance).

As shown in Figure 1.7, the realized frequency distribution of degrees is again similar to the Poisson approximation, although, as expected at this level of randomness, it is not a perfect match.

The degree distribution tells us a great deal about a network's structure. Let us examine this distribution in more detail, as it provides a first illustration of the

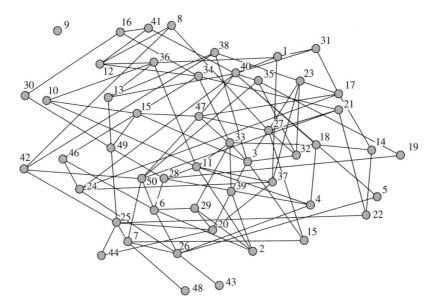

FIGURE 1.6   A randomly generated network with probability .08 on each link.

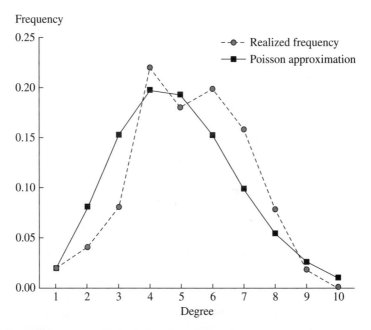

FIGURE 1.7   Frequency distribution of a randomly generated network and the Poisson approximation for a probability of .08 on each link.

concept of a *phase transition*, where the structure of a random network changes as the formation process is modified.

Consider the fraction of nodes that are completely isolated; that is, what fraction of nodes have degree $d = 0$? From (1.4) this number is approximated by $e^{-(n-1)p}$ for large networks, provided the average degree $(n - 1)p$ is not too large. To get a more precise expression, examine the threshold at which this fraction is just such that we expect to have one isolated node on average, or $e^{-(n-1)p} = \frac{1}{n}$. Solving this equation yields $p(n - 1) = \log(n)$, where the average degree is $(n - 1)p$. Indeed, this is a threshold for a phase transition, as shown in Section 4.2.2. If the average degree is substantially greater than $\log(n)$, then the probability of having any isolated nodes tends to 0, while if the average degree is substantially less than $\log(n)$, then the probability of having at least some isolated nodes tends to 1. In fact, as shown in Theorem 4.1, this threshold is such that if the average degree is significantly above this level, then the network is path-connected with a probability converging to 1 as n grows (so that any node can be reached from any other by a path in the network); below this level the network consists of multiple components with a probability converging to 1.

Other properties of random networks are examined in much more detail in Chapter 4. While it is clear that completely random networks are not always a good approximation for real social and economic networks, the analysis here (and in Chapter 4) shows that much can be deduced from such models and there are some basic patterns and structures that emerge more generally. As we build more realistic models, similar analyses can be conducted.

1.2.4 The Symmetric Connections Model

Although random network-formation models give some insight into the sorts of characteristics that networks can have and exhibit some of the features seen in the Add Health social network data, they do not provide as much insight into the Florentine marriage network. In that example marriages were carefully arranged. The last example discussed in this chapter is from the game-theoretic economics literature and provides a basis for the analysis of networks that form when links are chosen by the agents in the network. This example addresses questions about which networks would maximize the welfare of a society and which arise if the players have discretion in choosing their links.

This simple model of social connections was developed by Jackson and Wolinsky [361]. In it, links represent social relationships, for instance friendships, between players. These relationships offer benefits in terms of favors, information, and the like, and also involve some costs. Moreover, players also benefit from indirect relationships. Thus having a "friend of a friend" also results in some indirect benefits, although of lesser value than the direct benefits that come from having a friend. The same is true of "friends of a friend of a friend," and so forth. The benefit deteriorates with the distance of the relationship. This deterioration is represented by a factor δ that lies between 0 and 1, which indicates the benefit from a direct relationship and is raised to higher powers for more distant relationships. For instance, in a network in which player 1 is linked to 2, 2 is linked to 3, and 3 is linked to 4, player 1 receives a benefit δ from the direct

FIGURE 1.8 Utilities to the players in a three-link, four-player network in the symmetric connections model.

connection with player 2, an indirect benefit δ^2 from the indirect connection with player 3, and an indirect benefit δ^3 from the indirect connection with player 4. The payoffs to this network of four players with three links is pictured in Figure 1.8. For $\delta < 1$ there is a lower benefit from an indirect connection than from a direct one. Players only pay costs, however, for maintaining their direct relationships.[15]

Given a network g,[16] the net utility or payoff $u_i(g)$ that player i receives from a network g is the sum of benefits that the player gets for his or her direct and indirect connections to other players less the cost of maintaining these links:

$$u_i(g) = \sum_{\substack{j \neq i \,:\, i \text{ and } j \text{ are} \\ \text{path-connected in } g}} \delta^{\ell_{ij}(g)} - d_i(g)c,$$

where $\ell_{ij}(g)$ is the number of links in the shortest path between i and j, $d_i(g)$ is the number of links that i has (i's degree), and $c > 0$ is the cost for a player of maintaining a link.

Taking advantage of the highly stylized nature of the connections model, let us now examine which networks are "best" (most "efficient") from society's point of view, as well as which networks are likely to form when self-interested players choose their own links.

Let us define a network to be *efficient* if it maximizes the total utility to all players in the society. That is, g is efficient if it maximizes $\sum_i u_i(g)$.[17] It is clear that if costs are very low, it is efficient to include all links in the network. In particular, if $c < \delta - \delta^2$, then adding a link between any two agents i and j always increases total welfare. This follows because they are each getting at most δ^2 of value from having any sort of indirect connection between them, and since $\delta^2 < \delta - c$, the extra value of a direct connection between them increases their utilities (and might also increase, and cannot decrease, the utilities of other agents).

When the cost rises above this level, so that $c > \delta - \delta^2$ but c is not too high (see Exercise 1.3), it turns out that the unique efficient network structure is to have

---

15. In the most general version of the connections model the benefits and costs may be relation specific and so are indexed by ij. One interesting variation is when the cost structure is specific to some geography, so that linking with a given player depends on their physical proximity. That variation has been studied by Johnson and Gilles [367] and is discussed in Exercise 6.14.

16. For complete definitions, see Chapter 2.

17. This is just one of many possible measures of efficiency and societal welfare, which are well-studied subjects in philosophy and economics. How we measure efficiency has important consequences in network analysis and is discussed in more detail in Chapter 6.

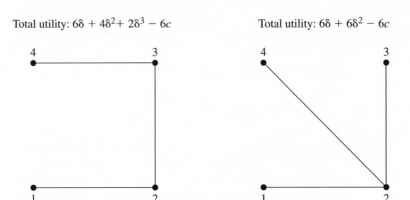

Total utility: $6\delta + 4\delta^2 + 2\delta^3 - 6c$ Total utility: $6\delta + 6\delta^2 - 6c$

FIGURE 1.9   Gain in total utility from changing a line into a star.

all players arranged in a "star" network. That is, there should be some central player who is connected to each other player, so that one player has $n - 1$ links and each of the other players has 1 link. A star involves the minimum number of links needed to ensure that all pairs of players are path connected, and it places each player within two links of every other player. The intuition behind why this structure dominates other structures for moderate-cost networks is then easy to see. Suppose for instance we have a network with links between 1 and 2, 2 and 3, and 3 and 4. If we change the link between 3 and 4 to be one between 2 and 4, a star network is formed. The star network has the same number of links as the starting network, and thus the same cost and payoffs from direct connections. However, now all agents are within two links of one another, whereas before some of the indirect connections involved paths of length three (Figure 1.9).

As we shall see, this result is the key to a remarkably simple characterization of the set of efficient networks: (1) costs are so low that it makes sense to add all links; (2) costs are so high that no links make sense; or (3) costs are in a middle range, and the unique efficient architecture is a star network. This characterization of efficient networks being stars, empty, or complete actually holds for a fairly general class of models in which utilities depend on path length and decay with distance, as is shown in detail in Section 6.3.

We can now compare the efficient networks with those that arise if agents form links in a self-interested manner. To capture how agents act, consider a simple equilibrium concept introduced in Jackson and Wolinsky [361]. This concept is called *pairwise stability* and involves two rules about a network: (1) no agent can raise his or her payoff by deleting a link that he or she is directly involved in and (2) no two agents can both benefit (at least one strictly) by adding a link between themselves. This stability notion captures the idea that links are bilateral relationships and require the consent of both individuals. If an individual would benefit by terminating some relationship that he or she is involved in, then that link would be deleted, while if two individuals would each benefit by forming a new relationship, then that link would be added, and in either case the network would fail to be stable.

In the case in which costs are very low ($c < \delta - \delta^2$), the direct benefit to the agents from adding or maintaining a link is positive, even if they are already indirectly connected. Thus in that case the unique pairwise-stable network is complete and is the efficient one. The more interesting case is when $c > \delta - \delta^2$, but c is not too high, so that the star is the efficient network.

If $\delta > c > \delta - \delta^2$, then a star network (that involves all agents) will be both pairwise stable and efficient. To see this we need only check that no player wants to delete a link, and no two agents both want to add a link. The marginal benefit to the center player from any given link already in the network is $\delta - c > 0$, and the marginal benefit to a peripheral player is $\delta + (n - 2)\delta^2 - c > 0$. Thus neither player wants to delete a link. Adding a link between two peripheral players only shortens the distance between them from two links to one and does not shorten any other paths, and since $c > \delta - \delta^2$ adding such a link would not benefit either of the players. While the star is pairwise stable, in this cost range so are some other networks. For example if $c < \delta - \delta^3$, then four players connected in a circle would also be pairwise stable. In fact, as we shall see in Section 6.3, many other (inefficient) networks can be pairwise stable.

If $c > \delta$, then the efficient (star) network is not pairwise stable, as the center player gets only a marginal benefit of $\delta - c < 0$ from any of the links. Thus in this cost range there cannot exist any pairwise-stable networks in which some player has only one link, as the other player involved in that link would benefit by severing it. For various values of $c > \delta$ there exist nonempty pairwise-stable networks, but they are not star networks: they must be such that each player has at least two links.

This model makes it clear that there are situations in which individual incentives are not aligned with overall societal benefits. While this connections model is highly stylized, it still captures some basic insights about the payoffs from networked relationships, and it shows that we can model the incentives that underlie network formation and see when the resultant networks are efficient.

This model also raises some interesting questions that are examined in the chapters that follow. How does the network that forms depend on the payoffs to the players for different networks? What are alternative ways of predicting which networks will form? What if players can bargain when they form links, so that the payoffs are endogenous to the network-formation process (as is true in many market and partnership applications)? How does the relationship between the efficient networks and those that form based on individual incentives depend on the underlying application and payoff structure?

1.3 ▪ Exercises

1.1 *A Weighted Betweenness Measure* Consider the following variation on the betweenness measure in (1.1). Any given shortest path between two families is weighted by the inverse of the number of intermediate nodes on that path. For instance, the shortest path between the Ridolfi and Albizzi involves two links; and the Medici are the only family that lies between them on that path. In contrast,

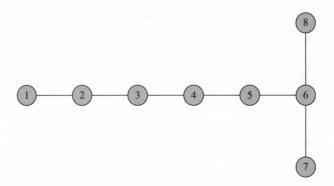

FIGURE 1.10   Differences in betweenness measures.

between the Ridolfi and the Ginori the shortest path is three links and there are
two families, the Medici and Albizzi, that lie between the Ridolfi and Ginori on
that path (see Figure 1.1).

More specifically, let $\ell(i, j)$ be the length of the shortest path between nodes i
and j and let $W_k(ij) = P_k(ij)/(\ell(i, j) - 1)$ (setting $\ell(i, j) = \infty$ and $W_k(ij) = 0$
if i and j are not connected). Then the weighted betweenness measure for a given
node k is defined by

$$W B_k = \sum_{ij:i \neq j, k \notin \{i,j\}} \frac{W_k(ij)/P(ij)}{(n - 1)(n - 2)/2},$$

where the convention is $\frac{W_k(ij)}{P(ij)} = 0/0 = 0$ if there are no paths connecting i and j.
Show that

(a) $W B_k > 0$ if and only if k has more than one link in a network and some of
     k's neighbors are not linked to one another;
(b) $W B_k = 1$ for the center node in a star network that includes all nodes (with
     $n \geq 3$); and
(c) $W B_k < 1$ unless k is the center node in a star network that contains all nodes.

Calculate this measure for the network pictured in Figure 1.10 for nodes 4 and 5.
Contrast this measure with the betweenness measure in (1.1).

1.2 *Random Networks*   Fix the probability of any given link forming in a Poisson
random network to be p, where $1 > p > 0$. Fix some arbitrary network g on k
nodes. Now, consider a sequence of random networks indexed by the number of
nodes n, as $n \to \infty$. Show that the probability that a copy of the k-node network
g is a subnetwork of the random network on the n nodes goes to 1 as n goes to
infinity.

[Hint: partition the n nodes into as many separate groups of k nodes as possible
(with some leftover nodes) and consider the subnetworks that form on each of
these groups. Using (1.2) and the independence of link formation, show that the
probability that none of these match the desired network goes to 0 as n grows.]

1.3 *The Upper Bound for a Star to Be Efficient*   Find the maximum level of cost, in terms of δ and n, for which a star is an efficient network in the symmetric connections model.

1.4 *The Connections Model with Low Decay* *   Consider the symmetric connections model with $1 > \delta > c > 0$.

   (a) Show that if δ is sufficiently close to 1 so that there is low decay and δ^{n-1} is nearly δ, then in every pairwise stable network every pair of players have some path between them and that there are at most $n - 1$ total links in the network.

   (b) In the case where δ is close enough to 1 so that any network that has $n - 1$ links and connects all agents is pairwise stable, what fraction of the pairwise-stable networks are also efficient networks?

   (c) How does that fraction behave as n grows (adjusting δ to be high enough as n grows)?

1.5 *Homophily and Balance across Groups*   Consider a society consisting of two groups. The set N_1 comprises the members of group 1 and the set N_2 comprises the members of group 2, with cardinalities n_1 and n_2, respectively. Suppose that $n_1 > n_2$. For individual i, let d_i be i's degree (total number of friends) and let s_i denote the number of friends that i has that are within i's own group. Let h_k denote a simple homophily index for group k, defined by

$$h_k = \frac{\sum_{i \in N_k} s_i}{\sum_{i \in N_k} d_i}.$$

Show that if h_1 and h_2 are both between 0 and 1, and the average degree in group 1 is at least as high as that in group 2, then $h_1 > h_2$. What are h_1 and h_2 in the case in which friendships are formed in percentages that correspond to the shares of relevant populations in the total population?

CHAPTER 2

Representing and Measuring Networks

This chapter presents some of the fundamentals of how networks are represented, measured, and characterized. It provides basic concepts and definitions that are the basis for the language of network research. Sprinkled throughout are observations from case studies that illustrate some of the concepts. More discussion about observed social and economic networks appears in Chapter 3.

2.1 ▪ Representing Networks

As networks of relationships come in many shapes and sizes, there is no single way of representing networks that encompasses all applications. Nevertheless, there are some representations that serve as a useful basis for capturing many applications. Here I focus on a few standard ways of denoting networks that are broad and flexible enough to capture a multitude of applications and yet sufficiently simple to be compact, intuitive, and tractable. As we proceed, I try to make clear what these concepts capture and what they omit.

2.1.1 Nodes and Players

The set $N = \{1, \ldots, n\}$ is the set of *nodes* that are involved in a network of relationships. Nodes are also referred to as "vertices," "individuals," "agents," or "players," depending on the setting. It is important to emphasize that nodes can be individual people, firms, countries, or other organizations; a node can even be something like a web page belonging to a person or organization.

2.1.2 Graphs and Networks

The canonical network form is an undirected graph, in which two nodes are either connected or they are not. For such graphs it cannot be that one node is related to a second without the second being related to the first. This behavior is generally true of many social and/or economic relationships, such as partnerships, friendships, alliances, and acquaintances. Such networks are central to most of the chapters that follow. However, there are other situations that are better modeled as directed networks, in which one node may be connected to a second without the second being connected to the first. For instance, a network that keeps track of which authors cite which other authors, or which web pages have links to which others would naturally take the form of a directed graph.

The distinction between directed and undirected networks is not a mere technicality. It is fundamental to the analysis, as the applications and modeling of the two types are quite different. In particular, when links are necessarily reciprocal, then it is generally the case that joint consent is needed to establish and maintain the relationship. For instance, to form a trading partnership, both partners need to agree to it. To maintain a friendship the same is generally true, as it is for a business relationship or an alliance. In the case of directed networks, one individual may direct a link at another without the other's consent, which is generally true in citation networks or in links between web pages. These distinctions result in some basic differences in the modeling of network formation, as well as different conclusions about which networks will arise, which are optimal, and so on.

In what follows the default is that the network is undirected, and I mention explicitly when directed networks are considered. Let us begin with the formal definitions of graphs that represent networks.

A *graph* (N, g) consists of a set of nodes $N = \{1, \ldots, n\}$ and a real-valued $n \times n$ matrix g, where g_{ij} represents the (possibly weighted and/or directed) relation between i and j. This matrix is often referred to as the *adjacency matrix*, as it lists which nodes are linked to each other, or in other words which nodes are adjacent to one another.[1] In the case in which the entries of g take on more than two values and can track the intensity level of relationships, the graph is referred to as a *weighted* graph. Otherwise, it is standard to use the values of either 0 or 1, and the graph is *unweighted*. In much of what follows, N will be fixed or given. Thus, I often refer to g as being a network or graph.

A network is *directed* if it is possible that $g_{ij} \neq g_{ji}$, and a network is *undirected* $g_{ij} = g_{ji}$ for all nodes i and j. Parts of the literature refer to directed graphs as *digraphs*.

For instance, if $N = \{1, 2, 3\}$, then

$$g = \begin{pmatrix} 0 & 1 & 0 \\ 1 & 0 & 1 \\ 0 & 1 & 0 \end{pmatrix} \tag{2.1}$$

is the (undirected and unweighted) network with a *link* between nodes 1 and 2, a link between nodes 2 and 3, but no link between nodes 1 and 3 (Figure 2.1). Nodes

---

1. There are more general graph structures that can represent the possibility of multiple relationships between different nodes; for instance, having different links for being friends, relatives, coworkers, and the like. These are sometimes referred to as a *multiplex networks*. One can also allow for relationships that involve more than two nodes at a time. For example, see Diestel [200] and Page and Wooders [518] for some more general representations.

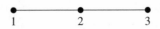

FIGURE 2.1   A network with two links.

are also often referred to as *vertices* and links as *edges* or *ties*; links are sometimes
called *arcs* in the case of directed graphs.

Self-links or *loops* often do not have any real meaning or consequence, and so
whether we set $g_{ii} = 1$ or $g_{ii} = 0$ as a default is usually (but not always!) irrelevant.
Unless otherwise indicated, assume that $g_{ii} = 0$ for all i.[2]

There are equivalent ways of representing a graph. Instead of viewing g as
an $n \times n$ matrix, it is sometimes easier to describe a graph by listing all links or
edges in the graph. That is, we can view a graph as a pair (N, g), where g is
the collection of links that are listed as a subsets of N of size 2. For instance,
the network g in Figure 2.1 can be written as $g = \{\{1, 2\}, \{2, 3\}\}$, or simplifying
notation a bit, $g = \{12, 23\}$. Thus ij represents the link connecting nodes i and
j. Then we can write $ij \in g$ to indicate that i and j are linked under the network
g; that is, writing $ij \in g$ is equivalent to writing $g_{ij} = 1$. I alternate between the
different representations as is convenient. It will also be useful to write $g' \subset g$, to
indicate that

$$\{ij : ij \in g'\} \subset \{ij : ij \in g\}.$$

Let the shorthand notation of $g + ij$ represent the network obtained by adding
the link ij to an existing network g, and, let $g - ij$ represent the network obtained
by deleting the link ij from the network g. We can represent directed networks in
an analogous manner, viewing ij as a directed link and distinguishing between ij
and ji. Let $G(N)$ be the set of all undirected and unweighted networks on N.

In some cases the specific identity of the node in a position in the network is
of interest, and in other situations only the structure of the network is important.
The idea that two networks or graphs have the same structure is captured through
the concept of an isomorphism. The networks (N, g) and (N', g') are *isomorphic*
if there exists a one-to-one and onto function (a bijection) $f : N \to N'$, such
that $ij \in g$ if and only if $f(i)f(j) \in g'$. Thus, f just relabels the nodes, and the
networks are the same up to that relabeling.

Given a subset of nodes $S \subset N$ and a network g, let $g|_S$ denote the network g
restricted to the set of nodes S, so that

$$[g|_S]_{ij} = \begin{cases} 1 & \text{if } i \in S, \ j \in S, \ g_{ij} = 1, \\ 0 & \text{otherwise.} \end{cases}$$

Thus $g|_S$ is the network obtained by deleting all links except those that are between
nodes in S. An example is pictured in Figure 2.2.

---

2. Sometimes graphs without any self-links (and without multiple links) are referred to as *simple
graphs*. Unless otherwise stated, the term graph refers to a simple graph in this book. If self-links
and multiple links between nodes are permitted, the resulting structure is termed a *multigraph*.

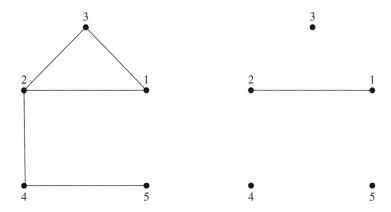

FIGURE 2.2   A network and the network restricted to $S = \{1, 2, 5\}$.

For any network g, let $N(g)$ be the set of nodes that have at least one link in the network g. That is, $N(g) = \{i | \exists j \text{s.t. } ij \in g, \text{ or } ji \in g\}$.[3]

2.1.3 Paths and Cycles

Much of the interest in networked relationships comes from the fact that individual nodes benefit (or suffer) from indirect relationships. Friends might provide access to favors from their friends, and information might spread through the links of a network. To capture the indirect interactions in a network, it is important to model paths through the network. In the case of an undirected network, a path is an obvious object. As there are multiple definitions in the case of a directed network, I return to those after providing definitions for an undirected network.

A *path* in a network $g \in G(N)$ between nodes i and j is a sequence of links $i_1 i_2, i_2 i_3, \ldots, i_{K-1} i_K$ such that $i_k i_{k+1} \in g$ for each $k \in \{1, \ldots, K-1\}$, with $i_1 = i$ and $i_K = j$, and such that each node in the sequence $i_1, \ldots, i_K$ is distinct.[4] A *walk* in a network $g \in G(N)$ between nodes i and j is a sequence of links $i_1 i_2, \ldots, i_{K-1} i_K$ such that $i_k i_{k+1} \in g$ for each $k \in \{1, \ldots, K-1\}$, with $i_1 = i$ and $i_K = j$. The distinction between a path and a walk is whether all involved nodes are distinct. A walk may come back to a given node more than once, whereas a path is a walk that never hits the same node twice.[5]

---

3. In this case it matters whether $g_{ii} = 1$, in which case $i \in N(g)$, or whether $g_{ii} = 0$, in which case $i \notin N(g)$.

4. A path may also be defined to be a subnetwork that consists of the set of involved nodes and the set of links between these nodes.

5. The definition of a path given here is the standard one from the graph theory literature. In some of the network literature, the term *path* is used more loosely and can refer to a walk, so that nodes can be visited more than once. This ambiguity can cause some confusion, which should be borne in mind when reading the literature.

A *cycle* is a walk $i_1 i_2, \ldots, i_{K-1} i_K$ that starts and ends at the same node (so $i_1 = i_K$) and such that all other nodes are distinct ($i_k \neq i_{k'}$ when $k < k'$ unless $k = 1$ and $k' = K$). Thus a cycle is a walk such that the only node that appears more than once is the starting/ending node. A cycle can be constructed from any path by adding a link from the end to the starting node; and conversely, deleting the first or last link of a cycle results in a path.

A *geodesic* between nodes i and j is a shortest path between these nodes; that is, a path with no more links than any other path between these nodes.

To summarize:

- A walk is a sequence of links connecting a sequence of nodes.
- A cycle is a walk that starts and ends at the same node, with all nodes appearing once except the starting node, which also appears as the ending node.
- A path is a walk in which a node appears at most once in the sequence.
- A geodesic between two nodes is a shortest path between them.

Note that for the convention of setting $g_{ii} = 0$, then $g^2 = g \times g$ tells us how many walks there are of length 2 between any two nodes.

For instance for the network

$$g = \begin{pmatrix} 0 & 1 & 1 & 0 \\ 1 & 0 & 0 & 1 \\ 1 & 0 & 0 & 1 \\ 0 & 1 & 1 & 0 \end{pmatrix},$$

g^2 is

$$g^2 = \begin{pmatrix} 2 & 0 & 0 & 2 \\ 0 & 2 & 2 & 0 \\ 0 & 2 & 2 & 0 \\ 2 & 0 & 0 & 2 \end{pmatrix}.$$

So, for instance there are two walks between 1 and 4 of length 2 (passing between 2 and 3, respectively). There are two walks from 1 back to 1 (passing through 2 and 3, respectively). For this network g^3 is

$$g^3 = \begin{pmatrix} 0 & 4 & 4 & 0 \\ 4 & 0 & 0 & 4 \\ 4 & 0 & 0 & 4 \\ 0 & 4 & 4 & 0 \end{pmatrix}.$$

There are four walks of length 3 between 1 and 2 (namely, (12,24,42), (13,34,42), (12,21,12), and (13,31,12)). Note that some walks have cycles in them (and hence the use of the term *walk* rather than *path*). The kth power of the network, g^k, keeps track of all possible walks of length k between any two nodes, including walks with many cycles within them.

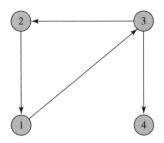

FIGURE 2.3 Directed path from 2 to 4 (via 1 and 3), directed cycle from 1 to 3 to 2 to 1, and directed walk from 3 to 2 to 1 to 3 to 4.

2.1.4 Directed Paths, Walks, and Cycles

In the case of directed networks, there are different possible definitions of paths and cycles. The definitions depend on whether we want to keep track of the direction of the links, and in various applications depend on whether communication is restricted to following the direction of the links or can move in both directions along a directed link, as for example in a network of links between web pages.

In the case in which direction is important, the definitions are just as stated above for undirected networks, but the ordering of the nodes in each link now takes on an important role. For instance, we might be interested in knowing whether one web page can be found from another by following (directed) links starting from one page and leading to the other. This case deals with directed paths, directed walks, and directed cycles.

A *directed walk* in a network $g \in G(N)$ is a sequence of links $i_1 i_2, \ldots, i_{K-1} i_K$ such that $i_k i_{k+1} \in g$ (that is, $g_{i_k i_{k+1}} = 1$) for each $k \in \{1, \ldots, K-1\}$.

A *directed path* in a directed network $g \in G(N)$ from node i to node j is a sequence of links $i_1 i_2, \ldots, i_{K-1} i_K$ such that $i_k i_{k+1} \in g$ (that is, $g_{i_k i_{k+1}} = 1$) for each $k \in \{1, \ldots, K-1\}$, with $i_1 = i$ and $i_K = j$, such that each node in the sequence $i_1, \ldots, i_K$ is distinct.

A *directed cycle* in a network $g \in G(N)$ is a sequence of links $i_1 i_2, \ldots, i_{K-1} i_K$ such that $i_k i_{k+1} \in g$ (that is, $g_{i_k i_{k+1}} = 1$) for each $k \in \{1, \ldots, K-1\}$, with $i_1 = i_K$.

These definitions are illustrated in Figure 2.3.

In cases where the direction of the link just indicates who initiated the link, but where links can conduct in both directions, we can keep track of undirected paths. There we think of i and j being linked if either $g_{ij} = 1$ or $g_{ji} = 1$. In that case, we can simply define the undirected network that comes from considering i and j to be linked if there is a directed link in either direction. In general, I refer to such paths, walks, and cycles as undirected.

To be more specific, given a directed network g let $\widehat{g}$ denote the undirected network obtained by allowing an undirected link for each directed one present in g. That is, let $\widehat{g}_{ij} = \max(g_{ij}, g_{ji})$. Then we say that there is an *undirected path* between nodes i and j in g if there is a path between them in $\widehat{g}$. An undirected cycle or walk is defined in a similar fashion. In Figure 2.3 there is no directed path from node 4 to any other node, but there is an undirected path from node 4 to each of the other nodes.

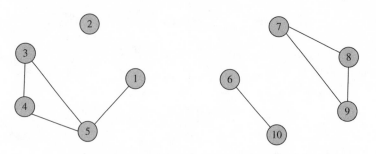

FIGURE 2.4   A network with four components.

2.1.5 Components and Connected Subgraphs

In many applications it is important to track which nodes can reach which other nodes through paths in the network. This tracking ability plays a critical role in phenomena like contagion, learning, and the diffusion of various behaviors through a social network. Looking at the path relationships in a network naturally partitions a network into different connected subgraphs that are commonly referred to as *components*. Again, definitions are first provided for undirected networks and later for directed ones.

A network (N, g) is *connected* (or path-connected) if every two nodes in the network are connected by some path in the network. That is, (N, g) is connected if for each $i \in N$ and $j \in N$ there exists a path in (N, g) between i and j.

A *component* of a network (N, g) is a nonempty subnetwork (N', g') such that $\emptyset \neq N' \subset N$, $g' \subset g$,

- (N', g') is connected, and
- if $i \in N'$ and $ij \in g$, then $j \in N'$ and $ij \in g'$.

Thus the components of a network are the distinct maximal connected subgraphs of a network. In the network shown in Figure 2.4 there are four components: the node 2 together with an empty set of links, the nodes $\{1, 3, 4, 5\}$ together with links $\{15, 35, 34, 45\}$, the nodes 6 and 10 together with the link $\{6-10\}$, and the nodes $\{7, 8, 9\}$ together with the links $\{78, 79, 89\}$. Note that under this definition of component, a completely isolated node that has no links is considered a component.[6]

The set of components of a network (N, g) is denoted $C(N, g)$. In cases for which N is fixed or obvious, I simply denote the components by $C(g)$. The component containing a specific node i is denoted $C_i(g)$.

Components of a network partition the nodes into groups within which nodes are path-connected. Let $\Pi(N, g)$ denote the partition of N induced by the network (N, g). That is, $S \in \Pi(N, g)$, if and only if $(S, h) \in C(N, g)$ for some $h \subset g$. For example, the network in Figure 2.4 induces the partition $\Pi(N, g) =$

6. This inclusion is a matter of convention, and one can also find definitions of components that only allow for subnetworks with links.

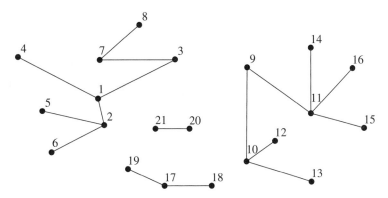

FIGURE 2.5 Four trees in a forest.

{{1, 3, 4, 5}, {2}, {6, 10}, {7, 8, 9}} over the set of nodes. Thus, a network is connected if and only if it consists of a single component (so that $\Pi(N, g) = \{N\}$).

A link ij is a *bridge* in the network g if $g - ij$ has more components than g.[7]

In the case of a directed network, there are again several different approaches to defining those terms. One way is to again ignore the directed nature of links and to consider the undirected network that has a link present if one is present in either direction. This method defines one notion of connection and components. In some applications in which direction is important, for instance in the transmission of information, we want to keep track of the directed nature of the network. In such cases, I refer to *strongly connected* graphs or subgraphs, so that each node can reach every other one by a directed path. Further definitions are specified as needed.

2.1.6 Trees, Stars, Circles, and Complete Networks

There are a few particular network structures that are commonly referred to.

A *tree* is a connected network that has no cycles.

A *forest* is a network such that each component is a tree. Thus any network that has no cycles is a forest, as in the example pictured in Figure 2.5.

A particularly prominent forest network is the star. A *star* is a network in which there exists some node i such that every link in the network involves node i. In this case, i is referred to as the *center* of the star.

Here are a few facts about trees that are easy to derive (see Exercise 2.2) and are worth mentioning.

- A connected network is a tree if and only if it has $n - 1$ links.
- A tree has at least two leaves, where leaves are nodes that have exactly one link.
- In a tree, there is a unique path between any two nodes.

---

7. There are variations on this definition, with some requiring that the components connected by the bridge both involve more than one node.

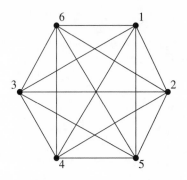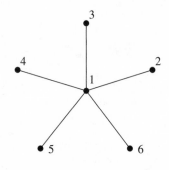

FIGURE 2.6   A complete network on six nodes and a star network on six nodes.

A *complete network* is one in which all possible links are present, so that $g_{ij} = 1$ for all $i \neq j$ (see Figure 2.6).

A *circle* (also known as a *cycle-graph*) is a network that has a single cycle and is such that each node in the network has exactly two neighbors.

In the case of directed networks, there can be many different stars involving the same set of nodes and having the same center, depending on which directed links are present between any two linked nodes. On occasion it is useful to distinguish between these stars, for instance, indicating whether links go in or out from the center node.

2.1.7 Neighborhood

The *neighborhood of a node i* is the set of nodes that i is linked to.[8]

$$N_i(g) = \{j : g_{ij} = 1\}.$$

Given some set of nodes S, the *neighborhood of S* is the union of the neighborhoods of its members. That is

$$N_S(g) = \bigcup_{i \in S} N_i(g) = \{j : \exists i \in S, \, g_{ij} = 1\}.$$

We can also talk about extended neighborhoods of a node, for instance of all the nodes that can be reached by walks of length no more than 2, and so on. The *two-neighborhood of a node i* is

$$N_i^2(g) = N_i(g) \cup \left(\bigcup_{j \in N_i(g)} N_j(g) \right).$$

---

8. Note that whether i is in i's neighborhood depends on whether $g_{ii} = 1$ is allowed. As I am following a default convention of $g_{ii} = 0$, i is generally not considered to be in i's neighborhood. This definition ensures that i's degree is the number of other nodes that i is linked to, which is then the cardinality of i's neighborhood.

Inductively, all nodes that can be reached from i by walks of length no more than k make up the *k-neighborhood* of i, which can be defined by

$$N_i^k(g) = N_i(g) \cup \left(\bigcup_{j \in N_i(g)} N_j^{k-1}(g) \right).$$

Similar definitions of k-neighborhoods hold for any set of nodes S, so that $N_S^k(g) = \cup_{i \in S} N_i^k(g)$ is the set of nodes that can be reached from some node in S by a walk of length no more than k. Generally the *extended neighborhood* of a node i is all of the nodes it is walk-connected to, or $N_i^n(g)$.

The above definitions also work for directed networks, in which case the nodes in $N_i^k(g)$ are those nodes that can be reached from i by a directed walk.

2.1.8 Degree and Network Density

The *degree* of a node is the number of links that involve that node, which is the cardinality of the node's neighborhood. Thus node i's degree in a network g, denoted $d_i(g)$, is

$$d_i(g) = \#\{j : g_{ji} = 1\} = \#N_i(g).$$

In the case of a directed network, the above calculation is the node's *in-degree*. The *out-degree* of node i is the corresponding calculation $\#\{j : g_{ij} = 1\}$. These definitions coincide for an undirected network (see Figure 2.7). The *density* of a network keeps track of the relative fraction of links that are present, and is simply the average degree divided by $n - 1$.

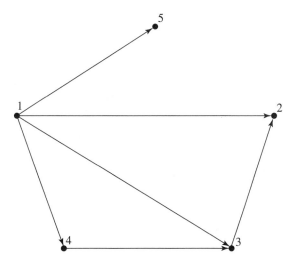

FIGURE 2.7   A directed network on five nodes. Node 1 has in-degree 2 and out-degree 4.

2.2 ▪ Some Summary Statistics and Characteristics of Networks

While a small network can be usefully described directly by its graph g and is easily illustrated in a figure, larger networks can be more difficult to envision and describe. Moreover, it is important to be able to compare networks and classify them according to their properties and thus to have a stable of summary statistics that provide meaningful insight into their structures.

2.2.1 Degree Distributions

A fundamental characteristic of a network is its degree distribution. The *degree distribution* of a network is a description of the relative frequencies of nodes that have different degrees. That is, $P(d)$ is the fraction of nodes that have degree d under a degree distribution P.[9]

For instance, a *regular* network is one in which all nodes have the same degree. A network is *regular of degree k* if $P(k) = 1$ and $P(d) = 0$ for all $d \neq k$. Such a network is quite different from the random network described in Section 1.2.3, in which there is a great deal of heterogeneity in the degrees of nodes, and the distribution is a Poisson distribution.

Beyond the degenerate degree distribution associated with a regular network, and the Poisson degree distribution associated with Poisson random networks discussed in Section 1.2.3, another prominent distribution is what is referred to as a *scale-free* degree distribution. These distributions date to Pareto [526], and they appear in a wide variety of settings including networks describing incomes, word usage, city populations, and degrees in networks (as is discussed in more detail in Chapter 3).[10]

A *scale-free distribution* (or power distribution) $P(d)$ satisfies[11]

$$P(d) = cd^{-\gamma}, \tag{2.2}$$

where $c > 0$ is a scalar (which normalizes the support of the distribution to sum to 1).[12] Thus if we increase the degree by a factor k, then the frequency is reduced by a factor of $k^{-\gamma}$. As this is true regardless of the starting degree d, the relative probabilities of degrees of a fixed relative ratio are the same independent of the scale of those degrees. That is, $P(2)/P(1)$ is the same as $P(20)/P(10)$. Hence the term *scale-free*. Scale-free distributions are often said to exhibit a *power law*, with reference to the power function $d^{-\gamma}$.

---

9. P can be a frequency distribution if we are describing data, or it can be a probability distribution if we are working with random networks.

10. For an informative overview, see Mitzenmacher [470].

11. One has to be careful about defining the value at $d = 0$, as it might not be well defined; so let us keep track of nodes with degree at least 1.

12. When the support is $\{1, 2, \ldots\}$, then the scalar is the inverse of what is known as the Riemann zeta function, $z(\gamma) = \sum_{d=1}^{\infty} \frac{1}{d^{\gamma}}$.

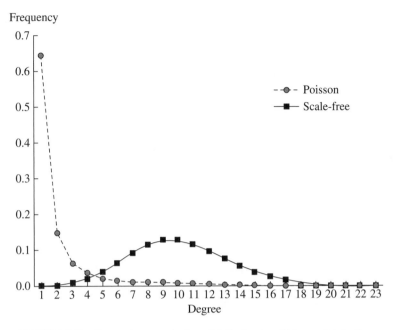

FIGURE 2.8   Comparing a scale-free distribution to a Poisson distribution.

Generally, given a degree distribution P, let $\langle d \rangle_P$ denote the expected value of d, and $\langle d^2 \rangle_P$ denote the expectation of the square of the degree, and so on. I often omit the $_P$ notation when P is fixed.

Scale-free distributions have "fat tails." That is, there tend to be many more nodes with very small and very large degrees than one would see if the links were formed completely independently so that degree followed a Poisson distribution. We can see this comparison in Figure 2.8, which shows plots of these degree distributions when the average degree is 10. The figure compares the Poisson degree distribution from (1.4) with the scale-free distribution from (2.2).

The fatter tail of the scale-free distribution is obvious in the lower tail (for lower degrees), while for higher degrees it is harder to see the differences. If we convert the plot to a log-log plot (i.e., log(frequency) versus log(degree) instead of the raw numbers), then the differences in the upper tail (for higher degrees) become more evident (Figure 2.9).

Figure 2.9 points out another interesting aspect of scale-free distributions: they are linear when plotted on a log-log plot. That is, we can rewrite (2.2) by taking logs of both sides to obtain:

$$\log(f(d)) = \log(c) - \gamma \log(d).$$

This form is useful when trying to estimate γ from data, as then a linear regression can be used.

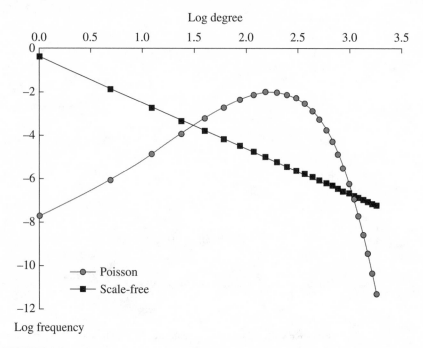

FIGURE 2.9 Comparing a scale-free distribution to a Poisson distribution: log-log plot.

2.2.2 Diameter and Average Path Length

The *distance* between two nodes is the length of (number of links in) the shortest path or *geodesic* between them. If there is no path between the nodes, then the distance between them is infinite. This concept leads us to another important characteristic of a network: its diameter. The *diameter* of a network is the largest distance between any two nodes in the network.[13]

To see how diameter can vary across networks with the same number of nodes and links, consider two different networks in which each node has on average two links, as in Figure 2.10. The first network is a circle, and the second is a tree. Even though both networks have approximately an average degree of 2, they are clearly very different in structure. The degree distribution reflects some aspect of the difference in that the circle is regular, so that every node has exactly two links, while in the binary tree almost half of the nodes have degree 3 and nearly half have degree 1 (the exception is the root node, which has degree 2). However, we need other measures to clearly distinguish these networks. For instance, the diameter of

---

13. Related measures, working with cycles rather than paths, are the girth and circumference of a network. The *girth* is the length of the smallest cycle in a network (set to infinity if there are no cycles), and the *circumference* is the length of the largest cycle (set to 0 if there are no cycles).

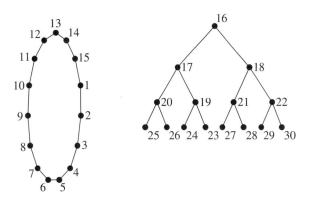

FIGURE 2.10   Circle and tree.

a circle of n nodes is either $n/2$ or $(n-1)/2$, while the diameter of a binary tree of n nodes is roughly $2\log_2(n+1) - 2$.[14]

The diameter is one measure of path length, but it only provides an upper bound. *Average path length* (also referred to as *characteristic path length*) between nodes is another measure that captures related properties. The average is taken over geodesics, or shortest paths. Clearly, the average path length is bounded above by the diameter; in some cases it can be much shorter than the diameter. Thus, it is often useful to see whether the diameter is being determined by a few outliers (literally), or whether it is of the same order as the average geodesic.

Many networks are not fully connected and may consist of a number of separate components. In such cases, one often reports the diameter and average path length in the largest component, being careful to specify whether that component is a giant component (containing a nontrivial fraction of the networks nodes).[15]

Recalling that raising the adjacency matrix g to a power k provides as its ijth entry the number of walks of length k between nodes i and j, we can easily calculate shortest path lengths. That is, the shortest path length between nodes i and j can be found by finding the smallest ℓ such that the ijth entry g^{ℓ} is positive: that entry is the number of shortest paths between those nodes. The same calculation provides shortest directed paths in the case of directed networks.

---

14. This measurement holds precisely if there is an integer K such that $n = 2^K - 1$ in the case of a binary tree.

15. There is a way to circumvent these problems. As Newman [503] suggests, the measure

$$\frac{n(n+1)}{2\sum_{ij}\frac{1}{\ell(i,j,g)}},$$

where $\ell(i, j, g)$ is the length of the shortest path between i and j in g and is set to infinity if the nodes are not connected. This measure can be calculated regardless of component structure. So rather than averaging path lengths, one looks at the reciprocal of the average of the reciprocal path lengths. Taking the reciprocal twice leads to something similar to averaging path lengths directly, but working with the reciprocals eliminates the influence of infinite path lengths.

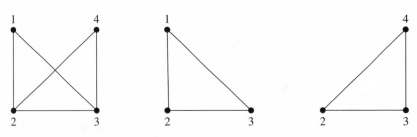

FIGURE 2.11 A network on four nodes and its two cliques.

Calculating shortest path lengths for all pairs of nodes, through successive powers of the adjacency matrix g, then provides a basic method of calculating diameter. There are more computationally efficient algorithms for calculating or estimating diameter,[16] and most network software programs include such calculations as built-in features.

2.2.3 Cliquishness, Cohesiveness, and Clustering

One fascinating and important aspect of social networks is how tightly clustered they are. For example, the extent to which my friends are friends with one another captures one facet of this clustering. There are a variety of concepts that measure how cohesive or closely knit a network is.

An early concept related to this is the idea of a clique. A *clique* is a maximal completely connected subnetwork of a given network.[17] That is, if some set of nodes $S \subset N$ are such that $g|_S$ is the complete network on the nodes S, and for any $i \in N \setminus S$ $g|_{S \cup \{i\}}$ is not complete, then the nodes S are said to form a clique.[18] Cliques are generally required to contain at least three nodes; otherwise each link could potentially define a clique of two nodes. Note that a given node can be part of several cliques at once. For example, in Figure 2.11 both nodes 2 and 3 are in two different cliques.

One measure of cliquishness is to count the number and size of the cliques in a network. One difficulty with this measure is that removing one link from a large clique can change the clique structure completely. For instance, removing one link from a complete network among four nodes changes the clique structure from having one clique involving four nodes to two cliques of three nodes. More generally, the clique structure is very sensitive to slight changes in a network.

---

16. For instance, there are more efficient ways of calculating powers of g when it is diagonalizable (see Section 2.4.1). Computational efficiency can be important when n is large.

17. Note the distinction between a clique and a component. A clique must be completely connected and not be a strict subset of any subnetwork that is completely connected, while a component must be path-connected and not be a strict subset of any subnetwork that is path-connected. Neither implies the other.

18. An early definition of this is from Luce and Perry [443].

The most common way of measuring some aspect of cliquishness is based on transitive triples or clustering.[19] Examining undirected networks, the most basic clustering measure is simply to perform the following exercise. Look at all situations in which two links emanate from the same node (e.g., ij and ik both involve node i) and ask how often jk is then also in the network. So if i has relationships with both j and k, how likely on average is it that j and k are related in the network? This clustering measure is represented by

$$Cl(g) = \frac{\sum_i \#\{jk \in g | k \neq j, j \in N_i(g), k \in N_i(g)\}}{\sum_i \#\{jk | k \neq j, j \in N_i(g), k \in N_i(g)\}} = \frac{\sum_{i;j \neq i;k \neq j;k \neq i} g_{ij} g_{ik} g_{jk}}{\sum_{i;j \neq i;k \neq j;k \neq i} g_{ij} g_{ik}}.$$

I will often refer to this as *overall* clustering to distinguish it from the other measures of clustering that follow.

Another measure that has also been used in the literature is similar to the clustering coefficient $Cl(g)$, except that instead of considering the fraction of fully connected triples out of the potential triples in which at least two links are present, the measure is computed on a node-by-node basis and then averaged across nodes. This measure is based on the following definition of *individual clustering for a node i*:

$$Cl_i(g) = \frac{\#\{jk \in g | k \neq j, j \in N_i(g), k \in N_i(g)\}}{\#\{jk | k \neq j, j \in N_i(g), k \in N_i(g)\}} = \frac{\sum_{j \neq i;k \neq j;k \neq i} g_{ij} g_{ik} g_{jk}}{\sum_{j \neq i;k \neq j;k \neq i} g_{ij} g_{ik}}.$$

Thus, $Cl_i(g)$ looks at all pairs of nodes that are linked to i and then considers how many of them are linked to one another.[20] Another way to write the individual clustering coefficient is then

$$Cl_i(g) = \frac{\#\{jk \in g | k \neq j, j \in N_i(g), k \in N_i(g)\}}{d_i(g)(d_i(g) - 1)/2}.$$

The *average clustering* coefficient is then

$$Cl^{Avg}(g) = \sum_i Cl_i(g)/n.$$

Note that this calculation is different from that for the overall clustering coefficient $Cl(g)$, where the average is taken over all triples. Under average clustering, one computes a clustering for each node and then averages across nodes. This method gives more weight to low-degree nodes than does the clustering coefficient method.

As an illustration of these two measures, let us compute them relative to the Florentine marriage network pictured in Figure 1.1. To compute the overall

---

19. Clustering in the sense used here comes from the recent random network literature (e.g., see Newman [503]). *Clustering* has an interesting history as a term, growing out of the earlier sociology literature and based on partitioning signed graphs into subsets in which nodes within elements of the partition have only positive relationships between them, and only negative relationships exist across elements of the partition (e.g., see Chapter 6 in Wasserman and Faust [650]).

20. A convention is to set $Cl_i(g) = 0$ if i has no more than one link.

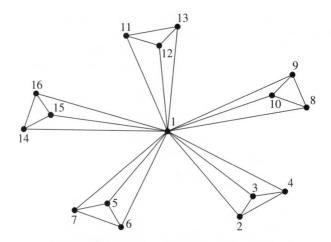

FIGURE 2.12   Differences in clustering measures.

clustering coefficient, first count how many configurations of the form ij, jk there are in the network. For instance, there is one with Medici-Barbadori and Barbadori-Castellan. There are three with the Ridolfi at the center (Strozzi-Ridolfi-Tornabuon, Strozzi-Ridolfi-Medici, and Tornabuon-Ridolfi-Medici). There are fifteen with the Medici at the center, and so forth. Totaling across all such configurations, we find that there are 47 such configurations in the network. Out of those 47 configurations, 9 of them are completed. Thus, $Cl(g) = 9/47$. In terms of the average clustering, compute the clustering for each family separately. For instance, the Barbadori have one possible pair and they are not connected, so the Barbadori have $Cl_{\text{Barbadori}}(g) = 0/1 = 0$. The Pazzi, Acciaiuol, Ginori, Lambertes, and Pucci are all 0 by convention (they each have one or no links). The Bischeri, Castellan, Ridolfi, and Tornabuon each have clustering $1/3$. The Strozzi are also $1/3$ (2 out of 6). The Peruzzi are $2/3$, the Medici $1/15$, and the Guadagni and Albizzi have at least two neighbors each, but still have clusterings of 0.[21] When we average across all of these, we get $Cl^{\text{Avg}}(g) = .15$ or $3/20$. This value is a bit less than the overall clustering $Cl(g)$, as a number of 0s are included in the average clustering.

The above calculations show that it is possible for these two common measures of clustering to differ. While in that example the average clustering is less than the overall clustering, it can also go the other way. Moreover, it is not uncommon to generate networks in which the two measures produce very different values. For instance, consider the following variation of a star network. Begin with a large number of triads (complete networks among three nodes), and then add a center node, to which every other node is connected (Figure 2.12).

As the number of nodes involved gets large, average clustering goes to 1, while overall clustering goes to 0! To see this note that all of the nodes other than the center node have individual clustering measures of 1. Thus when averaged the average clustering coefficient converges to 1. However, for the overall clustering

---

21. This relates back to the important role of the Medici, as many of their neighbors were not directly connected but were connected only indirectly through the Medici.

coefficient, each time that a new triad is added the number of possible pairs of links goes up by 3 times the number of links the center node already has (plus 12), while the number of those pairs that are completed only increases by 12. Thus overall clustering goes to 0. Clearly the two measures are capturing different aspects of clustering and so there is no "right" or "wrong" measure. This example shows that such simple coefficients cannot give a full picture of the interrelatedness of a network but only an impression of some aspect of it.

In the case of directed networks one has further choices for measuring clustering. One option is simply to ignore the direction of a link and consider two nodes to be linked if there is a directed link in either direction between them. Based on this derived undirected network, one can then apply the above measures of overall and average clustering. A different approach is to keep track of the percentage of *transitive triples*. This approach considers situations in which node i has a directed link to j, and j has a directed link to k, and then asks whether i has a directed link to k (i.e., the usual notion of transitivity of relationships).[22] The fraction of times in a network that the answer is "yes" is the *fraction of transitive triples*:

$$Cl^{TT}(g) = \frac{\sum_{i;j\neq i;k\neq j} g_{ij}g_{jk}g_{ik}}{\sum_{i;j\neq i;k\neq j} g_{ij}g_{jk}}.$$

The above fraction of transitive triples is a standard measure, but much of the empirical literature has instead simply ignored the directed nature of relationships.[23]

2.2.4 Centrality

Most of the measures discussed to this point are predominately macro in nature; that is, they describe broad characteristics of a network. In many cases, we might also be interested in micro measures that allow us to compare nodes and to say something about how a given node relates to the overall network. For instance, as we saw in the Florentine marriage example in Section 1.2.1, the idea of how central a node is can be very important. In particular, notions that somehow capture a node's position in a network are useful. As such, many different measures of centrality have been developed, and they each tend to capture different aspects of the concept, which can be useful when working with information flows, bargaining power, infection transmission, influence, and other sorts of important behaviors on a network.

Measures of centrality can be categorized into four main groups depending on the types of statistics on which they are based:[24]

1. degree—how connected a node is;
2. closeness—how easily a node can reach other nodes;

22. Alternatively, one could examine the percentage of times that k has a directed link to i so that a directed cycle emerges. This calculation can yield very different results, depending on the context.
23. There are also hybrid measures (mixing ideas of directed and undirected links) in which one counts the percentage of possible directed links among a node's direct neighbors that are present on average, as in, for example, Adamic's [2] study of the world wide web.
24. See Borgatti [94] for more discussion on categorizing measures of centrality.

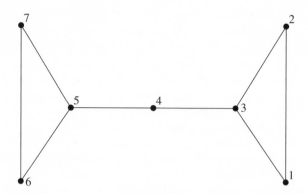

FIGURE 2.13   A central node with low degree centrality.

3. betweenness—how important a node is in terms of connecting other nodes; and

4. neighbors' characteristics—how important, central, or influential a node's neighbors are.

Given how different these notions are, even without looking at formal definitions it is easy to see that they capture complementary aspects of a node's position, and any particular measure will be better suited for some applications and less appropriate for others. Let me discuss some of the more standard definitions of each type.

Degree Centrality   Perhaps the simplest measure of the position of a given node in a network is simply to keep track of its degree. A node with degree $n - 1$ would be directly connected to all other nodes, and hence quite central to the network. A node connected to only two other nodes (for large n) would be, at least in one sense, less central. The *degree centrality* of a node is simply $d_i(g)/(n - 1)$, so that it ranges from 0 to 1 and indicates how well a node is connected in terms of direct connections.

Of course, degree centrality clearly misses many of the interesting aspects of a network. In particular, it does not measure how well located a node is in a network. It might be that a node has relatively few links, but lies in a critical location in the network. For many applications a centrality measure that is sensitive to a node's influence or marginal contribution to the network is important. For example, consider the network in Figure 2.13.

In this network the degree of nodes 3 and 5 are three, and the degree of node 4 is only two. Arguably, node 4 is at least as central as nodes 3 and 5, and far more central than the other nodes that each have two links (nodes 1, 2, 6, and 7). There are several senses in which we see a powerful or central role for node 4. If one deletes node 4, the component structure of the network changes. This change might be very important for applications involving information transmission, where node 4 is critical to path-connecting nodes 1 and 7. This aspect would be picked up by a measure such as betweenness. We also see that node 4 is relatively close to all other nodes in that it is at most two links away from any other node, whereas every other node has at least one node linked to it a distance of three or more. This aspect would be important in applications in which something is being conveyed or transmitted through the network (e.g., an opinion or favor), and there is a decay of the connection strength with distance. In that case, being closer can either help a node make use of other nodes (e.g., having access to favors) or can enhance its influence (e.g., conveying opinions). Thus we are brought to the next category of centrality measures.

Closeness Centrality This second class of measures tracks how close a given node is to any other node. One obvious closeness-based measure is just the inverse of the average distance between i and any other node j: $(n-1)/\sum_{j \neq i} \ell(i, j)$, where $\ell(i, j)$ is the number of links in the shortest path between i and j. There are various conventions for handling networks that are not connected, as well as other possible measures of distance, which leads to a family of closeness measures.

A richer way of measuring centrality based on closeness is to consider a decay parameter δ, where $1 > \delta > 0$ and then consider the proximity between a given node and every other node weighted by the decay. In particular, let the *decay centrality* of a node be defined as

$$\sum_{j \neq i} \delta^{\ell(i, j)},$$

where $\ell(i, j)$ is set to infinity if i and j are not path-connected. This centrality measure is related to the symmetric connections model of Jackson and Wolinsky [361], as it is just the benefit that a node receives in that model of a network. As δ approaches 1, it is easy to see that decay centrality measures how large a component a node lies in. As δ approaches 0, then decay centrality gives infinitely more weight to closer nodes than farther ones, so it becomes proportional to degree centrality. For intermediate values of δ, a node is rewarded for how close it is to other nodes, but in a way that very distant nodes are weighted less than closer nodes.

Betweenness Centrality A measure of centrality that is based on how well situated a node is in terms of the paths that it lies on was first proposed by Freeman [255]. We first discussed it in Section 1.2.1 in the context of the Florentine marriages (see Figure 1.1).

Let $P_i(kj)$ denote the number of geodesics (shortest paths) between k and j that i lies on, and let $P(kj)$ be the total number of geodesics between k and j. We can estimate how important i is in terms of connecting k and j by looking at the ratio $P_i(kj)/P(kj)$. If this ratio is close to 1, then i lies on most of the shortest paths connecting k to j, while if it is close to 0, then i is less critical to k and j. Averaging across all pairs of nodes, the *betweenness centrality* of a node i is

$$Ce_i^B(g) = \sum_{k \neq j: i \notin \{k, j\}} \frac{P_i(kj)/P(kj)}{(n-1)(n-2)/2}.$$

For the network in Figure 2.13, we find that $Ce_4^B(g) = 9/15$, $Ce_3^B(g) = 8/15$, and $Ce_1^B(g) = 0$. These values make it clear that nodes 3, 4, and 5 are much more central than the other nodes, and that 4 is the most central node in terms of connecting the other pairs of nodes.

Prestige-, Power-, and Eigenvector-Related Centrality Measures

Beyond these fairly direct measures of centrality, there are more intricate ones. One of the more elegant, both mathematically and in terms of the ideas that it captures, is a notion developed by Bonacich [91]. Bonacich's measure is based on ideas that trace back to Seeley [585], Katz [381], and earlier work of Bonacich [90], and it is useful to start by discussing those ideas. These measures are based on the premise that a node's importance is determined by how important its neighbors are. That is, we might like to account not only for the connectivity or closeness of a node to many other nodes, but also for its proximity to many other "important" nodes. This notion is central to such phenomena as citation rankings and Google page rankings. The difficulty is that such a measure becomes self-referential. The centrality of a node depends on how central its neighbors are, which depends on the centrality of their neighbors, and so forth. There are various approaches to dealing with this issue. The following is a nice application of some basic ideas from matrix algebra and fixed-point theory.

Define the *Katz prestige* of a node i, denoted $P_i^K(g)$, to be a sum of the prestige of i's neighbors divided by their respective degrees. Here I use the term *prestige* as in Katz [381], but it is also a measure of centrality. So i gains prestige from having a neighbor j who has high prestige. However, this measure is corrected by how many neighbors j has, so that if j has more relationships then i obtains less prestige from being connected to j, all else being equal. This correction for the number of relationships that j has might be thought of as correcting for the relative access or time that i spends with j. That is, the Katz prestige of a node i is

$$P_i^K(g) = \sum_{j \neq i} g_{ij} \frac{P_j^k(g)}{d_j(g)}. \tag{2.3}$$

This definition is self-referential, so it is not immediately obvious that it is uniquely (or always) defined. It does provide a series of equations and unknowns, so in principle it is solvable. We can see this as follows. Let $\widehat{g}_{ij} = g_{ij}/d_j(g)$ be the normalized adjacency matrix g so that the sum across any (nonzero) *column* is normalized to 1.[25] The relationship (2.3) can then be rewritten as

$$P^K(g) = \widehat{g} P^K(g), \tag{2.4}$$

or

$$(II - \widehat{g})P^K(g) = 0, \tag{2.5}$$

where P^K is written as a $n \times 1$ vector, and II is the identity matrix.

So, calculating the Katz prestige associated with the nodes of a given network reduces to finding the unit eigenvector of $\widehat{g}$, which is a standard calculation (see Section 2.4 for background on eigenvectors). Note that the Katz prestige is only determined up to a scale factor, so that if $P^K(g)$ solves (2.4) and (2.5), then so does cP^K for any scalar c.

---

25. Let $0/0 = 0$, so that if $d_j(g) = 0$, then set $\widehat{g}_{ij} = 0$.

Katz prestige turns out to be more novel in directed networks than in undirected ones. If in-degree is the same as out-degree for every node, then it is easy to check that the solution to (2.4) is the list of nodes' degrees (or any rescaling of them), so that $[P^K(g)]_i = d_i(g)$. This equality provides a justification for degree centrality but not for a new measure. In the case of a directed network, the normalization in $\widehat{g}_{ij} = g_{ij}/d_j(g)$ is generally by in-degree, so that columns still sum to 1, with the interpretation that directed links to a given node have equal access to that node. In that case, the measure of Katz prestige differs for in-degree and out-degree.[26]

When applied to the network in Figure 2.13, the Katz prestige measures are the $P_4^K(g) = 2$, $P_3^K(g) = 3$, and $P_1^K(g) = 2$. Thus more "prestige" is given to nodes 3 and 5 than to the middle node 4, which has the same prestige as nodes 1, 2, 6, and 7. Here we see the importance of the weighting in the Katz prestige calculation. The middle node 4 is linked to two prestigious nodes, but only receives 1/3 of their time each. So its prestige is $(3)/3 + (3)/3 = 2$. Nodes 3 and 5 are linked to three nodes each. Although each of these three nodes is less prestigious, 3 and 5 receive 1/2 of each of their weight: $2/2 + 2/2 + 2/2 = 3$.

In a variation on this idea that avoids reduction to degree centrality, one does not normalize the network of relations g. The measure is known as *eigenvector centrality*[27] and was proposed by Bonacich [90]. Let $C^e(g)$ denote the eigenvector centrality associated with a network g. The centrality of a node is proportional to the sum of the centrality of its neighbors: $\lambda C_i^e(g) = \sum_j g_{ij} C_j^e(g)$. In matrix notation:

$$\lambda C^e(g) = g C^e(g), \tag{2.6}$$

where λ is a proportionality factor. Thus from (2.6) $C^e(g)$ is an eigenvector of g, and λ is its corresponding eigenvalue. Given that it generally makes sense to look for a measure with nonnegative values, the standard convention is to use the eigenvector associated with the largest eigenvalue, which is nonnegative for the networks considered here (see Section 2.4).

Note that the definition of eigenvector centrality also works for weighted and/or directed networks, without any changes to the expressions. Thus the Katz prestige is a form of eigenvector centrality when we have adjusted the network adjacency matrix to be weighted.

Katz [381] introduced another way of tracking the power or prestige of a node. The idea presumes that the power or prestige of a node is simply a weighted sum of the walks that emanate from it. A walk of length 1 is worth a, a walk of length 2 is worth a^2, and so forth, for some parameter $0 < a < 1$. This scheme gives higher weights to walks of shorter distance, as in the connections model. So it is a method of looking at all walks from a given node and weighting them by distance.

Note that $g \mathbb{1}$ (where $\mathbb{1}$ is the $n \times 1$ vector of 1s) is the vector of degrees of nodes, which tells us how many walks of length 1 emanate from each node. Based on what we saw in Section 2.1.3, $g^k \mathbb{1}$ is the vector whose ith entry is the total

---

26. However, if one normalizes by out-degree, then the measure will be out-degree.
27. See Section 2.4 for background on eigenvectors.

number of walks of length k that emanate from each node. Thus the vector of the power of nodes, or prestige of nodes, can be written as

$$P^{K2}(g, a) = ag \mathbb{1} + a^2 g^2 \mathbb{1} + a^3 g^3 \mathbb{1} \cdots. \qquad (2.7)$$

We can rewrite (2.7) as

$$P^{K2}(g, a) = \left(1 + ag + a^2 g^2 \cdots \right) ag \mathbb{1}. \qquad (2.8)$$

For small enough $a > 0$, this is finite and can be expressed as[28]

$$P^{K2}(g, a) = (I\!I - ag)^{-1} ag \mathbb{1}. \qquad (2.9)$$

Another way to interpret (2.8) is to note that we can start by assigning some base value of $ad_i(g)$ to node i. This value is expressed as the vector $ag \mathbb{1}$. Then a given node receives its base value, plus a times the base value of each node it has a direct link to, plus a^2 times the base value of each node that it has a walk of length 2 to and weighted by the number of walks to the given node, plus a^3 times the base value of each node it has a walk of length 3 to, and so forth.

The measure introduced by Bonacich can be thought of as a direct extension of the above measure of power or prestige. It is often called *Bonacich centrality* and can be expressed as

$$Ce^B(g, a, b) = (I\!I - bg)^{-1} ag \mathbb{1}, \qquad (2.10)$$

where $a > 0$ and $b > 0$ are scalars, and b is sufficiently small so that (2.10) is well defined.[29]

Bonacich centrality can be thought of as a variation on the second prestige measure of Katz, where again we start with base values of $ad_i(g)$ for each node, but then we evaluate walks of length k to other nodes by a factor of b^k times the base value of the end node, allowing b to differ from a. So b is a factor that captures how the value of being connected to another node decays with distance, while a captures the base value on each node. When $b = a$, the two measures coincide.

Normalizing $a = 1$, we can calculate the Bonacich centrality of the network in Figure 2.13 for a couple of values of b, which are listed in Table 2.1 along with other centrality measures for the same network. Degree centrality favors nodes 3 and 5, but treats nodes 1, 2, 6, 7, and 4 similarly, and so misses some aspects of the structure of the network. Closeness differentiates the three types of nodes, favoring node 4, which is similar to betweenness centrality, but with less spread. Decay centrality treats nodes 3, 4, and 5 as being more central than nodes 1, 2, 6,

---

28. From (2.8), if $P^{K2}(g, a)$ is finite, then it follows that $P^{K2}(g, a) - ag P^{K2}(g, a) = ag \mathbb{1}$ or $(I\!I - ag) P^{K2}(g, a) = ag \mathbb{1}$. A sufficient condition for $P^{K2}(g, a)$ to be finite is that a be smaller than 1 divided by the norm of the largest eigenvalue of g; and for the latter to be true it is sufficient that a be smaller than 1 divided by the maximum degree of any agent.

29. Note that the scalar a is no longer relevant, as it simply multiplies all of the terms. It is only useful in comparing to the corresponding Katz measure. This is not to say that the Bonacich measure is the same as that of Katz, as being able to change b without forcing a to adjust in the same manner can lead to important differences.

TABLE 2.1
Centrality comparisons for Figure 2.13

Measure of centrality	Nodes 1, 2, 6, and 7	Nodes 3 and 5	Node 4
Degree (and Katz prestige P^K)	.33	.50	.33
Closeness	.40	.55	.60
Decay centrality ($\delta = .5$)	1.5	2.0	2.0
Decay centrality ($\delta = .75$)	3.1	3.7	3.8
Decay centrality ($\delta = .25$)	.59	.84	.75
Betweenness	.0	.53	.60
Eigenvector centrality	.47	.63	.54
Katz prestige-2 P^{K2}, $a = 1/3$	3.1	4.3	3.5
Bonacich centrality $b = 1/3$, $a = 1$	9.4	13.0	11.0
Bonacich centrality $b = 1/4$, $a = 1$	4.9	6.8	5.4

and 7 for any δ, but the relative rankings of 3 and 5 relative to 4 depend on δ. With a lower δ the results resemble those for like-degree centrality and favor nodes 3 and 5, while for higher δ they resemble those for closeness or betweenness and favor node 4. The eigenvector centralities and self-referential definitions of Bonacich and Katz prestige-2 all favor nodes 3 and 5, to varying extents. As b decreases the Bonacich favors closer connections and higher-degree nodes, while for higher b, longer paths become more important.

These measures are certainly not the only measures of centrality, and it is clear from the above that the measures capture different aspects of the positioning of the nodes. Given how complex networks can be, it is not surprising that there are many different ways of viewing position, centrality, or power in a network.

2.3 ▪ Appendix: Basic Graph Theory

Here I present some basic results in graph theory that will be useful in subsequent chapters.[30]

2.3.1 Hall's Theorem and Bipartite Graphs

A *bipartite* network (N, g) is one for which N can be partitioned into two sets A and B such that if a link ij is in g, then one of the nodes comes from A and the other comes from B. A bipartite network is pictured in Figure 2.14. Settings with two classes of nodes are often referred to as *matching* settings (and in some cases *marriage markets*), where one group is referred to as "women" and the other as

---

30. Excellent texts on graph theory are Bollobás [85] and Diestel [200].

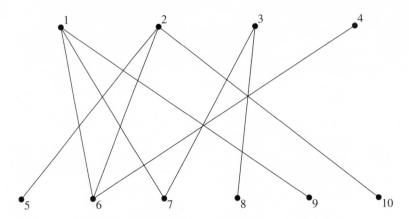

FIGURE 2.14   A bipartite network.

"men." It has applications to markets, where for instance one of the sets consists of buyers and the other of sellers, as well as to such processes as the assignment of students to schools, researchers to labs, and so forth. (See Roth and Sotomayor [568] for an overview of the matching literature.)

One interpretation of a bipartite graph in a matching setting is that it represents the potential relationships that might occur. The object is often to determine a matching for some set of nodes, say $S \subset A$, which is a pairing of the nodes in S with nodes in B such that each node in S is assigned to a distinct node of B and the pairings are feasible as defined by g. That is, a matching for $S \subset A$ relative to g is a mapping $\mu : S \rightarrow B$ such that $i\mu(i) \in g$ for each $i \in S$ and $j \neq i$ implies $\mu(j) \neq \mu(i)$.

It is clear that if we wish to assign each element of $S \subset A$ to a distinct element of B, then the number of neighbors of S in B must be at least as large as the size of S. Moreover, this condition must be true for any subset of S, since we wish to match each element to a different element from B. Hall's theorem states that this condition is not only necessary but also sufficient for such a matching to exist.

Theorem 2.1 (Hall)   *Consider a bipartite graph (N, g) with an associated bipartition of nodes $\{A, B\}$. There exists a matching of a set $S \subset A$ if and only if $|N_{S'}(g)| \geq |S'|$ for all $S' \subset S$.*

As we shall see in Chapter 10, this theorem is useful for working with networked models of markets, which are often bipartite in structure.

2.3.2 Set Coverings and Independent Sets

Given a network (N, g), an *independent* set of nodes $A \subset N$ is a set such that if $i \in A$ and $j \in A$ and $i \neq j$, then $ij \notin g$. An independent set of nodes A is maximal if it is not a proper subset of any other independent set of nodes as illustrated in Figure 2.15.

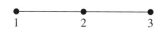

FIGURE 2.15   A three-node network. Independent sets: {1}, {2}, {3}, {1,3}; maximal independent sets: {1,3}, {2}.

The following observation (e.g., see Galeotti et al. [274]) is straightforward but useful when characterizing the equilibria of games played on networks.

Observation 2.1   *Consider a network* (N, g) *and a network* (N, g') *such that* $g \subset g'$. *Any independent set A of g' is an independent set of g, but if $g' \neq g$, then there exist (maximal) independent sets of g that are not (maximal) independent sets of g'.*

The proof of this observation is Exercise 2.8.

Independent sets are closely related to the equilibria of some games played on networks, as first pointed out by Bramoullé and Kranton [103]. To see how independent sets relate to equilibria, consider the following game played on a network.[31] Each player chooses whether to buy a product (e.g., a book). If a player does not buy the book, then he or she can freely borrow the book from any of his or her neighbors who bought it.[32] Indirect borrowing is not permitted, so a player cannot borrow the book from a neighbor of a neighbor. If none of a player's neighbors has bought the book, then the player would prefer to pay the cost of buying the book himself or herself rather than not having any access to the book. This problem is what is known as a classic *free-rider* problem, but defined on a network. A (pure strategy) equilibrium in this game is simply a specification of which players buy the book such that (1) no player who buys the book regrets it, and (2) no player who did not buy the book would rather buy the book. It is easy to see that the (pure strategy) equilibria of this game are precisely the situations in which the players who buy the book form a maximal independent set. This follows because (1) implies that if some player buys a book, then it must be that none of his or her neighbors buy the book, and (2) implies that any player who does not buy a book must have at least one neighbor who bought the book. Thus the first part implies that the set of people who buy the book must be independent, and the second part implies that the set must be maximal.

2.3.3 Colorings

Related to the concept of independent sets is that of colorings. One of the basic applications is to scheduling problems. For example, consider a network in which the nodes represent researchers who will attend a conference.[33] A link indicates

---

31. For more detailed definitions of game-theoretic concepts and a discussion of games played on networks see Chapter 9.

32. Assume that if some player buys the book, and several neighbors wish to borrow it, then they can coordinate on when they borrow it so that they can each borrow it without rivalry.

33. This example is from Bollobás [85].

that the two researchers wish to attend each other's presentations. The conference organizer wishes to know how many different time slots are needed (running parallel sessions within time slots) to ensure that each researcher can attend all of the presentations he or she would like to, and also present his or her own work. This problem is equivalent to coloring the associated graph. Suppose we have a different color to code each time slot of the conference. We want to color the nodes so that no two neighboring nodes have the same color. What is the minimum number of colors needed? That number is called the *chromatic number* of the graph.[34]

If we color the nodes of a network in k colors, then we have produced k independent sets. The coloring problem can then be thought of as finding the minimum number of independent sets needed to partition the set of nodes.[35] This challenging problem has resulted in some celebrated results. The most famous is probably the four-color theorem. That theorem concerns *planar* graphs. Without providing a formal topological definition, a planar graph is one that can be drawn on a piece of paper without having any two links cross each other (so that links can only intersect at one of their involved nodes). The four-color theorem states that every planar graph has a chromatic number of no more than four. This theorem was conjectured in the mid-1800s, and some false proofs were provided before it was proven in 1977 by Appel, Haken, and Koch [19].[36] An overview of coloring problems would take us beyond the scope of this text, but the problems are so central to graph theory and important in their applications that they at least deserve mention.

2.3.4 Eulerian Tours and Hamilton Cycles

The mathematician Leonhard Euler asked (and answered) a question that concerns paths in a graph. The puzzle traces back to a question concerning the old Prussian city of Königsberg, which lay on the Pregel River. The city was cut into four pieces by the river and had seven bridges. The question was whether it was possible to design a walk that started at some point in the city, crossed each bridge exactly once, and returned to the starting point. The four parts of the city can be thought of as the vertices or nodes of a graph, and the seven bridges as edges or links of the graph (Figure 2.17). The question then amounts to asking whether there exists

---

34. This problem is known as the vertex coloring problem. There are also edge coloring problems, and a recent generalization called list coloring problems. The edge coloring problem is to color the edges so that no two adjacent edges have the same color. The minimal number of colors needed has application, for instance, to having enough time slots for scheduling bilateral meetings of neighboring nodes, so that no node needs to be in more than one meeting at once. For an introduction to these problems, see Bollobás [85] or Diestel [200].

35. But note that the sets need not be maximal independent sets. For instance, node 1 is in its own element of the partition in Figure 2.16, but it is not a maximal independent set as it is not connected to 6. If we change the partition and color 6 to be the same color as 1, then we have another four-coloring. But then 2 is in its own element of the partition and does not form a maximal independent set.

36. That proof involved a computer verification that a series of 1,482 cases each reduces to being four-colorable. Shorter proofs have since been provided.

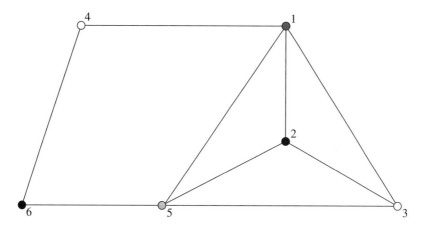

FIGURE 2.16   A planar network on six nodes with chromatic number 4.

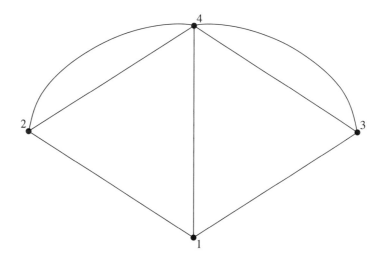

FIGURE 2.17   Multigraph for the Königsberg bridge problem.

a walk in the graph that contains each link in the graph exactly once and starts and ends at the same node.[37] Such a closed walk is said to be an *Eulerian tour* or circuit.

A walk is said to be *closed* if it starts and ends at the same node. It is clear that to have a closed walk that involves every link of a network exactly once, each node in

---

37. The graph here is actually a *multigraph,* as there is more than one link between some pairs of nodes. The general problem of finding Eulerian tours can be stated in either context.

the network must have an even degree.[38] This result holds because each time a node is "entered" by one link on the walk it must be "exited" by a different link, and each time the node is visited, it must be by a link that has not appeared previously on the walk. Euler's simple but remarkable theorem is that this condition is necessary *and sufficient* for such a closed walk to exist.

Theorem 2.2 (Euler) *A connected network g has a closed walk that involves each link exactly once if and only if the degree of each node is even.*

The proof is straightforward and appears as Exercise 2.9.

One can ask a related question for nodes rather than links: when is it possible to find a closed walk that involves each node in the network exactly once? Such a closed walk must be a cycle, and is referred to as a *Hamilton cycle* or a *Hamiltonian*. A further question is whether there exists a *Hamilton path* that includes each node exactly once. Clearly a network that has a Hamilton cycle also has a Hamilton path, while the converse is not true (consider a line).

Determining whether a network has a Hamilton cycle is a much more challenging question than whether it has an Euler tour; this has been an active area of research in graph theory for some time. It has direct applications to the traveling salesman problem, in which a salesman must visit each city on a trip exactly once, cities are nodes on a network, and the path must follow the links.

The seminal theorem on Hamilton cycles is due to Dirac [201]. Stronger theorems have since been developed, as we shall shortly see, but it is worth stating on its own, as it has an intuitive proof that illuminates the proofs of some of the later results.

Theorem 2.3 (Dirac) *If a network has $n \geq 3$ nodes and each node has degree of at least $n/2$, then the network has a Hamilton cycle.*

Proof of Theorem 2.3. First, the network must be connected, because if the minimum degree is $n/2$, then the smallest component has more than half the nodes, and so the network cannot consist of more than one component. Next, consider a longest path in this network, and if there is more than one longest path then pick any one of them. Let i be the starting node on the path and j the ending node. It must be that each of i's neighbors lies on the path, and so at least $n/2$ of the nodes in the path are neighbors of i. To see this, note that if it were not the case, then by starting with an omitted neighbor of i and then moving to i, we could find a longer path. Similarly, j has at least $n/2$ neighbors on the path. It is then easy to check that since i and j each have at least $n/2$ neighbors on the path, at least one of the nodes on the path that is a neighbor of i, say k, must have the previous node on the path be a neighbor of j. Thus consider the cycle formed as follows: $ik, k+1 \ldots j$, $jk-1, k-2 \ldots i$ (where the dots correspond to the original path). The claim is that this cycle is a Hamilton cycle. If this cycle does not include all nodes, then since the network is connected, there is some node outside the cycle connected to some node in the cycle. Then it is possible to make a path including that node and

---

38. Note that a closed walk is not necessarily a cycle (as it may visit some of the intermediate nodes more than once), but a cycle is a closed walk.

all nodes in the cycle, which contradicts the assumption that the original path was of maximal length. ■

An example of a strengthening of the Dirac theorem is the following theorem.

Theorem 2.4 (Chvátal [157]) *Order the nodes of a network of $n \geq 3$ nodes in increasing order of their degrees, so that node 1 has the lowest degree and node n has the highest degree. If the degrees are such that $d_i \leq i$ for some $i < n/2$ implies $d_{n-i} \geq n - i$, then the network has a Hamilton cycle.*

This theorem also has a converse. If a degree sequence does not have this property, then one can find a network that has a degree sequence with at least as high a degree in each entry that does not have a Hamilton cycle. While it is clear that there are networks that have low average degree and have Hamilton cycles (e.g., simply arrange nodes in a circle), this converse shows that guaranteeing the existence of Hamilton cycles either requires strong conditions on basic characteristics like degree sequences or requires much more information about the structure of the network.

2.4 • Appendix: Eigenvectors and Eigenvalues

Given an $n \times n$ matrix T, an *eigenvector v* is a nonzero vector such that

$$Tv = \lambda v, \qquad (2.11)$$

for some scalar λ, which is called the *eigenvalue* of v. Generally, we are interested in nonzero solutions to this equation (noting that a vector of 0s always solves (2.11)).

Eigenvectors come in two flavors: *left-hand* and *right-hand eigenvectors*, which are also known as *row* and *column eigenvectors,* respectively. These terms refer to whether the eigenvector multiplies the matrix T from the left-hand or right-hand side, and correspondingly whether it is a row or column vector. So a left-hand (row) eigenvector is a $1 \times n$ vector v such that

$$vT = \lambda v, \qquad (2.12)$$

whereas a right-hand (column) eigenvector is an $n \times 1$ vector v that satisfies (2.11) for some eigenvalue λ. As the definition at the start of this section suggests, *eigenvector* without a modifier usually refers to a right-hand eigenvector.

Basically, eigenvectors are vectors that, when acted upon by the matrix T, give back some rescaling of themselves, rather than being distorted to some new vector or new direction. So they serve as a sort of fixed point of the transformation T, and for many matrices (but not all), there will be as many eigenvector-eigenvalue pairs as there are dimensions: n.[39]

---

39. We have to be careful here to restrict attention to some normalization of each eigenvector, so it has norm 1, for instance. Otherwise, note that if v is an eigenvector of T, then so is kv for any scalar k, as (2.11), as well as (2.12), are satisfied if v is rescaled.

The usefulness of eigenvectors can be seen in some of their applications. We have already seen important applications—calculating centrality or power, and in particular calculating Katz prestige (and also the eigenvector centrality). The idea is that a given agent's prestige is a weighted average of his or her neighbors' prestige, where the weights correspond to weights from a social network. This measure then presents a self-referential problem, as the prestige has to be derived from the prestige. In this case, we look for an eigenvector with an eigenvalue of 1 since the prestige returns the prestige without rescaling. The existence of an eigenvector with an eigenvalue of 1 in this context is implied by the Perron-Frobenius theorem (see Meyer [466]).

The Perron-Frobenius theorem implies that if T is a nonnegative *(column) stochastic* matrix, so that the entries of each of its columns sum to 1, then there exists a nonnegative right-hand eigenvector v that solves (2.11) and has a corresponding eigenvalue $\lambda = 1$. The same is true of row stochastic matrices and left-hand eigenvectors. If in addition T^t has all positive entries for some t, then all other eigenvalues have a magnitude less than 1.[40]

Eigenvectors are also quite useful for examining the steady state or limit point of some system. Here we might think of T as a transition matrix. Starting with some column vector v, the system transitions to a new vector Tv. A steady state of such a system, or a convergence point, is often a point such that $v = Tv$, so that once the system reaches v, it stays there. Again, v is an eigenvector of T that has a unit eigenvalue. These systems play a central role in Markov chains (see Section 4.5.8), where the vs represent probabilities of being in different states of a system, and the entries of T represent probabilities of transferring from one state to another. This matrix is again stochastic (as probabilities sum to 1) and has a unit eigenvector.

Calculating the eigenvalues and corresponding eigenvectors of a matrix can be done using different methods, as the eigenvector calculation is basically a set of linear equations. If one knows λ, then (2.11) and (2.12) are systems of n equations in n unknowns. A useful way to solve for the eigenvalues associated with T is to rewrite (2.11) as

$$(T - \lambda I\!I)v = 0,$$

where $I\!I$ is the identity matrix (with 1 for each diagonal entry and 0 elsewhere). For this equation to have a nonzero solution v, $T - \lambda I\!I$ must be a singular (non-invertible) matrix.[41] Thus the *characteristic equation* of T is

$$\det(T - \lambda I\!I) = 0,$$

where $\det(\cdot)$ indicates determinant. The solutions to this equation are the eigenvalues of T.

---

40. The Perron-Frobenius theorem implies that the largest eigenvalue of any nonnegative matrix is real valued, and its corresponding eigenvector is nonnegative. Other eigenvalues can be complex valued.

41. This is a matrix in which some rows are linear combinations of other rows, or similarly for columns, which corresponds to having a determinant of 0.

2.4.1 Diagonal Decompositions

There are some particularly useful ways to rewrite a matrix T. To begin, let V be the matrix of left-hand eigenvectors—so that each row is one of the eigenvectors of T. Then we can write

$$VT = \Lambda V, \qquad (2.13)$$

where Λ is the matrix with the eigenvalues corresponding to each row of V on its diagonal:

$$\Lambda = \begin{pmatrix} \lambda_1 & 0 & 0 & 0 \\ 0 & \lambda_2 & 0 & 0 \\ \vdots & 0 & \ddots & \vdots \\ 0 & 0 & \dots & \lambda_n \end{pmatrix}.$$

From (2.13) it follows that if V is invertible, then

$$T = V^{-1}\Lambda V. \qquad (2.14)$$

Equation (2.14) is the *diagonal decomposition* of T. If it exists, then T is said to be *diagonalizable.*

It is sometimes useful to note that V^{-1}, if well-defined, is the matrix of right-hand (column) eigenvectors of T, and that they have the same matrix of eigenvalues as V. To see this, note that from (2.13) $VTV^{-1} = \Lambda VV^{-1} = \Lambda$. Thus $V^{-1}VTV^{-1} = V^{-1}\Lambda$, and so $TV^{-1} = V^{-1}\Lambda = \Lambda V^{-1}$, and V^{-1} is the vector of right-hand eigenvectors.

The decomposition (2.14) is useful for calculating higher powers of T (which, for instance recalling Section 2.1.3, is useful in calculating the walks of T if T has entries consisting of 0 or 1). From (2.14) it follows that

$$T^2 = V^{-1}\Lambda VV^{-1}\Lambda V = V^{-1}\Lambda^2 V,$$

and more generally that

$$T^t = V^{-1}\Lambda^t V,$$

which is useful for calculating speeds of convergence, as in Section 8.3.6.[42]

2.5 ■ Exercises

2.1 *Paths and Connectedness*   Given a network (N, g), define its complement to be the network (N, g') such that $ij \in g'$ if and only if $ij \notin g$. Show that if a network is

---

42. This formulation can help substantially from a computational perspective as well. Raising T to a power directly, for a large matrix, can be computationally intensive. Instead, raising Λ to a power is much easier since it involves raising only the diagonal entries to a power.

not connected, then its complement is. Provide an example of a four-node network that is connected and is such that its complement is also connected.

2.2 ***Facts about Trees*** Show the following:

    (a) A connected network is a tree if and only if it has $n - 1$ links.

    (b) A tree has at least two leaves, where leaves are nodes that have exactly one link.

    (c) In a tree, there is a unique path between any two nodes.

2.3 ***Diameter and Degree*** Consider a sequence of networks such that each network in the sequence is connected and involves more nodes than the previous network. Show that if the diameter of the networks is bounded, then the maximal degree of the networks is unbounded. That is, show that if there exists a finite number M such that the diameter of every network in the sequence is less than M, then for any integer K there exists a network in the sequence and a node in that network that has more than K neighbors.

2.4 ***Centrality Measures*** Consider a two-link network among three nodes. That is, let the network consist of links 12 and 23.

    (a) Calculate the Katz prestige (based on (2.5)) of each node, and compare it to the degree centrality and betweenness centrality for this network.

    (b) Calculate the second measure due to Katz (based on (2.9)) for each node, when $a = 1/2$, which is the Bonacich centrality of each node when $b = 1/2$ and $a = 1/2$. How does this compare to Bonacich centrality when $b = 1/4$ and $a = 1/2$? Which nodes are relatively favored when b increases and why? What happens as we continue to increase b to $b = 3/4$?

2.5 ***Average versus Overall Clustering*** Consider a network (g, N) such that each node has at least two neighbors ($n_i(g) \geq 2$ for each $i \in N$). Compare the average clustering measure of a network to the overall clustering measure in the following two cases:

    (a) $Cl_i(g) \geq Cl_j(g)$ when $d_i(g) \geq d_j(g)$, and

    (b) $Cl_i(g) \leq Cl_j(g)$ when $d_i(g) \geq d_j(g)$.

Hint: Write the average clustering as $\sum_i Cl_i(g) \left(\frac{1}{n} \right)$ and argue that overall clustering can be written as $\sum_i Cl_i(g) \left(\frac{d_i(g)(d_i(g)-1)/2}{\sum_j d_j(g)(d_j(g)-1)/2} \right)$. Then compare these different weighted sums.

2.6 ***Cohesiveness and Close-Knittedness*** There are various measures of how introspective or cohesive a given set of nodes is. Consider a set of nodes $S \subset N$. Given $1 \geq r \geq 0$ Morris [487] defines the set of nodes S to be *r-cohesive* with respect to a network g if each node in S has at least a fraction r of its neighbors in S. That is, S is r-cohesive relative to g if

$$\min_{i \in S} \frac{|N_i(g) \cap S|}{d_i(g)} \geq r, \tag{2.15}$$

where $0/0$ is set to 1.

Young [668] defines the set of nodes S to be *r-close-knit* with respect to a network g if each subset of S has at least a fraction r of its links remaining in S. Given S' and S, let $d(S', S, g) = |\{ij | i \in S', j \in S\}|$ be the number of links between members of S' and members of S. Then S is r-close-knit relative to g if

$$\min_{S' \subset S} \frac{d(S', S, g)}{\sum_{i \in S'} d_i(g)} \geq r,$$

where $0/0$ is set to 1.

Show that if a set of nodes S is r-close-knit relative to g then it is r-cohesive. Provide an example showing that the converse is false.

2.7 ***Independent Sets***   Show that there is a unique network on n nodes that is connected and is such that a maximal independent set of that network involves all nodes except node i. Show that there are two maximal independent sets of that network.

2.8 ***Independent Sets and Equilibria***   Prove Observation 2.1.

2.9 ***Euler Tours***   Prove Theorem 2.2. (Hint: First argue that any longest walk that does not involve any link more than once must be closed.)

CHAPTER 3

Empirical Background on Social and Economic Networks

There are numerous and extensive case studies for a variety of social and economic networks. Through such studies an immense amount has been learned about the structure of networks. In this chapter, I discuss some of the basic stylized facts and hypotheses that have come out of decades of empirical research on social and economic networks. As this literature is much too extensive to survey here, I focus on the fundamental characteristics of networks, mainly dealing with their structural aspects, and some of the hypotheses that we return to in later chapters.

I begin with two cautions regarding some of the stylized facts from the literature. First, examining the structure of any given social network is a formidable task that includes significant hurdles associated with how to define and measure links or relationships. For instance, a primary tool for estimating social networks is to use various sorts of surveys or interviews of the involved parties. Given that individuals have hundreds or even thousands of social relationships, getting them to recall the relevant ones with any desired accuracy is difficult.[1] In addition, it may be impossible to contact or observe all nodes in the network, and when contacted they may have reasons to distort or conceal relationships. Even recent studies of web pages, coauthorship, email, and citation networks, for which data are more easily obtained than for other social networks, have measurement idiosyncrasies. In addition, networks change over time and overlap in various ways. Close friends may fail to interact for long periods. Much of the information that we have about the structure of social networks comes from limited measurements of links that often describe a static and discrete view of something that is inherently dynamic and volatile. Second, as there are biases and idiosyncrasies associated with each

---

1. A sizable literature exists on techniques for measuring social networks, as well as on dealing with other measurement issues, such as missing data and biases in responses. For instance, see Marsden [451], Bernard [57], and Bernard et al. [58].

data set, and data are often collected and encoded in different ways, little has been done to systematically determine the prevalence of characteristics across ranges of social settings.[2] Thus much of what is discussed in this chapter is based on what might be termed *anecdotal evidence* gleaned from various case studies, and the stylized facts reported should be interpreted with the appropriate caution. There is a need for broader systematic studies and comparisons of networks across social settings.

3.1 ▪ The Prevalence of Social Networks

Social relationships play a critical role not only in daily life and behavior but also in determining long-run welfare. They affect the opinions we hold and the information we obtain, and are also often the key to accessing resources. While this is self-evident and gives sociology its foundation, quantifying the extent to which social relationships play roles in various aspects of life is an illuminating exercise.

One of the most robust and best-studied roles of social networks concerns obtaining employment. There have been a number of studies of how social contacts matter in obtaining information about job openings. Such studies began in the late 1940s, and a rich base of information now exists on this subject.[3] One of the earliest studies, by Myers and Shultz [495], was based on interviews with textile workers. They found that 62 percent heard about and applied for their first job through a social contact, in contrast with only 23 percent who directly applied, and 15 percent who found their job through an agency, ads, or the like. A study by Rees and Shultz [558] showed that these numbers were not particular to textile workers, but applied very broadly. For instance, the percentage of those interviewed who found their jobs through the use of social contacts as a function of their profession was: typist, 37.3 percent; accountant, 23.5 percent; material handler, 73.8 percent; janitor, 65.5 percent; and electrician, 57.4 percent. Moreover, the prevalent use of social contacts in finding jobs is robust across race and gender.[4]

The role of social networks is not limited to labor markets, but has been documented much more extensively. For example, networks and social interactions play a role in crime: Reiss [559], [560] finds that two-thirds of criminals commit crimes with others, and Glaeser, Sacerdote, and Scheinkman [290] find that social interaction is important in determining criminal activity, especially with respect to petty crime, youth activity in crime, and neighborhoods with unstable households. Networks have also been studied for various markets: Uzzi [632] finds that relation-specific knowledge is critical in the garment industry, and he documents how social networks play a key role in that industry; Weisbuch, Kirman, and Herreiner

---

2. There are authors, such as Watts [655] and Newman [503], who have looked across (a few) case studies to suggest some common features.

3. For a recent overview of research on social networks in labor markets see Ioannides and Datcher Loury [345].

4. See Corcoran, Datcher, and Duncan [177] for comparisons across race and gender, and Pellizzari [535] for data across countries.

[661] study repeated interactions in the Marseille fish market and discuss the importance of the network structure. Social networks also serve a vital role in the provision of social insurance. For instance, Fafchamps and Lund [237] show that social networks are critical to the understanding of risk sharing in rural Philippines, and De Weerdt [197] analyzes risk sharing in parts of Africa. The set of case studies is much more extensive than this list indicates and also includes extensive analyses of networks in disease transmission, the diffusion of language and culture, the collaboration on scientific research and invention, the citation of articles, the formation of opinions, political activity, choices of products to buy, and interactions of boards of firms—to name a few other applications.[5]

3.2 ▪ Observations on the Structure of Networks

I now discuss some of the regularities and stylized facts about social networks that these studies have revealed.

3.2.1 Diameter and Small Worlds

The stylized fact that large social networks exhibit features of small worlds (see Milgram [468]) is one of the earliest, best-known, and most extensively studied aspects of social networks. The term *small worlds* embodies the idea that large networks tend to have small diameters and small average path lengths.[6]

Milgram [468] pioneered the study of path length through a clever experiment in which people had to route a letter to another person who was not directly known to them. Letters were distributed to subjects in Kansas and Nebraska, who were told the name, profession, and some approximate residential details about a "target" person who lived in Massachusetts. The subjects were asked to pass the letter on to someone whom they knew well and would be likely to know the target or to be able to pass it on to someone else who did, and so on, with the objective of getting the letter to the target. Roughly a quarter of the letters reached their targets, and the median number of hops for a letter to reach its target was 5 and the maximum was 12. Given that the letters should not be expected to have taken the shortest path, these numbers are startlingly small. In addition, given the chains of interactions needed to get a letter from an initial subject to the target, the fact that a quarter of the letters reached their targets is also an impressive figure, especially since response rates in many voluntary surveys are on the order of 20–30 percent.[7]

To understand why many social networks exhibit small diameters, it is useful to think about neighborhood sizes. Most people have thousands of acquaintances.

---

5. The analysis also moves beyond social networks per se, to include things like analyses of the co-appearance of literary (comic-book) superheroes.

6. See Watts [655] for more discussion. This stylized fact is captured in the famous "six degrees of separation" of John Gaure's play, and actually dates to a 1929 play called *Chains* by Frigyes Karinthy.

7. This study has been replicated and extended a number of times. A recent example is research by Watts [657] (see also Dodds, Muhamad, and Watts [202]), who used email messages in a study involving nearly 50,000 subjects in 157 countries and found similar chain sizes.

Depending on whether one keeps track of strong relationships or casual acquaintances, the number might vary from the order of tens or hundreds to the order of thousands for a typical adult in a developed country (e.g., see Pool and Kochen [543], which is a key early study of small worlds). If we take a conservative estimate that a given individual has 100 relatives, friends, colleagues, and acquaintances with whom they are in somewhat regular contact, then a rough calculation (ignoring clustering and treating the network as if it were a tree) yields 100^2 or 10,000 friends of friends, and 1 million friends of friends of friends. With the inclusion of four links, we have covered a nontrivial portion of most countries. While this calculation overestimates the reach of a network (since it treats the network like a tree and ignores the clustering exhibited in most networks), it still provides a feeling for orders of magnitude. If we count more casual acquaintances and use a figure on the order of 1,000 acquaintances per person, then a tree network reaches a million nodes within a path distance of two and reaches a billion within a path distance of three.

Other examples provide similar impressions of path length and diameter measurements of observed networks. Watts and Strogatz [658] report a mean distance of 3.7 in a network among actors in which a link indicates that two actors have been in a movie together. Studies of networks of coauthorship in scientific journals also report relatively small path lengths and diameters on larger numbers of nodes. Here a link represents the coauthorship of a paper during some time period covered by the study. The well-known and prolific mathematician Paul Erdös had many coauthors, and as a fun distraction, many mathematicians (and economists, for that matter) have found the shortest path(s) from themselves to Erdös. For example, an author who coauthored a paper with Erdös has an Erdös number of 1. An author who never directly coauthored a paper with Erdös, but who coauthored with a coauthor of Erdös has an Erdös number of 2, and so forth. There are also some interesting patterns that emerge in such networks in terms of how they grow.[8] These networks are of scientific interest themselves, as they tell us something about how research is conducted and how information and innovation might be disseminated. Similar studies have now been conducted in various fields, including mathematics (Grossman and Ion [315], de Castro and Grossman [186]), biology and physics (Newman [503], [505]), and economics (Goyal, van der Leij, and Moraga-González [304]). Various statistics from these studies give some impression of the network structure, as shown in Table 3.1.[9]

Despite the differences among the networks along some dimensions (e.g., average degree, clustering, and size of the largest component), the average path

---

8. A web site (www.oakland.edu/enp/) maintained by Jerry Grossman, Patrick Ion, and Rodrigo de Castro provides a part of the Erdös graph. The American Mathematical Society web site also provides a platform that gives a shortest path between two authors. A similar analysis is of the Kevin Bacon network (see the web site for the computer science department at the University of Virginia, www.cs.virginia.edu/oracle/), in which a link indicates that two actors appeared in the same movie. In 2004, William Tozier auctioned (on eBay) a promise to coauthor an article, which would provide the purchaser with an Erdös number of 5 (Tozier's is 4). This auction led to a winning bid of more than $1,000 and a controversy, as well as several other such auctions (see *Science News Online,* June 12, 2004, vol. 165, no. 24).

9. As these networks are not connected (there are many isolated authors), the figures for average path length and diameter are reported for the largest component.

TABLE 3.1
Coauthorship networks

Measure	Biology	Economics	Math	Physics
Number of nodes	1,520,521	81,217	253,339	52,909
Average degree	15.5	1.7	3.9	9.3
Average path length	4.9	9.5	7.6	6.2
Diameter of the largest component	24	29	27	20
Overall clustering	.09	.16	.15	.45
Fraction of nodes in the largest component	.92	.41	.82	.85

lengths and diameters of the networks are comparable. Moreover, they are of an order substantially smaller than the number of nodes in the network, which gives an impression of the small-world nature of social networks.

To see how dramatic this effect can be, consider the average number of links it takes to move from one web page to another on the world wide web. Adamic [2] analyzed a sample of 153,127 web sites.[10] She found that there existed an undirected path starting at one page and ending at another in 85.4 percent of the possible cases, and in those cases the average minimum path length was only 3.1. In looking for directed paths, she found that of the 153,127 web sites, there was a strongly connected component of 64,826 sites (so that any web site in this component could be reached by a directed path from any other web site in the component). The average minimum directed path length in this component was 4.2. While not all pairs of sites are path-connected, the fact that it takes so few clicks to get from many of the sites to many others is impressive.[11]

3.2.2 Clustering

Another interesting observation about social networks is that they tend to have high clustering coefficients relative to what would emerge if the links were simply determined by an independent random process. Ideas behind clustering have been important in sociology since Simmel [594], who pointed out the interest in triads (triples of mutually connected nodes). A variety of large socially generated networks exhibit clustering measures much greater than would arise if the network were generated at random. For instance, let us reconsider the networks of researchers that have been analyzed in various fields of study. Newman [503] reports overall clustering coefficients of 0.496 for computer science and 0.45 for

---

10. This sample was based on a data set collected by Jim Pitkow of Xerox PARC. The initial data set contained 259,794 web sites and consisted of more than 50 million pages. The network was trimmed of any leaf nodes.

11. It is worth noting that the data were collected by an algorithm that followed links to locate nodes, and such web-crawling algorithms necessarily introduce some bias in the portion of the overall network that they identify, particularly with respect to path structure.

physics, while Grossman [314] reports an overall coefficient of 0.15 in mathematics. To get an idea of how these values compare to those for clustering that would appear in a purely random network, consider the physics network, which has 52,909 nodes and an average degree of 9.3. A purely random network with this average degree would have a probability of any given link forming of 9.27/52,908, or roughly .00018. For such a purely random network, the chance that link ik is present when ij and jk are present is simply the probability that ik is present, which is then .00018. Thus the clustering value of .45 is roughly 2,500 times greater than the clustering we would see in a random network of the same size and connectivity. We can also examine analogous numbers for a similar network constructed for researchers in economics. The data of Goyal, van der Leij, and Moraga-González [304] covering papers published in economics journals in the 1970s have a total of 33,770 nodes and an average degree of .894.[12] The clustering they report for that network is .193, whereas the corresponding clustering for a purely random network of the same degree is on the order of .894/33,770 or .000026. Here the observed clustering is almost 10,000 times larger than in the random network.[13]

Similarly high clustering has been observed in a variety of other contexts. For example, Watts [655] reports a clustering coefficient of 0.79 for the network consisting of movie actors linked by movies in which they have costarred. Several studies have also analyzed clustering in the world wide web. Adamic [2] gives a clustering measure of 0.1078 on the world wide web data set mentioned in Section 3.2.1. To get a feel for how large this clustering measure is, note that we expect a purely random graph with the same number of links to have a clustering coefficient of 0.00023, so that the observed network measure is about 469 times greater than expected if the links were formed independently.

3.2.3 Degree Distributions

Networks differ in their average numbers of links. For instance, in Table 3.1, the number of coauthors per paper varies dramatically across fields, and there are other differences in social structure across fields. In the economics data set, there are on average 1.6 authors per paper (and only 12 percent of papers have more than two authors), while in the biology data there are on average 3.8 authors per paper. Although the average degree of a network provides a rough feel for connectivity, there is much more information that we would like to know. For instance, how variable is the degree across the nodes of the network? We get a much richer feel for the structure of a social network by examining the full distribution of node degrees rather than just looking at the average.

---

12. The data in Table 3.1 are from the 1990s rather than 1970s, and have more nodes, higher average degree, and slightly lower clustering.
13. Note in such collaboration networks, as there may be many coauthors on any given paper, clustering partly reflects the fact that a multi-coauthored paper provides a complete set of connections between the authors. Given the large number of coauthors per paper in physics, this partly explains the high clustering number observed in this field. The economics data exhibit less of this tendency, as less than 4 percent of all economics papers had more than two coauthors, while roughly 25 percent of all papers had two coauthors.

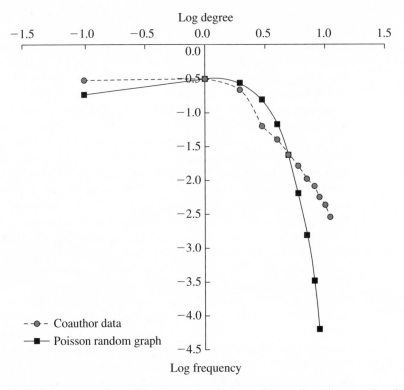

FIGURE 3.1 Comparison of the degree distributions of a coauthorship network and a Poisson random network with the same average degree.

Consider Figure 3.1, which provides a log-log plot of the frequency distribution of degrees using the economics coauthorship data from Goyal, van der Leij, Moraga-Gonzalez [304]. The degrees of economists in the data set range from 0 to more than 50. The distribution also has an interesting shape. It clearly exhibits some curvature. However, it also shows less curvature than the distribution of degrees in a network with the same number of links that are generated independently with identical probability (a Poisson random graph, as discussed in Section 1.2.3). Thus there are more economists with very high degree and more with very low degree than encountered in a network whose links were generated uniformly at random.

This fat-tailed property is not unique to this network, as discussed in Section 2.2.1. Much attention has been paid to the observation that the degree distributions of many observed large networks tend to exhibit fat tails. Price [546] was the first to document such distributions in a network setting, observing that citation networks among scientific articles seemed to follow a power law (both in terms of in- and out-degree). It has been said that these distributions approximate a scale-free or power-law distribution, at least in the upper tail. These terms refer to the frequency of a given degree being proportional to the degree raised to a power, so that the probability or frequency of a given degree can be expressed as

$$P(d) = cd^{-\gamma}, \tag{3.1}$$

where $c > 0$ and $\gamma > 1$ are parameters of the distribution. Hence the term *power-law*. The scale-free aspect refers to the fact that if we consider the probability of a degree d and compare it to that of degree d', then the ratio of $P(d)/P(d') = (d/d')^{-\gamma}$. Now suppose that we double the size of each of these degrees. We find that $P(2d)/P(2d') = (d/d')^{-\gamma}$. It is easy to see that rescaling d and d' by any factor still yields the same ratio of probabilities, and hence the relative probabilities of different degrees depend only on their ratio and not on their absolute size. Hence the term *scale-free*.[14]

For example, Figure 3.2 shows the degree distribution from the data set of Albert, Jeong, and Barabási [9], which is the distribution of in-degrees from the network of links among web pages on the Notre Dame domain of the world wide web in the late 1990s. For a scale-free distribution of the type in (3.1) from a log-log regression on these data, we find an estimate for γ of -2.56.[15]

Such scale-free distributions and fat tails appear well beyond network applications, such as in word usage (Estoup [231] and Zipf [675]), plant classifications (Willis and Yule [672]), city size (Auerbach [22] and Zipf [675]), and article citations (Price [546]).[16] There is a natural explanation for them (discussed in detail in Section 5.2).

There is a very important caution to be mentioned here. It is clear that many network degree distributions exhibit fat tails compared to a Poisson random graph, and that many of the other applications mentioned also have such fat tails. However, it is not so clear that these distributions are really power distributions. Most of these claims are made simply by examining log-log plots that appear approximately linear and then fitting a regression and finding a coefficient. On a plot of log frequency versus log degree, most of the data may end up occupying only a small portion of the figure. For example, in Figure 3.2, less than 10 percent of the data fall in the range below -4 on the vertical scale.[17] The few studies that have fit more than one distribution to a network have found that the degree distribution that best fits tended *not* to be a power distribution (e.g., Pennock et al. [538] and Jackson and Rogers [355]).

---

14. A discrete distribution, such as of citations, in which d can only take on the integer values $\{0, 1, 2, \ldots\}$, is sometimes hard to work with in terms of estimating expected values and conditional expectations. In some cases it is useful to use an approximation in the form of a continuous distribution in which d can take on noninteger values. The canonical continuous distribution satisfying a power law is a Pareto distribution, named for Vilfredo Pareto [526], who studied the distribution of wealth across individuals, among other things. The cumulative distribution function for a Pareto distribution with support $[1, \infty)$ and where $\gamma > 1$ is $F(d) = 1 - d^{-\gamma+1}$. The corresponding density is then $f(d) = (\gamma - 1)d^{-\gamma}$, which is of a similar form to the probability given in (3.1). An estimate of the cumulative distribution function corresponding to (3.1) is $\mathrm{Prob}[d \leq d'] = \sum_0^{d'} cd^{-\gamma}$, which does not have a nice closed form, but is approximately $1 - c'd^{-\gamma+1}$, where $c' > 0$ is a constant.

15. This value is from Jackson and Rogers [355].

16. See Mitzenmacher [470] for an overview of some of the literature on power laws.

17. In Figure 3.2 the points on the graph may be misleading, as there are more points below -4 on the scale. However, the frequency on the vertical scale provides the log weights, and so points higher on the scale represent orders of magnitude more data points. The point with log(degree) equal to 0 corresponds to more than 20 percent of the data!

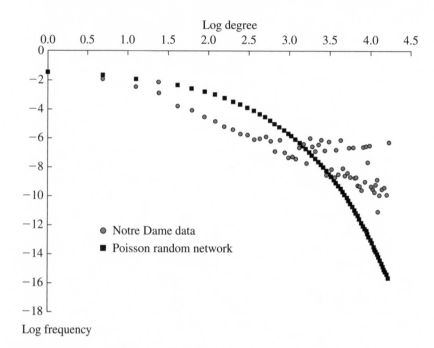

FIGURE 3.2 Distribution of in-degrees of Notre Dame web site domain from Albert, Jeong, and Barabási [9] compared to a Poisson random network.

Jackson and Rogers [355] provide a look at how close to scale-free versus independently random a network is. They examine a class of degree distributions in which the cumulative distribution function F is given by

$$F(d) = 1 - (rm)^{1+r} (d + rm)^{-(1+r)}, \tag{3.2}$$

where $m > 0$ is the average in-degree and r is a parameter that varies between 0 and ∞ and captures how randomly the links are formed. (See Section 5.3 for details.) In the extreme where r tends to 0 the distribution in (3.2) converges to a scale-free distribution, and in the extreme where r tends to ∞ it converges to a negative exponential distribution, which is the proper analog of the degree distribution of a purely random network that is growing over time.[18] Figures 3.3–3.5 show how the degree distribution changes as r is varied, changing from a scale-free distribution with fat tails to one with uniformly random attachment.

---

18. Instead of starting with a fixed number of nodes and randomly putting in links all at once, consider a process in which at each time a new node is born. That new node forms some links randomly with the existing nodes. As nodes age, they will gain links as more nodes are added, while newborn nodes have only their initial number of links. Rather than a Poisson distribution, this process leads to a negative exponential distribution of degrees, which fits some networks very well. This is discussed in more detail in Chapter 5.

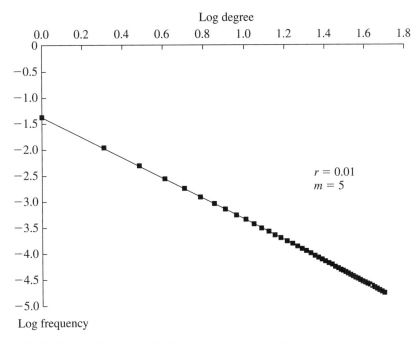

Log degree

FIGURE 3.3   Degree distribution with low r: essentially preferential attachment.

Fits to a few networks[19] give an idea of the range of network structures across applications, as shown in Table 3.2.[20] In Chapter 5, the derivation of a variation on this degree distribution as well as techniques for fitting it to data are detailed. In all cases, the degree distribution fits the data quite closely (the R^2 values on the corresponding regressions vary from .94 to .99).

19. The world wide web data are from an analysis of the links between web pages on the Notre Dame domain of the world wide web from Albert, Jeong, and Barabási [9]. The coauthorship data are from the study by Goyal, van der Leij, and Moraga-González [304]. The citation network consists of the network of citations among all papers that have cited Milgram's [468] paper or have the phrase "small worlds" in the title and is from Garfield [276]. The prison data record friendships among inmates in a study by MacRae [446], the ham radio data record interactions between ham radio operators from Killworth and Bernard [393], and the high school romance data collected romantic relationships between high school students over 1.5 years in a U.S. high school and are from Bearman, Moody, and Stovel [51]. The number of nodes, average degree, and clustering numbers are as reported by the studies. The estimates on randomness are from Jackson and Rogers [355]. The fits on these estimated r values are high, and R^2 ranges between 93 and 99 percent.

20. The clustering value for the coauthor data is actually for overall clustering, as the average number is not available but is likely to be higher, given that the clustering is decreasing in degree. The clustering for the high school romance network is special because that network is mainly heterosexual in its relationships, and so completed triads do not appear. Even if one looks for larger cycles, there are only five present in the whole network, which would be characteristic of a large network formed at random between two groups.

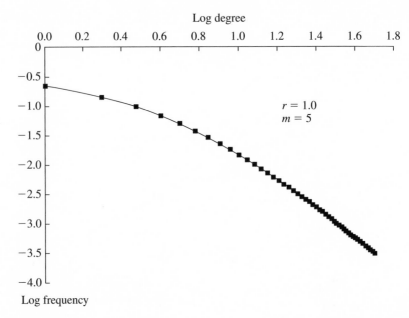

FIGURE 3.4 Degree distribution with medium r: a mixture of uniformly random and preferential attachment.

TABLE 3.2
Comparisons across applications

Measure	World wide web	Citations	Coauthor	Ham radio	Prison	High school romance
Number of nodes	325,729	396	81,217	44	67	572
Randomness r	0.57	0.63	4.7	5.0	∞	∞
Average in-degree m	4.6	5.0	.84	3.5	2.7	.83
Average clustering	.11	.07	.16	.47	.31	—

Note: — indicates not applicable.

The data of Albert, Jeong, and Barabási [9] for the Notre Dame web sites show that although the degree distribution appears to be scale-free to the naked eye, it is best fit by a model mixture of more than 1/3 part uniformly random to 2/3 part scale-free ($r = .57$ has a random/scale-free ratio of more than 1/2).[21] Some of the more purely social networks have parameters that indicate much higher levels of random link formation, which are very far from satisfying a power law. In fact,

21. For comparison, the statistical fit of this model (in terms of R^2) with $r = .57$ is .99 while the fit of a scale-free distribution is only .86.

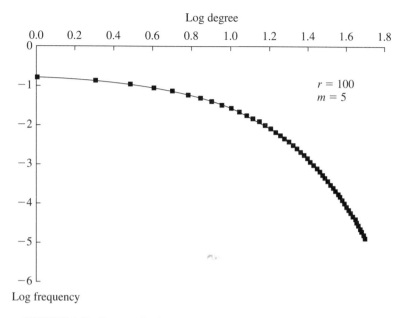

FIGURE 3.5   Degree distribution with high r: a growing random network.

the degree distribution of the romance network among high school students is essentially the same as that of a purely random network.

So there are two lessons here. The first is that many social networks exhibit fat tails in that there are more nodes with relatively high and low degrees than would tend to arise if links were formed independently. The second is that it is hard to find networks that actually follow a strict power law. Even networks that are often cited as exhibiting power laws (e.g., the world wide web) are better fit by distributions that differ significantly from a power distribution.

3.2.4 Correlations and Assortativity

Beyond the degree distribution of a network, we can also ask questions about the correlation patterns in the degrees of connected nodes. For instance, do relatively high-degree nodes have a higher tendency to be connected to other high-degree nodes? This tendency is termed *positive assortativity.*[22]

---

22. Even a finite network in which links are formed completely independently can exhibit correlation in degrees. For instance, consider the simplest possible setting with just two nodes, where the probability of a link is 1/2. The degree of the two nodes is perfectly correlated, as they either both have degree 1 or both have degree 0. So even when links are formed independently, the fact that a node has a high degree tells us something about which other nodes it is likely to be connected to based on their degrees. That correlation tends to disappear in a Poisson random network as the number of nodes grows, but this example shows that such correlations will be delicate.

TABLE 3.3
Correlations in degrees

	Math coauthorship	Physics coauthorship	Email network	Film actors	Internet wiring	Electric power grid	Neural network
Correlation	.12	.36	.09	.21	−.19	−.003	−.23

While there is little systematic study of assortativity, there is a hypothesis that positive assortativity is a property of many socially generated networks and contrasts with the opposite relationship, which is more prevalent in technological and biological networks. This hypothesis is put forth by Newman [503, p. 192] who examines the correlation in degree across linked nodes.[23] Table 3.3 lists correlations among degrees in seven different applications, as reported by Newman [503]. Newman refers to the first four networks as social and the latter three as technological, and remarks on the pattern of positive correlations for the social networks and negative ones for the technological networks.

The network of reported friendships among prison inmates of MacRae [446] shows a positive assortative relationship similar to the social networks in Table 3.3. It has a correlation between a node's in-degree and the average in-degree of its neighbors of .58 (as reported by Jackson and Rogers [355]). However, one can also find exceptions. For example, the ham radio network of interactions between amateur radio operators studied by Killworth and Bernard [393] has a negative correlation. The correlation between a node's in-degree and the average in-degree of its neighbors is −.26 (as reported by Jackson and Rogers [355]); however, since it is a small network, the correlation is not statistically significant. For such networks as those of trading relationships among countries, the structures can be thought of as primarily economic in nature and have some aspects of both social and technological relationships. For example, Serrano and Boguñá [588] find a negative correlation among the degrees of countries that trade with one another and suggest that the average degree of the neighbors of a given node is proportional to the inverse of the square root of that node's degree.[24] They describe the network as a hub-and-spoke system, in which smaller countries (the spokes) have few partners and trade with larger countries (the hubs), which tend to have many more partners. While many larger countries trade with one another, there is still a negative relationship overall.

---

23. Thus the calculation for a network g is simply

$$\frac{\sum_{ij\in g}(d_i - m)(d_j - m)}{\sum_{i\in N}(d_i - m)^2},$$

where m is the average degree and d_i is the degree of node i.

24. Their network has a directed link from one country to another if the first exports to the second. The relationship they examine is similar whether one examines out-degree, in-degree, an undirected version (with a link if there is a directed link in either direction), or an alternate undirected version in which only reciprocal links are examined (i.e., there are directed links both ways).

Related to assortativity, studies of some social networks have also suggested "core-periphery" patterns (e.g., see Brass [105]), for cases in which there is a core of highly connected and interconnected nodes and a periphery of less-connected nodes. Moreover, theories of structural similarity posit that people tend to use other people who are similar to themselves as a reference group (Festinger [247]). Studies building from this hypothesis (e.g., Burt [115]) have found that people with similar structural positions tend to have similar issues, which leads them to communicate with one another.

As the patterns of connections in a network can have a profound impact on processes like the diffusion of behavior, information, or disease, it is important to develop a better understanding of assortativity and other characteristics that describe who tends to be connected to whom in a network.

3.2.5 Patterns of Clustering

There are other patterns in networks that help characterize overall structure. Beyond degree distributions and correlations in degrees, one can also examine how clustering is distributed across a network. Clustering measures, such as average or overall clustering, are simple summary statistics. While they give some insight, we can look at much more detailed information about how clustering varies throughout a network.

For instance, in the example of the Florentine marriages in Section 1.2.1, we can keep track of the full distribution of individual clustering coefficients across nodes. In that network there are nine nodes with clustering coefficients of 0, five nodes with clustering coefficients of 1/3, one with a coefficient of 2/3, and one with a coefficient of 1/15. It is perhaps even more informative to see how the clustering relates to the degree of a node. In the Florentine marriage example, all nodes with degree two or less have individual clustering coefficients of 0. Degree-three nodes have on average a clustering of 4/15 (one has 0 and four have 1/3), degree-four nodes have on average a clustering of 1/6 (one has 2/6 and the other 0), and the degree-six node has a clustering of 1/15. A pattern emerges from this example. First, the low-degree nodes have clusterings of 0, most simply by convention. But more interestingly, the rate of clustering among the higher-degree nodes decreases with degree. This sort of pattern has been noted in other applications as well. That is, the neighbors of a higher-degree node are less likely to be linked to one another compared to the neighbors of a lower-degree node. For example, Goyal, van der Leij, and Moraga-González [304] observe that a network of coauthorship among 81,217 economists in the 1990s had an overall clustering coefficient of .157, while averaging over the 100 nodes with the highest degrees only yielded an average clustering of .043. Thus the highest-degree nodes tend to exhibit lower clustering than one sees on average across the whole network. A very simple way to see whether a network might exhibit such a pattern is to compare the overall clustering to the average clustering. Overall clustering can be thought of as a weighted averaging of clustering across nodes with weights proportional to the number of pairs of neighbors that the nodes have (so the weight on node i is $d_i(d_i - 1)/2)$, while average clustering weights all nodes equally. Thus overall clustering weights the higher-degree nodes much more than average clustering does; thus if the overall clustering is significantly lower than the average clustering, there is a sense in which the clustering is relatively lower for higher-degree nodes. We can see this trend by comparing the overall clustering numbers for the networks reported in Table 3.1 to the average clustering numbers for the same networks. For instance, the ratios of overall clustering to average clustering are .09/.60 for biology, .15/.34 for math, and .45/.56 for physics.[25] We can also simply examine the correlation between degree and the clustering in a node's neighborhood. Jackson and Rogers [355] calculate such correlations for the prison and ham radio networks mentioned above. They find a correlation of −.05 between a node's in-degree and the clustering in its neighborhood for MacRae's [446] friendship network among prison inmates, and a correlation of −.27 between a node's degree and the clustering in its neighborhood for Killworth and Bernard's [393] ham radio network. However, these are both small networks and so neither of these figures is statistically significant—they are only suggestive.

It is not obvious whether this pattern is general for social or other forms of networks, but it is at least exhibited in some observed networks, and we will see it exhibited by some models of network formation.

3.2.6 Homophily

Many social networks exhibit what was named "homophily" by Lazarsfeld and Merton [425]. As we saw in Section 1.2.2, this property refers to the fact that people are more prone to maintain relationships with people who are similar to themselves. It applies very broadly, as measured by age, race, gender, religion, or profession and is generally a robust observation (see McPherson, Smith-Lovin, and Cook [462] for an overview of research on homophily). It was first noted by Burton [118], who coined the phrase "birds of a feather." For example, based on a national survey Marsden [453] finds that only 8 percent of people discuss "important matters" with *any* people of another race. Homophily is an important aspect of social networks, since it means that some social networks may be largely segregated. For instance, homophily has profound implications for access to job information (e.g., see Calvó-Armengol and Jackson [130]). It can also have profound implications for the spread of other sorts of information, behaviors, and so forth.

There is an important distinction between different forms of homophily. One is due solely to opportunity, while the other is due to choice. For instance, it is not surprising that most children have closest friends who are of a similar age as themselves. Much of this is because they form friendships with other children with whom they regularly interact at school. This aspect is due to opportunity, which is constrained by the structure of classes within schools, among other things. But even when presented with opportunities to form ties across age, there is still a tendency to form a disproportionate fraction with own-age individuals. This tendency has been attributed to a number of factors, including maturity and interests. One also sees homophily with respect to other factors, such as race. For example, in middle school, less than 10 percent of "expected" cross-race friendships exist (Shrum, Cheek, and Hunter [592]). That is, given the composition of schools in terms

---

25. The number is not reported in the economics coauthorship data set, but some aspect of this relationship was just discussed.

TABLE 3.4

Friendship frequencies (in percent) compared to population percentages by ethnicity in a Dutch high school

	Ethnicity of students				
	Dutch $(n = 850)$	Moroccan $(n = 62)$	Turkish $(n = 75)$	Surinamese $(n = 100)$	Other $(n = 230)$
Percentage of the population (rounded)	65	5	6	8	17
Percentage of friendships with own ethnicity	79	27	59	44	30

Source: Based on data from Baerveldt et al. [27].

of race, if individuals form relationships in proportion to the relative numbers of people of various races that they encounter, there should be ten times more cross-race relationships than are observed. Thus, in addition to the substantial homophily one would expect because most schools are biased in their racial composition and thus there is a bias in opportunity toward own-race relationships, one also sees a very strong own-race bias in the relationships formed beyond that governed by relative population sizes.

To get a better impression of homophily, consider Table 3.4, which describes the frequency of friendships across different ethnicities in a Dutch high school. The data were collected by Baerveldt et al. [27]. For instance, the first column indicates that Dutch students form 79 percent of their friendships with other Dutch students, while they comprise 65 percent of the population. Here we see *inbreeding* homophily through the high percentages compared to the relative percentages in population. This inbreeding can be due to biases in the interactions within the school that provide the opportunities to form friendships; it can also be influenced by the choices made by the students.[26] There are other factors influencing these tendencies, such as religious and economic background.[27]

3.2.7 The Strength of Weak Ties

The role of social networks in finding jobs was at the heart of some of the most influential research in social networks, which was conducted by Granovetter [305], [308]. He interviewed people in Amherst, Massachusetts, across a variety of professions to determine how they found out about their jobs. He recorded not only whether they used social contacts in their employment searches but also

---

26. See Currarini, Jackson, and Pin [182] for evidence that both effects are present.

27. See Baerveldt et al. [27], Moody [482], Fong and Isajiw [252], Adamic and Adar [3], Fryer [262], and Currarini, Jackson, and Pin [182] for more discussion and background on the factors influencing friendship formation.

the strength of the social relationships as measured by frequency of interaction. He found that a surprising proportion of jobs were obtained through weak ties (as opposed to strong ones). There are various ways to measure the strength of a tie, but Granovetter's basic idea is that strength is related to the "amount of time, the emotional intensity, the intimacy (mutual confiding), and the reciprocal services which characterize the tie" (Granovetter [305, p. 1361]). His measure of the strength of a tie was by the number of times individuals had interacted in the past year (strong, at least twice a week; medium, less than twice a week but more than once a year; and weak, once a year or less). Out of the 54 people that Granovetter had detailed interviews with and who had found their most recent job through a social contact, he found that 16.7 percent had found their job through a strong tie, 55.7 percent through a medium tie, and 27.6 percent through a weak one.[28]

Granovetter's idea was that individuals involved in a weak tie were less likely to have overlap in their neighborhoods than individuals involved in a strong tie. Such ties then are more likely to form bridges across groups that have fewer connections to one another, and can thus play critical roles in the dissemination of information. Granovetter concludes that weak ties are the glue that holds communities together, and paradoxically, strong ties lead to more local cohesion but also to increased overall fragmentation.

There are numerous follow-up studies, including direct tests of some of the hypotheses put forth by Granovetter (e.g., see Friedkin [258]), as well studies of the roles of weak-ties hypotheses in a variety of settings, from the diffusion of technological information to patterns of immigration. There remain many interesting and basic questions about the relative use of weak ties that are not fully answered or whose answers vary across applications. For instance, given that individuals tend to have many weak ties (and for employed adults in large societies, conceivably orders of magnitude more weak than strong ties), how active should we expect weak ties to be in diffusing vital information? Are weak ties important solely because of their bridging behavior and the information they diffuse, or more generally because of other features they embody? And even more fundamentally, is it even generally true that individuals involved in stronger ties are more likely to have strong overlap in their neighborhoods? Regardless of the answers to these and other related questions, Granovetter's work on the strength of weak ties makes it clear that the abstraction to simple 0-1 networks is a crude approximation of interaction structures, and that developing richer models capturing additional nuances of interaction frequency, duration, and heterogeneity, is important.

3.2.8 Structural Holes

Another important concept regarding the structure of networks is due to Burt [115] and concerns what he named "structural holes." A structural hole is a void in a social structure, and in terms of social networks refers to an absence of connections between groups. As Burt points out, this absence does not mean that the groups are

---

28. These are raw numbers, and to keep these in perspective it is worth noting that we tend to have far more weak ties than strong ones.

unaware of one another, but instead that the lack of links between the groups leads to nonredundancies in the information between the groups and can also lead to a failure of diffusion between groups. One of Burt's main points is that individuals who fill structural holes, by offering connections between otherwise separated or sparsely interconnected groups, end up with power and control over the flow of information and favors between groups. For instance, Burt [116] offers evidence that filling structural holes leads to benefits in the form of promotion, bonuses, and other measures of performance networks of managers.

3.2.9 Social Capital

The term *social capital* has come to embody a number of different concepts related to how social relationships lead to individual or aggregate benefits in a society. The concept has been defined in many ways and applied in many contexts, so that there is no strict encompassing definition of social capital. For instance, in some incarnations it refers to the relationships that an individual has and the potential benefits that those relationships can bestow (e.g., Bourdieu [99]), and in others it refers to the aspects of broader social interaction (e.g., relationships, norms, trust) that facilitate cooperation (e.g., see Putnam [548], [549]).[29] Although there is resonance in the idea that social networks and social relationships can translate into individual and societal benefits, and although various definitions of social capital can be useful in identifying the relationship between social structure and welfare, it is important to be precise in the definition and application of social capital, as the term has been so broadly and differentially used that it is not always clear what it means. While I avoid the term *social capital,* many of the models that appear in this book—especially ones that directly relate social network structure to individual behavior and welfare—can be interpreted as capturing various aspects of social capital, operationally defined. There is much that remains to be done in terms of providing definitions and models of social capital that are incisive and yet still portable across applications.

Studies and theories of social capital have resulted in new conceptual understandings of social networks and their impact on behavior. For example, Coleman [163] emphasizes the role of *closure,* which roughly corresponds to high clustering, in the enforcement of social norms. He points out that closure allows agents to coordinate their sanctioning of individuals who deviate from a social norm and thus can help enforce prescribed behavior. Coleman also notes that closure helps in terms of spreading information about reputation from one neighbor of an individual to another.

3.2.10 Diffusion

One important role of social networks is as conduits of information. People often learn from one another, which has important implications not only for how they find

---

29. Sobel [610] provides an insightful overview of the objectives, as well as the shortcomings, of the literature, including some of the difficulties with various definitions.

employment, but also about what movies they see, which products they purchase, which technologies they adopt, whether they participate in government programs, whether they protest, and so forth. Many studies on the diffusion of innovation, including some classic early ones—such as Ryan and Gross's [573] study of the diffusion of hybrid corn seed among Iowa farmers and Hagerstrand's [317] examination of diffusion and the telephone—have shown how important social contacts are in determining behavior.[30]

A classic study in this area that illustrates the relation between social structure and adoption of a new technology is that of Coleman, Katz, and Menzel [165]. They examined the adoption of a new drug by doctors in four cities over a period of 17 months. The adoption of the drug means that it was prescribed to a patient by the doctor, as found through pharmacists' records. The drug was first used in trials by a few "innovators" and subsequently was adopted by almost all of the doctors by the end of the study. Along with the information about adoption times, Coleman, Katz, and Menzel interviewed the doctors to collect other information, including the type of each doctor's practice, his or her age, general prescription habits, as well as information about the doctor's social interaction. To determine the social network, Coleman, Katz, and Menzel asked each doctor questions: "To whom did he most often turn for advice and information?" "With whom did he most often discuss his cases in the course of an ordinary week?" "Who were the friends, among his colleagues, whom he saw most often socially?" The interviewed doctors were asked to provide three names in response to each question.

Using the answers to these questions, and the time series data about adoption rates, Coleman, Katz, and Menzel [165] were able to deduce some relations between the time of adoption and the social structure. For instance, they examined how the proportion of doctors who had adopted the drug depended on the number of social contacts the doctors had.[31]

As summarized in Table 3.5, after 6 months, among the 36 doctors who were not named as friends by any of the other doctors in their survey, only one-third had adopted the drug, while this ratio was just over one-half for the 56 doctors named as friends by one or two of the other doctors; the adoption ratio was more than 70 percent for the 33 doctors named as friends by three or more other doctors. By 10 months, the adoption rate among the doctors not named as friends was still just less than 50 percent, while it was roughly 70 percent among doctors named as friends by one or two others, and 94 percent for the doctors named as friends by three or more others.

As with any data, one has to be careful about inferring causation from correlations; but the differences in adoption rates do indicate that the level of social

---

30. Rogers [564] provides a detailed overview of much of the research on diffusion.

31. The particular explanation for this relationship is not obvious. It appears from the study that there was information about the drug widely available, and so one must rely on other sorts of explanations for such a peer effect, such as some sort of validation: one is more willing to prescribe if one knows a colleague who has prescribed, or that experience from colleagues is more trusted than studies and marketing information, or the like. See Section 13.1.1 for more discussion.

TABLE 3.5
Diffusion of drug adoption among doctors

	Named as friend		
Fraction of doctors adopting by	By no other doctors (36 subjects)	By one or two other doctors (56 subjects)	By three or more other doctors (33 subjects)
6 months	.31	.52	.70
8 months	.42	.66	.91
10 months	.47	.70	.94
17 months	.83	.84	.97

integration as measured through this survey is related to the speed of adoption. As we shall see in Chapters 7 and 8, this relation is to be expected for a variety of reasons involving position in a network. As one should intuitively expect, nodes with greater numbers of connections are more likely to obtain information more quickly and can serve as conduits of information. In terms of empirical work, to determine the role of social structure in influencing behavior, one has to carefully sort out other factors that might be correlated with position in a network and influence behavior. This task is often a challenge, as in any sort of empirical work in which critical variables are endogenous.

As an indicator of the variety of applications in which diffusion is important, consider a recent study by Christakis and Fowler [153]. They examined a network of 12,067 people from 1971 to 2003, based on data including both social relationships and health outcomes. Given that the data included weight at different times for the same individuals, they were able to examine whether weight gain by one individual correlated with weight gains of that individual's friends, while controlling for other factors that might have influenced weight gain. They reported a significant increase in the probability of a weight gain due to a friend's weight gain, which is not present when looking at close geographic proximity. While this result leaves many questions of causation and interpretation open, it does suggest that network structure is important in understanding various forms of diffusion.

With some definitions and empirical background in hand, let us now turn to modeling network formation.

PART II

MODELS OF NETWORK FORMATION

CHAPTER 4

Random-Graph Models of Networks

In this chapter, I discuss a few of the workhorse models of static random networks and some of the properties that they exhibit. As we saw in Chapter 1, randomly generated networks exhibit a variety of features found in the data, and through examining the properties of these models we can begin to trace traits of observed networks to characteristics of the formation process.

Models of random networks find their origin in the studies of random graphs by Solomonoff and Rapoport [611], Rapoport [553], and Erdös and Rényi [227], [228], [229]. The canonical version of such a model, a Poisson random graph, was discussed in Section 1.2.3. The next chapter is a "sister" to this one, in which I discuss a series of recent models of growing random networks that attempt to match more of the properties, such as those discussed in Chapter 3, that are exhibited by many observed networks. Indeed, random-graph models of networks have been a primary tool in analyzing various observed networks. For example the network of high school romances described in Section 1.2.2 has several features that are well described by a random-network model, such as a single giant component, a large number of much smaller components, and a few isolated nodes. Such random models of network formation help tie observed social patterns to the structure of the inherent randomness and the process of link formation.

Beyond their direct use in analyzing observed networks, random-network models also serve as a platform for modeling how behaviors diffuse through a network. For instance, the spread of a disease depends on the contacts among various individuals. That spread can be very different, depending on the average amount of interaction (e.g., interactions with a few others, or with hundreds of others) and its distribution in the population (e.g., everyone interacts with roughly the same number of people, or some people have contact with large numbers while others have contact with few). To understand how such diffusion works, one has to have a tractable model of the link structure within a society, and random-graph models provide such a base. These models are not only useful in understanding the diffusion of a disease, but also in modeling phenomena like the spread of information or decisions that are heavily influenced by peers (e.g., whether to go to college), as we shall see in more detail in Chapters 7 and 8.

Let me reiterate that random models of network formation are largely context-free, in that the nodes and processes for link formation are often simply governed by some given probabilistic rules. Some of these probabilistic rules have stories behind them, but this is not true of all such models. Thus these models are generally missing the social and economic incentives and pressures that underlie network formation, as discussed more fully in Chapter 6. Nevertheless, these models are still quite useful for the reasons mentioned above, and they also serve as benchmarks. By keeping track of the properties that random-graph models of networks exhibit, and which ones they fail to exhibit, we develop a reference point for building richer models and understanding the strengths and weaknesses of models that are tied to social and economic forces influencing individual decisions to form and maintain relationships.

The chapter starts with the presentation of a series of fundamental random-graph models that have been useful in various aspects of network analysis. They include variations on the basic Poisson random-graph model with correlations between links and allow richer degree distributions to be generated. I then present some properties of the resulting networks. These include understanding how small changes in underlying parameters can lead to large changes in the properties of the resulting graphs (thresholds and phase transitions), as well as understanding the conditions for connectedness and existence of a giant component, and other properties, such as diameter and clustering. The chapter concludes with an illustration of how random networks can be used as a basis for understanding the spread of contagious diseases or behaviors in a society.

4.1 ▪ Static Random-Graph Models of Random Networks

The term *static* refers to a typed model in which all nodes are established at the same time and then links are drawn between them according to some probabilistic rule. Poisson random graphs constitute one such static model. This class of static models contrasts with processes where networks grow over time. In the latter type of models new nodes are introduced over time and form links with existing nodes as they enter the network. Such growing processes can result in properties that are different from those of static networks, and they allow different tools for analysis. They are also naturally suited to different applications and are discussed in detail in Chapter 5.

4.1.1 Poisson and Related Random-Network Models

The Poisson random-graph model is one of the most extensively studied models in the static family. Closely related models are the ones mentioned in footnote 9 in Section 1.2.3, in which a network is randomly chosen from some set of networks. For instance, out of all possible networks on n nodes, one could simply pick one completely at random, with each network having an equal probability of being chosen. Alternatively, one could simply specify that the network should have M

4.1 Static Random-Graph Models of Random Networks   ■   79

links, and then pick one of those networks at random with equal probability (i.e., with each M link network having probability $\binom{N}{M}^{-1}$, where $N = \binom{n}{2}$ is the number of potential links among n nodes). Some of these models of random networks have remarkably similar properties. On an intuitive level, if in a network each link is formed with independent probability p, we expect to have $pn(n-1)/2$ links formed, where $n(n-1)/2$ is the potential number of links. While we might end up with more or fewer links, for a large number of nodes, an application of the law of large numbers ensures that the final number of links will not deviate too much from this expected number in terms of the percentage formed. This result guarantees that a model in which links are formed independently has many properties in common with a model network that is prescribed to have the expected number of links.[1]

While these networks are static in the way they are generated, much of the analysis of such random networks concerns what happens when n becomes large. It is easy to understand why most results for random graphs are stated for large numbers of nodes. For example, in the Poisson random-graph model, if we fix the number of nodes and some probability of a link forming, then every conceivable network has some positive probability of appearing. To talk sensibly about what might emerge, we want to make statements of the sort that networks exhibiting some property are (much) more likely to appear than networks that fail to exhibit that property. Thus most results in random-graph theory concern the probability that a network generated by one of these processes has a given property as n goes to infinity. For instance, what is the probability that a network will be connected, and how does this depend on how p behaves as a function of n? Many such results are proven by finding some lower or upper bound on the probability that a given property holds and then determining whether the bounds can be shown to converge to 0 or 1 as n becomes large. We shall examine some of these properties for a general class of static random networks later in the chapter.

Let me begin, however, by describing some variations of static random-graph models other than the Poisson model that provide a feel for the variety of such models and the motivations behind their development.

4.1.2 Small-World Networks

While random graphs can exhibit some features of observed social networks (e.g., diameters that are small relative to the size of the network when the average degree grows sufficiently quickly), it is clear that random graphs lack certain features that are prevalent among social networks, such as the high clustering discussed in Sections 3.2.2 and 3.2.5. To see this, consider the Poisson random-network model,

---

1. Let $G(n, p)$ denote the Poisson random-graph model on n nodes with probability p of any given link, and $G(n, M)$ denote the model network with M links chosen with a uniform probability over all networks of M links on n nodes. The properties of $G(n, p)$ and $G(n, M)$ are closely related for large n when M is near $pn(n-1)/2$. In particular, if $n^2 p(1-p) \to \infty$, and a property holds for each sequence of Ms that lie within $\sqrt{p(1-p)n}$ of $pn(n-1)/2$, then it holds for $G(n, p)$. The converse holds for a rich class of properties (called *convex properties*). See Chapter 2 in Bollobás [86] for detailed definitions and results.

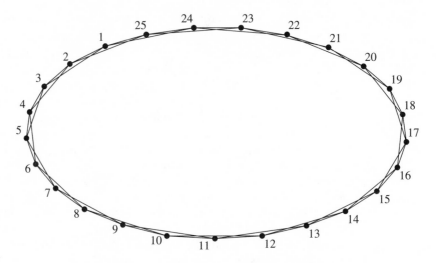

FIGURE 4.1   A ring lattice on 25 nodes with 50 links.

and let us ask what its clustering is. Suppose that i and j are linked and j and k are linked. What is the frequency with which i and k will be linked? Since link formation is completely independent, it is simply p. Thus, as n becomes large, if the average degree grows more slowly than n (which would be true in most large social and economic networks in which there are some bounds on the number of links that agents can maintain) then it must be that p tends to 0 and so the clustering (both average and overall) also tends to 0.

With this tendency in mind, Watts and Strogatz [658] developed a variation of a random network showing that only a small number of randomly placed links in a network are needed to generate a small diameter. They combined this model with a highly regular and clustered starting network to generate networks that simultaneously exhibit high clustering and low diameter, a combination observed in many social networks. Their point is easy to see. Suppose we start with a very structured network that exhibits a high degree of clustering. For instance, construct a large circle but connect a given node to the nearest four nodes rather than just its nearest two neighbors, as in Figure 4.1.

In such a network, each node's individual clustering coefficient is $1/2$. To see this, consider some sequence of consecutive nodes 1, 2, 3, 4, 5, that are part of the network for a large n. Consider node 3, which is connected to each of nodes 1, 2, 4, and 5. Out of all the pairs of 3's neighbors ({1, 2}, {1, 4}, {1, 5}, {2, 4}, {2, 5}, {4, 5}), half of them are connected: ({1, 2}, {2, 4}, {4, 5}). As n grows, the clustering (both overall and average) stays constant at $1/2$. By adjusting the structure of the local connections, we can also adjust the clustering.

While this sort of regular network exhibits high clustering, it fails to exhibit some of the other features of many observed networks, such as a small diameteter and at least some variance in the degree distribution. The diameter of such a network is on the order of $n/4$. The main point of Watts and Strogatz [658] is that by randomly rewiring relatively few links, we can create a network that has a much smaller diameter but still exhibits substantial clustering. The rewiring can be

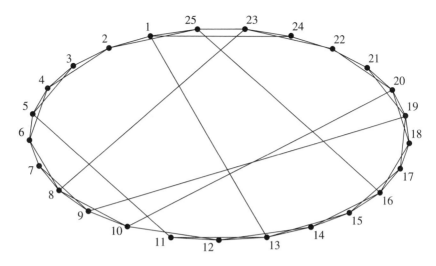

FIGURE 4.2   A ring lattice on 25 nodes starting with 50 links and rewiring seven of them.

done by randomly selecting some link ij and disconnecting it, and then randomly connecting i to another node k chosen uniformly at random from nodes that are not already neighbors of i. Of course, as more such rewiring is done, the clustering will eventually vanish. The interesting region is where enough rewiring has been done to substantially reduce (average and maximal) path length, but not so much that clustering vanishes.

After having rewired just six links, the diameter of the network has decreased from 6 in the network pictured in Figure 4.1 to 5 in the one shown in Figure 4.2, with minimal impact on the clustering. Note also that in Figure 4.1, every node is at distance 6 from three other nodes (e.g., node 1 and nodes 13, 14, and 15), so the rewiring has not simply shortened a few long paths, but rather these new links shorten many paths in the network: there are 39 pairs of nodes at a distance of 6 from each other in the original network that are all moved closer by the rewiring. This example is suggestive, and Watts and Strogatz perform simulations to provide an idea of how the process behaves for ranges of parameters.

This model makes an interesting point in showing how clustering can be maintained in the presence of enough random link formation to produce a low diameter. The model also has obvious shortcomings; in particular, the degree distribution is essentially a convex combination of a degenerate distribution with all weight on a single degree and a Poisson distribution. Such a degree distribution is fairly specific to this model and not often observed in social networks. I discuss alternative models that better match observed degree distributions in Section 5.3.

4.1.3 Markov Graphs and p^* Networks

In this section I describe a generalization of Poisson random graphs that has been useful in statistical analysis of observed networks and was introduced by Frank and Strauss [254]. They called this class of graphs *Markov graphs*. Such random-graph models were later imported to the social network literature by Wasserman and

Pattison [651] under the name of p^* *networks,* and further studied and extended.[2] The basic motivation is to provide a model that can be statistically estimated and still allows for specific dependencies between the probabilities with which different links form.

Again, one important aspect of introducing dependencies is related to clustering, since Poisson random networks with average degrees growing more slowly than the number of nodes have clustering ratios tending to 0, which are too low to match many observed networks. Having dependencies in the model can produce nontrivial clustering.

Conditional dependencies can be introduced so that the probability of a link ik depends on whether ij and jk are present. The obvious challenge is that such dependencies tend to interact with one another in ways that could make it impossible to specify the probability of different graphs in a tractable manner. For instance, if the conditional probability of a link ik depends on whether ij and jk are present but also on whether any other adjacent pairs are present, and the conditional probability of jk depends on other adjacent pairs being present, and so on, we end up with a complicated set of dependencies. The important contribution of Frank and Strauss [254] is to make use of a theorem by Hammersley and Clifford (see Besag [60]) to derive a simple log-linear expression for the probability of any given network in the presence of arbitrary dependencies.

One of the more useful results of Frank and Strauss [254] can be expressed as follows. Consider n nodes, and keep track of the dependencies between links by another graph, D, which is a graph among all of the $n(n-1)/2$ possible links.[3] So D is not a graph on the original nodes but one whose nodes are all the possible links. The idea is that if ij and jk are neighbors in D, then there is some sort of conditional dependency between them, possibly in combination with other links. Thus D captures which links are dependent on which others, possibly in quite complicated combinations. For example, D is empty for the Poisson random-graph model, as all links are independent. If instead we wish to capture the idea that there might be clustering, then the link ik should depend on the presence of ij and kj for each possible j. Thus D would have ik connected to every other link that contains either i or k (and possibly others, depending on the other dependencies).

Let $C(D)$ be all the cliques of D; that is, all of the completely connected subgraphs of D (where the singleton nodes are considered to be connected subgraphs). In the case of a Poisson random graph, $C(D)$ is simply the set of all links ij. With some dependencies, the set $C(D)$ includes all individual links and other cliques as well, for instance, all triads (sets of the form $\{ij, jk, ik\}$). Given a generic element $A \in C(D)$, let $I_A(g) = 1$ if $A \subset g$ (viewing g as a set of links), and $I_A(g) = 0$ otherwise. So if A is a triad $\{ij, jk, ik\}$, then $I_A(g) = 1$ if each of the links ij, jk, and ik are in g, and $I_A(g) = 0$ otherwise. Then Frank and Strauss use Hammersley and Clifford's theorem to show that the probability of a given network g depends only on which cliques of D it contains:

---

2. For instance, see Pattison and Wasserman [534] for an extension to multiple interdependent networks on a common set of nodes.

3. This method is easily adapted to directed links by making D a graph on the $n(n-1)$ possible directed links.

$$\log(\Pr[g]) = \sum_{A \in C(D)} \alpha_A I_A(g) - c, \tag{4.1}$$

where c is a normalizing constant, and the α_As are other free parameters.

In general, this structure can be used to specify the probabilities of different networks directly. Given that D can be very rich and the α_As can be chosen at will, (4.1) allows for an almost arbitrary probability specification. The difficulty and art in applying this type of model are in specifying the dependencies sparingly and imposing restrictions on the α_As so that the resulting probabilities are simple and practical. For certain kinds of dependencies, the expressions can be quite simple and useful (e.g., see Anderson, Wasserman, and Crouch [16]).

To see how the expressions can simplify, consider a case in which $C(D)$ is just the set of all links and all triads (triplets of the form $\{ij, jk, ik\}$). To simplify things further, suppose that there is a symmetry among nodes, so that the probability of any two networks that have the same architecture but possibly different labels on the nodes is identical. Then the α_As are the same across all As that correspond to single links, and the same across all As that correspond to triads. Thus (4.1) simplifies substantially. Let $n_1(g)$ be the total number of links in g, and let $n_3(g)$ be the total number of completed triads in g. Then there exist α_1, α_3, and c such that (4.1) becomes

$$\log(\Pr[g]) = \alpha_1 n_1(g) + \alpha_3 n_3(g) - c.$$

This expression provides a simple generalization of Poisson random graphs (which are the special case $\alpha_3 = 0$), which allows some control over the frequency of clusters. That is, we can adjust the parameters so that graphs that have more substantial clustering will be relatively more likely than graphs that have less clustering (for instance, by increasing α_3).[4]

While such a model can be cumbersome when attempting to capture more complicated dependencies, it still provides a powerful statistical tool for testing for the presence of a specific dependency.[5] One can test for significant differences between fits of a model where they are present and a model where they are absent. Obviously, the validity of the test depends on the appropriateness of the basic specification of the model, as it could be that the model is not a good fit with or without the dependencies, so that the comparison is invalidated.[6]

4.1.4 The Configuration Model

While the Markov model of random networks allows for general forms of dependencies, it is hard to track the general degree distribution and adjust it to match

---

4. See Park and Newman [527] for derivations of clustering probabilities for this example.

5. There are other such models designed for statistical analysis, as well as associated Monte Carlo estimation techniques, as for instance in Handcock and Morris [321].

6. There are some challenges in estimating such models. A useful technique is proposed by Snijders [607], based on the sampling of a Monte Carlo–style simulation of the model and then using an algorithm to approximate the maximum likelihood fit.

that of observed networks. To generate random networks with a given degree distribution, various methods have been proposed. One of the most widely used is the *configuration model,* as developed by Bender and Canfield [54]. The model has been further elaborated by Bollobás [86]; Wormald [667]; Molloy and Reed [473]; and Newman, Strogatz, and Watts [510], among others.

To see how the configuration model works, it is useful to use degree sequences rather than degree distributions. That is, given a network on n nodes, we establish a list of the degrees of different nodes: $(d_1, d_2, \ldots, d_n)$, which is the *degree sequence.*

Now suppose that we wish to generate the degree sequence $(d_1, d_2, \ldots, d_n)$ in a network of n nodes. The sequence is directly tied to the degree distribution, so that the proportion of nodes that have degree d in this sequence is $P^n(d) = \#\{i : d_i = d\}/n$.

Construct a sequence where node 1 is listed d_1 times, node 2 is listed d_2 times, and so on:

$$\underbrace{1, 1, 1, 1, 1, 1, 1, 1, 1, 1, 1, 1, 1}_{d_1 \text{ entries}} \quad \underbrace{2, 2, 2, 2, 2, 2}_{d_2 \text{ entries}} \quad \ldots$$

$$\underbrace{n, n, n, n, n, n, n, n, n, n, n, n}_{d_n \text{ entries}}.$$

Randomly pick two elements of the sequence and form a link between the two nodes corresponding to those entries. Delete those entries from the sequence and repeat.

Note the following about this procedure. First, it is possible to have more than one link between two nodes. The procedure then generates what is called a *multigraph* (allowing for multiple links) instead of a graph. Second, self-links are possible and may even occur multiple times; self-links have generally been ignored in our discussion of networks to this point. Third—and least significant— the sum of the degrees needs to be even or else an entry will be left over at the end of the process.

Despite these difficulties, this process has nice properties for large n. There are two different ways in which we can work around the multigraph problems. One is to work directly with multigraphs instead of graphs and then try to show that the multigraphs generated (under suitable assumptions on the degree sequence) have essentially the same properties as a randomly selected graph with the same degree sequence. The second is to generate a multigraph and delete from it self-links and duplicate links between two nodes. The result is a graph, and if the proportion of deleted links is suitably small, then the graph has a degree distribution close to the one we started with.[7]

---

7. For the purposes of proving results about such processes, one can alternatively simply consider that each graph with the desired degree sequence is generated with equal probability. The approach detailed in Section 4.5.10 is useful in operationalizing a variation of the configuration model.

4.1.5 An Expected Degree Model

Chung and Lu [155], [156] provide a different random model that also approximates a given desired degree sequence. The advantage of their process is that it forms a graph instead of a multigraph, although it still allows for self-loops and does not result in the exact degree sequence, even asymptotically.

Once more, start with n nodes and a desired degree sequence $\{d_1, \ldots, d_n\}$. Form a link between nodes i and j with probability $d_i d_j / (\sum_k d_k)$, where the degree sequence is such that $(\max_i d_i)^2 < \sum_k d_k$, so that each of these probabilities is less than 1.

It is clear that any node i's expected degree is indeed d_i (when a self-link ii is allowed to form with probability $d_i^2 / \sum_k d_k$).

To get a better feel for the differences between the configuration model and the Chung-Lu process, consider a degree sequence in which all nodes have the same number of links $k = \langle d \rangle$. First consider the configuration model, in which self- and duplicate links are deleted. As argued above, the probability that any given node has no self- or duplicate links, and hence has degree k, converges to 1. From here it is not difficult to conclude that with a probability tending to 1, the proportion of nodes with degree k will also converge to 1. Under the Chung-Lu process, although the expected degree of any given node is k (and approaches k if we exclude self-links), the chance that it ends up with exactly k links is bounded away from 1, regardless of whether self-links are allowed. To see this, note that the number of links to other nodes for any node follows a binomial distribution on $n - 1$ draws with a probability of k/n. As the probability of self-links vanishes, the probability that the degree is the same as the number of links excluding self-links approaches 1. However, even as n becomes large, a binomial distribution of $n - 1$ draws with probability k/n places a probability bounded away from 1 on having exactly k links. In fact, this process is effectively the same as having a Poisson random network! The probability of having exactly k links can be approximated from a Poisson approximation (recall (1.4)), and we find a probability on the order of $\frac{e^{-k}(k)^k}{k!}$, which is maximized at $k = 1$ and always less than $1/2$. Thus the realized degree distribution will differ significantly from the distribution of the expected degree sequence, which places full weight on degree k.

While the configuration process (under suitable conditions) leads to a degree distribution more closely tied to the starting one, the Chung-Lu expected degree process is still of interest and more naturally relates to the Poisson random networks. Both processes are useful.

4.1.6 Some Thoughts on Static Random-Network Models

The configuration model and the expected degree model are effectively algorithms for generating random networks with desired properties in terms of their degree sequences. They generally lack the observed clustering and correlation patterns that were discussed in Chapter 3, as the links are formed without regard to anything except relative degrees. A node forms links to two other nodes that are connected to each other purely by chance, and not because of their relation to each other. The two models are also severely limited as models of how social and economic networks form, since they do not account for the incentives and forces that influence the formation of relationships: the models describe a world governed completely and uniformly by chance. Why study such random-graph models? One of the biggest challenges in network analysis is developing tractable models. The combinatorial nature of networks that exhibit any heterogeneity makes them complex animals. Much of the theory starts by building up from simple models and techniques and seeing what can be carried further. These two models represent important steps in generalizing Poisson random graphs, and several of the basic properties of Poisson random graphs do generalize to some richer degree distributions. We also develop a better understanding of how degree distributions relate to other properties of networks. Although there are more refinements that we will introduce to the models, much can still be learned from looking at these relatively simple generalizations of the Poisson model. As we shall see, these models are the workhorses for providing foundations for understanding diffusion in a network, among other things.

4.2 ▪ Properties of Random Networks

If we fix some number of nodes n and then try to analyze the properties of a resulting random network, we run into difficulties. For instance, for the Poisson random-network model, each possible network has a positive probability of being formed. While some are much more likely than others, it is difficult to talk about what properties the resulting network will exhibit since everything is possible. We could try to sort out which properties are likely to hold and how this depends on the probability with which links are formed, but for a fixed n the likelihood of a given property holding is often a complicated expression that is difficult to interpret. One technique for dealing with this issue is to resort to computer simulations in which a large number of random networks are generated according to some model to estimate the probabilities of different properties being exhibited on some fixed number of nodes. Another technique is to examine the properties of the network at some limit, for instance as the number of nodes tends to infinity. If one can show that a property does (or does not) hold at the limit, then one can often conclude that the probability of it holding for a large network is close to 1 (or 0). Even with limiting properties, simulations are still useful in a number of ways. For instance, even limiting properties may be hard to ascertain analytically, and then simulations provide the only real tool for examining a property. Or perhaps we are interested in a relatively small network, or we want to see how the probability of a given property varies with parameters and the size of the population. As simulation techniques are more straightforward than analyses of limits, I illustrate the former at different points in what follows. The alternative approach of examining the limiting properties of large networks requires the development of some tools and concepts that I now discuss.

4.2.1 The Distribution of the Degree of a Neighboring Node

In a variety of applications, one is faced with the following sort of calculation. Start at some node i with degree d_i. Consider a neighbor j. How many neighbors do we expect j to have? This consideration is important for estimating the size of

i's expanding neighborhoods, keeping track of contagion and the transmission of beliefs, estimating diameters, and for many other calculations. Basically, any time we consider some process or calculation that navigates through the network and we wish to keep track of expansion rates, this is an important sort of calculation.

To understand such calculations, let us start by examining the following related calculation. Suppose that we randomly select a link from a network and then randomly pick one of the nodes at either end of the link. What is the conditional probability describing that node's degree? If the network has a degree distribution described by P, the answer is *not* simply P. To understand this, start with a simple case in which the network is such that $P(1) = 1/2 = P(2)$. So, half of the nodes have degree 1 and half have degree 2. For instance, consider a network on four nodes with links $\{12, 23, 34\}$. While the degree distribution is $P(1) = 1/2 = P(2)$, if we randomly pick a link and then randomly pick an end of it, there is a 2/3 chance of finding a node of degree 2 and a 1/3 chance of a node of degree 1. This just reflects the fact that higher-degree nodes are involved in a higher percentage of the links. In fact, their degree determines relatively how many more links they are involved with. In particular, if we randomly pick a link and a node at the end of it and consider two nodes of degrees d_j and d_k, then node k is d_k/d_j times more likely to be the one we find than is node j. Extrapolating, the distribution of degrees of a node found by choosing a link uniformly at random from a network that has degree distribution P and then picking either one of the end nodes with equal probability is

$$\widetilde{P}(d) = \frac{P(d)d}{\langle d \rangle},\qquad(4.2)$$

where $\langle d \rangle = E_P[d] = \sum_d P(d)d$ is the expected degree under the distribution P. Thus simply randomly picking a node from a network and finding nodes by randomly following the end of a randomly chosen link are two very different exercises. We are much more likely to find high-degree nodes by following the links in a network than by randomly picking a node.

Now let us return to the original problem: start at node i with degree d_i and examine the distribution of the degree of one of its randomly selected neighbors. If we consider either the configuration or expected degree models and let the number of nodes grow large and have the degree distribution converge (uniformly) to P, then the distribution of the degree of a randomly selected neighbor converges to the $\widetilde{P}$ described in (4.2). This is true since the degrees of two neighbors are approximately independently distributed for large networks provided the largest nodes are not too large.[8] It is also true in the Poisson random networks. We can also directly deduce that the distribution of the *expected* degree of the node at a given end of any given link (including self-links) under the Chung-Lu process is exactly given by $\widetilde{P}$. However, this distribution might not match that of the degree of the node at a given end of any given link. As an example, under the Chung-Lu

---

8. To see why this distribution is only approximate, consider any given degree sequence and the expected degree model. Say that there are n_d nodes with degree d. One of those nodes can only be connected to $n_d - 1$ nodes with degree d, while a node with degree $d' \neq d$ can be connected to n_d nodes with degree d. So there is actually a slight negative correlation in the degrees of neighboring nodes.

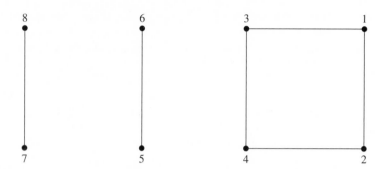

FIGURE 4.3   Forming networks with perfect correlation in degrees.

process if $P(2) = 1$, so that all nodes have an expected degree of 2, some nodes will have more than two links and others less. In this case, $\widetilde{P}$ places probability 1 on having degree 2. If we rewrite P to be the realized degree distribution, then for large n, (4.2) provides a good approximation of the degree of a neighbor.

However, it is important to note that (4.2) does not hold for many prominent models of growing random networks (e.g., those with preferential attachment) that are discussed in Chapter 5. In those random networks there is nonvanishing correlation between the degrees of nodes, so that higher-degree nodes tend to have neighbors with higher degrees than do lower-degree nodes.

To see how correlation can change the calculations, consider two methods of generating a network with a degree distribution such that half of the nodes have degree 1 and half have degree 2. First, generate such a network by operating the configuration model on a degree sequence (1, 1, 2, 2, 1, 1, 2, 2, 1, 1, 2, 2, 1, 1, 2, 2 ...).[9] In this case, it is clear that for any node, the degree of a randomly selected neighbor is 2 with a probability converging to 2/3 and 1 with a probability converging to 1/3.

Second, consider the following very different way of generating a network with the same degree distribution. Start with eight nodes. Connect four of them in a square, so that they each have degree 2 and have two neighbors each with degree 2. Connect the other four in two pairs, so that each has degree 1 and has a neighbor with degree 1, as in Figure 4.3. Now replicate this process. The same degree distribution results, so that half of the nodes are of degree 1 and half of degree 2, but nodes are segregated so that nodes with degree 1 are only connected to nodes of degree 1, and similarly nodes of degree 2 are only connected to nodes of degree 2. Then the degree of a node's neighbor is perfectly correlated with that node's degree. Note also that if we examine the degree of a neighbor of a randomly picked node in Figure 4.3, there is an equal probability that it has degree 1 or degree 2! That is, if we examine nodes 1 to 4, then any randomly selected neighbor has degree 2, while if we examine nodes 5 to 8, then any randomly selected neighbor has degree 1. So the distribution of the degree of a node's neighbor is quite different

---

9. For this calculation, let us work with the resulting multigraph, so that self- and duplicate links are accounted for and so that the degree distribution is exactly realized when the number of nodes is a multiple of four.

from $\widetilde{P}$, regardless of whether we condition on the starting node's degree or we simply pick a node uniformly at random and then examine one of its neighbors' degrees.

While this example is stark, it illustrates that we do need to be careful in tracking how a network was generated, and not only its degree distribution, to properly calculate properties like the distribution of degrees of neighboring nodes.[10]

4.2.2 Thresholds and Phase Transitions

When we examine random networks on a growing set of nodes for some given parameters or structure, often properties hold with either a probability approaching 1 or a probability approaching 0 in the limit. So while it may be difficult to determine the precise probability that a property holds for some fixed n, it is often much easier to discern whether that probability is tending to 1 or 0 as n approaches infinity.

To see how this works, consider the Poisson random-network model on a growing set of nodes n, where we index the probability of a link forming as a function of n, denoted $p(n)$. It is quite natural to have the probability of a link forming between two nodes vary with the size of the population. For example, people have on average several thousand acquaintances, then p needs to be on the order of 1 percent for a network that includes a few hundred thousand nodes, but p should be on the order of a fraction of a percent for a population of millions of nodes. So to keep the average degree constant as the number of nodes grows $p(n)$ should be proportional to $1/n$. With a $p(n)$ in hand, we can ask what the probability is that a given property holds as $n \to \infty$. Interestingly, many properties hold with a probability that approaches either 0 or 1 as the number of nodes grows, and the probability that a property holds can shift sharply between these two values as we change the underlying random-network process. For example, we can ask what the probability is that a network will have some isolated nodes. For some random-network-formation processes, if the network is large, then it will be almost certain that some isolated nodes exist, while for other network-formation processes it will be almost certain that the resulting network will not have any isolated nodes. This sharp dichotomy is true of a variety of properties, such as whether the network has a giant component, or has a path between any two nodes, or has at least one cycle. There are also many exceptions, in terms of properties that do not exhibit such convergence patterns. For instance, consider the property that a network has an even number of links. For many random-network processes, the probability of this property holding will be bounded away from 0 and 1.

There are different methods for specifying a property, but an easy way is just to list the networks that satisfy it. Thus properties are generally specified as a set of networks for each n, and then a property is satisfied if the realized network is in the set. Then a property is a list of $A(N) \subset G(N)$ of the networks that have the

---

10. Some of the literature proceeds with calculations as if there were no correlation between neighboring nodes, even though some of the models (like preferential attachment discussed in Chapter 5) used to motivate the analysis generate significant correlation. Using a variation on the configuration model is one approach to avoiding such problems, but it does limit the scope of the analysis.

property when the set of nodes is N. For instance, the property that a network has no isolated nodes is

$$A(N) = \{g \mid N_i(g) \neq \emptyset \; \forall i \in N\}.$$

Most properties that are studied are referred to as *monotone* or *increasing properties*. Those are properties such that if a given network satisfies the property, then any supernetwork (in the sense of set inclusion) satisfies it. So a property $A(\cdot)$ is monotone if $g \in A(N)$ and $g \subset g'$ implies that $g' \in A(N)$. The property of having an even number of links is obviously not a monotone property, while the property of being connected is monotone.

For the Poisson model, the model is completely specified by $p(n)$, where n is the cardinality of the set of nodes N. In that case, a *threshold function* for some given property is a function $t(n)$ such that

$$\Pr[A(N)|p(n)] \to 1 \text{ if } p(n)/t(n) \to \infty,$$

and

$$\Pr[A(N)|p(n)] \to 0 \text{ if } p(n)/t(n) \to 0.$$

When such a threshold function exists, it is said that a *phase transition* occurs at that threshold.[11] Even when there are no sharp threshold functions, we can still often produce lower or upper bounds so that a given property holds for value of $p(n)$ above or below those bounds.

This definition of a threshold function is tailored to the Erdös-Rényi or Poisson random-network setting, as it is based on having a function $p(n)$ describe the network-formation process. We can also define threshold functions for other sorts of random-network models, but they will be relative to some other description of the random process, generally characterized by several parameters.

To get a better feel for a threshold function, consider a relatively simple one. Let us consider the property that node 1 has at least one link; that is, $A(N) = \{g \mid d_1(g) \geq 1\}$. In the Poisson model, the probability that node 1 has no links is $(1 - p(n))^{n-1}$, and so the probability that $A(N)$ holds is $1 - (1 - p(n))^{n-1}$. To derive a threshold function, we need to determine for which $p(n)$ this tends to 0 and for which $p(n)$ it tends to 1. If $t(n) = \frac{r}{n-1}$, then by definition of the exponential

---

11. There are different sorts of probabilistic statements that one can make, analogous to differences between the weak and strong laws of large numbers. That is, it can be that as n grows, the probability of a property holding tends to 1. This is the weak form of the statement. The stronger form reverses the order between the probability and the limit, stating that the probability that the property holds in the limit is 1. This is also stated as having something hold *almost surely*. For many applications this difference is irrelevant, but in some cases it can be an important distinction. In most instances in this text, I claim or use the weaker form, as that is generally much easier to prove and one can work with a series of probabilities, which keeps the exposition relatively clear, rather than having a probability defined over sequences. Nevertheless, many of these claims hold in their stronger form.

function (see Section 4.5.2), the limit of the probability that node 1 has no links is

$$\lim_{n}(1 - t(n))^{n-1} = \lim_{n}\left(1 - \frac{r}{n-1}\right)^{n-1} = e^{-r}. \tag{4.3}$$

So if $p(n)$ is proportional to $\frac{1}{n-1}$, then the probability that node 1 has at least one link is bounded away from 0 and 1 in the limit. Thus $t^*(n) = \frac{1}{n-1}$ is a function that could potentially serve as a threshold function. Let us check that $t^*(n) = \frac{1}{n-1}$ is in fact a threshold function. Suppose that $p(n)/t^*(n) \to \infty$, which implies that $p(n) \geq \frac{r}{n-1}$ for any r and large enough n. From (4.3) it follows that $\lim_{n}(1 - p(n))^{n-1} \leq e^{-r}$ for all r, and so $\lim_{n}(1 - p(n))^{n-1} = 0$. Similarly, if $p(n)/t^*(n) \to 0$, then an analogous comparison implies that $\lim_{n}(1 - p(n))^{n-1} = 1$. Thus $t^*(n) = \frac{1}{n-1}$ is indeed a threshold function for a given node having neighbors in the Poisson random-network model.

Note that the threshold function is not unique here, as $t(n) = a/(n+b)$ for any fixed a and b also serves as a threshold. Moreover, threshold functions provide conclusions only about how large or small $p(n)$ has to be in terms of its limiting order, and conclusions only hold in the limit. How large n has to be for the property to hold with a high probability depends on more detailed information. For instance, $p(n) = e^{-n}$ and $p(n) = 1/n^{1.0001}$ both lead to probabilities of 0 that node 1 has any neighbors in the limit, but the second function gets there much more slowly. Determining properties for smaller n requires examining the probabilities directly, which is feasible in this example, but more generally may require simulations.

Much is known about the properties and thresholds of the Poisson random-network model. A brief summary is as follows.

▪ At the threshold of $1/n^2$ the first links emerge, so that the network is likely to have no links in the limit for $p(n)$ of order less than $1/n^2$, while for $p(n)$ of order larger than $1/n^2$ the network has at least one link with a probability going to 1.[12] (The proof of this is Exercise 4.6.)

▪ Once $p(n)$ is at least $n^{-3/2}$ there is a probability converging to 1 that the network has at least one component with at least three nodes.

▪ At the threshold of $1/n$ cycles emerge, as does a giant component, which is a unique largest component that contains a nontrivial fraction of all nodes (at least cn, for some factor c).[13]

▪ The giant component grows in size until the threshold of $\log(n)/n$, at which the network becomes connected.

---

12. Note that this does not contradict the calculations above, which were for the property that a single node did/did not have any neighbors. The property here is that none of the nodes have any neighbors.

13. Below the threshold of $1/n$, the largest component includes no more than a factor times $\log(n)$ of the nodes; at the threshold the largest component contains a number of nodes proportional to $n^{2/3}$. See Bollobás [85] for details.

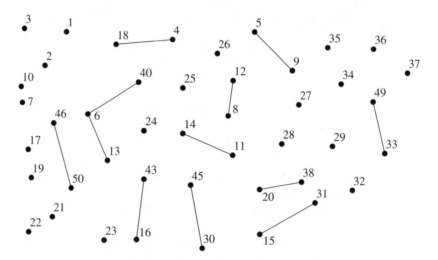

FIGURE 4.4 A first component with more than two nodes: a random network on 50 nodes with $p = .01$.

These various thresholds are illustrated in Figures 4.4–4.7. Shown are Poisson random networks generated on 50 nodes (using the program UCINET). For $n = 50$, the first links emerge at $p = 1/n^2 = .0004$. The threshold for the first component with more than two nodes to emerge is at $p = n^{-3/2} = .003$. Indeed, the first network with $p = .01$ has a component with three nodes, but the network is still very sparse, as seen in Figure 4.4.

At the threshold of $p = \frac{1}{n} = .02$ cycles start to emerge. As an example, note that in Figure 4.4 (the network with $p = .01$) no cycles appear, while in Figures 4.5–4.7 the networks with $p = .03$ or more cycles are present. Moreover, the first signs of a giant component also appear at the threshold $p = .02$, as pictured in Figure 4.5.

As p increases the giant component swallows more and more nodes, as pictured in Figure 4.6. Eventually, at the threshold of $p = \frac{log(n)}{n} = .08$ the network should become connected, as pictured in Figure 4.7.

To better understand how these thresholds work, let us start by examining the connectedness of a random network.

4.2.3 Connectedness

Whether or not a network is connected—and more generally, its component structure—significantly affects the transmission and diffusion of information, behaviors, and diseases, as we shall see in Chapter 7. Thus it is important to understand how these properties relate to the network-formation process.

The phase transition from a disconnected to a connected network was one of the many important discoveries of Erdös and Rényi [227] about random networks. Exploring this phase transition in detail is not only useful for its own sake, but also because it helps illustrate the idea of phase transitions and provides some basis for extensions to other random-network models.

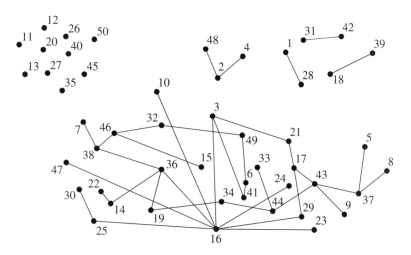

FIGURE 4.5   Emergence of cycles: a random network on 50 nodes with $p = .03$.

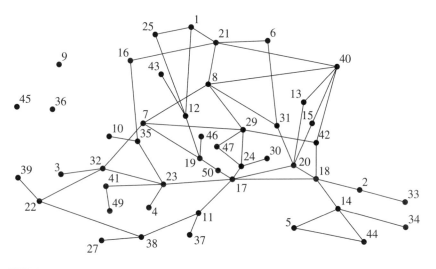

FIGURE 4.6   Emergence of a giant component: a random network on 50 nodes with $p = .05$.

Theorem 4.1 (Erdös and Rényi [229])   *A threshold function for the connectedness of the Poisson random network is $t(n) = \log(n)/n$.*

The theorem states that if the probability of a link is larger than $\log(n)/n$, then the network is connected with a probability tending to 1, while if it is smaller than $\log(n)/n$, then the probability that it is not connected tends to 1. This threshold corresponds to an expected degree of $\log(n)$.

The ideas behind Theorem 4.1 are relatively easy to understand, and a complete proof is not too long, even though the conclusion of the theorem is profound. To

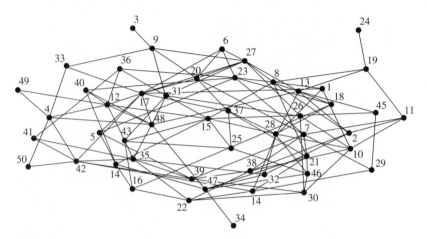

FIGURE 4.7   Emergence of connectedness: a random network on 50 nodes with $p = .10$.

show that a network is not connected, it is enough to show that there is some isolated node. It turns out that $t(n) = \log(n)/n$ is not only the threshold for a network being connected, but also for there not to be any isolated nodes. To see why, note that the probability that a given node is completely isolated is $(1 - p(n))^{n-1}$ or roughly $(1 - p(n))^n$. When working with $p(n)$ near the threshold, $p(n)/n$ converges to 0, and so we can approximate $(1 - p(n))^n$ by $e^{-np(n)}$. Thus the probability that any given node is isolated tends to $e^{-p(n)n}$, which at the threshold is $1/n$. For n nodes, it is then not too hard to show that this is the threshold of having some of the nodes be isolated, as below the threshold the chance of any node being isolated is significantly less than $1/n$, while above the threshold it is significantly greater than $1/n$. The proof then shows that above this threshold it is not only that there are no isolated nodes, but also no components of size less than $n/2$. The intuition behind this logic is that the probability of having a component of some small finite size is similar (asymptotically) to that of having an isolated node: there need to be no connections between any of the nodes in the component and any of the other nodes. Thus either some of the nodes are isolated or else the smallest components must be approaching infinite size. However, the chance of having more than one component of substantial size goes to 0, as there are many nodes in each component and there cannot be any links between separate components. So components roughly come in two flavors: very small and (uniquely) very large.

I now offer a full proof of Theorem 4.1 to give a rough idea of how some of the many results in random-graph theory have been proven: basically by bounding probabilities and expectations and showing that the bounds have the claimed properties.

Proof of Theorem 4.1.[14]   Let us start by showing that $t(n) = \log(n)/n$ is the threshold for having isolated nodes. First, we show that if $p(n)/t(n) \to 0$, then

---

14. This proof is adapted from two different proofs by Bollobás (Theorem 7.3 in [86] and Theorem 9 on page 233 of [85]).

the probability that there are isolated nodes tends to 1. This clearly implies that the network is not connected.

The probability that a given node is completely isolated is $(1 - p(n))^{n-1}$ or roughly $(1 - p(n))^n$ if $p(n)$ is converging to 0. Given that $p(n)/n$ converges to 0, we can approximate $(1 - p(n))^n$ by $e^{-np(n)}$. Thus the probability that any given node is isolated goes to

$$e^{-p(n)n}.$$

We can write $p(n) = \frac{\log(n) - f(n)}{n}$, where $f(n) \to \infty$ and $f(n) < \log(n)$, and then $e^{-p(n)n}$ becomes

$$\frac{e^{f(n)}}{n}.$$

The expected number of isolated nodes is then $e^{f(n)}$, which tends to infinity.

While expecting a divergent number of isolated nodes in the limit is suggestive that there will be some isolated nodes, it does not prove that the probability of there being at least one isolated node converges to 1. We show this via Chebyshev's inequality.[15] Let X^n denote the number of isolated nodes. We have shown that $E[X^n] \to \infty$. If we can show that the variance of X^n, $E[(X^n)^2] - E[X^n]^2$, is no more than twice $\mu = E[X^n]$, then we establish the claim by applying Chebyshev's inequality. In particular, we then can conclude that $\Pr[X^n < \mu - r\sqrt{2\mu}] < 1/r^2$ for all $r > 0$, which since $\mu \to \infty$ implies that the probability converges to 1 that X^n will be arbitrarily large, and so there will be an arbitrarily large number of isolated nodes. To obtain an upper bound on $E[(X^n)^2] - E[X^n]^2$, note that $E[X^n(X^n - 1)]$ is the expected number of ordered pairs of isolated nodes, which is $n(n - 1)(1 - p)^{2n-3}$, since a pair of nodes is isolated from all other nodes if none of the $2(n - 2)$ links from either of them is present and the link between them is not present. Thus

$$E\left[(X^n)^2\right] - E\left[X^n\right]^2 = n(n - 1)(1 - p)^{2n-3} + E[X^n] - E[X^n]^2$$

$$= n(n - 1)(1 - p)^{2n-3} + E\left[X^n\right] - n^2(1 - p)^{2n-2}$$

$$\leq E\left[X^n\right] + pn^2(1 - p)^{2n-3}$$

$$= E\left[X^n\right]\left(1 + pn(1 - p)^{n-2}\right)$$

$$\leq E\left[X^n\right]\left(1 + (\log(n) - f(n))e^{-\log(n)+f(n)}(1 - p)^{-2}\right)$$

$$\leq 2E\left[X^n\right].$$

To complete the proof that $t(n) = \log(n)/n$ is the threshold for having isolated nodes, we need to show that if $p(n)/t(n) \to \infty$, then the probability that there

---

15. Chebyshev's inequality (see Section 4.5.3) states that for a random variable X with mean μ and standard deviation σ, $\Pr[|X - \mu| > r\sigma] < 1/r^2$ for every $r > 0$.

are isolated nodes tends to 0. It is enough to show this for $p(n) = \frac{\log(n)+f(n)}{n}$, where $f(n) \to \infty$ but $f(n)/n \to 0$.[16] By a similar argument to the one above, we conclude that the expected number of isolated nodes is tending to $e^{-f(n)}$, which tends to 0. The probability of having X^n be at least one then has to tend to 0 as well for $E[X^n] \to 0$.

To complete the proof of the theorem, we need to show that if $p(n)/t(n) \to \infty$, then the chance of having any components of size 2 to size $n/2$ tends to 0. Let X_k denote the number of components of size k, and write $p(n) = \frac{\log(n)+f(n)}{n}$, where $f(n) \to \infty$ and $f(n)/n \to 0$.[17] It is enough to show that $E[\sum_{k=2}^{n/2} X_k] \to 0$:

$$E\left[\sum_{k=2}^{n/2} X_k\right] \le \sum_{k=2}^{n/2} \binom{n}{k}(1-p)^{k(n-k)}$$

$$= \sum_{k=2}^{n^{3/4}} \binom{n}{k}(1-p)^{k(n-k)} + \sum_{k=n^{3/4}}^{n/2} \binom{n}{k}(1-p)^{k(n-k)}$$

$$\le \sum_{k=2}^{n^{3/4}} \left(\frac{en}{k}\right)^k e^{-knp} e^{k^2 p} + \sum_{k=n^{3/4}}^{n/2} \left(\frac{en}{k}\right)^k e^{-knp/2}$$

$$\le \sum_{k=2}^{n^{3/4}} e^{k(1-f(n))} k^{-k} e^{2k^2\log(n)/n} + \sum_{k=n^{3/4}}^{n/2} \left(\frac{en}{k}\right)^k e^{-knp/2}$$

$$\le 3e^{-f(n)} + n^{-n^{3/4}/5},$$

which tends to 0 in n. ∎

The above proof used the specific structure of the Poisson random-network model fairly extensively. How far can we extend it to other random-network models? It is fairly clear that the argument used to prove Theorem 4.1 is not well suited to the configuration model. For the configuration model, under some reasonable bounds on degrees, each node acquires its specified degree with a probability approaching 1, which renders the above approach inapplicable. If the limiting degree distribution has a positive mass on nodes of degree 0 in the limit, then the network will not be connected, but otherwise it is not clear what will happen. For instance, if the associated $\widetilde{P}$ has mass on nodes of some bounded degree in the limit, then there is a nonvanishing probability that the network will not be connected. However, requiring that the mass on nodes of some bounded

---

16. Having no isolated nodes is clearly an increasing property, so that it holds for larger $p(n)$. The reason for working with $f(n)/n \to 0$ is to ensure that the approximation of $(1 - p(n))^n$ by $e^{-np(n)}$ is valid asymptotically.

17. Here again, we work with $p(n)$ "near" the threshold, as this will establish that the resulting network is connected with a probability going to 1 for such p values, and then it holds for larger values of p.

degree vanish is not enough, as it is still possible that the network has a nontrivial probability of being disconnected.

The expected-degree model of Chung and Lu [156] is better suited for an analysis of connectivity. At least we can make some progress with regard to the threshold for the existence of isolated nodes, since the model is essentially a generalization of the Poisson random-network model that allows for different expected degrees across nodes (with the possibility of self-loops).

Recall that in the expected-degree model there are degree sequences of an expected degree for each node $d_1, \ldots, d_n$. Let

$$\text{Vol}^n = \sum_{i=1}^{n} d_i \qquad (4.4)$$

denote the total expected degree of the network on n nodes. The probability of a link between nodes i and j is then $\frac{d_i d_j}{\text{Vol}^n}$, and so the probability that node i is isolated is

$$\prod_j \left(1 - \frac{d_i d_j}{\text{Vol}^n}\right).$$

By (4.4) the probability that a given node i is isolated is then approximately $e^{-d_i \sum_j d_j / \text{Vol}^n} = e^{-d_i}$ for large n (under the assumption that $\max_i \frac{d_i^2}{\text{Vol}^n}$ converges to 0, which is maintained under the expected-degree model). The probability that no node is isolated is then

$$\prod_i (1 - e^{-d_i})$$

or approximately

$$e^{-\sum_i e^{-d_i}}.$$

This expression suggests a threshold such that if $\sum_i e^{-d_i} \to 0$, then there will be no isolated nodes, while if $\sum_i e^{-d_i} \to \infty$, then isolated nodes will occur.[18] As a double-check of this hypothesis, let $d_i = d(n) = \log(n) + f(n)$ for each i in the Poisson random-network setting (where $p(n) = d(n)/n$). Then $\sum_i e^{-d_i} = e^{-f(n)}$, so if $f(n) \to \infty$, there are no isolated nodes (and a connected network) and if $f(n) \to -\infty$, then the probability tends to 1 that there are isolated nodes. Indeed, this threshold corresponds to the one we found for the Poisson random-network model.

---

18. I am not aware of results on this question or the connectedness of the network under the expected-degree model. While it seems natural to conjecture that the threshold for the existence of isolated nodes is the same as the threshold for connectedness, the details need to be checked.

4.2.4 Giant Components

In cases for which the network is not connected, the components structure is of interest, as there will generally be many components. In fact, we have already shown that if the network is not connected in the Poisson random-network model, then there should be an arbitrarily large number of components. We also know from Section 4.2.2, that in this case there may still exist a giant component. Let us examine this in more detail and for a wider class of degree distributions.

In defining the size of a component, a convention is to call a component *small* if it has fewer than $n^{2/3}/2$ nodes, and *large* if it has at least $n^{2/3}$ nodes (e.g., see Chapter 6 in Bollobás [86]). The term *giant component* refers to the unique largest component, if there is one. This component may turn out to be small in some networks, but we are generally interested in giant components that involve nonvanishing fractions of nodes, which are necessarily "large" components.

The idea of there being a unique largest component is fairly easy to understand, in the case where these are large components. It relates to the proof of Theorem 4.1: for any two large sets of nodes (each containing at least $n^{2/3}$ nodes) it is very unlikely that there are no links between them, unless the overall probability of links is very small. For instance, in the Poisson random-network model the probability of having no links between two given large sets of nodes is no more than $(1 - p)^{n^{4/3}}$. If $pn^{4/3} \to 0$, then this expression is positive, but otherwise it tends to 0. Proving that the probability of not having two separate large components goes to 0 involves a bit more proof, but is relatively straightforward (see Exercise 4.7).

4.2.5 Size of the Giant Component in Poisson Random Networks

As already seen, it is not even clear whether each node is path-connected to every other node. Unless p is high enough relative to n, it is likely there exist pairs of nodes that are not path-connected. As a result, diameter is often measured with respect to the largest component of a network.[19] But this also raises a question as to what the network looks like in terms of components. The answer is one of the deeper and more elegant results of the work of Erdös and Rényi.

To gain some impression of the size of the giant component, let us do a simple heuristic calculation.[20] Form a Poisson random network on $n - 1$ nodes with a probability of any given link being $p > 1/n$. Now add a last node, and again connect this node to every other node with an independent probability p. Let q be the fraction of nodes in the largest component of the $n - 1$ node network. As a fairly accurate approximation for large n, q will also be the fraction of nodes in the largest component of the n-node network. (The only possible exception is if the added node connects two large components that were not connected before. As argued above, the chance of having two components with large numbers of

---

19. This practice can result in some distortions; for instance, a network in which each node has exactly one link has a diameter much smaller than a network that has many more links.
20. The heuristic argument is based on Newman [503], but a very different and complete proof of the characterizing equation above the threshold for the emergence of the giant component can be found in Bollobás [86].

nodes that are not connected to each other goes to 0 in n, given that $p > 1/n$.) The chance that this added node lies outside the giant component is the probability that none of its neighbors are in the giant component. If the new node has degree d_i this probability converges to $(1 - q)^{d_i}$, as n becomes large. As we can think of any node as having been added in this way, in a large network the expected frequency of nodes of degree d_i that end up outside the giant component is approximately $(1 - q)^{d_i}$.[21] So the overall fraction of nodes outside the giant component, $1 - q$, can then be found by averaging $(1 - q)^{d_i}$ across nodes:[22]

$$1 - q = \sum_d (1 - q)^d P(d). \tag{4.5}$$

When we apply (4.5) to the Poisson degree distribution described by (1.4), the fraction of nodes outside the giant component is then approximated by the solution of

$$1 - q = \sum_d \frac{e^{-(n-1)p}((n-1)p)^d}{d!}(1 - q)^d.$$

Since

$$\sum_d \frac{((n-1)p(1-q))^d}{d!} = e^{(n-1)p(1-q)},$$

an approximation is described by the solution to

$$q = 1 - e^{-q(n-1)p}. \tag{4.6}$$

There is always a solution of $q = 0$ to (4.6). When the average degree is larger than 1 (i.e., $p > 1/(n-1)$), and only then, there is also a solution for q that lies between 0 and 1.[23] This case corresponds to a phase transition, in that the appearance of such a giant component comes above the threshold of $(n-1)p = 1$. That is, there is a

---

21. There are steps omitted from this argument, as for any finite n the degrees of nodes in the network are correlated, as are their chances of being in the largest component conditional on their degree. For example, a node of degree 1 is in the giant component if and only if its neighbor is. If that neighbor has degree d, then it has $d - 1$ chances to be connected to a node in the giant component. Thus the calculation approaches $(1 - q)^{d-1}$ for the neighbor to be in the giant component. To see a fuller proof of this derivation, see Bollobás [86].

22. Here take the convention that $0^0 = 1$, so that if $q = 1$, then the right-hand side of this equation is $P(0)$.

23. To see this, let $f(q) = 1 - e^{-q(n-1)p}$. We are looking for points q such that $f(q) = q$ (known as *fixed points*). Since $f(0) = 1 - e^0 = 0$, $q = 0$ is always a fixed point. Next, note that f is increasing in q with derivative $f'(q) = (n-1)pe^{-q(n-1)p}$ and is strictly concave, as the second derivative is negative: $f''(q) = -((n-1)p)^2 e^{-q(n-1)p}$. Since $f(1) = 1 - e^{-(n-1)p} < 1$, f is a function that starts at 0 and ends with a value less than 1, and f is increasing and strictly concave. The only way in which it can ever cross the 45-degree line is if it has a slope greater than 1 when it starts, otherwise it will always lie below the 45-degree line, and 0 will be the only fixed point. The slope at 0 is $f'(0) = (n-1)p$, and so there is a $q > 0$ such that $q = f(q)$ if and only if $(n-1)p$ is greater than 1.

marked difference in the structure of the resulting network depending on whether the average degree is greater than or less than 1. If the average degree is less than 1, then there is essentially no giant component: instead the network consists of many components that are all small relative to the number of nodes. If the average degree exceeds 1, then there is a giant component that contains a nontrivial fraction of all nodes (approximately described by (4.6)).

Note that if we let $p(n-1)$ grow (so that the average degree is unbounded as n grows), then the solution for q tends to 1. Of course, that requires the average degree to become large. In a random network for which there is some bound on average degree, so that $p(n-1)$ is bounded, then q is between 0 and 1. For example, for a solution to $q = 1 - e^{-q(n-1)p}$ when $n = 50$ and $p = .08$, q roughly satisfies $q = 1 - e^{-4q}$, or q is about .98. So an estimate for the size of the giant component is 49 nodes out of 50—which happens to match the realized network in Figure 1.6 exactly.

4.2.6 Giant Components in the Configuration Model

Understanding giant components more generally is especially important, as they play a central role in various problems of diffusion, and a giant component gives an idea of the most nodes that one might possibly reach starting from a single node. Let us examine giant components for more general random networks, using the configuration model as a basis.[24] We work with randomly formed networks according to the configuration model on n nodes and will examine the limiting probability that the resulting networks have a giant component when n grows large. Consider a sequence of degree sequences, ordered by the number of nodes n, with corresponding degree distributions described by $P^n(d)$. Assume that these satisfy some conditions:

1. the degree distributions converge uniformly to a limiting degree distribution P that has a finite mean,
2. there exists ε such that $P^n(d) = 0$ for all $d > n^{\frac{1}{4}-\varepsilon}$,
3. $(d^2 - 2d)P^n(d)$ converges uniformly to $(d^2 - 2d)P(d)$, and
4. $E_{P^n}[d^2 - 2d]$ converges uniformly to its limit (which may be infinite).

An important aspect of such sequences is that the probability of having cycles in any small component tends to 0. Let us examine properties of the degree distribution that indicate when such networks exhibit a giant component. The following is a simple and informal derivation. A somewhat more complete derivation appears in Section 4.5.

The idea is to look for the threshold at which, starting at a random node, there is some chance of finding a nontrivial number of other nodes through tracing out expanding neighborhoods. Indeed, if a node is in a giant component, then exploring longer paths from the node should lead to the discovery of more and more nodes,

---

24. Similar results hold for the expected degree model of Chung and Lu [155], and under weaker restrictions on the set of admissible degree distributions, but they follow from a less intuitive argument.

while if it is in a small component, then expanding neighborhoods will not result in finding many more nodes.

At or below the threshold at which the giant component just emerges, the components are essentially trees, and so each time we search along a link that has not been traced before, we will find a node that has not been previously visited. This fact allows us to analyze the component structure up to the point where the giant component emerges as if the network were a collection of trees. The following argument (due to Cohen et al. [161]) provides the idea behind there being negligible numbers of cycles below the threshold.[25] Consider any link in the configuration model on n nodes. The probability that the link connects two nodes that were already connected in a component with s nodes (where s is the size of some component ignoring that link) is the probability that both of its end nodes lie in that component, which is proportional to $\left(\frac{s}{n}\right)^2$. Thus the fraction of links that end up on cycles is on the order of $\sum_i \left(\frac{s_i}{n}\right)^2$, where s_i is the size of component i in the network. This sum is less than $\sum_i \frac{s_i S}{n^2}$, where S is the size of the largest component. Thus, since $\sum_i s_i = n$, we find that the proportion of links that lie on cycles is of an order no more than S/n. If we are at or below the threshold at which the giant component is just emerging, then with probability 1, S/n is vanishing for large n. Thus when we consider a sequence of degree distributions at or below the threshold of the emergence of the giant component, the components are essentially trees.[26]

To develop an estimate of component size as the network grows, let ϕ denote the limiting number of nodes that can be found on average by picking a link uniformly at random, picking with equal chance one of its end nodes, and then exploring all of the nodes that can be found by expanding neighborhoods from that end node. Given an absence of cycles, the number of new nodes reached by a link is the first node reached plus that node's degree minus 1 (as one of its links points back to the original node) times ϕ, which indicates how many new nodes can be expected to be reached from each of the first node's neighbors. Thus

$$\phi = 1 + \sum_{d=1}^{\infty} (d-1)\widetilde{P}(d)\phi = 1 + \sum_{d=1}^{\infty} \frac{P(d)d}{\langle d \rangle}(d-1)\phi.$$

We can rewrite this equation as

$$\phi = 1 + \frac{(\langle d^2 \rangle - \langle d \rangle)\phi}{\langle d \rangle},$$

or

$$\phi = \frac{1}{2 - \frac{\langle d^2 \rangle}{\langle d \rangle}}. \tag{4.7}$$

---

25. This is part of the informality of the derivation, and I refer the interested reader to Molloy and Reed [473] for a more complete proof.

26. The more rigorous result proven by Molloy and Reed [473] establishes that almost surely no component has more than one cycle.

Now we deduce the threshold at which a giant component emerges. If ϕ has a finite solution, then for a node picked uniformly at random in the network, we expect to find a finite number of nodes that can be reached from one of its links. This places the node in a finite component. If ϕ does not have a finite solution, then we expect at least some nodes that are found uniformly at random to be in components that are growing without bound, which should occur at the threshold for the emergence of a giant component. For ϕ to have a finite solution it must be that $0 > \langle d^2 \rangle - 2 \langle d \rangle$. Thus if

$$\langle d^2 \rangle - 2 \langle d \rangle > 0, \tag{4.8}$$

there is a giant component, and so the threshold is where $\langle d^2 \rangle = 2 \langle d \rangle$.

In the case of a Poisson distribution $\langle d^2 \rangle = \langle d \rangle + \langle d \rangle^2$, and so the giant component emerges when $\langle d \rangle^2 > \langle d \rangle$, or $\langle d \rangle > 1$. Indeed the threshold for the existence of a giant component for the Poisson random-network model is $t(n) = 1/n$, which corresponds to an average degree of 1.

In the case of a regular network, for which the degree sequences have full weight on some degree k, if we solve for $\langle d^2 \rangle = 2 \langle d \rangle$ we find $k^2 = 2k$, and so a threshold for a giant component is $k = 2$. Clearly, for $k = 1$ we just have a set of dyads (paired nodes) and no giant component.

For a scale-free network, where the probability of degree d is of the form $P_n(d) = cd^{-\gamma}$, we find that $\langle d^2 \rangle$ diverges when $\gamma < 3$, and so generally there is a giant component regardless of the specifics of the distribution.

The arguments underlying the derivation of (4.5) were not specific to a Poisson distribution, and so for the configuration model when the probability of loops is negligible we still have as the approximation for the size of the giant component the largest q that solves

$$1 - q = \sum_d (1 - q)^d P(d). \tag{4.9}$$

Using this expression, there is much that we can deduce about how the size of the giant component changes with the degree distribution. For instance, if we change the distribution to place more weight on higher nodes (in the sense of first-order stochastic dominance; see Section 4.5.5), then the right-hand side expectation goes down for any value of q, and the new value of $1 - q$ that solves (4.9) has to decrease as well, which corresponds to a larger giant component, as detailed in Exercise 4.11. This trend makes sense, since we can think of such a modification as effectively adding links to the network, which should increase the size of the giant component. Interestingly, providing a mean-preserving spread in the degree distribution has the opposite effect, decreasing the size of the giant component. This result is a bit more subtle, but has to do with the fact that $(1 - q)^d$ is a convex function of d. So spreading out the distribution leads to some higher-degree nodes that have a higher chance of being in a giant component, but also produces some lower-degree nodes with a much lower chance of being in the giant component. The key is that the convexity implies that there is more loss in probability from moving to lower-degree nodes than gain in probability from the high-degree nodes.

4.2.7 Diameter Estimation

Another important feature of a network is its diameter. This characteristic, as well as other related measures of distances between nodes, is important for understanding how quickly behavior or information can spread through a network, among other things.

Let us start by calculating the diameter of a network that makes such calculations relatively easy. Suppose that we examine a tree component so that there are no cycles. A method of obtaining an upper bound on diameter is to pick some node and then successively expand its neighborhood by following paths of length ℓ, where ℓ is increased until the paths are sufficiently long to reach all nodes. Then every node is at distance of at most ℓ from our starting node and no two nodes can be at a distance of more than 2ℓ from each other. Thus the diameter is bounded below by ℓ and above by 2ℓ.[27] What this diameter works out to be will depend on the shape of the tree.

Let us explore a particularly nicely behaved class of trees. Consider a tree such that every node either has degree k or degree 1 (the "leaves"), and such that there is a "root" node that is equidistant from all of the leaves. Start from that root node.[28] If we then move out by a path of 1, we have reached k nodes. Now, by traveling on all paths of length 2, we will have reached all nodes in the immediate neighborhoods of the nodes in the original node's neighborhood: $k + k(k-1)$ or k^2 nodes. Extending this reasoning, by traveling on all paths of length ℓ, we will have reached

$$k + k(k-1) + k(k-1)^2 + \ldots + k(k-1)^{\ell-1}.$$

This expression can be rewritten (see Section 4.5.1) as

$$k\frac{(k-1)^\ell - 1}{k - 1 - 1} = \left(\frac{k}{k-2}\right)\left((k-1)^\ell - 1\right).$$

We can thus find the neighborhood size needed to reach all nodes by finding the smallest ℓ such that

$$\left(\frac{k}{k-2}\right)\left((k-1)^\ell - 1\right) \geq n - 1.$$

Approximating this equation provides a fairly accurate estimate of the neighborhood size needed to reach all nodes of $(k-1)^\ell = n - 1$, or

$$\ell = \frac{\log(n-1)}{\log(k-1)},$$

and then the estimated diameter for this network is

$$2\frac{\log(n-1)}{\log(k-1)}.$$

---

27. Note that we can easily see that both of these bounds can be reached. If the network is a line with an odd number of nodes, and we do this calculation from the middle node, then the diameter is exactly 2ℓ, while if we start at one of the end nodes, then the diameter is exactly ℓ.

28. Trees for which all nodes have either degree k or degree 1 are known as *Cayley trees*.

Newman, Strogatz, and Watts [510] follow similar reasoning to develop a rough estimate of the diameter of more general sorts of random networks by examining the expansion in the neighborhoods. The calculation presumes a tree structure, which in the Poisson random network setting we know not to be valid beyond the threshold at which the giant component emerges, and so it only gives an order of magnitude approximation near the threshold. Generally, obtaining bounds on diameters is a very challenging problem. We will encounter other situations where there are potential problems with the calculation as we proceed.

A randomly picked node i has an expected number of neighbors of $\langle d \rangle$ (recalling the $\langle \cdot \rangle$ notation for the expectation operator). If we presume that nodes' degrees are approximately independent, then each of these nodes has a degree described by the distribution $\widetilde{P}(d)$ from (4.2). Thus each of these nodes has an expected number of neighbors (besides i) of $\sum_d (d-1)\widetilde{P}(d)$, or $\frac{\langle d^2 \rangle - \langle d \rangle}{\langle d \rangle}$. So the expected number of i's second neighbors (who are at a distance of 2 from i) is *very* roughly $\langle d \rangle \frac{\langle d^2 \rangle - \langle d \rangle}{\langle d \rangle}$.[29] Iterating, the expected number of kth neighbors is estimated by

$$\langle d \rangle \left(\frac{\langle d^2 \rangle - \langle d \rangle}{\langle d \rangle} \right)^{k-1},$$

so that expanding out to an ℓth neighborhood reaches

$$\sum_{k=1}^{\ell} \langle d \rangle \left(\frac{\langle d^2 \rangle - \langle d \rangle}{\langle d \rangle} \right)^{k-1} \tag{4.10}$$

nodes. When this sum is equal to $n - 1$, it gives some idea of how far we need to go from a randomly picked node to hit all other nodes. This number gives us a crude estimate of the diameter of the largest component. Substituting for the sum of the series in (4.10) (see Section 4.5.1 for some facts about sums of series), we obtain an estimate of diameter as the ℓ that solves

$$\langle d \rangle \left(\frac{\left(\frac{\langle d^2 \rangle - \langle d \rangle}{\langle d \rangle} \right)^{\ell} - 1}{\left(\frac{\langle d^2 \rangle - \langle d \rangle}{\langle d \rangle} \right) - 1} \right) = n - 1,$$

or

$$\ell = \frac{\log\left[(n-1)\left(\langle d^2 \rangle - 2\langle d \rangle \right) + \langle d \rangle^2 \right] - \log\left[\langle d \rangle^2 \right]}{\log\left[\langle d^2 \rangle - \langle d \rangle \right] - \log\left[\langle d \rangle \right]}. \tag{4.11}$$

---

29. Thus there are several approximations. The tree structure is implicit in the assumption that each of these second neighbors are not already first neighbors. There is an assumption about the correlation in neighbors' degrees implicit in the use of $\widetilde{P}$ to calculate these degrees. And the expected number of second neighbors is found by multiplying first neighbors times the expected number of their neighbors, again embodying some independence to have the expectation of a product equal the product of the expectations.

In cases for which $\langle d^2 \rangle$ is much larger than $\langle d \rangle$, (4.11) is approximately

$$\ell = \frac{\log[n] + \log[\langle d^2 \rangle] - 2\log[\langle d \rangle]}{\log[\langle d^2 \rangle] - \log[\langle d \rangle]} = \frac{\log[n/\langle d \rangle]}{\log[\langle d^2 \rangle/\langle d \rangle]} + 1, \quad (4.12)$$

although (4.12) should be treated with caution, as cycles are ignored, and when we are above the threshold for a giant component to exist (e.g., when $\langle d^2 \rangle$ is much larger than $2\langle d \rangle$), then there can be nontrivial clustering for some degree sequences.

If we examine (4.11) for the case of a Poisson random network, then $\langle d^2 \rangle - \langle d \rangle = \langle d \rangle^2$, and then

$$\ell = \frac{\log\left((n-1)\frac{\langle d \rangle - 1}{\langle d \rangle} + 1\right)}{\log(\langle d \rangle)}.$$

When $\langle d \rangle$ is substantially greater than 1, this is roughly $\log(n)/\log(\langle d \rangle)$, which is very similar to the result for the regular tree example. If p is held constant, then as n increases, ℓ decreases and converges to 1 from above. In that case we would estimate the diameter to be 2. In fact, it can be shown that for a constant p, this crude approximation is right on the mark: the diameter of a large random graph with constant p is 2 with a probability tending to 1 (see Corollary 10.11 in Bollobás [86]). Next let us consider the case in which the average degree is not exploding but instead is held constant so that $p(n-1) = \langle d \rangle > 1$. Then our estimate for diameter is on the order of $\log(n)/\log(\langle d \rangle)$. Here the estimate is not as accurate.[30] Applying this estimate to the network generated in Figure 1.6, where $n = 50$, $p = .08$, and the average degree is roughly $\langle d \rangle = 4$, yields an estimated diameter of 2.8. While the calculation is only order of magnitude, this value is not far off for the largest component in Figure 1.6.

Developing accurate estimates for diameters, even for such completely random networks, turns out to be a formidable task that has been an active area of study in graph theory for the past four decades.[31] Nevertheless, the above approximations reflect the fact that the diameter of a random network is likely to be "small" in the sense that it is significantly smaller than the number of nodes, and one can work with specific models to develop accurate estimates.

4.3 • An Application: Contagion and Diffusion

To develop a feel for how some of the derivations from random networks might be useful, consider the following application. There is a society of n individuals. One of them is initially infected with a contagious virus (possibly even a computer virus). Let the network of interactions in the society be described by a Poisson random network with link probability p.

---

30. In this range of p, the network generally has a giant component but is most likely not completely connected.

31. See Chapter 10 in Bollobás [86] for a report on some of the results and references to the literature.

The initially infected person interacts with each of his or her neighbors. Some of the neighbors are immune to the virus, while others are not. Let any given individual be immune with a probability π. For instance, π might represent the fraction of indivduals with natural immunity, a percentage of people who have been vaccinated, or the percentage of people whose computers are not susceptible to the virus. This example is a variation on what is known as the *Reed-Frost model* in the epidemiology literature (see Bailey [28], as the work of Reed and Frost was never published), and is discussed in more detail in Section 7.2. The eventual spread of the disease can then be modeled by:

- generating a Poisson random network on n nodes with link probability p,
- deleting πn of the nodes (uniformly at random) and considering the remaining network, and
- identifying the component that the initially infected individual lies in on this subnetwork.

This calculation is equivalent to examining a network on $(1 - \pi)n$ nodes with a link probability of p and then examining the size of the component containing a randomly chosen node. Thus, given that the threshold for the emergence of a giant component is at $p(1 - \pi)n = 1$, then if $p(1 - \pi)n < 1$, we expect the disease to die out and only infect a negligible fraction of the population. In contrast, if $p(1 - \pi)n > 1$, then there is a nontrivial probability that it will spread to some fraction of the originally susceptible population. In particular, from (4.6) we know that for large n, if an agent in the giant component of the susceptible population is infected, then the expected size of the epidemic as a percentage of the nodes that are susceptible is approximated by the nonzero q that solves

$$q = 1 - e^{-q(1-\pi)np}. \tag{4.13}$$

Furthermore, from Theorem 4.1 if $p > \frac{\log((1-\pi)n)}{(1-\pi)n}$, then with a probability approaching 1 (as n grows) the network of susceptible agents will be connected, and so all of the susceptible population will be infected.

While (4.13) is difficult to solve directly for q, we can rewrite the equation as

$$(1 - \pi)np = \frac{\log(1 - q)}{q}. \tag{4.14}$$

Then for different values of q on the right-hand side of (4.14), we find the corresponding levels of $(1 - \pi)np$ that lead to those q values, which leads to Figure 4.8. The figure displays an initial threshold of $(1 - \pi)np = 1$. Nearly the entire population of susceptible individuals is connected as $(1 - \pi)np$ approaches 5. So, for instance, if half the population is susceptible, and the average degree is greater than 10, then nearly all of the susceptible agents are interconnected, and the probability of them all becoming infected from a tiny initial seed is quite high.

It is also worth emphasizing that this model can also capture diffusion of various behaviors. For instance, define as susceptible someone who would buy a certain product if made aware of it. Then $(1 - \pi)$ can be interpreted as the percentage of the population who would buy the product if everyone was aware of it. The size of the giant component from these calculations indicates the potential

Fraction of susceptible nodes in giant component

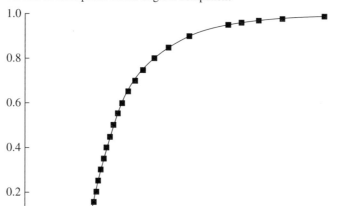

FIGURE 4.8   Fraction of the susceptible population in the largest component of a Poisson random network as a function of the proportion of susceptible nodes $1 - \pi$ times the link probability p times the population size n.

impact of informing a few agents in the population about the product, when they communicate by word of mouth with others and each individual is sure to learn about the product from any neighbor who buys it.

This analysis is built on contagion taking place with certainty between any infected and susceptible neighbors. When the transmission is probabilistic, which is the case in some applications, then the analysis needs to account for that. Such diffusion is discussed in greater detail in Chapter 7.

4.4 ▪ Distribution of Component Sizes*

The derivations in Section 4.2.6 provide an idea of when a giant component will emerge, and its size, but we might be interested in more information about the distribution of component sizes that emerge in a network. Again, we will see how important this is when we examine network-based diffusion in more detail in Chapter 7. Following Newman, Strogatz, and Watts [510], we can use probability generating functions to examine the component structure in more detail. (For readers not familiar with generating functions, it will be useful to read Section 4.5.9 before preceding with this section.)

This analysis presumes that adjacent nodes have independent degrees, and so it is best to fix ideas by referring to the configuration model, in which approximate independence holds for large n. Let the degree distribution be described by P.

Consider the following question. What is the size of the component of a node picked uniformly at random from the network? We answer this by starting at a node, picking one of its edges and examining the neighboring node, and then following the edges from that neighboring node and determining how many additional nodes we find. Then summing across the edges leaving the initial node, we have an idea of the expected size of the component. This method presumes a tree structure and is thus only a good approximation when the degree distribution is such that the number of cycles in the network is negligible.

Let Q denote the distribution of the number of nodes that can be found by picking an edge uniformly at random from the network, picking one of its nodes uniformly at random, and then counting that node plus all nodes that are found by following all paths from that node that do not use the original link. Let $G_Q(x)$ denote the generating function associated with this distribution.

Note that Q can be thought of in the following way. There is probability $\widetilde{P}(d)$ that the node at the end of the randomly selected edge has degree d. In that case, it has $d - 1$ edges emanating from it. The number of additional nodes that can be found by starting from each such edge is a random variable Q.[32] We now use some facts about generating functions to deduce the generating function of Q. In Section 4.5.9 (see (4.23)) it is shown that the generating function of the sum of $d - 1$ independent draws from the distribution of Q is the the generating function of Q raised to the power $d - 1$, so the generating function of additional nodes found through the node if it happens to have degree d is $\left[G_Q(x)\right]^{d-1}$. The overall distribution of the number of nodes found through the additional node is then given by a mixture of distributions: first pick some random d according to $\widetilde{P}(d)$, and then draw a random variable from a distribution having generating function $[G_Q(x)]^{d-1}$ (see (4.24)). So the generating function of the distribution of the additional nodes found after the first one is $\sum_d \widetilde{P}(d) \left[G(q)\right]^{d-1}$. Finally, we need to add one node for the first one found, and the generating function of a distribution of a random variable plus 1 is just x times the generating function of the random variable (see (4.25)). Thus the distribution function of the number of nodes found from one side of an edge picked uniformly at random is

$$G_Q(x) = x \sum_d \widetilde{P}(d)[G_Q(x)]^{d-1}.$$

Noting that $G_{\widetilde{P}}\left(G_Q(x)\right) = \sum_d \widetilde{P}(d)[G_Q(x)]^d$, we rewrite the above as[33]

$$G_Q(x) = x \frac{G_{\widetilde{P}}\left(G_Q(x)\right)}{G_Q(x)}$$

or

$$G_Q(x) = \left(x G_{\widetilde{P}}\left(G_Q(x)\right)\right)^{1/2}. \tag{4.15}$$

---

32. There must be a large number of nodes for this approximation to be accurate, as otherwise there are fewer potential nodes to explore.

33. This equation appears to be different from (25) in Newman, Strogatz, and Watts [510], but in fact $\frac{G_{\widetilde{P}}(\cdot)}{G_Q(x)}$ is the same as their G_1 and allows for an easy derivation of (4.15).

As Newman, Strogatz, and Watts [510] point out, finding a solution to (4.15) is in general impossible without knowing something more about the structure of P. However, we can solve for the expectation of Q, as that is simply $G'_Q(1)$:

$$G'_Q(x) = \tfrac{1}{2} \left(x G_{\widetilde{P}} \left(G_Q(x) \right) \right)^{-1/2} \left(G_{\widetilde{P}} \left(G_Q(x) \right) + x G'_{\widetilde{P}} \left(G_Q(x) \right) G'_Q(x) \right).$$

Thus recalling that $G(1) = 1$ for any generating function we find that

$$G'_Q(1) = \tfrac{1}{2} \left(1 + G'_{\widetilde{P}}(1) G'_Q(1) \right). \tag{4.16}$$

Since $G'_{\widetilde{P}}(1) = E_{\widetilde{P}}[d] = \langle d^2 \rangle / \langle d \rangle$ it follows from (4.16) that when the expectation of Q does not diverge, it must be that

$$G'_Q(1) = \frac{1}{2 - \frac{\langle d^2 \rangle}{\langle d \rangle}}. \tag{4.17}$$

If $\langle d^2 \rangle \geq 2 \langle d \rangle$, then the expectation of Q diverges, and so (4.17) is no longer valid. Indeed this expression grows as $\langle d^2 \rangle$ approaches $2 \langle d \rangle$. This behavior is consistent with (4.7).

Now we can calculate the average size of a component. Let H be the distribution of the size of the component of a node picked uniformly at random. Starting from a node picked uniformly at random, the degree is governed by $P(d)$, the extended neighborhood size has generating function $[G_Q(x)]^d$, and we have to account for the initial node as well.[34] Thus the generating function for H is

$$G_H(x) = x \sum_d P(d)[G_Q(x)]^d = x G_P \left(G_Q(x) \right). \tag{4.18}$$

From (4.18) it follows that the average size of the component that a randomly selected node lies in is (if the average over Q does not diverge)

$$G'_H(1) = 1 + G'_P(1)G'_Q(1) = 1 + \frac{\langle d \rangle^2}{2 \langle d \rangle - \langle d^2 \rangle}. \tag{4.19}$$

For Poisson random networks, $\langle d \rangle = (n-1)p < 1$ if the network is to remain disconnected, and so this must also hold to prevent the average size of the component for diverging. Indeed, in the Poisson random-network model substituting $\langle d \rangle = (n-1)p$ and $\langle d^2 \rangle = \langle d \rangle^2 + \langle d \rangle$, we find that average component size of a node picked uniformly at random is

$$G'_H(1) = 1 + \frac{1}{1 - (n-1)p}.$$

For instance, if $(n-1)p = 1/2$ then the average component size is 3; if $(n-1)p = 9/10$, then the average is 11.

---

34. The derivation of the distribution for Q was based on randomly picking a node at either end of an edge. Here, we are working out from a given node, but given (approximate) independence in degrees, the calculation is still valid.

For scale-free networks, $\langle d^2 \rangle$ is generally large relative to $\langle d \rangle$ and diverges as n grows. In that case, the expected component size diverges. The intuition behind this result is as follows. Even though the average degree might be low, if a given node has a neighbor, that neighbor is likely to have a high degree (as $\widetilde{P}(d) = P(d)d/\langle d \rangle$, which in a scale-free network places great weight on the highest-degree nodes). It is then even more likely to have additional high-degree neighbors, and so forth.

4.5 ▪ Appendix: Useful Facts, Tools, and Theorems

This appendix contains a few mathematical definitions, formulas, theorems, and approximations that are useful in working with random networks.

4.5.1 Sums of Series

A geometric series is one in which a series of powers of x is summed, where $x \neq 1$:

$$\sum_{i=m}^{n} ax^i = a\frac{x^m - x^{n+1}}{1 - x}.$$

Thus

$$\sum_{i=0}^{n} ax^i = a\frac{1 - x^{n+1}}{1 - x}.$$

For $x < 1$ it follows that

$$\sum_{i=1}^{\infty} ax^i = \frac{ax}{1 - x},$$

and

$$\sum_{i=0}^{\infty} ax^i = \frac{a}{1 - x}.$$

Another series of interest (especially for scale-free degree distributions) is

$$\sum_{i=1}^{\infty} a\frac{1}{i^\gamma}.$$

This is the Riemann zeta function, $z(\gamma) = \sum_1^\infty \frac{1}{\gamma}$, which is convergent when γ is greater than 1.

A special case is $\gamma = 1$, when we can look at a truncated series

$$\sum_{i=1}^{n} \frac{1}{i} = H_n, \tag{4.20}$$

which is known as a *harmonic number* and has various approximations. For large n, an approximation of H_n is $\gamma + \log(n)$, where the γ of roughly .577 is the Euler-Mascheroni constant. The difference between this approximation and H_n tends to 0. Equation (4.20) is useful in approximating some sequences such as

$$\frac{1}{i+1} + \frac{1}{i+2} + \cdots + \frac{1}{t}, \tag{4.21}$$

which can be written as $H_t - H_i$. For large t, (4.21) is approximately $\log(t) - \log(i)$, or $\log(t/i)$.

4.5.2 The Exponential Function e^x and Stirling's Formula

The exponential function e^x can be defined in various ways that provide useful formulas. Fixing x (at any positive, negative or complex value),

$$\lim_{n \to \infty} \left(1 + \frac{x}{n}\right)^n = e^x.$$

Another definition of e^x is given by

$$e^x = \sum_{i=0}^{\infty} \frac{x^n}{n!}.$$

Stirling's formula for large n is

$$n! \sim \sqrt{2\pi n} \left(\frac{n}{e}\right)^n.$$

4.5.3 Chebyshev's Inequality and the Law of Large Numbers

Chebyshev's inequality states that for a random variable X with mean μ and standard deviation σ,

$$\Pr[|X - \mu| > r\sigma] < 1/r^2$$

for every $r > 0$. This inequality is easy to prove directly from the definition of standard deviation. Letting $r = \frac{x}{\sigma}$, we can also write this as

$$\Pr[|X - \mu| > x] < \sigma^2/x^2$$

for every $x > 0$. Chebyshev's inequality leads to an easy proof of a version of the weak law of large numbers:

Theorem 4.2 (The weak law of large numbers) *Let $(X_1, X_2, \ldots)$ be a sequence of independently distributed random variables such that $E[X_i] = \mu$ for all i and there is a finite bound B so that $\mathrm{Var}(X_i) \le B$ for all i. Then*

$$\Pr\left[\left|\frac{\sum_{i=1}^{n} X_i}{n} - \mu\right| > \varepsilon\right] \to_n 0$$

for all $\varepsilon > 0$.

Proof of Theorem 4.2.   Let $S_n = \sum_{i=1}^{n} \frac{X_i}{n}$. Then

$$\text{Var}(S_n) = \sum_i \frac{\text{Var}(X_i)}{n^2} \leq \frac{B}{n}.$$

Thus $\text{Var}(S_n) \to 0$. By Chebyshev's inequality, fixing any $\varepsilon > 0$,

$$\Pr\left[\left|\sum_{i=1}^{n} \frac{X_i}{n} - \mu\right| > \varepsilon\right] \leq \frac{\text{Var}(S_n)}{\varepsilon^2} \to 0,$$

which establishes the claim. ▪

A stronger conclusion is also possible. The weak law of large numbers just states that the probability that a sequence of observed sample means deviates from the true mean of the process tends to 0. This does not directly imply that there is a probability 1 that the sequence will converge. The strong law of large numbers provides this stronger conclusion. For a proof, see Billingsley [69].

Theorem 4.3 (The strong law of large numbers)   *Let $(X_1, X_2, \ldots)$ be a sequence of independently and identically distributed random variables such that $E[X_i] = \mu$ for all i. Then*

$$\Pr\left[\lim \frac{\sum_{i=1}^{n} X_i}{n} = \mu\right] = 1.$$

4.5.4 The Binomial Distribution

There are many situations in which we need to make use of the binomial distribution. This section defines the distribution and also provides an illustration of the law of large numbers.

Consider flipping a coin repeatedly, when the coin is not a "fair" coin, but instead has a probability of p of coming up heads and of $1 - p$ of coming up tails. A single flip is called a *Bernoulli trial*. In many instances we are interested in a set of flips. If we ask about the probability of a particular sequence of heads and tails being realized, say heads, tails, tails, heads, heads, . . ., where there are m heads out of n flips, then its probability is $p^m(1 - p)^{n-m}$. Noting that there are $\binom{n}{m}$ (read "n choose m," where $\binom{n}{m} = \frac{n!}{m!(n-m)!}$) different orderings that have m heads out of n flips of the coin, the probability that there are m heads if we flip it n times is

$$\binom{n}{m} p^m (1 - p)^{n-m}.$$

The expected number of heads out of n flips is simply pn, while the standard deviation is $\sqrt{np(1 - p)}$.

Note also that the expected fraction of flips of the coin that come up heads is simply p, and the standard deviation of this fraction out of n flips is $\sqrt{\frac{p(1-p)}{n}}$.

Then applying Chebyshev's inequality and letting X be the realized fraction of flips that come up heads,

$$\Pr\left[|X - p| > r\sqrt{\frac{p(1-p)}{n}}\right] < \frac{1}{r^2}.$$

If $r = n^{1/4}$, then

$$\Pr\left[|X - p| > \frac{\sqrt{p(1-p)}}{n^{1/4}}\right] < \frac{1}{n^{1/2}},$$

and so with a large number of flips of the coin it is very unlikely that the realized fraction of heads will differ from p by very much, just as the law of large numbers states.

4.5.5 Stochastic Dominance and Mean-Preserving Spreads

Consider discrete distributions $\widehat{P}$ and P with support on $\{0, 1, 2, \ldots\}$.[35] The concept of first-order stochastic dominance captures the idea that P is obtained by shifting mass from $\widehat{P}$ to place it on higher values. The following conditions are equivalent:

- $\sum f(d)P(d) \geq \sum f(d)\widehat{P}(d)$ for all nondecreasing functions f,
- $\sum_0^x P(d) \leq \sum_0^x \widehat{P}(d)$ for all x,
- $\sum_x^\infty P(d) \geq \sum_x^\infty \widehat{P}(d)$ for all x,

and if they hold we say that P *first order stochastically dominates* $\widehat{P}$.

The dominance is *strict* if the inequalities hold strictly for some x (or f). Note that if strict dominance holds, then $\sum f(d)P(d) > \sum f(d)\widehat{P}(d)$ for any strictly increasing f. An example of a degree distribution that (strictly) first-order stochastic dominates another is pictured in Figure 7.6.

The last two conditions are clearly equivalent and capture the idea that P places less weight on low values and thus more weight on higher values than does $\widehat{P}$. The idea that stochastic dominance provides higher expectations for all nondecreasing functions is not difficult to prove, as P shifts weight to higher values of the function f. The converse is easily seen using the third condition and a simple step function that has value 0 up to x and then 1 from x onward. Often referring to first-order stochastic dominance, the "first-order" is omitted and it is simply said that P stochastically dominates $\widehat{P}$.

The idea of second-order stochastic dominance is a less demanding relationship than first-order stochastic dominance, and so it orders more pairs of distributions. It is implied by first-order stochastic dominance. Instead of requiring a higher expectation relative to all nondecreasing functions, it only requires a higher expectation relative to all nondecreasing functions that are also concave. This concept has deep

---

35. The extension of these definitions is straightforward to the case of more general probability measures: simply substitute $\int \cdot dP$ in the place of sums with respect to P.

roots in the foundations of decision making and risk aversion, although for us it is quite useful in comparing degree distributions of different networks.

Theorem 4.4 (Rothschild and Stiglitz [569]) *The following are equivalent:*

- $\sum f(d) P(d) \geq \sum f(d) \widehat{P}(d)$ *for all nondecreasing, concave functions* f,
- $\sum f(d) P(d) \leq \sum f(d) \widehat{P}(d)$ *for all nonincreasing, convex functions* f,
- $\sum_{z=0}^{x} \sum_{d=0}^{z} P(d) \leq \sum_{z=0}^{x} \sum_{d=0}^{z} \widehat{P}(d)$ *for all* x,

and when they hold we say that P *second-order stochastically dominates* $\widehat{P}$. *If* P *and* $\widehat{P}$ *have the same mean then the above are also equivalent to*

- $\widehat{P}$ *is a* mean-preserving spread *of* P,[36]
- $\sum f(d) P(d) \geq \sum f(d) \widehat{P}(d)$ *for all concave* f.

Again, the dominance (or mean-preserving spread) is strict if the inequalities listed hold strictly for some f (or x). In that case, $\sum f(d) P(d) > \sum f(d) \widehat{P}(d)$ for any strictly increasing and strictly concave functions f.

So if P and $\widehat{P}$ have the same average, then P second-order stochastically dominates $\widehat{P}$ if and only if $\widehat{P}$ is a mean-preserving spread of P. This condition implies that $\widehat{P}$ has a (weakly) higher variance than does P, but also requires a more structured relationship between the two. Having a higher variance and identical mean is not sufficient for one distribution to be a mean-preserving spread of another.

4.5.6 Domination

There are also definitions of domination for distributions on several dimensions.

Consider two probability distributions μ and ν on $\mathbb{R}^n$: μ *dominates* ν if

$$E_{\mu}\left[f\right] \geq E_{\nu}\left[f\right]$$

for every nondecreasing function $f : \mathbb{R}^n \to \mathbb{R}$. The domination is *strict* if strict inequality holds for some nondecreasing f.

Domination captures the idea that "higher" realizations are more likely under μ than under ν. For $n = 1$, domination reduces to first-order stochastic dominance.

4.5.7 Association

Beyond comparing two different distributions, we will also be interested in knowing when a joint distribution of a set of random variables exhibits relationships between the variables. Concepts like *correlation* and *covariance* can be applied to two random variables, but when working with networks we often deal with groups

---

36. This indicates that the random variable described by $\widehat{P}$ can be written as the random variable described by P plus a random variable with mean 0.

of variables. A notion that captures such relationships is *association*. The definition is due to Esary, Proschan, and Walkup [230].

Let μ be a joint probability distribution describing a random vector $\mathbf{S} = (S_1, \ldots S_n)$, where each S_i is real-valued. Then μ is *associated* if

$$\operatorname{Cov}_\mu (f, g) = E_\mu[f(\mathbf{S})g(\mathbf{S})] - E_\mu[f(\mathbf{S})]E_\mu[g(\mathbf{S})] \geq 0$$

for all pairs of nondecreasing functions $f : \mathbb{R}^n \to \mathbb{R}$ and $g : \mathbb{R}^n \to \mathbb{R}$.

If $S_1, \ldots, S_n$ are the random variables described by a measure μ that is associated, then we say that $S_1, \ldots, S_n$ are *associated*. Association of μ implies that S_i and S_j are nonnegatively correlated for any i and j, and it entails that all dimensions of $\mathbf{S}$ are nonnegatively interrelated.[37]

To establish strictly positive relationships, as opposed to nonnegative ones, Calvó-Armengol and Jackson [130] define a strong version of association. A partition Π of $\{1, \ldots, n\}$ captures which random variables are positively related (e.g., the components of nodes in a network). A probability distribution μ describing the random variables $(S_1, \ldots S_n)$ is *strongly associated* relative to the partition Π if it is associated, and for any $\pi \in \Pi$ and nondecreasing functions f and g

$$\operatorname{Cov}_\mu (f, g) > 0$$

when f is increasing in s_i for all s_{-i}, g is increasing in s_j for all s_{-j}, and i and j are in π.

An implication of strong association is that S_i and S_j are positively correlated for any i and j in π.

4.5.8 Markov Chains

There are many settings in which one considers a random process over time, the world can be described by a state, and the transition from one state to another depends only on the current state of the system and not how we got there. For the applications in this book, we are mainly concerned with finite-state systems. For instance, a network that describes a society at a given time can be thought of as a state. Alternatively, a state might describe some action of the agents in a network.

Let the finite set of states be denoted by S. If the state of the system is $s_t = s$ at time t, then there is a well-defined probability that the system will be in state $s_{t+1} = s'$ at time $t + 1$ as a function only of the state s_t. Let Π be the $n \times n$ matrix describing these *transition probabilities* with entries

$$\Pi_{ss'} = \operatorname{Pr}(s_{t+1} = s' \mid s_t = s).$$

---

37. This concept is weaker than that of affiliation, which requires association when conditioning on various events. The weaker concept is useful in many network settings in which the states of nodes of a network are associated but are not affiliated. See Calvó-Armengol and Jackson [130] for an example and discussion.

This matrix results in what is known as a (finite-state) *Markov chain,* where "Markov" refers to the property that the distribution of what will happen in the future of the system only depends on its current state, and not on how the current state came about. Markov chains have a number of applications and very nice properties, so they have been studied extensively. There are some basic facts about Markov chains that are quite useful.

The Markov chain is said to be *irreducible* when for any two states s' and s, if the system starts in state s' in some period, then there is a positive probability that it will reach s at some future date. Irreducibility corresponds to the strong connectedness of the associated directed graph in which the nodes are the states and s points to s' if $\Pi_{s's} > 0$.

An irreducible Markov chain is *aperiodic* if the greatest common divisor of its cycle lengths is 1, where the cycles are in the associated directed graph just described. Checking whether a system is aperiodic is equivalent to asking the following. Start in some state s at time 0 and list all the future dates for which there is a positive probability of being in this state again. If the answer for a state s is a list of dates with the greatest common divisor greater than 1, then that state is said to be *periodic*. If no state is periodic, then the Markov chain is aperiodic.[38]

Noting that the probability of starting in state s and ending in state s' in two periods is simply Π^2, we see by similar reasoning that the probability of starting in state s and ending in state s' in t periods is Π^t. If Π^t has all entries greater than 0 for some t, then it is clearly both irreducible and aperiodic, as it will then have all positive entries for all times thereafter. In contrast, if it never has all positive entries, then either it fails to be irreducible or it is periodic for some states.

An important theorem about Markov chains states that an irreducible and aperiodic finite-state Markov chain has what is known as a *steady-state* distribution (e.g., see Billingsley [69]). The steady state of the Markov process is described by a vector μ with dimension equal to the number of states, where μ_s is the probability of state s. The steady-state condition is that if the process is started at time 0 by randomly drawing the state according to the steady-state distribution, then the distribution over the state at time 1 will be given by the same distribution. That is,

$$\mu_{s'} = \sum_s \mu_s \Pi_{ss'},$$

or

$$\mu = \mu\Pi.$$

We can find the steady-state distribution as a left-hand unit eigenvector, noting that Π is a row-stochastic matrix (i.e., the elements of each row sum to 1).

Other useful facts about the steady-state distribution of a finite-state, irreducible, and aperiodic Markov chain include that it provides the long-run limiting average fraction of periods that the process will spend in each state regardless of

---

38. For those readers who want to master all of the definitions, verify that if a Markov chain has a finite number of states and is irreducible, then one state is periodic if and only if all states are periodic, and in that case they all have the same period (greatest common divisor of dates at which they have a probability of recurring).

the starting state; and regardless of where we start the system the probability of being in state s at time t as t increases goes to μ_s. Thus, when behavior can be described by a Markov chain, we have sharp predictions about behavior over the long run.

4.5.9 Generating Functions

Generating functions (also known as *probability generating functions*[39]) are useful tools for encapsulating the information about a discrete probability distribution and for calculating moments of the distribution and various other statistics associated with the distribution.

Let $\pi(\cdot)$ be a discrete probability distribution, which for our purposes has support in $\{0, 1, 2\ldots\}$. The *generating function* associated with π, denoted G_π, is defined by

$$G_\pi(x) = \sum_{k=0}^{\infty} \pi(k)x^k = E_\pi\left[x^k\right]. \tag{4.22}$$

Note that since $\pi(\cdot)$ is a probability distribution, $G_\pi(1) = 1$. Moreover, G_π has a number of useful properties. Taking various derivatives of it helps to recover the various expectations with respect to π:

$$G'_\pi(x) = \sum_{k=0}^{\infty} \pi(k)kx^{k-1}.$$

Thus

$$G'_\pi(1) = \sum_{k=1}^{\infty} \pi(k)k = E_\pi[k] = \langle k \rangle.$$

More generally,[40]

$$\left(x\frac{d}{dx}\right)^m G_\pi = \sum_{k=1}^{\infty} \pi(k)k^m x^k,$$

and so

$$E\left[k^m\right] = \langle k^m \rangle = \left(x\frac{d}{dx}\right)^m G_\pi \big|_{x=1}.$$

Next, suppose that we consider two independent draws of the random variable k and we want to know the sum of them. The probability that the sum is k is given

---

39. These are distinct from *moment generating functions*, which are defined by $\sum_{k=0}^{\infty} \pi(k)e^{xk} = E_\pi\left[e^{xk}\right]$.

40. The notation $\left(x\frac{d}{dx}\right)^m G_\pi$ indicates taking the derivative of G_π with respect to x and then multiplying the result by x, and then taking the derivative of the new expression and multiplying it by x, and so forth, for m iterations.

by $\sum_{i=0}^{k} \pi(i)\pi(k-i)$. This new distribution of the sum, denoted by π_2, is then such that $\pi_2(k) = \sum_{k=0}^{k} \pi(i)\pi(k-i)$. It has an associated generating function

$$G_{\pi_2}(x) = \sum_{k=0}^{\infty} \pi_2(k)x^k = \sum_{k=0}^{\infty} \sum_{i=0}^{k} \pi(i)\pi(k-i)x^k.$$

Note that

$$[G_\pi(x)]^2 = \left[\sum_{k=0}^{\infty} \pi(k)x^k\right]^2 = \sum_{i,j} \pi(i)\pi(j)x^{i+j} = G_{\pi_2}(x).$$

This result extends easily to higher powers (simply iterating gives even powers) and so the generating function associated with the distribution π_m of a sum of m independent draws of k from π is given by

$$G_{\pi_m}(x) = [G_\pi(x)]^m. \tag{4.23}$$

Another useful observation is the following. Consider a distribution π that is derived by first randomly selecting a distribution from a series of distributions $\pi_1, \pi_2, \ldots, \pi_i, \ldots$, picking each with corresponding probability γ_k, and then drawing from the chosen distribution. Then it follows almost directly that

$$G_\pi = \sum_i \gamma_i G_{\pi_i}. \tag{4.24}$$

Finally, there are many situations in which we have a variable k with distribution P and we want to work with the distribution of $k + 1$. The distribution $\overline{P}$ of $k + 1$ is described by $\overline{P}(k) = P(k - 1)$, where $k \geq 1$. Thus it has a generating function of

$$G_{\overline{P}}(x) = \sum_{k=1}^{\infty} P(k-1)x^k = xG_P(x). \tag{4.25}$$

To use generating functions in the context of degree distributions, let us begin with a degree distribution P. Let it have an associated generating function G_P, defined as in (4.22). Suppose that we are also interested in the generating function $G_{\widetilde{P}}$ associated with the distribution of neighboring degrees under the configuration model, denoted $\widetilde{P}$. Recalling from (4.2) that $\widetilde{P}(d) = \frac{d\,P(d)}{\langle d\rangle}$, it follows that

$$G_{\widetilde{P}}(x) = \sum_{k=0}^{\infty} \widetilde{P}(d)x^d = \sum_{k=0}^{\infty} \frac{P(d)d}{\langle d\rangle}x^d = \frac{xG'_P(x)}{G'_P(1)}.$$

4.5.10 Multiple and Self-Links in the Configuration Model

Let us make the idea of growing the sequence more explicit. Begin with an infinite degree sequence $(d_1, d_2, d_3, \ldots)$ and then consider increasing portions of the sequence. Let q_i^n denote the number of self- or duplicate links for node i when the configuration model is applied to the first n nodes. Let Q_i^n denote the probability that under the configuration model, node $i \leq n$ has at least one self- or duplicate link, so that $Q_i^n = \Pr[q_i^n > 0]$. We can then show the following.

Proposition 4.1 *If a degree sequence* $(d_1, d_2, d_3, \ldots)$ *is such that* $\max_{i \leq n} d_i / n^{1/3} \to 0$, *then* $\max_{i \leq n} Q_i^n \to 0$.

This proposition is not true if we drop the restriction that $\max_{i \leq n} d_i / n^{1/3} \to 0$ (see Exercise 4.2). The reason is that if some nodes have degrees that are too large relative to n, then nontrivial portions of the links involve these nodes, and the probability of self-links and/or multiple links can be nontrivial. Thus while the configuration model is useful when the degrees of nodes do not grow too large relative to the number of nodes, the degree sequences must be chosen carefully for the resulting multigraph to approximate a graph.

The proposition establishes that if $\frac{\max_{i \leq n} d_i}{(n \langle d \rangle)^{1/3}}$ tends to 0, then the chance that any given node (including the largest ones) has a duplicate or self-link tends to 0. From this proposition, we can deduce that if self- and multiple links are deleted from the multigraph, then the proportion of nodes with the correct degree approaches 1 as the number of nodes grows, and the degree distribution converges to the desired distribution (pointwise, if there is an upper bound on degrees). However, convergence does not imply that the resultant multigraph will be a graph. When one aggregates across many nodes, there tend to be some duplicate and self-links, except under more extreme assumptions on the degree sequences; but there will not be many of them relative to the total number of links. To explore this idea in more detail, consider different statements about the probability of self- or multiple links under the configuration model:

1. Fixing a node and its degree, as the number of nodes grows the probability that the given node has any self- or multiple links vanishes. That is, $Q_i^n \to 0$.

2. The maximum probability across the nodes of having any self- or multiple links vanishes. That is, $\max_{i \leq n} Q_i^n \to 0$.

3. The fraction of nodes that have self- or duplicate links tends to 0. That is, for any $\varepsilon > 0$, $\Pr\left[\#\{i \leq n : q_i^n > 0\}/n > \varepsilon \right] < \varepsilon$ for large enough n.

4. The fraction of links that are self- or duplicate links tends to 0. That is, for any $\varepsilon > 0$, $\Pr\left[\sum_{i \leq n} q_i^n / \sum_{i \leq n} d_i > \varepsilon \right] < \varepsilon$ for large enough n.

5. The probability of seeing any self- or multiple links vanishes. That is, $\Pr\left[\sum_{i \leq n} q_i^n > 0 \right] \to 0$.

It is easy to see that (1) is true for any degree sequence, presuming that the sequence includes an infinite number of nodes with positive degree. Statement (2) is what is shown in Proposition 4.1, which then implies (3) based on the argument that when the probability across nodes of having self- or duplicate links goes to 0 uniformly across nodes, then it is impossible to expect a nontrivial fraction of nodes to have self- or multiple links (see Exercise 4.1). A similar argument establishes (4). The statement that would make our lives easiest in terms of ending up with a graph instead of a multigraph, (5), is only true under extreme conditions. To see why (5) generally fails, consider a degree sequence of $(2, 2, 2, \ldots)$, which is about as well behaved as one could want in terms of having a good chance of avoiding self- and duplicate links. But even for this regular degree sequence there is still a nontrivial limiting probability of having at least one self-link. The probability that any given link is not a self-link is $1 - \frac{1}{2n-1}$. To see this, start by connecting one end of the link to some node, and then there are $2n - 1$ equally likely entries in the full sequence of points to attach the other end of this link under the configuration model (see the sequence displayed at the beginning of Section 4.1.4). Only one of these choices leads to a self-link. Continue the process of randomly picking an entry to be one end of the link and picking a second entry for the other end. As we proceed, there will be at least $n/2$ links for which the initial node for the first end of the link does not yet have any link attached to it. For each of these links, an upper bound on the probability of not ending up with a self-link is $1 - \frac{1}{2n-1}$. So we have an upper bound on the probability of not ending up with any self-links in the whole process, $\left(1 - \frac{1}{2n}\right)^{n/2}$, which converges to $e^{-1/4}$.

The useful implication of statement (3) is that for a degree sequence that has a nice limiting degree distribution $P(d)$,[41] if we delete self- and duplicate links, then the proportion of nodes that have degree d converges almost surely to $P(d)$ (so $\Pr\left[\lim_n |p^n(d) - P(d)| = 0\right] = 1$, where $p^n(d)$ is the realized proportion of nodes with degree d after the deletion of duplicate and self-links).

Proof of Proposition 4.1. Let $\widehat{d}^n = \max_{i \leq n} d_i$ be the maximum degree up to node n and $\langle d \rangle^n = \frac{\sum_{i \leq n} d_i}{n}$ be the average degree through node n. We can find a bound for the probability that any given node has a self- or a duplicate link. First, instead of thinking of the configuration process as picking two entries at random and matching them and then iterating, start by picking the first entry of the first element and randomly choosing a match for it, and then doing the same for the second remaining entry, and so forth. It is not hard to see that this process leads to the same distribution over matchings and thus of links. Consider the first node and its first link (isolated nodes can be discarded). The chance that the link is not a self- or duplicate link (so far) is $1 - (d_1 - 1)/(n\langle d\rangle^n - 1)$, as only self-links need be considered. This probability is greater than $1 - \widehat{d}^n/(n\langle d\rangle^n - \widehat{d}^n)$. The chance that the second link (if it has degree greater than 1) is not a self- or duplicate link (so far), presuming the first one is not a self-link, is then

$$1 - \frac{d_1 - 2}{n\langle d\rangle^n - 2} - \frac{d_i - 1}{n\langle d\rangle^n - 2},$$

where d_i is the degree of the node that the first link went to. This expression is greater than $1 - (2\widehat{d}^n)/(n\langle d\rangle^n - \widehat{d}^n)$. Continuing in this manner, we establishg a lower bound on the probability of self- or duplicate links:

---

41. There are various definitions of a limiting distribution that are useful. For instance, it could be that $P_n(d)$ converges to $P(d)$ for each d, but that it takes much longer to reach the limit for some ds compared to others. To make the above statement precise, consider a form of uniform convergence where $\max_d |P^n(d) - P(d)| \to 0$. We can also work with other (weaker) definitions of convergence, such as pointwise convergence, weak convergence, or convergence in distribution (e.g., see Billingsley [68]).

$$\prod_{j=1,\ldots,\widehat{d^n}} \left(1 - \frac{j\widehat{d^n}}{n\langle d\rangle^n - \widehat{d^n}}\right),$$

which is larger than

$$\left(1 - \frac{(\widehat{d^n})^2}{n\langle d\rangle^n - \widehat{d^n}}\right)^{\widehat{d^n}}.$$

If $\widehat{d^n}/((n\langle d\rangle^n - \widehat{d^n})^{1/3})$ tends to 0, then we can approximate the above expression by[42]

$$e^{-(\widehat{d^n})^3/(n\langle d\rangle^n - \widehat{d^n})},$$

which tends to 1 if (and only if) $\frac{\widehat{d^n}}{(n\langle d\rangle^n)^{1/3}}$ tends to 0. ■

4.6 ▪ Exercises

4.1 *Self- and Multiple Links in the Configuration Model** Show that in the configuration model if $\max_{i\leq n} Q_i^n \to 0$, then the fraction of nodes that experience self- or multilinks vanishes as the population size n grows; or more specifically, for any $\varepsilon > 0$, $\Pr\left[\#\{i \leq n : q_i^n > 0\}/n > \varepsilon\right] < \varepsilon$ for large enough n.

4.2 *A Degree Sequence That Always Has Large Nodes* Consider the degree sequence $(1, 1, 2, 4, 8, 16, \ldots)$. Show that in the configuration model given node has a probability of self- or multiple links that tends to 0 as n becomes large, but for each $n \geq 2$ there is some node with a significant probability of having a self-link or multiple links. That is, show that $Q_i^n \to 0$, but that $Q_n^n \to 1$ for all n.

4.3 *A Degree Sequence for the Power Distribution in the Configuration Model* Find a degree sequence that converges to a power distribution and for which $\frac{\widehat{d^n}}{(n\langle d\rangle)^{1/3}}$ tends to 0.

4.4 *The Distribution of Neighbors' Degrees in the Configuration and Expected-Degree Models* Consider a constant degree sequence $(d, d, d \ldots)$. Form a random network by applying the configuration model and form another random network by applying the expected-degree model. Provide an expression for the resulting degree distributions in the limit as n grows (working with the resulting multigraph in the configuration model). Provide an expression for the limiting distribution $\widetilde{P}$ of the degree of a node found at either end of a uniformly randomly chosen link.

4.5 *The Distribution of Neighbors' Degrees*

   (a) Consider the Poisson random-network model on n nodes with a link probability of p. Consider a node i and a node j, which are fixed in advance.

---

42. We can approximate $(1 - \frac{r}{x})^x$, when $r \to 0$ and x does not decrease, by e^{-r}. See Section 4.5.2 for approximating expressions.

Conditional on the link ij being present, what is the distribution of j's degree?

(b) Uniformly at random choose a node i out of those having at least one link, presuming that the network is nonempty. Randomly pick one of its neighbors (with equal probability on each neighbor). Argue that the conditional distribution of the node's degree is different from the conditional distribution for j's degree that you found in part (a). What does this distribution converge to as n grows if p is set to keep the average degree constant (so that $p = m/(n-1)$ for some fixed $m > 0$)?

(c) Explain the difference between these two distributions.

4.6 A Threshold for Links in the Poisson Random-Network Model   Show that $t(n) = 1/n^2$ is a threshold function for there being at least one link in a Poisson random network.

4.7 There Is at Most One Giant Component in the Poisson Random-Network Model[*]   Consider the Poisson random-network model when p (as a function of n) is such that there exists $m > 0$ such that $pn \geq m$ for all n. Show that the probability of having more than one giant component vanishes as n grows.

4.8 Size of the Giant Component   (a) Show that there is a solution to (4.9) of $q = 1$ if and only if $P(0) = 0$. (b) Find a nonzero solution to (4.9) when $P(0) = 1/3$ and $P(2) = 2/3$.

4.9 Estimating the Extent of an Infection in an Exponential Random-Network Model[*]   Consider a degree distribution given by

$$P(d) = \frac{e^{\frac{-d}{(1-\pi)m}+1}}{m}$$

with support from $(1 - \pi)m$ to ∞, which has a mean of $2(1 - \pi)m$ (which is derived in Section 5.1 as the distribution corresponding to a uniformly random network in which the number of nodes grows over time). Use (4.9) to estimate the percentage of susceptible nodes that will be infected when a random selection π of nodes are immune. Hint: See Section 4.5.1 for helpful formulas for sums of series.

4.10 Estimating the Diameter in an Exponential Random-Network Model   Consider a degree distribution given by

$$P(d) = \frac{e^{\frac{-d}{m}+1}}{m}$$

with support from m to ∞. Use (4.11) to estimate the diameter.

4.11 First-Order Stochastic Dominance and Increasing Giant Components

(a) Consider two degree distributions $\widehat{P}$ and P, such that P first-order stochastically dominates $\widehat{P}$ (see Section 4.5.5 if this definition is unfamiliar). Show that if q' and q are interior solutions to (4.9) relative to $\widehat{P}$ and P, respectively, then $q \geq q'$.[43]

(b) If $\widehat{P}$ is a mean-preserving spread of P, and q' and q are interior solutions to (4.9) relative to $\widehat{P}$ and P, respectively, how are q' and q ordered?

4.12 ***Mean-Preserving Spreads and Decreasing Diameters*** Consider two degree distributions $\widehat{P}$ and P, such that $\widehat{P}$ is a mean-preserving spread of P (see Section 4.5.5 if this definition is unfamiliar). Show that the solution to (4.11) under $\widehat{P}$ is lower than that under P. Show that if we change "is a mean-preserving spread of" to "first-order stochastically dominates," then the solutions to (4.11) cannot be ordered.

4.13 ***First-Order Stochastic Dominance and Decreasing Diameters*** * Consider two finite-degree sequences in the expected-degree model of Section 4.1.5, with corresponding distributions $\widehat{P}$ and P, such that P first-order stochastically dominates $\widehat{P}$. Show that the random networks associated with $\widehat{P}$ have higher diameters in the sense of first-order stochastic dominance of the realized network diameters compared to those associated with P.

4.14 ***Component Sizes for a Family of Degree Distributions*** *

(a) Calculate $\langle d^2 \rangle$ using the degree distribution that has a distribution function of

$$F(d) = 1 - (rm)^{1+r} \, (d+rm)^{-(1+r)} \, ,$$

from (3.2), using this continuous distribution as an approximation for distributions with large n.

(b) Show that $\langle d^2 \rangle$ diverges when $r < 1$. Use the expression for $\langle d^2 \rangle$ and (4.19) to estimate the expected component size in large networks with such a degree distribution when $r > 1$ and for $m = \langle d \rangle$ such that $\langle d^2 \rangle < 2\langle d \rangle$.

---

43. To offer a complete proof to this statement, note that (4.9) can be written as a function $1 - q = H(1 - q)$, where you can show that $H(\cdot)$ is increasing and strictly convex and $H(1) = 1$. Thus, you can show that it has at most one solution other than $q = 0$. Drawing a picture is helpful.

CHAPTER 5

Growing Random Networks

In contrast to state networks, there is another prominent class of models of random networks that consists of those in which new nodes are born over time and form attachments to existing nodes when they are born. As an example, consider the creation of a new web page. When the web page is designed, it will often include links to existing pages. The web page might be updated over time; nonetheless, a nontrivial portion of its links will have been included when the page was first created. Over time, an existing page will be linked to by new web pages. The same is true of people entering a school, a new job, or new neighborhood. Thus there is a fundamental difference between a growing network and a static one. Time introduces a natural heterogeneity to nodes based on their age in a growing network. This heterogeneity is important for two reasons. First, it is present in many applications in which individuals or nodes enter (and leave) networks over time. Such networks include the web page example mentioned above, as well as many networks of friendships, acquaintances, citations, and professional relationships. People enter and leave networks and may accumulate connections over time. Second, the added heterogeneity comes in a simple form that allows the model to move beyond the Poisson random networks of Erdös and Rényi and to have extra layers of richness, yet still be amenable to analysis.

In this chapter, I discuss growing random-network models and show some of the properties that emerge. An important aspect is that the resulting degree distributions are richer than those of Poisson random networks, and at one extreme will provide an explanation for how scale-free distributions might naturally emerge. This contrasts with such models as the configuration model, in which we essentially program in the desired degree distribution and then generate a network. The configuration model is a workhorse for studying diffusion and other properties on a random network, but it is used precisely because it is simple and lacks the richness of more foundational or grounded models of networks. Growing networks allow for several key factors that generate network characteristics that match some of the observations about social networks discussed in Chapter 3. When nodes are born we can consider different ways in which they attach to existing nodes. At one extreme, when newborn nodes uniformly randomly select nodes to link to, the result is just a growing variation on an Erdös-Rényi random network. At the other extreme, when nodes are selected in proportion to the current degrees of the existing nodes (named "preferential attachment" by Barabási and Albert [42]), nodes that are older and have had a chance to grow in degree will grow faster than younger ones, which have lower degrees. This sort of rich-get-richer process leads to scale-free distributions. Preferential attachment has a nice interpretation in that if we randomly pick a node and then start searching through the network, we end up finding nodes in proportion to their degree. We can also consider hybrid models, in which the attachment probabilities vary between these extremes. Despite claims in the literature that many social networks exhibit power laws (having scale-free degree distributions), by fitting such hybrid models, it becomes clear that most social networks lie between these extremes.

Beyond degree distributions, growing random networks provide insight into other observed characteristics. By the nature of the process, older nodes have higher degrees on average, and, since older nodes have a greater proportion of their connections to older nodes than do younger nodes, a natural positive correlation in degrees emerges. Actually, not only do we see correlation in degrees, but also age-based homophily, which is also consistent with many observed social networks. To the extent that the process involves some form of preferential attachment, we also see large hub nodes emerge in the networks, which produces a lower diameter than in a Poisson random network. Finally, certain variations of hybrid models produce the high clustering observed in many social networks. This clustering emerges in hybrid models in which newborn nodes find the nodes that they link to by navigating the network itself. For instance, to the extent that people are introduced to a new friend through an old one, then natural clustering emerges. The same is true in citation networks in which one finds new articles to cite by examining the bibliography of other relevant articles; and similarly in finding new web pages to link to by following links of pages one is already linked to, and so forth. In short, growing random networks introduce new aspects to network formation that generate structures that more closely match many observed characteristics of social networks. Beyond seeing characteristics of observed networks and understanding how those characteristics emerge, we examine models of growing random networks in this chapter. I also introduce some versatile techniques for network analysis.

5.1 ▪ Uniform Randomness: An Exponential Degree Distribution

To illustrate a process in which a network grows over time, I start by discussing a dynamic variation on the Poisson random-network model, in which nodes are born over time and form links to existing nodes at the time of their birth.[1] I begin by showing how we can deduce the degree distribution that results as we grow a

---

1. One of the deficiencies of such models is already evident in this description, as the only links formed over time are between a new node and an existing one. There are no new links formed between existing nodes over time. I return to this feature later.

random network, and in Section 5.4 I return to discuss other characteristics of such networks.

As nodes are born over time, index them by the order of their birth. Thus node i is born at date i, where $i \in \{0, 1, 2 \ldots\}$. I postpone the question of how to interpret the scale of time until we investigate situations in which nodes are born in clusters. For now, think of each period of time as indicating that a new node has been born, regardless of how much physical time has passed since the last node entered the system.

A node forms links to existing nodes when it is born. To start, examine the case in which links are undirected. Let $d_i(t)$ be the degree of node i (born at time i) at a time t. So $d_i(i)$ is the number of links formed at the node's birth, and $d_i(t) - d_i(i)$ is the number of links that node i forms to the new nodes that were born between time i and time t.

Consider a variation of the Poisson random setting, in which each newborn node uniformly randomly selects m of the existing nodes and links to them. For definiteness, start the network with $m + 1$ nodes born at times $\{0, 1, \ldots, m\}$, all connected to one another. The specifics are not of great consequence when we look at the limiting properties of the system, but it is helpful in order to properly analyze the system. Thus the first newborn node that we consider is the one born at time $m + 1$.

At the end of time $m + 1$, m of the older nodes will have new links and one older node will not, while the newest node will have m links. Each of the pre-existing nodes expects to gain $m/(m + 1)$ links (or one link with a probability of $m/(m + 1)$). At the end of time $m + 2$, there are different possibilities: m of the $m + 2$ pre-existing nodes will have gained a new link, while two of them will not. Depending on which two do not gain a link, there are different possibilities for degree distributions. As time progresses the number of possible realizations of the degree distribution grows. While it is hard to keep track of the potential realizations and their relative probabilities, we can do some more direct calculations. If we look at time t, a node i born at time $m \leq i < t$ has an *expected* degree at time t of

$$m + \frac{m}{i + 1} + \frac{m}{i + 2} + \cdots + \frac{m}{t},$$

or

$$m \left(1 + \frac{1}{i + 1} + \frac{1}{i + 2} + \cdots + \frac{1}{t}\right). \tag{5.1}$$

For large t, this is approximately[2]

$$m \left(1 + \log\left(\frac{t}{i}\right)\right). \tag{5.2}$$

---

2. $\sum_{k=1}^{n} \frac{1}{k} = H_n$ is a harmonic number and has various approximations, as discussed in Section 4.5.1. For large n, an approximation of H_n is $\gamma + \log(n)$, where γ is the Euler-Mascheroni constant, which is roughly 0.577; the difference between this approximation and H_n tends to 0. Thus we can rewrite (5.1) as $m(1 + H_t - H_i)$, which then leads to the stated approximation.

Therefore, although it is difficult to deduce the actual degree distribution of this network-formation process, it is relatively straightforward to deduce the distribution of *expected* degrees. For a large t, the nodes that have expected degree less than d are (using the approximation) those such that

$$m \left(1 + \log\left(\frac{t}{i}\right)\right) < d. \tag{5.3}$$

We rewrite (5.3) as the nodes i such that

$$i > t e^{1-\frac{d}{m}}.$$

Thus the nodes with expected degree less than d (where $d < m \left(1 + \log\left(\frac{t}{m}\right)\right)$) are those born after time $t e^{1-\frac{d}{m}}$. This is a fraction of $1 - e^{1-\frac{d}{m}}$. Thus we have deduced an approximation to the distribution function of the expected degrees at time t. For $d < m \left(1 + \log\left(\frac{t}{m}\right)\right)$, the fraction of nodes with expected degrees less than d is

$$F_t(d) = 1 - e^{-\frac{d-m}{m}}. \tag{5.4}$$

This expression is a variation of an exponential distribution. In particular, each node starts out with m links, and the expected links that a random node gains over time have an exponential distribution with expected value m.

Note that the distribution is in fact independent of time t because the fraction of nodes with no more than some degree d is actually constant over time. As more nodes are born, more of them are also below any given level. Because the expected degree of a given node in (5.2) is dependent on t/i, the degree depends on when i was born relative to the set of all nodes, and not on i's absolute date of birth. The time independence of the distribution is true for some processes and not others.

5.1.1 Mean-Field Approximations

In the previous section, we calculated the distribution of expected degrees after time t. How close is this distribution to that of actual degrees after time t? It turns out to be a good approximation in this particular model. Proving this statement is beyond the scope of this text, but let me outline the issues. First, the derived distribution can easily be shown to be an accurate approximation of the distribution of expected degrees for large t. The only approximations in this respect came in rounding the sum of the harmonic series and in neglecting the differences between the first m nodes and other nodes, which is an issue for a vanishingly small fraction of nodes and would enter the calculation only if we examined the very high tail of the distribution. However, the approximation might face difficulties in the difference between a distribution of expected degrees and actual degrees. Here we benefit from the fact that the relative distance of the degree of a node from the expected degree (actual degree minus expected degree divided by expected degree) tends to 0 for this process. The degree of a given node can be viewed as the realization of a sum of independent random variables (whether or not each newborn node happens to link to it). While these independent random variables have different probabilities and expectations, their summed expectation relative to the expected value of any given one tends to infinity, and so by a variation on the law of large numbers (e.g., see Landers and Rogge [422]) we can deduce that the ratio of the actual degree relative to the expected degree of any given node tends to 1. There is still much work to be done to establish that the distribution over nodes then converges, as one has to aggregate across nodes. If we fix some degree d and then ask how many nodes end up on the other side of it compared to the node's degree, this probability vanishes for any given node, since its expected degree only increases and its ratio over the expected degree tends to 1. The key to aggregating is then showing that this convergence has some uniformity to it, so that the fraction of nodes whose ratios of realized to expected degrees is off by more than a given amount tends to 0.

When we compare the network that emerges from this growing system to a Poisson random network, two differences are apparent. First, in this process, each node starts with a given number m of links. Then it is only the additional links that are random. Thus an appropriate benchmark random network would not be the Poisson random network of Erdös and Rényi, but instead a variation in which there are t nodes, and each links at random to m other nodes.[3] Each node would approximately have a degree of m plus a Poisson random variable with expectation m. The main difference between the distribution for a Poisson random network and exponential distribution from the growing random network is that the exponential distribution has more of a spread to it: The older nodes tend to have higher degrees and the younger nodes have lower degrees.

The previous section introduced what is known as a *mean-field approximation*. The full randomness of the process was quite complex, and so rather than try to deduce the degree distribution of the process directly, we found an approximation to the distribution of expected degrees. I then argued why, at least in that specific case, the distribution of expected degrees is a valid approximation of the actual degree distribution.

Random graph processes tend to be complex; especially those in which heterogeneity enters the system through time so that nodes have different distributions for their degree of connectivity. As such, deriving the degree distribution at any point in time can be quite difficult. A standard technique, borrowed from the statistical physics literature, for solving such complex dynamic systems is to use a mean field approximation. That is, one assumes that the system evolves so that events occur at the average level rather than randomly. For instance, suppose that there are already 100 nodes and a new node appears and is supposed to form links to existing nodes independently with probability 1/10. Under a mean field approximation, we instead suppose that the node forms exactly 10 links. This assumption, coupled with a continuous time approximation, allows us to model the change in time of a given node's degree at a fixed rather than a stochastic rate. All of the variation that comes into the system under such an approximation is not due to the stochastics, but rather because of other forms of heterogeneity in the initial conditions; for instance, here the heterogeneity is in terms of dates of birth. While there are obvious departures from reality in such an approximation, in many situations these techniques provide remarkably accurate estimates.

---

3. Here we would have to either admit duplicate links between pairs of nodes or ignore them.

Analytically, we know distressingly little about when such approximations are good and when they are not. The analysis above provides an argument for why this approximation is accurate for large growing networks with uniform probabilities on attachment, but such arguments become increasingly difficult with more complex processes and are generally overlooked entirely.[4] A standard (but not fully satisfactory) technique for verifying the accuracy of approximation is to compare it to simulations of the actual process for some range of parameter values. While we might prefer to calculate things like degree distributions analytically, this usually turns out to be intractable for all but the starkest of models; hence turning to mean-field approximations is a next-best alternative.

5.1.2 Continuous Time Approximations of Degree Distributions

Let us now re-derive the degree distribution associated with the growing random network analyzed in Section 5.1, but do so with an alternative technique: working with a continuous time mean-field approximation.

A new node is born at time t. It forms m links by uniformly randomly picking m out of the t existing nodes.

Node i's degree is thus described by a starting condition of $d_i(i) = m$ and an approximate change over time of

$$\frac{dd_i(t)}{dt} = \frac{m}{t},$$

for each $t > i$. This expression is due to the new node born at each time spreading its m new links randomly over the t existing nodes at time t.

This differential equation has a solution

$$d_i(t) = m + m \log\left(\frac{t}{i}\right).$$

From this solution we can derive an approximation of the degree distribution. We again note that the degrees of nodes are increasing over time. So for instance, if we ask how many nodes have degree of no more than 100, and we see that a node born at time τ has degree of exactly 100, then we are equivalently asking how many nodes were born on or after time τ. So if it is currently time t, then the fraction of nodes having degree of no more than 100 would be $(t - \tau)/t$. In this manner we derive a degree distribution.

Thus for any d and time t, we find the node $i(d)$ such that $d_{i(d)}(t) = d$. The nodes that have degree of less than d are then those born after $i(d)$. The resulting cumulative distribution function is then $F_t(d) = 1 - i(d)/t$.

Applying this technique to the uniformly random growing network process, we solve for $i(d)$ such that

$$d = m + m \log\left(\frac{t}{i(d)}\right),$$

---

4. For one such analysis, see Benaïm and Weibull [53].

which implies that

$$\frac{i(d)}{t} = e^{-\frac{d-m}{m}}.$$

Such a network would have a distribution function described by

$$F_t(d) = 1 - e^{-\frac{d-m}{m}}.$$

This is a negative exponential distribution with support from m to infinity and a mean degree of $2m$ (as it intuitively should be, because each link involves two nodes and each new node brings m links with it).

Note that this result matches (5.4) in Section 5.1, where we worked with discrete time rather than a continuous time approximation. For the mean-field approximation, the continuous time approximation is relatively minor, smoothing things out and allowing the use of differential equations, which can substantially simplify calculations. Again, the main approximation of concern is working with expected rather than realized values.

5.2 ▪ Preferential Attachment

Now that we have developed some understanding of techniques for tackling growing processes, let us enrich the process a bit.

As discussed in Section 3.2.3, degree distributions of a number of observed networks exhibit "fat tails." Price [546], [547] initiated the study of power distributions in networks. He adapted the ideas of Simon [595] to scale-free degree distributions in a setting of growing citation networks. His empirical focus was on citations among scientific papers. His idea was that an article would gain citations over time in a manner proportional to the number of citations the paper already had. This observation is consistent with the idea that researchers randomly find some article (e.g., via searching for key words on the internet) and then search for additional papers by tracing through the references of the first article. The more citations an article has, the larger the likelihood that it will be found and cited again. So, ignoring other issues guiding citation decisions, the probability that an article gets cited is proportional to the number of citations it already has. In the recent literature, such a link formation process was named "preferential attachment" by Barabási and Albert [42] and shown to apply to networks more generally.[5]

---

5. Barabási and Albert [42] developed a similar model to that of Price [547] except that theirs is undirected, while Price's was directed. There are a number of subsequent studies generating power or scale-free degree distributions based on variations of preferential attachment. These include, for example, Kumar et al. [420], whose copying method is akin to preferential attachment, as well as Dorogovtsev and Mendes [205], Levene et al. [431], and Cooper and Frieze [174]. See Newman [503] and Mitzenmacher [470] for more discussion of such processes and their development. There are also other models generating scale-free distributions by appealing directly to the fitness of nodes and proposing that links depend on fitness and that fitness has a power distribution (e.g., see Caldarelli et al. [121]).

It is important to note that fat-tailed distributions have been found in a wide variety of applications, and so the basic ideas behind generating such distributions have a long history. Some of the first work done on this was by Pareto [526], for whom the canonical power distribution is named. In the 1890s, Pareto was studying wealth distributions and noticed that they had scale-free features, where there were many more individuals who had large or small amounts of wealth than would appear in a normal or other purely random distribution. Such features were also observed by Zipf [675] in the frequency of word usage and of city sizes (known as *Zipf's law*). Explanations for why systems should exhibit such a distribution were first put forth by Yule in 1925 [672] and Simon in 1955 [595]. Most processes that generate scale-free distributions are essentially (and sometimes unknowingly) variations on the ideas first formalized by Simon, with roots tracing back to Yule. The two basic ingredients that lead to scale-free distributions are (1) that the system grows over time so that new objects continue to enter (e.g., nodes in network applications, people in wealth applications, cities in city-size applications) and (2) that existing objects grow at rates that are proportional to their size. This second feature ensures that the rich get richer (i.e., faster than the poor) and is essential to obtaining such a distribution. This proportional growth feature is also central to a close cousin of scale-free distributions: lognormal distributions. It is the specifics of the system's growth that result in the scale-free features rather than the lognormal nature of the distribution.

Let us explore a basic preferential-attachment model in more detail. Nodes are born over time and indexed by their date of birth $i \in \{0, 1, 2 \ldots, t, \ldots\}$. Just as in the model in the previous section, upon birth each new node forms m links with pre-existing nodes. The difference is the way in which a new node selects existing nodes to link to. Instead of selecting m of the nodes uniformly at random, it attaches to nodes with probabilities proportional to their degrees. For example, if one existing node has three times as many links as some other existing node, then it is three times as likely to be linked to by the newborn node. Thus the probability that an existing node i receives a new link to the newborn node at time t is m times i's degree relative to the overall degree of all existing nodes at time t, or

$$ m \frac{d_i(t)}{\sum_{j=1}^{t} d_j(t)}. $$

As there are tm total links in the system at time t, it follows that $\sum_{j=1}^{t} d_j(t) = 2tm$. Therefore, the probability that node i gets a new link in period t is

$$ \frac{d_i(t)}{2t}. $$

Again, there are some details to consider in starting such a process. So, start with a pre-existing group of m nodes all connected to one another. Now we have a well-defined stochastic process. Following the discussion above, we examine the mean-field approximation. The mean-field, continuous-time approximation of this process is described by

$$ \frac{dd_i(t)}{dt} = \frac{d_i(t)}{2t}, $$

with initial condition $d_i(i) = m$. This equation has a solution

$$d_i(t) = m \left(\frac{t}{i}\right)^{1/2}. \tag{5.5}$$

Thus nodes are born over time and then grow. Just as before, the degrees of nodes can be ordered by their ages, with the oldest nodes being the largest. To find the fraction of nodes with degrees that exceed some given level d at some time t, we just need to identify which node is at exactly level d at time t, and then we know that all nodes born before then are the larger nodes. Let $i_t(d)$ be the node that has degree d at time t, or such that $d_{i_t(d)}(t) = d$. From (5.5) it follows that

$$\frac{i_t(d)}{t} = \left(\frac{m}{d}\right)^2.$$

The fraction of nodes that have degree smaller than d at time t is the proportion born after node $i_t(d) = t(m/d)^2$. At time t, this is the fraction m^2/d^2. Thus the distribution function is

$$F_t(d) = 1 - m^2 d^{-2},$$

with a corresponding density or frequency distribution (for $d \geq m$) of

$$f_t(d) = 2m^2 d^{-3}.$$

Thus the degree distribution (of expected degrees) is a power distribution with an exponent of -3.

This distribution has the same time independence that we saw with the exponential distribution that arises from a growing random network in which new links are formed to existing nodes uniformly at random. This time independence again results because the relative degrees of nodes are determined by their relative birth dates.

To understand why the proportional density distribution has an exponent of -3 and not some other exponent, let us examine the growth process. Recall from (5.5) that the degree of node i as a function of time can be written as $d_i(t) = m(t/i)^{1/2}$. Thus nodes are growing over time at a rate that is proportional to the square root of the time measured relative to their birthdate. It is this square root that translates into the -2 in the distribution function (and then the -3 in the density or frequency distribution). So why is it a square root here? It was a particular aspect of process that resulted in nodes growing at a specific rate. To get a better feel for this, suppose a node's degree grows at a rate of

$$\frac{dd_i(t)}{dt} = \frac{d_i(t)}{\gamma t}.$$

Following the same steps as before, a node's degree would be $d_i(t) = m(t/i)^{1/\gamma}$ and the degree distribution would be described by $F_t(d) = 1 - m^\gamma d^{-\gamma}$, or a freqency of

$$f_t(d) = \gamma m^\gamma d^{-\gamma-1}.$$

A slower growth rate of any given node's degree over time (corresponding to a higher γ) leads to a distribution of degrees with a steeper fall-off in its frequency. That is, degrees become relatively more bunched at lower levels as γ increases. How would one interpret γ? In the model we examined, m links were formed at each date and to nodes in proportion to their relative degrees. So the sum of degrees is $2mt$ in the network at time t, and probability is $md_i/(2mt)$ of a node i getting a new link in period t. The value of γ is then 2.

How could one justify a probability of $md_i/(\gamma mt)$ for node i getting a new link in period t? The γ here is not an easy parameter to justify altering. One possibility is as follows. Suppose that instead of a single node being born at time t, a group of new nodes comes in at time t. They form a fraction of links among themselves and the remaining fraction to existing links. For instance if they form a fraction α of links to existing nodes and $1 - \alpha$ among themselves, then the probability of node i (born before t) getting a new link in period t is $\alpha md_i/(2mt)$. Therefore, $\gamma = 2/\alpha$. If $\alpha = 1$ and new nodes form all of their links with pre-existing ones, then this scenario describes the preferential attachment model with $\gamma = 2$. As α decreases and new nodes form more of their links among themselves, then γ increases, which corresponds to slower growth in the degrees of pre-existing nodes and a distribution with more concentration on relatively lower degrees.

We are again dealing with expectations and a continuous time approximation. But now there is an added complication. In the case in which links were uniformly random, as in Section 5.1, the realization of links was independent of time. In the present case, if a node happens to receive more links at an early stage, that can snowball into more links at later stages. Moreover, this effect is nonlinear. Thus not only is it difficult to see whether the above approximation is a good approximation for the degree distribution, but in fact it is not even clear that it is a reasonable one for the distribution of expected degrees. The fact that a preferential attachment process does lead to the stated degree distribution has been verified by Bollabás et al. [88], but by using an approach to keep track of the degree distribution directly (explicitly tracking the possible degree sequences that can emerge over time) rather than by showing that mean-field approximations are accurate.

To get some feel for such a network and how it might differ from the previous random graph models, consider Figure 5.1, which shows a 25-node network generated using such a preferential attachment process. Each new node forms two links. To get this process started, node 1 forms no links at birth, node 2 forms only a link to 1, and the process is subsequently well defined.

This network looks very different from the earlier models, which had similar average degree (Figures 4.6 and 4.7). As the nodes are indexed by their birth dates, the older nodes tend to have much higher degrees. For instance, node 2 has degree 11, while nodes 22–25 have only degree 2—the links they form at birth. In the mean-field approximation, this bias is taken to an extreme, where older nodes always have higher degrees. Even observed networks that display higher degrees for older nodes (e.g., citation networks, coauthorship networks) have degree distributions that are not so purely age dependent. Some older nodes might have few links, and some younger ones might have more links. By adding a weighting of "fitness" parameter, it is easy to extend the model so that some younger nodes can overtake older nodes because they are more attractive to link to, as outlined by Bianconi and Barabási [62]. For instance, suppose the probability

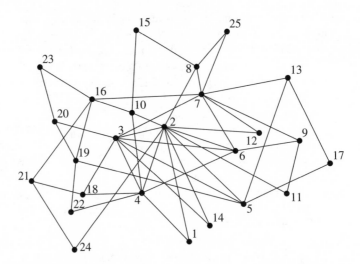

FIGURE 5.1   A network of 25 nodes formed by preferential attachment.

of a node getting a new link is changed by preferential attachment from $d_i/(2t)$ to $d_i v_i/(2tv)$, where v_i is the node's inherent attractiveness or fitness, and v is the average fitness in the population. In this case a node can grow more quickly because of having a large d_i or a large v_i.

5.3 ▪ Hybrid Models

While network formation by preferential attachment leads to a degree distribution that is scale-free and consistent with a power law, many observed degree distributions match neither the exponential process of Section 5.1 nor the preferential attachment process of Section 5.2. For example, consider the degree distribution pictured in Figure 5.2 from the coauthorship network that was discussed in Section 3.2.3.

In Figure 5.2 the degree distribution lies between the two extremes of uniformly random link formation and preferential attachment. This result suggests that a more general network-formation model is needed to match observed degree distributions.

5.3.1 Mean-Field Analyses of Growing Network Processes

A variety of models are hybrids of random and preferential attachment (e.g., Kleinberg et al. [403], Kumar et al. [420], Dorogovtsev and Mendes [205], Pennock et al. [538], Levene et al. [431], Cooper and Frieze [174], Vazquez [637], and Jackson and Rogers [355]). Interestingly, most of these ignore the fact that the resulting degree distributions are not scale-free, but instead try to show that the distribution is at least approximately scale-free for high degrees. The explicit interest in examining all aspects of the degree distribution and in matching the features that are

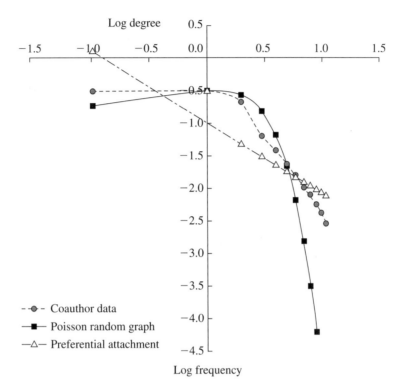

FIGURE 5.2 A degree distribution of a coauthorship network that is indetermediate in degree distribution between a uniformly random network and one formed by preferential attachment.

not scale-free first appears in Pennock et al. [538] and is also a feature of Jackson and Rogers [355]. Let us examine this matter in more detail.

To work with more general processes, I start by noting the basic techniques underlying the mean-field analyses in the previous sections.

Consider any growing network in which nodes are indexed in the order of their birth and node i's degree at time t can be represented as

$$d_i(t) = \phi_t(i),$$

where $\phi_t(i)$ is a decreasing function of i. The fact that ϕ_t is an decreasing function of i indicates that younger nodes have lower degrees. It also means that ϕ_t is an invertible function, so that if some degree d is specified, then we can determine which node has degree d at time t. Because degree increases with age, the fraction of nodes with degree at least d are precisely those older than the node i satisfying $\phi_i(t) = d$; that is, those nodes older than $\phi_t^{-1}(d)$. Thus the degree distribution at time t is

$$F_t(d) = 1 - \frac{\phi_t^{-1}(d)}{t}. \tag{5.6}$$

Thus if we can derive an expression for $d_i(t)$ that is decreasing in i, so that older nodes have more links, then we can easily derive the associated degree distribution.

5.3.2 Mixing Random and Preferential Attachment

Now let us examine a hybrid model of link formation. Suppose, for instance, that a newborn node links to existing nodes through two different processes: a combination of linking by uniformly at random and preferential attachment. Each newborn node forms m links, with a fraction $\alpha < 1$ of them formed to existing nodes selected uniformly at random and a fraction $1 - \alpha$ of them formed to existing nodes by preferential attachment. The mean-field expression for the change in the degree of a node over time can be written as

$$\frac{dd_i(t)}{dt} = \frac{\alpha m}{t} + \frac{(1-\alpha)md_i(t)}{2mt} = \frac{\alpha m}{t} + \frac{(1-\alpha)d_i(t)}{2t}, \tag{5.7}$$

where the second expression has $(1 - \alpha)m$ links being formed by preferential attachment and node i has a probability of $d_i(t)/(2mt)$ of receiving any one of them, and the first expression represents the chance of receiving one of the αm links being formed by picking uniformly at random from the t existing nodes.

Equation (5.7) is a differential equation that has as its solution

$$d_i(t) = \phi_t(i) = \left(d_0 + \frac{2\alpha m}{1-\alpha}\right)\left(\frac{t}{i}\right)^{(1-\alpha)/2} - \frac{2\alpha m}{1-\alpha}, \tag{5.8}$$

where d_0 is the initial number of links that a node has when it is born. From (5.8) we deduce that

$$\phi_t^{-1}(d) = t\left(\frac{d_0 + \frac{2\alpha m}{1-\alpha}}{d + \frac{2\alpha m}{1-\alpha}}\right)^{2/(1-\alpha)}. \tag{5.9}$$

Thus setting $d_0 = m$ in (5.6), from (5.9) we conclude that

$$F_t(d) = 1 - \left(\frac{m + \frac{2\alpha m}{1-\alpha}}{d + \frac{2\alpha m}{1-\alpha}}\right)^{2/(1-\alpha)}. \tag{5.10}$$

When $\alpha = 0$ this degree distribution becomes $1 - (m/d)^2$, which is the power distribution in the case of pure preferential attachment. When $\alpha \to 1$, then the limit is harder to determine, but it approaches the exponential distribution in (5.4) for the model in which links were formed uniformly at random. To see this, let $x = 2\alpha/(1 - \alpha)$ and then note that

$$\left(\frac{m + xm}{d + xm}\right)^x = \left(1 + \frac{m-d}{d + xm}\right)^x,$$

which for large x is approximately $\left(1 + \frac{m-d}{xm}\right)^x$, which tends to $e^{\frac{m-d}{m}}$.

5.3.3 Simulations as a Check on the Degree Distribution

We are again faced with the difficulty that we have not shown that the continuous time mean-field process, in which nodes grow deterministically over time, matches the actual distribution of degrees in a large random network. While this is a challenging open problem, we can perform a rough check that has become a standard technique when faced with such problems.[6] That is, we can simulate the process for some parameter choices and check that the resulting degree distribution is well approximated by the analysis above. While not a guarantee, this procedure can provide some reassurance that the process is not too far off for some parameter values.

Simulating such a process is actually quite easy, especially if we just want to keep track of the resulting degree distribution and not the network as a whole. Let $D(t)$ be the sequence such that if node i has $d_i(t)$ links at time t, then the label i appears $d_i(t)$ times in the entries of the vector $D(t)$:

$$D(t) = \underbrace{1, 1, 1, 1, 1, 1, 1, 1, 1, 1, 1, 1}_{d_1(t) \text{ entries}} \quad \ldots \quad \underbrace{i, i, i, i}_{d_i(t) \text{ entries}} \quad \ldots .$$

If each new node forms m links, then the vector is of length $2mt + m^2$, since each link counts for two different nodes plus whatever we started the system with (say, m^2 additional entries).

Now consider what happens at time $t + 1$. The new node forms αm links uniformly at random and $(1 - \alpha)m$ links by preferential attachment. Take these numbers to be integers. The αm links can be found simply by drawing αm numbers from the set of integers from 1 to t. The $(1 - \alpha)m$ links made by preferential attachment can be chosen by picking an entry out of $D(t)$ with equal weight on each entry. If more than one link forms to an existing node, we have three options: keep duplicate links, delete a duplicate link (and not redraw), or redraw according to the appropriate part of the process until the link is not a duplicate. The third option is most in the spirit of the network-formation process, at least for cases in which links keep track of whether two individuals have some social or economic relationship.[7]

---

6. Berger et al. [55] have recently verified distributions for hybrid models based on Polya urn models.

7. Although it is tempting to proclaim that the other two options should lead, asymptotically, to the same distribution by appealing to arguments such as those in Section 5.1.1 for the configuration model, we must be careful. That argument worked with a prespecified degree sequence that was already on a large number of nodes. One of the main reasons for running simulations in the present case is to check how the process evolves when we keep track of the full evolution of the system and are not working with an approximation or a limiting argument that starts at some late time. It is conceivable, at least with pure preferential attachment, that some node will grow so large that it completely dominates the system, having almost all of the links and thus gaining more and more links. In this case if we do not redraw, lower-degree nodes are placed at a relative disadvantage with regard to gaining new links, compared to a situation in which we redraw and are forced to form new links to different nodes in each iteration. Thus, without a careful argument that applies to the full evolution of the system and takes into account starting conditions, we cannot be sure whether these details will matter.

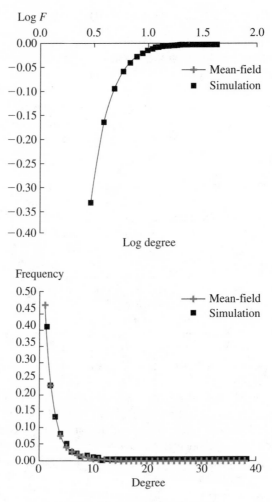

FIGURE 5.3   Comparision of the degree distribution from a simulation on 1,000 nodes and from the mean-field approximation in (5.10).

To be sure that the link formation process is fully specified, add the uniformly random links first (without replacement, thus redrawing until we have αm links to distinct nodes) and then add the preferential attachment links (again redrawing if any links are duplicated).

Figure 5.3 shows the results from a simulation for the case $m = 2$ and $\alpha = 1/2$, so that each newborn node forms one link uniformly at random and one by preferential attachment.

5.3.4 Fitting Hybrid Degree Distributions to Data

The degree distribution described by (5.10) is useful for fitting to data, because when α varies it ranges between the extremes of uniformly random attachment

TABLE 5.1

Degree sequence of the network of ham radio operators network studied by Killworth and Bernard [393]

Degree	Number of operators	Degree	Number of operators	Degree	Number of operators
0	3	9	3	18	0
1	11	10	2	19	0
2	5	11	2	20	1
3	2	12	1	21	1
4	1	11	1	22	1
5	0	14	0	23	0
6	2	15	0	24	0
7	2	16	1	25	1
8	3	17	0	26	0
				27	1

and preferential attachment. By estimating α from fitting the hybrid model to an observed network, we can get some feel for how links might have been formed.

Such analyses have been carried out by Pennock et al. [538] and Jackson and Rogers [355] for different variations on such a process. Let us examine a couple of examples to see how such an estimation works.

Consider a degree distribution from a network of amateur radio operators,[8] in which a link represents the fact that two operators had a radio conversation during a one-month period. These data were collected by Killworth and Bernard [393]. The degree distribution is listed in Table 5.1.

First we directly calculate m. Since m is the number of new links formed each period, it is half of the added degree in each period. The overall degree is $2tm$, and so m is half of the average degree. In this network the average degree is 6.95, and so m is roughly 3.5.

Next we need to derive the parameter α, which determines the proportion of links that are formed uniformly at random versus by preferential attachment. Recall that the continuous time mean-field approximation to the degree distribution was described in (5.10) as

$$F(d) = 1 - \left(\frac{m + \frac{2\alpha m}{1-\alpha}}{d + \frac{2\alpha m}{1-\alpha}} \right)^{2/(1-\alpha)}.$$

This equation is nonlinear in α. While there are different approaches to estimating α in such a situation, let us take a simple iterative least squares regression approach

---

8. With the advent of modern computer-based communications, the era of such ham radio operators has largely come to a close.

TABLE 5.2
Initial parameter estimates α_0 and resulting fits α_1

α_0	0.1	0.2	0.3	0.4	0.5	0.6	0.7	0.8	0.9	0.99	0.999	0.9999
α_1	−0.63	−0.40	−0.21	−0.04	0.12	0.28	0.44	0.61	0.79	0.89	0.98	0.9998

that provides fairly accurate estimates.[9] We can rewrite (5.10) in the following form:

$$\log\left(1 - F(d)\right) = \frac{2}{1-\alpha} \log\left(m + \frac{2\alpha m}{1-\alpha}\right) - \frac{2}{1-\alpha} \log\left(d + \frac{2\alpha m}{1-\alpha}\right).$$

Starting with an initial guess of α—say, α_0—we can regress $\log(1 - F(d))$ on $\log\left(d + \frac{2\alpha_0 m}{1-\alpha_0}\right)$ to estimate $2/(1-\alpha)$ and hence obtain an estimate α_1. We can either iterate this process until the estimate converges to some α^* or examine a grid of values for α_0 and find those estimates α_1 that result from a grid of values for α_0.

Table 5.2 shows that the estimate for α is nearly 1, as we are looking for an α_0 that gives itself back as an estimate α_1. Thus the best fit for the degree distribution of the ham radio operator network in the hybrid model is the extreme in which links are formed uniformly at random.

To compare this result to another application, we can also fit such a hybrid distribution to a data set of coauthorship relationships among economists from Goyal, van der Liej, and Moraga-González [304]. This data set consists of researchers who published an article in a journal listed in EconLit (a bibliography of economics literature) during the 1990s. A link indicates that two researchers were coauthors on at least one article during this period. The degree distribution is listed in Table 5.3.

Again, we can estimate the parameter α by starting with an initial estimate, examining the fit of the estimate, and searching for a fixed point of this process. Table 5.4 provides the grid for the coauthor data. In this case α is estimated to be about .56. This value corresponds to a ratio of links formed uniformly at random to links formed by preferential attachment of about 1.27.[10]

---

9. As discussed by Jackson and Rogers [355], we could also perform a maximum likelihood estimation. However, that requires deriving the probability of observing any given degree distribution as a function of the parameters of the model, which appears to be difficult analytically, and so simulations would be required. The process outlined here is relatively straightforward, although finding a fixed point in this manner only checks a necessary condition and we still need to check the fit of the resulting distribution.

10. This value differs significantly from the Jackson and Rogers [355] estimates reported in Table 3.2, as they worked with a directed process and ours is an undirected one. In their setting, the preferential attachment aspect is dependent on in-degree, and they estimate in-degree to be half of the total degree. Here the preferential attachment is based on the total degree. Dividing the degree by 2 changes the size of the degrees in the upper tail, so that less preferential attachment is needed (roughly a factor of 4 less).

TABLE 5.3
Frequency distribution of degrees from a coauthorship network studied by Goyal, van der Liej, and Moraga-González [304]

Degree	Number of authors	Degree	Number of authors	Degree	Number of authors
0	24,578	21	16	41	2
1	25,078	22	9	42	0
2	17,139	23	11	43	0
3	5,069	24	15	44	0
4	3,089	25	7	45	1
5	1,853	26	7	46	0
6	1,232	27	6	47	0
7	815	28	7	48	0
8	631	29	1	49	0
9	443	30	2	50	0
10	329	31	3	51	0
11	226	32	1	52	0
12	156	33	2	53	0
13	134	34	2	54	1
14	87	35	0	55	0
15	73	36	0	56	0
16	60	37	0	57	0
17	51	38	0	58	0
18	35	39	0	59	0
19	19	40	0	60	0
20	27			61	1

TABLE 5.4
Initial parameter estimates α_0 and resulting fits α_1

α_0	0.1	0.2	0.3	0.4	0.5	0.6	0.7	0.8	0.9	0.56
α_1	0.43	0.45	0.48	0.51	0.54	0.58	0.62	0.69	0.79	0.56

5.4 ▪ Small Worlds, Clustering, and Assortativity

Growing random-network models are useful beyond their versatility in fitting observed degree distributions and tracing those back to variations in the link-formation process. Such models provide insight into the emergence of other observed aspects of social networks.

5.4.1 Diameter

The diameter of a growing random network can be quite different from that of, say, a Poisson random network, because the growing network can have very-high-degree nodes emerge that can serve as hubs and decrease distances. Generally, diameters and average path lengths are very difficult to calculate in networks more advanced than Poisson random networks or variations on the configuration model, for which the independence of the randomness provides something of a calculational toehold.

A result has been worked out by Bollobás and Riordan [87] for the special case of a preferential attachment network-formation process in which each node forms a single link (see also Reed [556]). They show that the network consists of a single component with diameter proportional to $\log(n)$ almost surely; if more than one link is formed by each new node, the diameter is proportional to $\frac{\log(t)}{\log\log(t)}$. Their proof of the following proposition is quite long and is omitted here.

Proposition 5.1 (Bollobás and Riordan [87]) *In a preferential attachment model in which each newborn node forms $m \geq 2$ links, as n grows the resulting network consists of a single component with diameter proportional to $\frac{\log(n)}{\log\log(n)}$ almost surely.*

Thus the diameter of a pure preferential attachment process is lower than that of a Poisson random network (which is proportional to $\log(n)$ when the average degree is held constant), and as pointed out above, the intuition for this result comes from the presence of high-degree nodes, which serve as hubs in the network.

One might conjecture that such a result provides an upper bound on more general growing network processes in which at least two links are formed by preferential attachment. The validity of such a conjecture is not clear, and we cannot simply extend the Bollobás and Riordan [87] argument, which is particular to pure preferential attachment. The difficulty is that as we alter the process, the mix of the degrees at any date is changed, which then alters the probabilities of attachment and the rates at which different nodes grow.

5.4.2 Positive Assortativity and Degree Correlation

As mentioned in Section 3.2.4, a number of social networks have positive correlation in their degree distribution. Such correlation is absent in the Poisson random-networks model, and generally in the configuration model by design. It emerges quite naturally in growing random networks, and the growth of the network might help explain why we see such correlation in social networks, a point first made by Krapivsky and Redner [415].[11] The following result is a (nondirected) variation on a (directed) result from Jackson and Rogers [355]. It shows a very strong form of correlation in that older nodes have a distribution of neighbors' degrees that first-order stochastically dominates that of younger nodes. This conclusion is stronger

---

11. For an alternative explanation, relating to the relevance of nodes, see Capocci, Caldarelli, and De Los Rios [136].

than positive correlation in degree, showing that the whole distribution of neighbors' degrees is shifted up for higher-degree nodes and not just the mean of the distribution.

Let $F_i^t(d)$ denote the fraction of node i's neighbors at time t who have degree d or less.

Proposition 5.2 (Jackson and Rogers [355]) *Consider a growing hybrid random network-formation process as described in Section 5.3. Under the mean-field estimate, a node i's degree is larger than a node j's degree at time t after both are born if and only if i is older than j. In that case, if $\alpha > 0$, then the estimated distribution of i's neighbors' degrees strictly first-order stochastically dominates that of j's at each time $t > j$ relative to younger nodes; that is, $F_i^t(d) < F_j^t(d)$ for all $d < d_j(t)$.*

The first part of the proposition is obvious under the mean-field estimate, as degrees grow deterministically and faster for higher degrees. The second part of the condition follows from the form of the distribution functions, as shown in the proof that follows.

It is worth stressing that the extent to which networks exhibit positive degree correlation, or have degree correlated with age, depends on the application and seems more typical of social networks than other forms.[12]

Proof of Proposition 5.2. Under the mean-field approximation, the degree of node i at time t is described by (5.8):

$$d_i(t) = \left(m + \frac{2\alpha m}{1 - \alpha}\right)\left(\frac{t}{i}\right)^{(1-\alpha)/2} - \frac{2\alpha m}{1 - \alpha}. \qquad (5.11)$$

If $d_i(t) > d_j(t)$, then since $d_i(t)$ is decreasing in i, it must be that $i < j < t$. For $d < d_i(t)$,

$$F_i^t(d) = 1 - \frac{d_i(t^*(d, t))}{d_i(t)},$$

where $t^*(d, t)$ is the date of birth of a node that has degree d at time t. So consider $d < d_j(t)$. It is thus enough to show that for any $i < j < t' < t$

$$\frac{d_i(t')}{d_i(t)} > \frac{d_j(t')}{d_j(t)}. \qquad (5.12)$$

Then (5.11) implies that

$$\frac{d_i(t')}{d_i(t)} = \frac{\left(m + \frac{2\alpha m}{1-\alpha}\right)\left(t'\right)^{(1-\alpha)/2} - \left(\frac{2\alpha m}{1-\alpha}\right) i^{(1-\alpha)/2}}{\left(m + \frac{2\alpha m}{1-\alpha}\right) t^{(1-\alpha)/2} - \left(\frac{2\alpha m}{1-\alpha}\right) i^{(1-\alpha)/2}}.$$

This expression is decreasing in i, which establishes the result. ▪

---

12. See, for instance, Adamic and Huberman [4] for evidence against the correlation of age and degree on the world wide web.

As is clear from the proof, the result also holds for many other growing processes, including exponential random networks (see Exercise 5.7). However, the result does not hold if the rate of growth of degree accelerates substantially beyond being proportional to degree, as then older nodes grow so fast that (5.12) no longer holds.

5.4.3 Clustering in Growing Random Networks

As discussed in Chapter 3, a distinguishing feature of observed networks is that they exhibit high clustering, as well as low diameters and a positive correlation between neighbors' degrees. While the hybrid model in Section 5.3 does a good job of matching observed degree distributions and degree assortativity, it lacks some other properties of observed networks, as do the extreme models of preferential attachment and uniformly random attachment. In particular, regardless of the choices of α and m, as the network grows the average and overall clustering both converge to 0.

As an illustration, consider the exponential random-network model in which links are formed to existing nodes uniformly at random. The only way a cluster can form is when a newborn node links to both ends of an existing link. Consider a node born at time $t + 1$, and consider two of its newly formed links. What is the chance that a link will be present? Given that the two linking nodes are chosen uniformly at random, and there are $t(t - 1)/2$ such pairs to pick from and tm existing links in the network, there is a probability of $2m/(t - 1)$ that a link is present, which converges to 0 as t grows. For the hybrid model or for pure preferential attachment, this calculation is a bit more complicated, as now the attachment probability favors high-degree nodes, and as just seen in Proposition 5.2, they are more likely to be connected to one another than to lower-degree nodes. Nevertheless, the process still has a clustering that tends to 0, as the likelihood that any two high-degree nodes are connected to each other is still vanishing, because the number of high-degree nodes is growing sufficiently rapidly (the fat tail of the power distribution). This convergence to 0 is difficult to show, but can be established in a mean-field analysis.

There are several simple processes of network formation that are hybrid processes that tend to have some clustering in the limit. The simplest of these is described in Vázquez [637] in the context of finding web pages to link to. A new node first randomly chooses some existing node. Next, with a probability q the new node follows a link out from that existing one, and with probability $1 - q$ jumps to another node chosen uniformly at random. It continues this process until it has formed some set number of links. First, note that if $q = 0$, then this process simply reduces to the growing network formed by uniformly random attachment. At the other extreme, $q = 1$, only the first node is identified uniformly at random, while the others are found by the network structure itself. Nodes that have higher degrees are more likely to be neighbors of the first node, and thus more likely to be chosen as the second node. This process lends a sort of preferential attachment aspect to the network formation. If node i has degree d_i, then it can be found by this network-based search process if any d_i of its neighbors are found in the first step. Having a higher degree, holding all else constant, leads to a proportionately greater chance of being found by this search process, just as in preferential attachment. We can see a reason for clustering to emerge, as nodes find other nodes to link to by following the existing link structure of the network; thus they link to neighboring nodes on a regular basis.

To work out the details, it helps to use a directed version of this network, for reasons explained in the next section.

5.4.4 A Meetings-Based Network-Formation Model

Consider a variation on a network-formation process proposed and analyzed by Jackson and Rogers [355], for which it is easy to estimate clustering expressions. Each new node meets $m_r > 0$ nodes uniformly at random and forms directed links to them. Then the new node randomly chooses m_n of the out-links from the first group of nodes and follows those links to meet new nodes and form additional links. We start this process with some initial network with enough nodes so that any group of m_r nodes has at least m_n additional neighbors (but beyond this requirement, the precise form of the initial network does not matter for asymptotic behavior). Let $m = m_r + m_n$ be the total number of out-links formed by each newborn node and let $d_i^{in}(t)$ be a node's in-degree at date t. Let m, m_r, and m_n be integers. According to an approximation of this process at large time t, a node has probability m_r/t of being linked to uniformly at random, and probability $m_n d_i^{in}(t)/(mt)$ of being found through a search along links of the first group of nodes. In this second expression, node i can be found if any of its $d_i^{in}(t)$ neighbors are found in the first step (which occurs with probability $m_r d_i^{in}(t)/t$) and then, conditional on one of its neighbors being found, it is found with probability $m_n/(m_r m)$ (where m_n is the number of out-links followed from the initial m_r nodes out of their $m_r m$ total out-links).[13]

Thus the expected change in the number of links at time t for a node with in-degree $d_i^{in}(t)$ is

$$\frac{dd_i^{in}(t)}{dt} = \frac{m_r}{t} + \frac{m_n d_i^{in}(t)}{mt}. \tag{5.13}$$

If we compare (5.13) to the hybrid process (5.7), it is as if $\alpha = m_r/m$ and $1 - \alpha = m_n/m$, except that a factor of 2 is missing in the denominator of the second term in (5.13). The difference results from the use of a directed rather than an undirected search. This change is minor, and following the same steps as before leads to the continuous time mean-field approximation of the distribution of in-degrees described by

$$F(d^{in}) = 1 - (rm)^{1+r} \left(d^{in} + rm \right)^{-(1+r)},$$

where $r = m_r/m_n$, which matches (3.2).[14]

---

13. These calculations ignore the chance of finding more than one neighbor, or being found by multiple paths, which are second-order effects and of relatively negligible probability for large t.
14. This substitution sets $d_0 = 0$, so that new nodes have no in-degree.

These processes model network formation well when the network has a natural direction to it, such as following links among web pages or locating scientific articles and then locating articles that are cited by those articles, and so on. However, for processes that are more purely social, in which an individual meets the friends of a new friend, it is be more natural to examine an undirected process, because friendships tend to be reciprocal. But for an undirected version of the above process, the probability that a node with degree $d_i(t)$ is linked to by a new node at time t is more complicated. There is still a probability of $\frac{m_r}{t}$ of being linked to uniformly at random. It is also true that i may be found if any of its $d_i(t)$ neighbors are found in the first step, which occurs with probability $\frac{m_r d_i(t)}{t}$. But then conditional on one of its neighbors being found, the probability that i ends up with a new link depends on the total degrees of its neighbors rather than on just the out-degrees of the neighbors. In the directed process we knew the out-degrees of any neighbor of i to be m. However, in the undirected case, the degree of a neighbor of i is actually positively correlated with $d_i(t)$ and this correlation varies over time. Now a node with a higher degree has a greater chance of one of its neighbors being found than a node with a lower degree; but then conditional on a neighbor being found, the higher-degree node actually has a lower chance of being met through the network search process, because its neighbors tend to have more neighbors than the neighbors of a lower-degree node. The first effect still dominates, but the second correction can be substantial.

5.4.5 Clustering

Let us examine the clustering in this meetings-based network-formation model. Given the directed nature of the process, we begin by examining the fraction of transitive triples defined in Section 2.2.3. Recall that this measure is

$$Cl^{TT}(g) = \frac{\sum_{i;j\neq i;k\neq j} g_{ij}g_{jk}g_{ik}}{\sum_{i;j\neq i;k\neq j} g_{ij}g_{jk}}.$$

This transitive triple measure examines directed links ij and jk and counts the fraction of these in which the link ik is present.

We immediately see why this process leads to nontrivial clustering, since if node i finds j at random, and j has a directed link to k, then there is a nontrivial chance that i will find k through the search of j's out-links and then i will form a directed link to k. Thus it is precisely because some nodes are met through meeting the "friends of friends" that clustering occurs in the process. We can easily derive a lower bound for this clustering. First, note that the denominator of the fraction of transitive triples is simply the m^2 potential triples that are generated for each distinct i (counting across its out-degree m of js and then each j's out-degree m of ks), and so at time t the denominator is tm^2. The numerator has the cardinality t times the number of situations for each i such that i connects to both j and k and subsequently those two are linked to each other. Each newborn node i has at least m_n instances in which i found k by following a link from j. Thus a lower bound on Cl^{TT} is

$$\frac{tm_n}{tm^2} = \frac{m_n}{m^2} = \frac{1}{(r+1)m}.$$

This expression is accurate for the fraction of transitive triples when $r \geq 1$, but is only a lower bound otherwise, as there are also possibilities that j and k are neighbors of each other and both found by the network search process when $r < 1$ (as then $m_n > m_r$). Let us explore these derivations in more detail.

To develop clustering estimates, consider a special case of the process such that when $r \geq 1$, then at most one link is formed in the neighborhood of each node found uniformly at random, and otherwise exactly m_n/m_r (letting m_n/m_r be a positive integer) links are formed in the neighborhood of each node found uniformly at random.

Proposition 5.3 (Jackson and Rogers [355]) *Under a mean-field approximation the fraction of transitive triples, Cl^{TT}, tends to*

$$\begin{cases} \frac{1}{(r+1)m} & \text{if } r \geq 1, \\ \frac{(m-1)r}{m(m-1)(1+r)r-m(1-r)} & \text{if } r < 1. \end{cases}$$

Let us examine how the fraction of transitive triples behaves as a function of r, the relative weight on uniformly random versus network-based meetings. As r grows, then the fraction of transitive triples tends to 0, just as we would expect, given that the model then operates almost uniformly at random and clustering in such models goes to 0. At the other extreme, as r becomes small we have to be a bit careful. There is a lower bound on r in this case since there is always at least one node found uniformly at random so that the newborn node can search its neighborhoods. So, for r to be low, m must be large. For instance, fixing $m_r = 1$ implies that $r = 1/m_n = 1/(m-1)$. Then Cl^{TT} simplifies to $\frac{m-1}{2m}$, which tends to $1/2$ as m grows (and r shrinks).[15] To see this explicitly, recall from the lower bound discussion that the denominator of the fraction of transitive triples is tm^2. Then the numerator is t times the number of ij, ik pairs that a typical i has that end up with a link between j and k. For $m_r = 1$, the newborn i finds one other node j at random, attaches to it, and then attaches to all but one of its neighbors. So i has $m - 1$ completed triples of the form ij and jk. Then out of the $(m - 1)(m - 2)/2$ pairs of js and ks in j's neighborhood that i has linked to, those completed triples are linked at the same rate as j's neighbors are linked to one another. That happens at a rate of $\frac{Cl^{TT}m^2}{m(m-1)/2}$.[16] So we have

$$Cl^{TT} = \frac{(m-1) + \frac{Cl^{TT}m^2}{m(m-1)/2}(m-1)(m-2)/2}{m^2},$$

---

15. This process differs from the one in Jackson and Rogers [355], in which a probability of linking to nodes that are found is allowed. In that case, one can lower the probability of linking to the uniformly randomly found node without requiring that m grow. At the extreme of preferential attachment, the fraction of transitive triples, overall clustering, and average clustering all tend to 0.

16. $Cl^{TT}m^2$ provides the total number of pairs of linked neighbors that a typical node will have, and then $m(m - 1)/2$ is the number of such pairs.

which simplifies to $Cl^{TT} = \frac{m-1}{2m}$. The proof of the overall proposition proceeds similarly.

With the estimate of transitive triples in hand, overall clustering (ignoring the direction of links) in the meetings-based network-formation model can be estimated to tend to (see Exercise 5.8):[17]

$$\begin{cases} 0 & \text{if } r \le 1, \\ \frac{6(r-1)}{(1+r)[3(m-1)(r-1)+4mr]} & \text{if } r > 1. \end{cases}$$

Here the expression is 0 at or below $r = 1$ and then tends to 0 again as r becomes very large, so clustering is only significantly positive in an intermediate range where $r > 1$ but r is not too large. The critical aspect that requires $r > 1$ is that total clustering does not account for the direction of links in the same way that transitive triples do. When $r < 1$, then very-high-degree nodes start to appear, as the preferential aspect of the attachment becomes prevalent. High-degree nodes have large numbers of pairs of neighbors, and a vanishing fraction of them are connected to one another, because many of them are nodes that found the high-degree node by different paths. These high-degree nodes dominate the calculation. They do not dominate the calculation if we account for the directions of links (as in the transitive triples calculation) or average across nodes, so that the effect of low clustering seen among the high-degree nodes is offset by the nonvanishing clustering among low-degree nodes. The latter have many fewer pairs of neighbors and are the initiators of the links to a nontrivial fraction of high-degree nodes. This model thus also illustrates how careful one has to be in terms of which definition of clustering one uses.

Sketch of a Proof of Proposition 5.3. To derive the expression for $Cl^{TT}(g)$, note that the denominator of Cl^{TT} is tm^2, and then consider that the numerator is then t times the expected number of situations in which some i has links ij and ik and there is also a link between j and k. So we need to find this expectation for a given i and then divide by m^2. The situations in which there is a pair of links ij and ik for which either jk or kj is present break into three cases relative to how node i found j and k:

1. both j and k were found at random;
2. one of j and k (say, j) was found at random and the other by a network-based meeting; or
3. both j and k were found by network-based meetings.

Under case 1, the probability of j and k being connected tends to 0 as t becomes large, just as in the uniformly random case. Under case 2, j and k tend to be connected if k was found through j, but not if k was found by search of some $j' \ne j$'s neighborhood. There are a total of m_n situations in which k was found by j's neighborhood. Under case 3, if j and k are found by the search of different nodes' neighborhoods, then the probability that they are linked tends to 0. It is only

17. Average clustering is more cumbersome, and the interested reader is referred to Jackson and Rogers [355] for details.

when they are found by a search of the same node's neighborhood that they have a nonvanishing probability of being linked. Under the process described above, this can only occur when $m_n \geq m_r$; and so let us examine that case. There are $\frac{m_n}{m_r}$ links formed by a new node in the neighborhood of any one of the nodes that were found uniformly at random, and there are m_r such neighborhoods, and so there are $m_r \frac{m_n}{m_r} (\frac{m_n}{m_r} - 1)/2$ such pairs in total. As the initial node and these links are independently and uniformly chosen, these potential clusters are completed with probability $\frac{Cl^{TT} m^2}{m(m-1)/2}$, since the initial node i' has approximately $Cl^{TT} m^2$ completed triples of $m(m-1)/2$ possible pairs of outward links. Thus there are approximately

$$\frac{Cl^{TT} m m_n}{m-1} \left(\frac{1}{r} - 1\right)$$

completed triples from case 3 if $m_n \geq m_r$ and 0 otherwise. Summing across the three cases we expect a given newborn node to have

$$m_n + \frac{Cl^{TT} m m_n}{m-1} \left(\frac{m_n}{m_r} - 1\right)$$

clusters out of m^2 possibilities if $m_n \geq m_r$, and m_n clusters otherwise. Thus

$$Cl^{TT} = \frac{m_n}{m^2} + Cl^{TT} \frac{m_n}{m(m-1)} \left(\frac{m_n}{m_r} - 1\right), \qquad (5.14)$$

if $m_n \geq m_r$, and

$$Cl^{TT} = \frac{m_n}{m^2}$$

otherwise. Solving for Cl^{TT} in (5.14) yields the claimed expression. ∎

While a meetings-based model offers one explanation for how clustering might emerge, there are at least two other reasons for its emergence. One is that nodes might be connected based on some cost and benefit structure, and we then expect clustering among groups of nodes that share low connection costs because of geographical or other characteristics. This possibility is explored in Section 6.5. Another is that groups of nodes might be born in waves. For example, Klemm and Eguíluz [404] have a variation of a preferential attachment model in which nodes are declared to be either "active" or "inactive." A new node enters as active, and then one existing active node is randomly deactivated (with a probability inversely proportional to its degree). New nodes attach to each active node. Then with a probability μ, each of these links is rewired to a random node in the population chosen according to preferential attachment. This process thus has a fixed number of active nodes, and each entering node ends up linked to a proportion $1 - \mu$ of nodes that were active when the new node was born. This, coupled with the fact that the list of active nodes only changes by one each period, results in significant clustering.

5.5 ▪ Exercises

5.1 *Growing Objects and Degree Distributions* Suppose we start with a population of an object of size 1. Suppose also that a new object of size 1 is born at each date, and that existing objects double in size in each period. Over time, the sequence of populations as listed by their sizes will look like (1), (1,2), (1,2,4), (1,2,4,8), (1,2,4,8,16), and so forth.

   (a) Show that fraction of objects that have size less than d at date t is $\log(d)/(t-1)$ for $d \in \{1, 2, 4, \ldots, 2^{t-1}\}$.
   (b) What is the difference between this and the preferential attachment system described in Section 5.2? (How many links are added each period if we interpret this as a system in which the sizes are degrees?)

5.2 *Three Types of Link Formation* In the appendix of Jackson and Rogers [355] the following sort of growing network-formation process is considered. Newborn nodes form links to existing nodes. An existing node receives links from a newborn in three different ways:

   - some links are formed with a probability relative to the size of the existing node (as in preferential attachment),
   - some links are formed with a probability depending on the total time that has already evolved (as in the growing variation of the purely random network), and
   - some links are formed with a constant probability.

We can think of the second and third ways of forming links as different extensions of the idea of purely random Poisson networks applied to a growing set of nodes. The difference between these two is only in terms of how the probability of a link scales with the size of the society. In both cases, each existing node at some time has an equal chance of receiving a new link from a newborn node. The difference lies in what the probability of a link is: Are we keeping the average degree of newborn nodes constant (which necessitates a probability of link formation that decreases with the size of the society) or are we holding the probability of a link between any two nodes constant (which necessitates a growing average degree)? The analysis in Section 5.3 relies on holding the average degree constant, but the other approach is also natural in some applications.

   Allowing for an arbitrary combination of all three of these different methods of forming links leads to the following expression for the change in node i's degree over time at time t:

$$\frac{dd_i(t)}{dt} = \frac{ad_i(t)}{t} + \frac{b}{t} + c, \qquad (5.15)$$

where a, b, and c are scalars.

   Solve for the degree distribution using a continuous time mean-field approximation of the process under the condition that $a > 0$ and either $c = 0$ or $a \neq 1$, and with an initial condition of $d_i(i) = d_0$. Note that in those cases, the solution to (5.15) is

$$d_i(t) = \phi_t(i) = \left(d_0 + \frac{b}{a} - \frac{c}{1-a}\right)\left(\frac{t}{i}\right)^a - \frac{b}{a} + \frac{ct}{1-a}.$$

5.3 *Dying Links* Consider a growing network process such that a newborn node forms m links with a portion α (with $1 > \alpha > 0$) uniformly at random and a portion $1 - \alpha$ by preferential attachment. Also, in any given period qm links are destroyed, where $(1 - \alpha)/2 \geq q \geq 0$ and the links are selected uniformly at random out of all links that exist at the beginning of the period. Solve for the degree distribution using a continuous time mean-field approximation.

5.4 *Degree Distributions with Groups of Self-Attaching Newborn Nodes* Suppose that nodes are born in groups of n in each period. Suppose that they attach a fraction f of their links uniformly at random to other newborn nodes, and a fraction $1 - f$ to older nodes by preferential attachment. Using a continuous time mean-field approximation, develop an expression for the degree distribution.

5.5 *Stochastic Dominance in Hybrid Growing Network Models** Consider the distribution function given in (5.10) corresponding the hybrid growing random network model, which has support for degrees of m and higher. Show that for any fixed α, the distribution associated with m strictly first-order stochastically dominates an alternative distribution with $m' < m$. Show that for any fixed m, the distribution associated with α strictly second-order stochastically dominates an alternative distribution with $\alpha' > \alpha$.

5.6 *Degree Distributions with Growth in the Numbers of Newborn Nodes over Time* The models in this chapter generally had a single node born at each point in time. The systems are generally unchanged if a fixed number of nodes are born at each date. However, if the number of newborn nodes grows over time, then the degree distribution changes. Consider an extension of the hybrid model such that the number of nodes entering at each date grows over time. Let the number of new nodes entering at time t be gn_t, where n_t is the number of nodes at time t and $g > 0$ is a growth rate. Derive an estimated degree distribution using a continuous time mean-field approximation.

5.7 *Positive Assortativity in Exponential Growing Random Networks* Show that Proposition 5.2 also holds for exponential growing random networks of the sort described in Section 5.1.

5.8 *Overall Clustering in the Meeting-Based Network-Formation Model** Building on the sketch of a proof of Proposition 5.3, show that overall clustering in the meetings-based network-formation model, treating directed links as if they were undirected, tends to

$$\begin{cases} 0 & \text{if } r \leq 1, \\ \frac{6(r-1)}{(1+r)[3(m-1)(r-1)+4mr]} & \text{if } r > 1. \end{cases}$$

Hint: First argue that the overall clustering at time t can be approximated by

$$\frac{3m^2 Cl^{TT}}{m(m-1)/2 + m^2 + \frac{1}{t}\sum_i d_i(d_i - 1)/2},$$

and then calculate $\frac{1}{t}\sum_{i=1}^{t} d_i(d_i - 1)/2$ using the mean-field approximation.

TABLE 5.5
Frequency distribution of degrees for a network of friendships among prison inmates as collected by MacRae [446]

Degree	Number of prisoners
0	7
1	17
2	11
3	9
4	12
5	3
6	4
7	3
8	1

5.9 *Fitting a Degree Distribution from a Hybrid Model* Consider the degree distribution for a network of friendships among prison inmates as collected by MacRae [446] and listed in Table 5.5. Using the techniques given in Section 5.3.4, fit the degree distribution given by (5.10) to the data in Table 5.5.

CHAPTER 6

Strategic Network Formation

While the random-network models discussed in Chapters 4 and 5 are useful in growing large and complex networks that exhibit certain features, they are still lacking some important aspects. In particular, there are many settings in which not only chance but also choice plays a central role in determining relationships. Social settings generally involve sentient actors who have discretion in which relationships they form and maintain and generally have discretion in how much effort, time, or other resources they devote to different relationships. Examples include trading relationships, political alliances, employer-employee relationships, marriages, professional collaborations, citations, email correspondence, and friendships.

There are two central challenges in modeling networks from a strategic point of view. The first is to explicitly model the costs and benefits that arise from various networks. Doing this not only enables us to model how networks develop in the face of individual incentives to form or sever links, but also provides well-defined measures of overall societal welfare. Thus, we not only have predictions about which networks might form, but we also have measures of which networks are "best" from society's point of view. The second challenge of modeling strategic network formation is to make a prediction of how individual incentives translate into network outcomes. In this chapter I focus on an equilibrium method, and I come back to discuss a variety of possible methods as well as dynamic models in Chapter 11.

From the outset, it is important to emphasize what is or is not embodied in a strategic model. Individuals need not be Machiavellian and calculate their potential benefits and costs from each potential relationship. What is critical is that they have a tendency to form relationships that are (mutually) beneficial and to drop relationships that are not. The forces behind such incentives can be quite strong and can operate without individuals realizing that they are being influenced in this way. The term *strategic* thus carries with it connotations that are not necessary to its application.

Some of the important conclusions from the literature on strategic network formation regard the comparison between the networks that form based on individual incentives and those networks that maximize overall societal welfare. There is often some disparity and generally a tension between the individual incentives and societal welfare. This is not surprising, given that there are externalities. For example, one of my trading partners in a market might care about whom I choose as my other trading partners, as the choice could affect the prices or other terms of trade, even though he or she may have little direct influence over my choice. The interesting aspect of this tension is how extensive and resilient it is. In particular, as we shall see, even if there exist transfers so that individuals can be subsidized to maintain relationships that are in society's interest but are not in their own interests, it can still be impossible to maintain socially efficient networks, under some reasonable restrictions on transfers.

Another important aspect of strategic models of network formation is that they provide answers to *why* networks take particular forms, rather than just *how* they take those forms. For example, growing random-network models allowed us to trace certain aspects of networks, such as the shape of the degree distribution and clustering, to specific types of network formation, such as the extent to which nodes are formed uniformly at random versus by preferential attachment, and whether new nodes are met by navigating the network or searching at random. While such analyses relate features of the network to the formation process, they do not provide an understanding of why people tend toward preferential attachment in some settings and not in others. The strategic approach ties explanations to fundamental aspects of the setting. For example, as discussed below, the explanation behind the combination of high clustering and low diameter comes from a strategic analysis that relates high clustering to low costs of connecting to nodes that are close in social or geographical distance, and low diameter to the benefits of accessing the information held by distant nodes. This process brings us to a related point. In a situation in which information diffuses through a network, agent payoffs depend on access to information that they have. As a result, this access shapes their incentives regarding which relationships to form or maintain and ultimately affects the network structure.

6.1 ▪ Pairwise Stability

To model network formation in a way that accounts for individual incentives, we first need to model the net payoffs or utility that each agent receives as a function of the network. In this setting, the nodes of the network $N = \{1, \ldots, n\}$ are often referred to as players. The overall benefit net of costs that a player enjoys from a network is modeled by a *utility function* or *payoff function*. That is, the payoff to player i is represented by a function $u_i : G(N) \to \mathbb{R}$, where $u_i(g)$ represents the net benefit that i receives if network g is in place.[1]

The utility function captures all benefits net of costs that a given player experiences as a function of the network. Depending on the setting, these can include very different things, such as the value of trading opportunities in a trading network or the value of information that might be obtained in a job-contact network. The

---

1. This concept can be viewed as a special case of a richer object called an *allocation rule,* as defined in Jackson and Wolinsky [361], and described in more detail in Chapter 12.

extent to which the players in a network know their own or other player's utility functions is very much context dependent. What is most critical for the approach described here is that they be aware of changes in their own utility as they add or delete links, or at least react in terms of adding relationships that increase payoffs and deleting those that decrease payoffs.

To capture the fact that forming a relationship or link between two players usually involves mutual consent, while severing a relationship only involves the consent of one player, we need an equilibrium or stability concept that differs from an off-the-shelf adaptation of a noncooperative game-theoretic solution, such as a Nash equilibrium (see Section 9.9 for a primer on game theory). Nash equilibrium–based solution concepts fail to capture the possibility that if two players each want to engage in a relationship then we should expect them to.

To develop a feel for this issue, let us consider a basic example (see Chapter 11 for a more detailed discussion). Consider just two individuals and a choice of whether or not to form a link. A natural inclination is to model this game as one in which players (simultaneously) announce whether they wish to be linked to each other. If they both announce that they wish to form the link, then it is formed, while if either says that they do not wish to then the link is not formed. Thus mutual consent is needed to form a relationship. Suppose the link is beneficial to both players. One might try to use the concept of Nash equilibrium. A Nash equilibrium is a choice of action by each player, such that no player would benefit by changing his or her action, given the actions of the other player(s). Unfortunately, that is not a very useful concept here. There are two equilibria: one where both players say they wish to form the link and it is formed, and another where both players say they do not wish to form the link and it is not formed. The second pair of actions form a Nash equilibrium since neither player has an incentive to change his or her action, given the (correct) anticipation that the other player will say that he or she does not want to form the link. This second equilibrium does not make much sense in a social setting, where we would expect the players to talk to each other and form the link if it is in their mutual interest. However, standard game-theoretic concepts do not take this possibility into account.[2] Thus some simple games and standard game-theoretic equilibrium notions are not well suited for the study of network formation, as they do not properly account for the communication and coordination that is important in the formation of social relationships in networks.

A die-hard game theorist might respond that this failure occurs simply because the game has not been properly defined. We could explicitly model all communications that are available between the individuals, and then the actions that they might take in response, and so on. While on the face of it this strategy might seem reasonable, it is impractical for at least two reasons. One is that modeling the possible communication is very cumbersome. A game that incorporates all of the back and forth that might go on in forming a social relationship is complex, and yet the added complexity only captures a very simple idea: that two individuals should be

---

2. There are some refinements of the Nash equilibrium (such as undominated Nash or trembling hand perfect equilibrium) that select the "natural" equilibrium of forming the link in this particular example, but fail to handle other examples. These refinements are discussed in more detail in Chapter 11.

able to coordinate on forming a link when it is in their mutual interest. Moreover, once such a game is modeled, it might have multiple equilibria and need special refinements on beliefs and other aspects of equilibrium to make fairly obvious predictions. An alternative is to directly define an equilibrium notion on networks that incorporates mutual consent.

A very simple stability concept that captures mutual consent is pairwise stability, as defined by Jackson and Wolinsky [361],[3] which we previewed in Section 1.2.4. A network g is *pairwise stable* if

1. for all $ij \in g$, $u_i(g) \geq u_i(g - ij)$ and $u_j(g) \geq u_j(g - ij)$, and
2. for all $ij \notin g$, if $u_i(g + ij) > u_i(g)$ then $u_j(g + ij) < u_j(g)$.

A network is pairwise stable if no player wants to sever a link and no two players both want to add a link. The requirement that no player wishes to delete a link that he or she is involved in implies that a player has the discretion to unilaterally terminate relationships that he or she is involved in. The second part of the definition captures the mutual consent needed to form a link and can be stated in various ways. For a network to be pairwise stable, it is required that if some link is not in the network and one of the involved players would benefit from adding it, then the other player would suffer from the addition of the link. Alternatively if a network g is such that the creation of some link would benefit both players involved (with at least one of them strictly benefiting), then g is not stable.

While pairwise stability is natural and easy to work with, there are limitations to the concept that deserve discussion (and are discussed at more length in Chapter 11). First, pairwise stability is a weak notion in that it only considers deviations on a single link at a time. Although this restriction makes it easy to apply, if other sorts of deviations are viable and attractive, then pairwise stability could be too weak a concept. For instance, it could be that a player would not benefit from severing any single link but would benefit from severing several links simultaneously, and yet the network could still be pairwise stable. Second, pairwise stability considers only deviations by at most a pair of players at a time. It might be that some group of players could all be made better off by some more complicated reorganization of their links, which is not accounted for under pairwise stability. To the extent that larger groups can coordinate their actions in making changes in a network, a stronger solution concept might be needed. While such coordinated actions might sound artificial, these group actions can capture behaviors like the expulsion or ostracism of an individual. In both of these regards, pairwise stability might be thought of as a necessary but not sufficient requirement for a network to be stable over time. Nevertheless, pairwise stability still turns out to be quite useful and often provides tight predictions about the set of stable networks without the need to consider richer deviations.

---

3. This concept should not be confused with a similarly named one that has been used in the "marriage market" literature following Gale and Shapley [267]. Although related, there are distinctions as the Gale and Shapley notion allows a pair of individuals to simultaneously divorce their previous partners and marry each other. The pairwise stability notion defined on networks considers only one link at a time.

6.2 ▪ Efficient Networks

Next let us turn our attention to the evaluation of the overall benefits that society sees from a given network. Payoffs not only provide an individual's perspective on the network but also enable us to at least partially order networks with regard to their overall societal benefits. Given that we have well-defined payoffs to players as a function of the network, there are two obvious and standard notions of welfare.

6.2.1 Efficiency

One way of evaluating societal welfare is by a utilitarian principle, which is to say the best network is the one that maximizes the total utility of the society. This notion was referred to as "strong efficiency" by Jackson and Wolinsky [361], but I simply refer to it as efficiency as in much of the subsequent literature. A network g is *efficient* relative to a profile of utility functions $(u_1, \ldots, u_n)$ if $\sum_i u_i(g) \geq \sum_i u_i(g')$ for all $g' \in G(N)$. It is clear that there will always exist at least one efficient network, given that there are only finitely many networks.

6.2.2 Pareto Efficiency

Another standard tool used by economists for examining overall societal welfare is that of Pareto efficiency, as first defined by Pareto [526]. A network g is *Pareto efficient* relative to $(u_1, \ldots, u_n)$ if there does not exist any $g' \in G$ such that $u_i(g') \geq u_i(g)$ for all i, with strict inequality for some i.

We say that one network *Pareto dominates* another if it leads to a weakly higher payoff for all individuals and a strictly higher payoff for at least one. A network is then Pareto efficient if it is not Pareto dominated by any other network. Pareto domination indicates unanimity in the ordering between two networks and thus is a quite compelling argument in favor of the dominating network compared to the dominated network, at least from the perspective of judging networks by the welfare they confer. The difficulty of course is that such a unanimous ordering can be quite rare, and so while Pareto domination can help rule out some networks, we are often faced with a large set of Pareto efficient networks, and so the concept may not be sufficiently discriminating.

Figure 6.1 illustrates these definitions in the context of a four-player setting. The numbers next to the nodes are the payoffs to the respective player for each network.

There are many networks that are not pictured in Figure 6.1. Let any permutation of the pictured networks have correspondingly permuted payoffs to the players, and any networks that are not permutations of the pictured ones lead to payoffs of 0 for all players. The arrows in the figure point away from networks that are unstable in that some player would benefit by deleting a link, or two players would each benefit by adding a link. The arrows point to the networks that would result if the player(s) who benefit from the action take that action. In this figure there is just one efficient network, which is to match the players into two pairs and have each

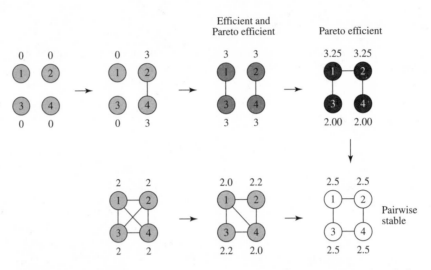

FIGURE 6.1 An example of efficient, Pareto efficient, and pairwise stable networks in a four person society.

player have one link. This is also a Pareto efficient network, as any other network leads to lower payoffs for some player. The efficient network here is not pairwise stable, as two disconnected players would benefit from adding a link. If such an action is taken, then the new network is the top-right one in Figure 6.1. The players who have formed the link have increased their payoffs from 3 to 3.25, which has led to a lowering of the payoffs for the other two players. This network is also Pareto efficient, as there is no other network that gives all players a weakly higher payoff with some a strictly higher payoff. However, the network is not pairwise stable. Here, the two players who have only one link would benefit by adding a link to each other. This action leads to the bottom-right network, which is the only pairwise stable network of those pictured. No player would gain by adding or severing a link here. We already see a conflict between stability and efficiency here, as the only pairwise stable network (or networks, if we count the permutations) is Pareto dominated by the efficient network.

To better understand the relationship between efficiency and Pareto efficiency, note that if g is efficient relative to $(u_1, \ldots, u_n)$ then it must also be Pareto efficient relative to $(u_1, \ldots, u_n)$. However, the converse is not true, as seen in Figure 6.1. What is true is that g is efficient relative to $(u_1, \ldots, u_n)$ if and only if it is Pareto efficient relative to all payoff functions $(\widehat{u}_1, \ldots, \widehat{u}_n)$ such that $\sum_i \widehat{u}_i = \sum_i u_i$.

Thus efficiency is a more discriminating notion and is the more natural notion when there is some freedom to change the way in which utility is allocated throughout the network, for instance by reallocating value through transfers (e.g., taxes and subsidies). It can also be justified in settings in which the utility functions are fixed but one is willing to make interpersonal comparisons of utility and take a utilitarian perspective on welfare. Pareto efficiency is a much less restrictive notion, often admitting many networks, but it might be more reasonable in contexts where the payoff functions are fixed, no transfers are possible, and the inherent difficulties in comparing utilities across individuals are taken into account.

Beyond these notions of efficiency, one may want to consider others. For instance it may be that some reallocation of value is possible, but only under the constraints that the allocations are balanced on each component. Such constraints lead to definitions of constrained efficiency, as considered in Exercises 6.8 and 6.9. With definitions of efficiency in hand, we can start to take a hard look at the relationship between stability and efficiency of networks.

6.3 • Distance-Based Utility

I begin with a generalization from Bloch and Jackson [79] of the symmetric connections model discussed in Section 1.2.4. The basic idea is that players obtain utility from their direct connections and also from their indirect connections, and the utility deteriorates with the distance between individuals. So, all else held equal, being closer to another player brings higher benefits. The rate at which benefit decreases with distance is captured by a benefit function.

Let $b : \{1, \ldots, n-1\} \to \mathbb{R}$ denote the net benefit that a player receives from (indirect) connections as a function of the distance between the players. The *distance-based utility model* is one in which a player's utility can be written as

$$u_i(g) = \sum_{j \neq i : j \in N^{n-1}(g)} b(\ell_{ij}(g)) - d_i(g)c,$$

where $\ell_{ij}(g)$ is the shortest path length between i and j. Let $b(k) > b(k+1) > 0$ for any k and $c \geq 0$. These inequalities embody the idea that a player sees higher benefits for having a shorter distance to other players.

The benefit function is fairly general, allowing the benefits to vary with distance in a wide variety of ways. It could be that being at a distance of two links rather than one is almost as beneficial as being directly connected; or it might be that the benefits fall off dramatically. This variation would depend on the application. The symmetric connections model is a special case of this model, in which the benefits fall off exponentially with distance, so that $b(k) = \delta^k$.

This distance-based utility model has two critical aspects: similar utility functions for different players, and benefits from indirect connections that only depend on minimum path length. While these characteristics are clearly special, they still capture some basic aspects of the costs and benefits of many social and economic situations. The utility also serves as a useful benchmark, so that as we add heterogeneity or benefits that depend on other aspects of the network structure, we can understand how they change the analysis.

The following proposition shows that efficient networks in the distance-based utility model share the same features as the special case of the symmetric connections model.

Proposition 6.1 *The unique efficient network structure in the distance-based utility model is*

1. *the complete network if $b(2) < b(1) - c$,*

2. *a star encompassing all nodes if $b(1) - b(2) < c < b(1) + \frac{(n-2)}{2}b(2)$, and*

3. *the empty network if $b(1) + \frac{(n-2)}{2}b(2) < c$.*

So efficient networks take simple and intuitive forms in a broad class of settings. If link costs are high relative to benefits, then it does not make sense to form any links, and so the empty network is the only efficient network (3). If link costs are sufficiently low $(c < b(1) - b(2))$, then it makes sense to form all links, as the cost of adding a link is less than the gain from shortening a geodesic of length at least two into a path of length one, and so the unique efficient network is the complete network (1). The more interesting case arises for intermediate costs of links relative to benefits, such that the only efficient network structure is a star (2).

Proof of Proposition 6.1.   To see (1), note that adding a link ij cannot decrease the utility of any $k \notin \{i, j\}$, and so if the utility to i and j increases as the result of adding a link, then total utility increases. Thus it suffices to show that adding any link benefits the two nodes involved in the link regardless of the starting network. Note that adding a link ij cannot increase distances between them and any other nodes, and it decreases the distance between i and j. Thus adding a link between any i and j increases each of their utilities by at least $b(1) - c - b(2)$, which is greater than 0 in (1). Thus adding the link increases total utility and the unique efficient network is the complete network.

Next, let us verify (2) and (3). To connect any k nodes involves at least $k - 1$ links. A star network involves exactly $k - 1$ links. A star network with $k - 1$ links leads to a total utility of

$$2(k - 1)(b(1) - c) + (k - 1)(k - 2)b(2). \tag{6.1}$$

Note that if a component has k nodes and $m \geq k - 1$ links, then the value of the direct connections due to the links is $2m(b(1) - c)$. This leaves $\frac{k(k-1)}{2} - m$ pairs of players who are at a distance of at least 2 from each other. The value of each such indirect connection is at most $b(2)$. Therefore the overall value of the component is at most

$$2m(b(1) - c) + (k(k - 1) - 2m)b(2). \tag{6.2}$$

The difference between (6.1) and (6.2) is

$$2(m - (k - 1))(b(2) - (b(1) - c)).$$

Since $b(2) > b(1) - c$, this expression is greater than 0 when $m > k - 1$. So the value can only equal the value of the star when $m = k - 1$. Any network other than a star with $k - 1$ links connecting k nodes leads to a total utility that is

$$2(k - 1)(b(1) - c) + X,$$

where $X < (k - 1)(k - 2)b(2)$, since if it is not a star and has only $k - 1$ links among k nodes, then some of the nodes are at a distance of more than 2, and at most $k - 1$ pairs of nodes are directly connected. Thus if one chooses to involve k nodes and have exactly $k - 1$ links, then a star is the most efficient architecture. Thus when $b(2) > b(1) - c$, efficient networks must involve some combinations of stars and disconnected nodes.

Now let us show that if two stars, involving $k_1 \geq 1$ and $k_2 \geq 2$ nodes, respectively, each lead to nonnegative utility, then a single star among $k_1 + k_2$ nodes leads to strictly higher total utility. This follows from (6.1), noting that the total utility from a star of $k_1 + k_2$ nodes is

$$(k_1 + k_2 - 1)[2(b(1) - c) + (k_1 + k_2 - 2)b(2)],$$

which is larger than

$$(k_1 - 1)[2(b(1) - c) + (k_1 - 2)b(2)] + (k_2 - 1)[2(b(1) - c) + (k_2 - 2)b(2)],$$

when both terms in this latter expression are nonnegative.

Thus we can conclude that if $b(2) > b(1) - c$, then an efficient network is either a star involving all nodes or an empty network. The condition differentiating between (2) and (3) is exactly the calculation of whether the value of a star involving all n players is positive or negative (which is given in (6.1), setting $k = n$). ∎

We can now compare the efficient networks with those that arise if players form links in a self-interested manner. The pairwise stable networks in the distance-based utility model have similar properties to those in the symmetric connections model.

Proposition 6.2 *In the distance-based utility model:*

1. *A pairwise stable network has at most one (nonempty) component.*

2. *For $b(2) < b(1) - c$, the unique pairwise stable network is the complete network.*

3. *For $b(1) - b(2) < c < b(1)$, a star encompassing all players is pairwise stable, but for some n and parameter values in this range it is not the unique pairwise stable network.*

4. *For $b(1) < c$, in any pairwise stable network each node has either no links or else at least two links. Thus every pairwise stable network is inefficient when $b(1) < c < b(1) + \frac{(n-2)}{2}b(2)$.*

The proof appears as Exercise 6.2.

As one might expect, for high and low costs to links, efficient networks coincide with pairwise stable networks. Disparities occur in cases of intermediate link costs relative to benefits. In the range of costs and benefits such that $b(1) - b(2) < c < b(1) + \frac{(n-2)}{2}b(2)$, a star involving all players is the unique efficient network architecture, but is only sometimes pairwise stable and even then not uniquely so.

Moreover, there are situations in which all pairwise stable networks are Pareto inefficient. To see this, consider the case of $n = 4$ and $b(1) < c < b(1) + \frac{b(2)}{2}$, and so a star network is the unique efficient structure. Here the only pairwise stable network is the empty network. We can argue this as follows. If a player has three links, then severing one leads to an increase in payoff of $c - b(1)$. If a player has two links, there are two possibilities: the player is in a component of all players, or the player is in a component of just three players. In the second case, severing one of the links leads to an increase in payoff of $c - b(1)$. In the first case, the player is directly connected to two players and then has a path length of 2 to the third player. One of the two links can be severed without increasing the path length to the third player, which leads to an increase in payoff of $c - b(1)$. Thus each player in the network has at most one link. In that case, any player who has one link would increase his or her payoff by severing the link since it does not lead to any indirect payoffs. Although the empty network is the unique pairwise stable network, it is not even Pareto efficient. The empty network is Pareto dominated by a line (e.g., $g = \{12, 23, 34\}$). To see this, note that for the line, the payoff to the end players (1 and 4) is $b(1) + b(2) + b(3) - c$, which is greater than 0; to the middle two players (2 and 3) the payoff is $2b(1) + b(2) - 2c$, which is also greater than 0 since $c < b(1) + \frac{b(2)}{2}$.

Thus there exist cost ranges for the distance-based utility model (and hence the symmetric connections model) for which all pairwise stable networks are Pareto inefficient, and other cost ranges for which all pairwise stable networks are efficient. There are also cost ranges for which some pairwise stable networks are efficient and other pairwise stable networks are not even Pareto efficient.

6.3.1 Externalities

The inefficiency of pairwise stable networks in the distance-based utility model stems from the externalities that are present. Externalities occur when the utility or payoffs to one individual are affected by the actions of others, although those actions do not directly involve the individual in question. In the distance-based utility model, beyond his or her own links, a player in this setting can have an increase in payoffs (hence positive) as his or her neighbors form more links or even if indirectly connected players form more links.

Let us say that there are *nonnegative externalities* under $u = (u_1, \ldots, u_n)$ if

$$u_i(g + jk) \geq u_i(g)$$

for all $i \in N$, $g \in G(N)$, and jk such that $j \neq i \neq k$. There are *positive externalities* under $u = (u_1, \ldots, u_n)$ if there are nonnegative externalities under $u = (u_1, \ldots, u_n)$ and the inequality is strict in some instances. It is easy to see that the distance-based utility model is one of positive externalities, as added links can only bring players closer together.

Let us say that there are *nonpositive externalities* under $u = (u_1, \ldots, u_n)$ if

$$u_i(g + jk) \leq u_i(g)$$

for all $i \in N$, $g \in G(N)$, and jk, such that $j \neq i \neq k$. There are *negative externalities* under $u = (u_1, \ldots, u_n)$ if there are nonpositive externalities under $u = (u_1, \ldots, u_n)$ and the inequality is strict in some instances. We shall see an example of a model with negative externalities shortly.

6.3.2 Growing Networks and Inefficiency

As there can be many pairwise stable networks, even when some are efficient we might not expect that those would be the ones to arise. How can we predict which networks are likely to emerge from a multitude of pairwise stable networks? There are a variety of approaches focusing on either refining the equilibrium concept or examining some dynamic process. To get an impression of one such dynamic, let us examine a natural and intuitive process that was introduced by Alison Watts [652].

Consider a random ordering over links, in which at any point in time any link is as likely as any other to be identified. If the link has not yet been added to the network, and at least one of the two players involved would benefit from adding it and the other would be at least as well off given the current network (and not accounting for what might happen in the future), then the link is added. If the identified link has already been added, then it is deleted if either player would (myopically) benefit from its deletion. If this process comes to rest on a fixed configuration, then it must be at a pairwise stable network. It is also possible for the process to cycle. (I discuss the full range of possibilities in more detail in Chapter 11.)

Based on this process, we can deduce which pairwise stable networks will be reached in the symmetric distance-based utility model. If the empty network is the only pairwise stable network, then the process will halt there.[4] This happens whenever $c > b(1)$, even in cases for which there are nonempty networks that are strictly preferred by all players to the empty network. When $b(1) - c > b(2)$, then it is clear that all links will form, and the efficient complete network will be reached. More subtle cases arise when $b(2) > b(1) - c > 0$. In this range a star is the efficient network, but players are willing to add a link to players with whom they do not have any indirect connection (or have only a sufficiently distant one). For a star to form, the links must be identified in an order that always includes some particular player (who becomes the center) until all possible $n - 1$ links to that player have been formed. So, for instance, if the first link that pops up is ij, then the next one (other than ij) has to be of the form ik or jk. If ik is the next one that arises, then the subsequent links to be added all have to involve i until the star forms. If any other link pops up first, the star network will not be formed (this statement requires proof, which is given below). As n grows, the chance that the star forms clearly tends to 0.

While Alison Watts's [652] result was stated for the symmetric connections model, it extends to the symmetric distance-based utility model.

Proposition 6.3 *Consider the symmetric distance-based utility model in the case $b(1) - b(2) < c < b(1)$. As the number of players grows, the probability that the above described dynamic process leads to an efficient network (star) converges to 0.*

Proof of Proposition 6.3. First, note that if a player forms a link at some point in the process, then that player will henchforth always be linked to at least one player. This follows from the observation that when $b(1) > c$, no player would ever sever a link to a player who has no other connections (nor would that player sever his or her only link).

---

4. Here it is clear that the myopic nature of the process is critical. If players anticipate further additions to the network, they may form links that are initially costly but could later lead to net benefits. For a discussion of forward-looking behavior, see Chapter 11.

With this observation in hand, let us show that forming a star involves specific orders of links being identified, and that the probability of such an order being realized tends to 0. Consider a star forming with some center player, without loss of generality labeled as 1. Order the other players in terms of the last time they linked to 1, and without loss of generality, label them as $2, \ldots, n$.

Note that 1 will only link to n if n is not linked to any other player when they meet, as otherwise 1 is already at a distance of two links to n, and $b(2) > b(1) - c$. By the first observation above, for 1 to link to n, it must be that n has not met any other players before meeting 1. Similar reasoning then implies that when 1 meets $n - 1$, it must be that $n - 1$ has not met any other player previously. Based on this reasoning, the only way for a star to form is for some link ij to form, then ik or jk to form, and then the center player to meet every other player before any other two players meet each other. Suppose that ij meet first. The chance that the next two players who meet and have not met before are other than ij is $(n - 2)(n - 3)/[n(n - 1) - 2]$.[5] This probability goes to 1, and so the probability that the star results goes to 0. ■

6.3.3 The Price of Anarchy and the Price of Stability

Beyond simply knowing that the network formation might lead to inefficiencies, we might also be interested in the extent to which the emergent networks are inefficient. That is, the situation is somehow worse if the stable networks are "very" inefficient compared to if they are "nearly" efficient. This issue of quantifying the social inefficiency that results from selfish individuals acting in a system is not only of interest in network settings but is also critical to a variety of settings. The issue has become known as the *price of anarchy*.[6]

To get an idea of the price of anarchy, consider a special case of the distance-based utility model in which preferences are directly proportional to distance. That is, let

$$u_i(g) = \sum_{j \neq i} -\ell_{ij}(g) - d_i(g)c, \tag{6.3}$$

where ℓ_{ij} is set to ∞ if i and j are not in the same component. Such a model was considered by Fabrikant et al. [233].

These payoffs are written so that they are always negative. They can be interpreted as a sort of cost of communication. Here, the price of anarchy is the ratio largest total cost (in absolute value) generated by any pairwise stable network compared to the cost of the efficient network. A ratio of 1 indicates that all pairwise stable networks are efficient, while a ratio greater than 1 indicates that there are

---

5. There are $(n - 2)(n - 3)/2$ such (unordered) pairs. The total number of possible pairs is $n(n - 1)/2$, and one has already formed. So the probability is $(n - 2)(n - 3)/2$ divided by $[n(n - 1)/2] - 1$.

6. For example, the problem was studied in selfish-routing settings by Roughgarden and Tardos [571], and was named the *price of anarchy* by Papadimitriou [524].

higher costs (lower payoffs) associated with some pairwise stable networks than the efficient network.

We can distinguish between the best and worst possible pairwise stable networks. This distinction is between the price of stability and the price of anarchy (e.g., see Tardos and Wexler [620]). The *price of stability* is the ratio of the lowest total cost (in absolute value) generated by any pairwise stable network to the cost of the efficient network. Clearly the price of anarchy always exceeds the price of stability, as one is a worst-case scenario and the other is the best-case scenario. A price of stability of 1 indicates that the efficient network is stable, while a price of anarchy of 1 indicates that all stable networks are efficient. A price of stability greater than 1 indicates that all stable networks are inefficient, while a price of stability of 1 and a price of anarchy greater than 1 indicates that some stable networks are efficient while others are not. These prices can differ substantially, and we can keep track of a *anarchy-stability gap*.

An easy variation on Proposition 6.1 shows that in the model described by (6.3) the unique efficient network structure is

1. the complete network if $c < 1$, and
2. a star encompassing all nodes if $c > 1$.

We also see that a pairwise stable network in this model includes all players in one component, given that there is an infinite cost of not being connected to some other player. If $c < 1$, then it is clear that the unique pairwise (Nash) stable network is the complete network, and so in this case the price of anarchy and that of stability are both 1. When $c \geq 1$, in this model a star is pairwise (Nash) stable, and so the price of stability remains 1. However, for these higher costs, there are other pairwise (Nash) stable networks and so the price of anarchy increases to greater than 1. Fabrikant et al. [233] provide an upper bound on this price of anarchy. The bound is fairly easy to derive by bounding the diameter of a pairwise stable network and the number of links it can contain.

Proposition 6.4 (Fabrikant et al. [233]) *The diameter of any pairwise stable network in the model described by (6.3) is at most $2\sqrt{c} + 1$, and such a network contains at most $n - 1 + 2n^2/\sqrt{c}$ links. Thus, the price of anarchy is no more than $17\sqrt{c}$.*

Proof of Proposition 6.4. First let us bound the diameter of a pairwise stable network. Suppose that the diameter is at least $2D$ and no more than $2D + 1$, where D is a positive integer. If we show that D cannot exceed $\sqrt{c}$, then it follows that the diameter cannot exceed $2\sqrt{c} + 1$. Consider players i and j at a maximal distance from each other, which is at least $2D$. If they link to each other, the gain in payoff for each one will be at least

$$(2D - 1) + (2D - 3) + \cdots + 1 = D^2,$$

since they reduce the distance between them from at least $2D$ to 1 and the distance from the next closest player on the path between them from $2D - 1$ to 2 (a gain of $2D - 3$), and so forth. Thus, given pairwise stability, the players cannot gain from adding this link, and so $D^2 \leq c$ and $D \leq \sqrt{c}$, as claimed.

Next argue that the number of links in a pairwise stable network is at most $n-1+3n^2/\sqrt{c}$. There are at most $n-1$ edges that are bridges (so that the network would have more than one component if the link were removed).[7] So we need to argue that there are at most $3n^2/\sqrt{c}$ links that are not bridges. Consider a link ij that is not a bridge. Let A_{ij} be the set of nodes (including j) for which the unique shortest path to i goes through j. If ij is deleted, the distance between i and a node in A_{ij} can become at most double the diameter of the network (denoted by d). Thus deleting the link increases the distance costs by at most $2d|A_{ij}|$ and so this expression must be at least c, as otherwise i should sever the link. Thus $|A_{ij}| \geq c/2d$, which implies that any given node i can have at most $2dn/c$ nonbridge links, so there are at most dn^2/c nonbridge links in total. Since $d \leq 3\sqrt{c}$, it follows that there are at most $3n^2/\sqrt{c}$ nonbridge links.

To derive the price of anarchy, we simply need to bound the cost of pairwise stable networks. A crude upper bound on the cost is $n(n-1)$ times the diameter plus two times c times the number of links. The cost of the efficient network when $c > 1$ is that of a star, which is $2(n-1)[n-1+c]$. Therefore the price of anarchy is at most

$$\frac{\left(2\sqrt{c}+1\right)n(n-1)+2(n-1)c+6n^2\sqrt{c}}{2(n-1)[n-1+c]},$$

which (noting that $1 < \sqrt{c}$ and $n \geq 2$) is less than $17\sqrt{c}$. ▪

This bound can be loose, but it shows that the price of anarchy is no more than the order of $\sqrt{c}$ in this setting, and so is the price of stability and the anarchy gap.

In this extreme model, the price of stability is 1, as there is always some pairwise stable and efficient network, but there is a price of anarchy, as some pairwise stable networks are inefficient. More generally, in the distance-based utility model all stable networks can be inefficient, and so the price of stability can exceed 1.

Such calculations of the prices of anarchy and stability are important as they provide a magnitude to the inefficiency of selfish network formation. Because such calculations can be challenging outside simple settings, the prices of stability and anarchy are still unknown for most models, especially those with any heterogeneity across players.

6.4 ▪ A Coauthor Model and Negative Externalities

The analyses in the previous section for the distance-based model show that self-centered incentives can lead to inefficient networks. That model has a specific form of positive externality in it: individuals benefit from indirect connections. That is, one individual can benefit because another individual has connections. The tension arising between stability and efficiency results because individuals do not account for the indirect benefits that their connections bring to their neighbors. That is, an individual considers whether his or her payoff will increase when forming a link

---

7. Each bridge that is removed breaks the network into one more component than it had initially, and there are at most n components, so there can be at most $n-1$ bridges (noting that the fact that one link is a bridge is not affected by the removal of another bridge).

but does not pay attention to whether the link would increase the payoffs of other players in the network.

Let us now consider another simple model of network payoffs that has a different sort of externality. Consider a situation in which there are negative externalities due to links. That is, a given individual would rather that his or her neighbors have fewer connections rather than more. This situation corresponds to one in which an individual is in competition with other indirect connections for access to the individual's neighbors. It contrasts with the connections and distance-based models, in which individuals benefit from indirect connections.[8]

This model is called the *coauthor model*, as introduced by Jackson and Wolinsky [361]. The story that accompanies the payoff structure is that individuals benefit from interacting with others, for instance, in collaborating on a research project. Beyond the benefit of having the other player put time into the project, there is also a form of synergy. The synergy is proportional to the product of the amounts of time the two researchers devote to the project. If they spend more time together, they generate more synergy. This scenario leads to the negative externality. If an individual's collaborator increases the time spent on other projects, then the individual sees less synergy with that collaborator. Effectively, each player has a fixed amount of time to spend on projects, and the time that researcher i spends on a given project is inversely related to the number of projects, $d_i(g)$, that he or she is involved in. The synergy between two researchers depends on how much time they spend together and is captured by $\frac{1}{d_i(g)d_j(g)}$. The more projects a researcher is involved in, the lower the synergy that is obtained per project. Player i's payoff is represented by

$$u_i(g) = \sum_{j:ij\in g} \left(\frac{1}{d_i(g)} + \frac{1}{d_j(g)} + \frac{1}{d_i(g)d_j(g)} \right)$$

for $d_i(g) > 0$, and $u_i(g) = 1$ if $d_i(g) = 0$. So the value generated by any given research project is proportional to the sum of the time that i puts into the project, the time that j puts into it, and a synergy that is dependent on an interaction between the time the two researchers put into the project. Note that in the coauthor model there are no directly modeled costs to links. Costs are implicit in the diluted synergy when efforts are spread among more coauthors.

Proposition 6.5 (Jackson and Wolinsky [361]) *In the coauthor model, if n is even, then the efficient network structure consists of n/2 separate pairs. If a network is pairwise stable and n ≥ 4, then it is inefficient and can be partitioned into fully intraconnected components, each of which has a different number of members. Moreover, if m is the number of members of one component of a pairwise*

---

8. There are also models for which either positive or negative externalities can result from indirect links depending on the network configuration and the players in question. For instance, the payoff of a player can be related to a centrality measure, such as betweenness centrality, as in Buechel and Buskens [114]. Adding a link to a network could increase some player's centrality by placing him or her between new pairs of players, even if that player is not involved in the link. So there could be positive externalities in some cases. Adding a link could also decrease a player's centrality, as it could result in new paths between other players that circumvent the given player. Thus adding a link could exhibit negative externalities.

stable network and $\widehat{m}$ is the number of members of a different component that is no larger than the first, then $m > \widehat{m}^2$.

Proof of Proposition 6.5.   To verify efficiency, note that

$$\sum_{i \in N} u_i(g) = \sum_{i:d_i(g)>0} \sum_{j:ij \in g} \left[\frac{1}{d_i(g)} + \frac{1}{d_j(g)} + \frac{1}{d_i(g)d_j(g)} \right],$$

so that

$$\sum_{i \in N} u_i(g) \leq 2N + \sum_{i:d_i(g)>0} \sum_{j:ij \in g} \frac{1}{d_i(g)d_j(g)},$$

and equality can hold only if $d_i(g) > 0$ for all i. Then the result follows since

$$\sum_{i:d_i(g)>0} \sum_{j:ij \in g} \frac{1}{d_i(g)d_j(g)} \leq n,$$

with equality only if $d_i(g) = 1 = d_j(g)$ for all i and j, and $3n$ is the value of $n/2$ separate pairs.

To characterize the pairwise stable networks, consider i and j who are not linked. It follows from the formula for $u_i(g)$ that i will strictly want to link to j in a given network g if and only if

$$\frac{1}{d_j(g)+1}\left(1 + \frac{1}{d_i(g)+1}\right) > \left[\frac{1}{d_i(g)} - \frac{1}{d_i(g)+1}\right] \sum_{k:k \neq j, ik \in g} \frac{1}{d_k(g)},$$

(substitute 0 on the right-hand side if $d_i(g) = 0$), which simplifies to

$$\frac{d_i(g)+2}{d_j(g)+1} > \frac{1}{d_i(g)} \sum_{k:k \neq j, ik \in g} \frac{1}{d_k(g)}. \tag{6.4}$$

The following facts are then true of a pairwise stable network:

1. If $d_i(g) = d_j(g)$, then $ij \in g$.

To prove (1), it is enough to show that if $d_j(g) \leq d_i(g)$, then i would benefit from linking to j. Note that if $d_j(g) \leq d_i(g)$, then $\frac{d_i(g)+2}{d_j(g)+1} > 1$ while the right-hand side of (6.4) is at most 1 (the average of d_i fractions). Therefore, i benefits from linking to j.

2. If $d_h(g) \leq \max\{d_k(g) | ik \in g\}$, then i benefits from a link to h.

To prove (2), let j be such that $ij \in g$ and $d_j(g) = \max\{d_k(g) | ik \in g\}$. If $d_i(g) \geq d_j(g) - 1$ then $\frac{d_i(g)+2}{d_h(g)+1} \geq 1$. If $\frac{d_i(g)+2}{d_h(g)+1} > 1$ then (6.4) clearly holds for i's link to h. If $\frac{d_i(g)+2}{d_h(g)+1} = 1$, then $d_h(g) \geq 2$ and so $d_j(g) \geq 2$. Thus the right-hand side of (6.4) is strictly less than 1 when calculated for adding the link h. Thus (6.4) holds. If $d_i(g) < d_j(g) - 1$, then

$$\frac{d_i(g)+1}{d_j(g)} < \frac{d_i(g)+2}{d_j(g)+1} \leq \frac{d_i(g)+2}{d_h(g)+1}.$$

Since $ij \in g$ and g is pairwise stable, it follows from (6.4) that

$$\frac{d_i(g) + 1}{d_j(g)} \geq \frac{1}{d_i(g) - 1} \sum_{k:k \neq j, ik \in g} \frac{1}{d_k(g)}.$$

Also,

$$\frac{1}{d_i(g) - 1} \sum_{k:k \neq j, ik \in g} \frac{1}{d_k(g)} \geq \frac{1}{d_i(g)} \sum_{k:ik \in g} \frac{1}{d_k(g)},$$

since the extra element on the right-hand side is $1/d_j(g)$, which is smaller than (or equal to) all terms in the sum. Thus

$$\frac{d_i(g) + 2}{d_h(g) + 1} > \frac{1}{d_i(g)} \sum_{k:ik \in g} \frac{1}{d_k(g)}.$$

Facts (1) and (2) imply that all players with the maximal number of links are connected to one another and to nobody else. (By (1), they must all be connected to one another. By (2), anyone connected to a player with a maximal number of links would like to connect to all players with no more than that number of links, and hence all those with that number of links.) Similarly, all players with the next to maximal number of links are connected to one another and to nobody else, and so on.

The only thing that remains to be shown is that if m is the number of members of one (fully intraconnected) component and $\widehat{m}$ is that of the next largest component in size, then $m > \widehat{m}^2$. Notice that for i in the next largest component to be unwilling to link to j in the largest component, it must be that

$$\frac{d_i(g) + 2}{d_j(g) + 1} \leq \frac{1}{d_i(g)}$$

(using (6.4), since all nodes to which i is connected also have $d_i(g)$ connections). Thus $d_j(g) + 1 \geq d_i(g)(d_i(g) + 2)$. It follows that $d_j(g) > d_i(g)^2$. ∎

The coauthorship model, while very different in structure from the distance-based utility model, exhibits similar features: it has a simple structure to its efficient networks, and yet the pairwise stable networks tend to be inefficient. In both models the inefficiencies are tied to externalities, but these are of different sorts. In the distance-based settings, when stars are efficient, the center may not have an incentive to maintain links with solitary players. The externality is that when the center forms a link, it benefits other players, since it brings them valuable indirect connections. The failure, or tension between efficiency and stability, comes about because the indirect value that the center generates is inadequately reflected in the payoffs that the center sees from a direct connection. In the coauthorship model, by forming additional connections, a player dilutes the time he or she spends with his or her original partners, which harms them. Here the inefficiency of stable networks stems from the fact that (up to a point) a given player sees more benefit from adding a new link than harm in terms of dilution of value from existing partnerships, while those existing partners see only harm. In both models, social and private incentives are not aligned, but for different reasons stemming from opposite sorts of externalities.

6.5 ▪ Small Worlds in an Islands-Connections Model

Before moving on to discuss the tension between stability and efficiency more generally, let us examine one more model that is another variation on the connections model. The model shows how some of the observed features of real-world networks, such as small-world properties, can be explained from a strategic point of view. It provides a very different perspective on why we observe small worlds than the random-network perspective.

The reasoning behind small worlds in this model is that high clustering stems from a distance-based cost structure. Nodes that are closer (or more similar) find it cheaper to maintain links to one another, which generates high clustering. Short overall path length occurs because if there were no sufficiently short paths between two given nodes, then even if there were a high cost to adding a link, that link would bridge distant parts of the network and bring high benefits to that pair of nodes.

This argument highlights an important distinction between strategic models and purely random models. The random models can identify processes that generate certain features but do not explain why those processes might arise. In a strategic model, the explanation for a specific characteristic of a network is instead traced back to more primitive elements, such as costs and benefits from social relationships. Thus, in a sense, the strategic model can be thought of as explaining "why," whereas the random-graph models can be thought of as explaining "how." This is not to say that strategic models are "better." Each modeling technique has its strengths and weaknesses, and they are quite complementary. For instance, with random-graph models it is easy to produce processes that match any given degree distribution, something that is difficult (at least analytically) to do with a fully strategic model. Yet strategic models allow us to evaluate networks in terms of overall welfare and trace structure back to underlying primitives.

Consider a simple islands variation of a truncated version of the connections model from Jackson and Rogers [353]. There are two modifications to the symmetric connections model discussed in Section 1.2.4. First, if the minimum path length between two players is more than D links, then they do not receive any value from each other. Second, there is a geographic structure to costs. That is, there are K islands, each of which has J players on it. Forming a link between players i and j costs i and j each c if they are on the same island, and C otherwise, where $C > c > 0$. So it is cheaper to link to nearby players.

This geography provides a very simple way of introducing heterogeneity among players or nodes. It is important to emphasize that the geographic structure need not be interpreted literally, but instead can describe differences among players in terms of social or political attributes, research interests, compatibility of research and development programs, and the like. In a richer model, players would be coded by lists of attributes, and linking costs would depend on the vectors of attributes.[9] However, this simple formulation already captures some essential aspects of social interaction and provides substantial insight into small-world phenomena.

---

9. See Johnson and Gilles [367], as discussed in Exercise 6.14, for an alternative geographic cost structure based on distance on a line.

The overall utility to a player i in network g is

$$u_i(g) = \sum_{j \neq i:\ell(i,j) \leq D} \delta^{\ell(i,j)} - \sum_{j:ij \in g} c_{ij},$$

where $c_{ij} = c$ if i and j are on the same island and C otherwise.[10] For large enough D, the truncation is irrelevant. For smaller D, truncation captures the idea that benefits fall off quite dramatically beyond some threshold connection distance. For instance, it is impossible to ask for favors from the friend of a friend of a friend of a friend. The results in this model extend to the distance-based utility model under a suitable formulation, but the truncation makes the present model particularly transparent.

Recall from Chapter 3 that many social networks exhibit small-world characteristics embodied by a relatively small diameter, short average path length, and high clustering (compared to an independent random network). The following proposition from Jackson and Rogers [353] shows that for suitable parameter values, an islands version of the truncated connections model exhibits small-world characteristics. It makes clear how costs and benefits can explain small-world phenomena, a point first made by Carayol and Roux [138].

Proposition 6.6 *If $c < \delta - \delta^2$ and $C < \delta + (J-1)\delta^2$, then any network that is pairwise stable or efficient is such that*

1. *the players on any given island are completely connected to one another,*

2. *the diameter and average path length are no greater than $D + 1$, and*

3. *if $\delta - \delta^3 < C$, then a lower bound on individual, average, and overall clustering is $(J-1)(J-2)/(J^2K^2)$.*[11]

The intuition behind the proposition is relatively straightforward. Low costs of connections to nearby players (those on the same island) lead to high clustering. The high value of linking to other islands (accessing many other players) leads to low average path length. The high cost of linking across islands means that there are only a few links across islands.

These properties are illustrated in Figure 6.2, which illustrates the case $c < .04$, $1 < C < 4.5$, and $\delta = .95$. Players are grouped in sets of five that are completely connected to and lie on the same island, and there are five separate islands.

This economic analysis of small worlds gives complementary insights to those of rewiring analysis of Watts and Strogatz [658] discussed in Section 4.1.2. The random rewiring analyzed by Watts and Strogatz shows that it is possible to have

---

10. This cost structure is similar to that of the insiders-outsiders model of Galeotti, Goyal, and Kamphorst [273], while the benefits structure is quite different: the insiders-outsiders model has almost no decay in value with distance, while the islands-connection model matches the truncated version of the connections model (see Jackson and Wolinsky [361]). The difference in benefit structure between the islands-connections and the insider-outsider models leads to very different conclusions regarding clustering.

11. For the bounds on clustering it is assumed that $\delta - \delta^2 \neq C$. If $\delta - \delta^2 = C$, then there is a great deal of indifference over links, and the set of pairwise stable networks mushrooms.

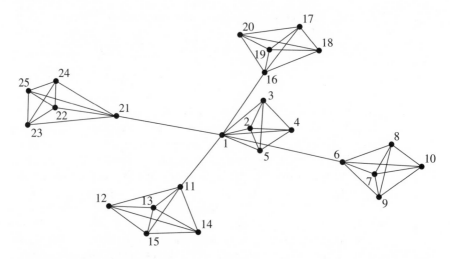

FIGURE 6.2   A pairwise stable "small world" in an islands version of the connections model.

both high clustering and short path length, whereas the above model gives more insight into why we should expect these traits to be exhibited by social networks.

Another feature distinguishing an economic modeling and a random modeling of these network characteristics concerns *shortcut* links (i.e., those which link distant parts of the network and, if deleted, would substantially alter the distance between the connected nodes). In a random rewiring model, shortcut links would at least occasionally occur in close proximity to each other. Under the strategic approach, the cost of building a second shortcut link next to an existing one would outweigh the benefit.[12]

Proof of Proposition 6.6.   Let us first show (1). If two players on the same island are not connected in some network, then they would each gain at least $\delta - \delta^2 - c > 0$ by adding the link, and this action would only help other players, so the network cannot be pairwise stable or efficient.

For (2), suppose that there are two players (on distinct islands), say i and j, such that $\ell(i, j) \geq D + 2$. As just argued, in any pairwise stable or efficient network, j is directly connected to all members of his or her island and so is i. Thus i is at a distance of at least $D + 1$ from each member of j's island and so enjoys no benefit from any of these players; the same is true for j from i's island. Thus by linking to j, i would gain at least $\delta + (J - 1)\delta^2 - C > 0$ (and vice versa), so this network cannot be pairwise stable or efficient.

Next, derive a lower bound for an individual's clustering (and thus average clustering). Consider an individual with L interisland links. All of the player's pairs of intraisland neighbors are themselves neighbors in either an efficient or pairwise stable network. Thus there are at least $(J-1)(J-2)/2$ pairs of i's neighbors that are linked out of a maximal total of $(J+L-1)(J+L-2)/2$ pairs of neighbors, which leads to a lower bound of $(J-1)(J-2)/[(J+L-1)(J+L-2)]$. Since $L \leq J(K-1)$, we have a loose lower bound of $(J-1)(J-2)/(J+J(K-1))^2$, resulting in the claimed expression.

The lower bound on the overall clustering coefficient is established as follows. For a given network, write i's clustering coefficient as a_i/b_i, where a_i is the number of links among neighbors in i's neighborhood and b_i is the number of pairs of neighbors in i's neighborhood. We have established a lower bound for a_i/b_i. Note that overall clustering is $(\sum_i a_i)/(\sum_i b_i)$, which is clearly greater than $\min_i(a_i/b_i)$.[13] ▪

Proposition 6.6 identifies small-world properties in a strong sense. The diameter is bounded above by $D+1$, and average path lengths are smaller, since each island is fully connected. Next observe that average clustering is approximated by K^{-2}. Thus clustering can remain large when n is very large, provided that the per-island population is not too small. In cases in which C is large enough so that the number of interisland links is lower (bounded by KJ), then the lower bound for clustering is even higher (on the order of $[J/(J+K)]^2$); and then even for large K relative to J, the clustering is much larger than that observed for an independently random network (which goes to 0 as the population grows, holding the probability of links constant, as discussed in Section 4.1.2).

Proposition 6.6 applies to networks that are either pairwise stable or efficient, and thus it shows that there are some similarities between these sets of networks. Understanding the exact relationship between pairwise stable networks and efficient networks is complex in the islands-connection model. Jackson and Rogers [353] characterize efficient networks when the intraisland costs are low. From that characterization they conclude that for some range of interisland costs the pairwise stable and efficient networks coincide, whereas for other cost ranges the set of pairwise stable networks, though always exhibiting small-world features, can be quite varied. These results are discussed in Exercises 6.11 and 6.12.

These results apply to cases in which the intraisland cost of connections is low enough that players are completely connected within their own islands. The analysis becomes more complex when the intraisland connection cost rises, so that not all players within an island are connected. Exercise 6.13 concerns diameters in such a situation.

6.6 ▪ A General Tension between Stability and Efficiency

In the above models, we have seen that there are settings in which all pairwise stable networks are inefficient and sometimes all pairwise stable networks are even Pareto inefficient. This contrast between efficieny and stability raises a number of interesting issues that we shall examine here and further in Chapter 11.

---

13. It is straightforward to check that $(a_1+a_2)/(b_1+b_2) \geq \min(a_1/b_1, a_2/b_2)$. The result then follows by induction.

6.6.1 Transfers: Taxing and Subsidizing Links

One question is to what extent this problem of reconciling stability and efficiency can be dealt with by transfers among the players. These might take different forms. It could be that a government or other entity intervenes to tax and subsidize different links (e.g., subsidizing research and development partnerships) if it decides that there are positive externalities and individual players might form too few partnerships. Or the players themselves could bargain over some payments to maintain links. For instance, the center of a star could negotiate with the other players to receive some payments or favors for maintaining his or her links with the other players. In fact, intuition from the sociology literature would suggest that a player in such a central position should receive a high payoff (e.g., see Burt [115]), which could come from the implicit power that the player receives from the implicit threat of severing links, or the favors and benefits that come along with the indirect connections that the player provides. If we start to account for such reallocations, can efficiency and stability be reconciled? And, more generally, what characterizes the settings in which there is a tension and when can transfers help?

Let us start with the basic question of whether it is possible to make transfer payments among the players so that at least some efficient networks are stable. Jackson and Wolinsky [361] showed that there are very simple and natural settings in which it is not possible to make transfers to align incentives and efficiency without violating some basic principles about how transfers should or would be structured. To make this idea precise, we need a few definitions.

A *transfer rule* is a function $t : G \to \mathbb{R}^N$ such that $\sum_i t_i(g) = 0$ for all g. A transfer rule can thus capture any reallocation of payoff in a given network. These payments could subsidize or tax certain links or collections of links and could be due to intervention by an outside authority or to bargaining by the players. What matters to players is the net payoff they receive as a function of the network. The requirement that transfers sum to 0 is usually termed a *balance condition,* and it embodies the idea that the system neither depends on any outside infusion of capital to operate nor destroys value.

In the presence of transfers, player i's net payoff becomes $u_i(g) + t_i(g)$, and this is used by the player in decisions regarding the addition or deletion of links. That is, pairwise stability is then applied where the payoffs to the players from a network g are $u_i(g) + t_i(g)$ rather than $u_i(g)$.

First note that there is a transfer rule that aligns individual and societal incentives. That is the *egalitarian transfer rule* (denoted by t^e) such that

$$u_i(g) + t_i^e(g) = \frac{\sum_j u_j(g)}{n}$$

or

$$t_i^e(g) = \frac{\sum_j u_j(g)}{n} - u_i(g).$$

This transfer rule completely equalizes all players' payoffs on any given network. Under this rule, any network that is efficient will also maximize each individual's net payoff, as each individual equally shares in the overall societal value. While the application of this rule is one way to realign individual incentives, there are reasons that such rules would not tend to arise. These are captured in the following conditions.

6.6.2 Component Balance

One condition that we would expect transfers to satisfy in a bargaining or voluntary process and when secession is a concern has to do with component balance.

A transfer rule t is *component balanced* if there are no net transfers across components of the network; that is, $\sum_{i \in S} t_i(g) = 0$ for each network g and component of players $S \in \Pi(N, g)$. Component balance requires that the value of a given component of a network is allocated to the members of that component. This condition is one that a planner or government would respect to avoid secession by components of the network or to reallocate value only among the individuals who generated it.

Whether component balance of the transfers is a compelling condition depends on the context, and in particular on the utility functions. If the utility functions exhibit externalities across components, so that the payoffs in one component depend on how other components are organized, then it may be important to make transfers across components. This is the case, for instance, when links are cooperative ventures among firms and firms are in competition with one another. For example, if links refer to code-sharing among airlines, then one group of airlines might care to what extent airlines to whom they are not linked are linked to one another. Component balance is relevant for those applications in which, for instance, links represent friendships and it does not matter to a given player how players in separate components are organized. Component balance is also a condition that one might expect to arise naturally if the transfers come about by a bargaining process. For example, individuals in one component might not be willing to make transfers to another component of individuals if the second component's organization has no effect on the first component.

It is important to emphasize that the following result only requires that component balance be applied in situations with absolutely no externalities across components.[14] In particular, there are no externalities across components when u is component-decomposable, which is defined as follows.

A profile of utility functions u is *component-decomposable* if $u_i(g) = u_i(g|_{N_i^n(g)})$ for all i and g.[15]

6.6.3 Equal Treatment of Equals

Another basic condition regarding transfers is an equal treatment condition. The condition of equal treatment has a rich tradition in social choice (e.g., see Thomson [628]). It requires that two players who are identical according to all criteria

---

14. The definition here is adapted from a condition defined on allocation rules, where it can be made explicit in the actual definition that it only be applied when the allocation rule is component additive. For more on that formulation, see Chapter 12.

15. Recall that $N_i^n(g)$ is the set of all players at a distance of no more than n from i, where n is the number of players, and so is the set of all players in i's component.

should end up with the same transfers or allocations. It is one of the most basic fairness criteria. In the context of social networks, the condition can be formulated as follows. Given two players i and j and a network g, let g^{ij} denote the network derived from switching the positions of i and j (and switching each of their connections).[16]

Two players i and j are *complete equals* relative to a profile of utility functions u and a network g if

- $ik \in g$ if and only if $jk \in g$,
- $u_k(\hat{g}) = u_k(\hat{g}^{ij})$ for all $k \notin \{i, j\}$ and for all $\hat{g} \in G(N)$, and
- $u_i(\hat{g}) = u_j(\hat{g}^{ij})$ and $u_j(\hat{g}) = u_i(\hat{g}^{ij})$ for all $\hat{g} \in G(N)$.

Thus two players are complete equals relative to a network and a profile of utility functions if they are completely symmetric relative to all players in the network, all other players see them as interchangeable in forming a network, and the two players have the same utility function as a function of the structure of the network.

A transfer rule satisfies *equal treatment of equals* relative to a profile of utility functions u if $t_i(g) = t_j(g)$ when i and j are complete equals relative to u and g. Equal treatment is the weakest possible anonymity condition, stating that two players should get the same transfers when they are identical in terms of their position in the network and interchangeable in the eyes of all players, including themselves.

This condition has several justifications. From the normative side, if one is designing transfers, it captures the most basic fairness principle: one should treat identical individuals equally. From the positive (descriptive) side, in cases for which the transfers might arise endogenously, it captures the idea that identical individuals would have similar bargaining positions, which lead them to similar outcomes.

6.6.4 Incompatibility of Pairwise Stability and Efficiency

The following proposition is a variation on a result of Jackson and Wolinsky [361].[17] The proposition can be strengthened to replace efficiency with a weaker form of efficiency, as outlined in Exercise 6.8.

Proposition 6.7 *There exist component-decomposable utility functions such that every pairwise stable network relative to any component-balanced transfer rule satisfying equal treatment of equals is inefficient.*

Proof of Proposition 6.7. The proof is by example. It is presented for $n = 3$, but is easily adapted to any n. Suppose that the utility of each player in the complete network is 4, the utility of each connected player in a linked pair is 6, and that of

---

16. Thus $g_{kh} = g_{kh}^{ij}$ when $k \notin \{i, j\}$ and $h \notin \{i, j\}$; and $g_{jk} = g_{ik}^{ij}$ and $g_{ik} = g_{jk}^{ij}$ for all k.

17. Their formulation was stated in terms of allocation rules rather than transfer rules, which is essentially equivalent (see the discussion in Chapter 12). They also required a stronger anonymity condition rather than the equal treatment condition, but their proof works with the equal treatment condition.

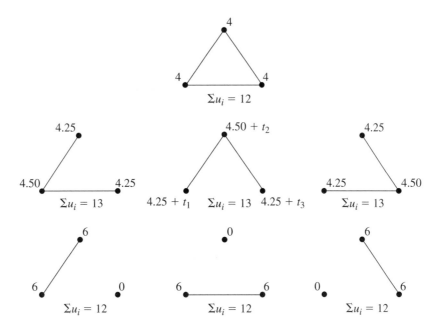

FIGURE 6.3 Payoffs such that no pairwise stable network is efficient regardless of transfers.

players who are disconnected is 0. The efficient networks are those with two links, which have a total utility of 13, with the central player having a utility of 4.5 and the other two players one of 4.25 each. This example is pictured in Figure 6.3.

In the absence of any transfers, the pairwise stable networks all fail to be efficient. The pairwise stable networks are limited to those involving a single link. In any other network some player(s) has an incentive to sever a link (each player has such an incentive in the complete network, and the center player has an incentive to do so in each of the two-link networks).

Consider introducing transfers to ensure that at least one efficient network is pairwise stable. Equal treatment of equals implies that the transfers on the complete network must be 0, and equal treatment plus component balance implies the same for the single-link and empty networks. So consider the middle network out of the two-link networks in Figure 6.3, and let us try to introduce transfers to make it pairwise stable. Payoffs with transfers are $4.25 + t_1$, $4.5 + t_2$ and $4.25 + t_3$, where t_1, t_2, and t_3 are the transfers on this network to the three players. Given the complete symmetry between the first and third players, equal treatment implies that their transfers must be equal so that $t_1 = t_3$. For the middle two-link network to be pairwise stable, the first and third players must not gain from adding the missing link. So $t_1 = t_3 \geq -.25$. However, to have the network be pairwise stable, the second (or center) player must be willing to keep both links that are in place. As that player receives a payoff of 6 if either link is deleted, it must be that $t_2 \geq 1.5$. However, now the sum of the transfers exceeds 0, and hence component balance is violated (in fact, even a basic feasibility condition is violated). ▪

This example extends to weaker notions of efficiency and a variety of notions of stability, as for example, in Exercises 6.9 and 6.10.

While this proof is based on an example, the example has natural properties in that some intermediate-size network is efficient. It is clear that the argument holds for a range of examples and that it is easily modified to hold for larger societies.

Both of the conditions of component balance and equal treatment of equals are required for the result to hold. The importance of component balance is easily seen. The egalitarian transfer rule satisfied equal treatment of equals and is such that all efficient networks are pairwise stable. Individuals only have incentives to form or sever links when these actions increase total utility. If we drop equal treatment of equals, but keep component balance, then a careful and clever construction of transfers by Dutta and Mutuswami [213] ensures that some efficient network is strongly stable for a class of utility functions. This result is stated in the following proposition.

Proposition 6.8 (Dutta and Mutuswami [213]) *If the profile of utility functions is component-decomposable and all nonempty networks generate positive total utility, then there exists a component-balanced transfer rule such that some efficient network is pairwise stable. Moreover, while transfers will sometimes fail to satisfy equal treatment of equals, they can be structured to treat completely equal players equally on at least one network that is both efficient and pairwise stable.*[18]

While the details of the proof of Proposition 6.8 are intricate, we can see some of the intuition motivating the proof by noting how it would work in the example in Figure 6.3. Here we set transfers so that if player 2 is in a single-link network, then he or she pays the other player at least 1.5. Under such transfers, the two-link network with player 2 in the middle is pairwise stable, as he or she no longer obtains a higher utility from severing one of the links.

Next, note that if we weaken efficiency to Pareto efficiency, Proposition 6.7 no longer holds. This statement takes a bit of proof to show generally, as one needs to work with specific transfers and then find an algorithm to identify pairwise stable networks that are Pareto efficient. That method is described in Exercise 6.15. It is easy to see why the proposition fails to extend to Pareto efficiency in the network in Figure 6.3, as without any transfers the single-link networks are pairwise stable and Pareto efficient. Although this case is of some interest, when admitting transfers, Pareto efficiency is arguably not the right notion of efficiency. Instead, an efficiency notion that considers the admitted transfers is more appropriate, and then Proposition 6.7 can be extended as there is an incompatibility of stability and efficiency allowing for component-balanced transfers that satisfy equal treatment of equals (as outlined in Exercise 6.9).

---

18. Dutta and Mutuswami work with a variation of strong stability that is not quite a strengthening of pairwise stability, as it only considers one network to defeat another if there is a deviation by a coalition that makes all of its members strictly better off; while pairwise stability allows one of the two players adding a link to be indifferent. However, the construction of Dutta and Mutuswami extends to pairwise stability as well.

So reconciling the tension between stability and efficiency requires relinquishing one of our desired conditions of equal treatment of equals, component balance, and (constrained) efficiency; this tension is characteristic of many network games.

There are many related questions associated with this tension that are addressed in subsequent chapters. For instance, which settings (in terms of the structure of costs and benefits) naturally lead efficient networks to be stable? In which settings can transfers help reconcile efficiency and stability? How efficient or ineffectient are the networks that form if players bargain over payoffs during the network-formation process?

6.7 ▪ Exercises

6.1 *Efficiency versus Pareto Efficiency* Provide an example of a society of three individuals and corresponding utility functions such that there are several Pareto efficient networks and yet only a single efficient network.

6.2 *Pairwise Stability in the Distance-Based Utility Model* Prove Proposition 6.2.

6.3 *Diameters in Large Pairwise Stable Networks in the Connections Model**
(a) Consider the symmetric connections model when $c > \delta$. Show that as n grows, the diameter of any nonempty pairwise stable network is bounded above.
(b) Show that the same is true in the distance-based utility model for any specification of $c > b(1) > b(2) > b(3) > \cdots > b(k) > b(k+1) > \cdots$, provided that these are fixed and independent of n.

6.4 *The Diameter in the Symmetric Connections Model* Using techniques similar to those in the proof of Proposition 6.4, show that if the diameter of a nonempty component of a pairwise stable network in the symmetric connections model is between $2D$ and $2D + 1$ for a positive integer D, then the increase in payoff to forming a link to each of a pair of nodes in that component that are at maximum path distance from each other is at least $\delta(1 - \delta^D)^2/(1 - \delta)$. Using this lower bound on the increase in payoffs from link formation, show that an upper bound on the diameter of a nonempty component of a pairwise stable network when $c < \delta(1 - \delta)$ is

$$2\frac{\log\left(1 - \sqrt{c(1-\delta)/\delta}\right)}{\log(\delta)} + 1.$$

6.5 *An Asymmetric Connections Model* Consider an asymmetric version of the connections model, in which all players have a common δ parameter where $0 < \delta < 1$, and the only asymmetry is that individuals might have different costs per link. In particular, suppose that each individual's cost for a link is the same for all links, so that individual i has a cost c_i for each link that i is involved with, but it is possible that these costs differ across players.
(a) Provide an example where the unique efficient network is not a star network, nor a complete network, nor the empty network.

(b) Show that every efficient network has a subnetwork that is a star network (possibly only involving a subset of the players) and such that the center player in that star has a minimal cost (that is, $c_i \leq c_j$ for all j).

(c) Show that if individuals have different values for δ_i, then it is possible to have an efficient network that is not empty and does not have a star network as a subnetwork.

6.6 *Growing Strategic Networks* Consider the network growth process in the distance-based utility model described in Section 6.3.2 when $n = 4$. Suppose that $b(1) - b(2) < c < b(1)$. Find the probability that an efficient network forms.

6.7 *Pareto Inefficiency in the Coauthor Model* Show that if $n \geq 4$ is even, then any pairwise stable network in the coauthor model of Section 6.4 is not only inefficient but is, in fact, Pareto dominated by some efficient network. (Hint: show that any player in any component of a pairwise stable network [including being alone] that does not involve exactly two players, receives a lower payoff than he or she would in any efficient network.)

6.8 *Constrained Efficiency* The following notion of efficiency (from Jackson [347]) falls between efficiency and Pareto efficiency. A network g is *constrained efficient* relative to a profile of component-decomposable utility functions u if there does not exist any $g' \in G(N)$ and a component balanced t satisfying equal treatment of equals relative to u such that $u_i(g') + t_i(g') \geq u_i(g) + t_i(g)$ for all i with strict inequality for some i.

(a) Show that for any profile of component-decomposable utility functions u, the set of efficient networks is a subset of the constrained efficient networks.

(b) Let $n = 5$. Consider a component-decomposable utility for which all individuals are interchangeable such that the complete network generates utility of 2 per player, a component consisting of a pair of individuals with one link gives each of the pair a payoff of 1, and a completely connected component among three individuals gives each player in the component a payoff of 3. All other networks generate utility of 0. Identify the efficient networks, and show that the completely connected network is constrained efficient but not efficient.

(c) Let $n = 3$. Consider a profile of utility functions u such that the complete network has a payoff of 3 for each player, any network with two links leads to a payoff of 4 to the center player and 2 to each of the other players, and all other networks lead to payoffs of 0 to all players. Find a Pareto efficient network relative to this u that is not constrained efficient.

6.9 *Proposition 6.7 with Constrained Efficiency* Show that Proposition 6.7 holds when efficiency is replaced by constrained efficiency.

6.10 *Side Payments and the Incompatibility of Efficiency and Stability* Consider a stronger definition of stability than pairwise stability due to Jackson and Wolinsky [361]. A network g' *defeats* a network g *allowing for side payments* if

- $g' = g - ij$ and $u_i(g) < u_i(g')$ or $u_j(g) < u_j(g')$, or
- $g' = g + ij$ and $u_i(g') + u_j(g') > u_i(g) + u_j(g)$.

The network g is *stable relative to side payments* if it is not defeated by another network allowing for side payments. Show that Proposition 6.7 holds with stability relative to side payments replacing pairwise stability, and without the need for condition of equal treatment of equals.

6.11 *Efficient Networks in the Islands Model* Jackson and Rogers [353] provide the following partial characterization of efficient networks in the islands model. Let $c < \delta - \delta^2$. In any efficient network, each island is internally completely connected and interisland links are as follows.

- If $C < \delta - \delta^2$, then the unique efficient network is the completely connected network.
- If $\delta - \delta^2 < C < \delta - \delta^3$ and $K = 2$, then the efficient networks are those such that there are exactly J links between the two islands, and on at least one island each player is involved in exactly one of the J links.
- If $\delta - \delta^3 + 2(J - L - 1)(\delta^2 - \delta^3) < C < \delta - \delta^3 + 2(J - L)(\delta^2 - \delta^3)$ and $K = 2$, then the efficient networks are those such that there are exactly $1 \le L < J$ links between the two islands and no player is involved in more than one of these links.
- If $\delta - \delta^3 + 2(J - 2)(\delta^2 - \delta^3) < C$ and $K = 2$, then the efficient networks have at most one link between the two islands.

Prove these claims. Show also that when K is very large and $C < \delta - \delta^3$, completely connecting all players within each island and then connecting every player on every island other than island 1 directly to the same player on island 1 can be more efficient than having intraisland links that do not all pass through the same island.

Hint for the proof of the proposition: Suppose there are just two islands, and call the number of links between them L. Use the fact that

$$V(x_1, x_2, L) = 2L\delta + 2\left[Jx_1 + Jx_2 - x_1x_2\right]\left(\delta^2 - \delta^3\right) - 2L\delta^2 + 2J^2\delta^3$$

is the utility obtained by the members of island 1 from connections to island 2 plus the reverse, where $x_i \le L$ is the number of players on island i having links to the other island.

6.12 *Inefficiency of Pairwise Stable Networks in the Islands Model* Consider the islands model when $c < \delta - \delta^2$ and show an example with at least three islands and three players per island, such that all pairwise stable networks are distinct from all efficient networks.

6.13 *Diameter in the Islands Model* Consider the islands model when $\delta - \delta^2 < c < \delta$ and $C < \delta + (J - 1)\delta^D$. Show that the diameter of all pairwise stable networks is no greater than $2D$.

6.14 *The Spatial Connections Model* A version of the connections model studied by Johnson and Gilles [367] introduces geography to costs. Let there be more than three players. The benefits take the same form as in the symmetric connection model, but there is a geography to player locations, and the cost of forming a link between players i and j, denoted by c_{ij}, is related to physical distance. Let players be spaced equally on a line and i's location be at the point i, and let c_{ij} be proportional to $|i - j|$.

(a) Provide an example for which all efficient networks are nonempty, incomplete, and not stars.

(b) Provide an example for which all pairwise stable networks are ineffcient.

6.15 *Pairwise Stable and Pareto Efficient Networks*   There are component-balanced transfer rules, satisfying equal treatment of equals relative to utility functions that are component-based, relative to which there always exists at least one pairwise stable network that is Pareto efficient.[19]

Given a profile of component-decomposable utility functions u and a set of nodes $S \subset N$, let

$$g(u, S) = \text{argmax}_{g \in G(S)} \frac{\sum_{i \in S} u_i(g)}{\#N(g)}$$

denote the network with the highest per capita value to individuals who have links in the network out of all networks that can be formed by any subset of players in S. Given a component-decomposable utility function u, find a network g^u through the following algorithm. Pick some $h_1 \in g(u, N)$ with a maximal number of links. Next select $h_2 \in g(u, N \setminus N(h_1))$ with a maximal number of links. Iteratively, at stage k pick a new component $h_k \in g(u, N \setminus N(\cup_{i \leq k-1} h_i))$ with a maximal number of links. Once there are only empty networks left, stop. The union of the components selected in this way defines a network g^u. Consider component-balanced transfers such that if i and j are in the same component of g, then $u_i(g) + t_i(g) = u_j(g) + t_j(g)$.

(a) Show that such transfers satisfy equal treatment of equals.

(b) Show that if u is component-decomposable, then g^u is pairwise stable and Pareto efficient (relative to the payoffs including transfers).

(c) Show that if the selection algorithm did not have the quantifier "with a maximal number of links," then the resulting g^u could fail to be pairwise stable.

---

19. This case is a variation of a result due to Banerjee [39], based on an algorithm that is adapted to work for pairwise stability by Jackson [348].

PART III

IMPLICATIONS OF NETWORK STRUCTURE

CHAPTER 7

Diffusion through Networks

Consider the following fundamental scenario in which networks play a central role. There is a society of individuals whose relationships are described by a network, indicating who has contact with whom on a regular basis. This network might be one of email correspondents, friendships, acquaintances, or even sexual relationships or some sort of blood-to-blood contacts. To keep things simple at this point, let us presume that individuals are either linked or not, so that we abstract away the fact that some relationships might involve more frequent or intense interaction than others. Consider the introduction of a disease. Individuals can catch the virus if they interact with one of their neighbors who is infected with it. Transmission occurs somewhat randomly, as the chance of interaction might be random and it might also take specific conditions for the disease to be transmitted. Let us also suppose that the chance of a given healthy individual becoming infected increases with the number of neighbors who are infected. Under what conditions will an initial outbreak spread to a nontrivial portion of the population? What percentage of the population will eventually become infected? How does this depend on the network of contacts in the society? How does it depend on the contagion probabilities and recovery rates? What if a number of individuals are immune to the disease, or require different frequencies of interaction to be infected? How would immunizing the population affect the spread of the disease?

This problem is of obvious importance. Beyond the direct application to disease transmission, it is closely related to the diffusion through a network of information, opinions, product purchasing, fashions, participation in programs, and various other behaviors.[1] For example, network structure is central to "viral marketing" (e.g., see Leskovec, Adamic, and Huberman [430]) as well as "word-of-mouth

---

1. See Rogers [564] and Strang and Soule [617] for overviews of some of the literature on diffusion and its many applications.

marketing" (e.g., see Silverman [593]).[2] Moreover, it is clear that the network structure can be very important in determining the outcome. If there are relatively few contacts between individuals, then the disease might never spread. If some pockets of the society are isolated from others, they might remain relatively uninfected (or uninformed, depending on the application). Networks with a few highly connected individuals might develop different patterns of infection than networks in which all individuals have roughly the same degree.

In this chapter I discuss a number of different models of such diffusion processes. The chapter begins with a basic model of disease transmission and then discusses some variations on it. Beyond these issues of diffusion, there are also related questions of navigation on networks. That is, individuals are searching through the network to find specific nodes. How difficult is it for an individual to find a specific web page by following links on the web? How can we understand the ease (in terms of small number of hops) with which subjects in Milgram's [468] small-world experiments were able to send something to an individual they did not know? Finally, there are issues of how fragile networks are to interruption. If we delete some set of nodes, does diffusion or navigation change dramatically? How does this change depend on which nodes are removed? The topics of this chapter dovetail with the learning and games on networks that are discussed in the next two chapters, and we will see related themes and tools emerging.

An outline of the chapter is as follows. First I discuss the Bass model, which is a parsimonious and widely used model of diffusion that incorporates some ideas of imitation but without explicit network structure. Next I turn to models of diffusion that explicitly incorporate social structure, and discuss results on component size in random networks and how that relates to the potential extent of a diffusion in a society. Then standard models from the epidemiology literature are discussed; in their models, contagion across nodes is random and the probability of becoming infected depends on the number of neighbors that a node has and the state of those neighbors. The analysis depends on the extent to which nodes can recover from an infection, whether they can become infected again, and the ease with which the disease or information is transmitted from one neighbor to another. The chapter then turns to a related but distinct question of navigation of a network, or following paths in a network to find specific nodes; it examines how this ability depends on the structure of the network and how much knowledge about the network and the node is used in navigation. The chapter also includes a look at how robust diffusion and navigation results are to the removal of nodes.

---

2. In word-of-mouth marketing, referrals, direct conversations, and recommendations among consumers are used to spread information about a product (e.g., providing initial access to select individuals and then allowing them to communicate with others, as in the way that the email service Gmail was marketed). Part of this strategy builds on the idea of the high weight that people might place on a friend's recommendation for a product. Viral marketing is a more general concept that includes word of mouth as a special case, but also includes other uses of interactions among consumers. For example, the email service Hotmail was initially marketed by including information about its availability in outgoing emails of Hotmail users. This transmission was not voluntary word of mouth, but still took advantage of the social structure for the spread of information.

7.1 ▪ Background: The Bass Model

An early model of diffusions (that is still a workhorse) is the Bass [46] model. The model is lean and tractable and incorporates social aspects into its structure. Although it does not have any explicit social network structure, it still incorporates imitation. The model is built on two key parameters: one captures the rate at which agents innovate or spontaneously adopt, and the other captures the rate at which they imitate other agents or adopt because others have.[3] One can also interpret the innovation as a response to outside stimuli, such as media or advertising, while the imitation aspect captures social and peer effects.

Consider discrete time periods t, and let $F(t)$ be the fraction of agents in a society who have adopted a new product or behavior by time t. The Bass model is then described by the difference equation

$$F(t) = F(t-1) + p\,(1 - F(t-1)) + q\,(1 - F(t-1))\,F(t-1),$$

where p is the rate of innovation and q that of imitation. The expression $p(1 - F(t-1))$ is the rate of innovation times the fraction of people who have not yet adopted. The expression $q(1 - F(t-1))F(t-1)$ captures the imitation process, where the rate of imitation is multiplied by two factors. The first factor, $(m - F(t-1))$, is again the fraction of people who have not yet adopted, and the second expression, $F(t-1)$, is the fraction of agents who have adopted and can therefore be imitated. A continuous time version of this model is described by

$$\frac{dF(t)}{dt} = (p + qF(t))\,(1 - F(t)). \qquad (7.1)$$

Solving (7.1) when $p > 0$ and setting $F(0) = 0$ leads to

$$F(t) = \frac{1 - e^{-(p+q)t}}{1 + \frac{q}{p}e^{-(p+q)t}}. \qquad (7.2)$$

As the parameters p and q are varied, the Bass model can be made to fit a wide variety of diffusion curves, and as such has been used in both forecasting diffusion and in empirical analyses of diffusion, where one can estimate q and p from fitting the model to data. There are also several extensions of the model to include such things as pricing and advertising.[4] Note that in the expression for $F(t)$ in (7.2), the levels of p and q simply multiply time, and the ratio of q to p is the critical parameter that determines the overall shape of the curve. Thus, up to rescalings of time, the curves can be thought of as generated by a single parameter that captures the relative ratio of imitators to innovators.

Part of the insight that emerges from the Bass model is into the process behind some widely observed patterns in diffusion studies. Beginning with one of the first studies of diffusion by Tarde [619], who emphasized the importance of imitation

---

3. The version here omits a third parameter m, which simply represents the portion of the population which could potentially be adopters. Here $m = 1$.

4. Examples of recent models of diffusion, which push in various directions beyond the Bass model, are Leskovec, Adamic, and Huberman [430] and Young [668].

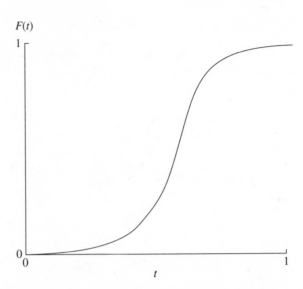

FIGURE 7.1   An S-shaped diffusion/adoption curve.

in diffusion, many studies have found diffusion curves that are S-shaped. Perhaps the best known such studies are those of Ryan and Gross [573] and Griliches [311], which documented S-shaped diffusion curves in the adoption of hybrid corn seeds among Iowa farmers. Figure 7.1 provides an example of an S-shaped adoption curve, where diffusion begins slowly, then accelerates, and eventually slows down and asymptotes.

To see how the Bass model works, and why it can exhibit the S-shape, note that initially there are no agents in the population to imitate. Thus the first adopters are almost entirely those who adopt from their own spontaneous "innovation." In (7.1), when $F(t)$ is close to 0 the equation is approximated by p. As the process progresses, there are more agents around to be imitated, which leads to an increase in the rate of diffusion, since now agents adopt through imitation as well as innovation. A balance occurs as time progresses, since there are more agents around to imitate, but fewer around to do the imitating. Eventually the process has to slow down simply because there are no longer any agents left who might innovate or imitate.

Many features that are stationary in the original Bass model can be enriched. For example, for spontaneous innovation it is not clear that the rate p should be constant over time. As Rogers [564] describes, the earliest adopters can often differ substantially in their characteristics from others who adopt later (even without imitation). One can enrich the model to incorporate such features as heterogeneous costs of adoption among the population, as well as different sorts of imitation principles. For example, is it raw numbers or the percentage of neighbors who have adopted that influence an agent to adopt? I discuss some of the variations of models in what follows and also in Chapter 9. For now, let us examine settings in which social structure is modeled more explicitly.

7.2 ▪ Spread of Information and Disease

The component structure of a network naturally partitions a society into separate groups who do not interact or communicate with one another. Different components might be subjected to the same outside stimuli, and it will take some such external consideration to foster diffusion across components of a network. In many instances, the component structure serves as a natural first limit on the extent of diffusion or contagion in a social network. The component structure is also important to understand with respect to navigation: if one can only follow paths in the network, then components are natural barriers.

7.2.1 Percolation, Component Size, Immunity, and Diffusion

The general problems of modeling contagion, the spread of information, and navigation through a society all involve determining when paths exist that connect different nodes of a network and properties of those paths, including the components that they generate. The variations on the problem incorporate nodes that might be immune to infection or averse to listening to or passing information along, and situations in which different links transmit or function probabilistically. This behavior might reflect the fact that individuals interact only sporadically, or that special conditions need to be met for transmission to take place. For many applications, models that provide insight into diffusion and navigation have to be rich enough to capture such variations.

This model enrichment also leads to overlap with what is known in the physics and math literatures as *percolation theory*.[5] A canonical scenario from percolation theory deals with some substance, which is porous, and a central question is whether a liquid on one side of the substance will penetrate to the other side. One might think of the substance as having holes or pores in it, and these possibly being connected to one another. The holes or pores can naturally be modeled as nodes with links, and thus as a network or in many cases even as a lattice. The question of percolation is then equivalent to whether there exists a path from one side of the network to the other. The analysis of diffusion and navigation on networks has thus drawn on tools and insights from percolation theory, and in turn enriched that theory.

To begin, let us revisit the scenario studied in Section 4.3, which examined component size when some members of the population are completely immune. This model also corresponds to percolation through a network when some nodes are removed. To recall the setting, there is a society of n individuals. One of them is initially infected with a disease. Members of the society are immune with a probability π. The question of whether the disease can spread to a nontrivial fraction of the population amounts to whether the initially infected individual lies in a component containing a nontrivial fraction of the population once the immune individuals are removed from the network. In Section 4.3 we considered Poisson

---

5. See Stauffer and Aharony [615] or Grimmett [313] for background.

random networks, but the analysis generalizes easily to other degree distributions. Let us work with the configuration model.

We can generalize beyond Poisson degree distributions as follows. Recall from (4.8) that the threshold for the emergence of a giant component in the configuration model is[6]

$$\left\langle d^2 \right\rangle_\pi = 2 \left\langle d \right\rangle_\pi . \tag{7.3}$$

The subscript π indicates that we are operating on a network in which a proportion π of the nodes are removed from the network uniformly at random, and the expectations are then taken with respect to the resulting degree distribution. Thus (7.3) determines the threshold of percolation in this system, or the point at which a single infected individual has a probability of infecting a number of others that becomes infinite as n grows.

Before analyzing this expression for different values of π and the resultant degree distributions, let me point out a different argument supporting (7.3). This threshold was originally determined by calculating neighborhood sizes and examining when the expected extended neighborhood sizes became infinite as n grows, but (7.3) can also be explained by a simple (heuristic) argument due to Cohen et al. [161]. Consider starting at a given node and then randomly picking one of its neighbors. For the neighborhoods to keep expanding, the expected degree of this neighbor should be at least 2 (the connection to the original node plus a continuation to some other nodes). The expected degree of a neighboring node under the configuration model is simply $E_{\widetilde{P}_\pi}[d]$ (recall that $\widetilde{P}_\pi(d) = P_\pi(d)d/\langle d \rangle_\pi$ is the conditional distribution of a neighbor's links, adjusting for the fact that the neighbor was found by a link in the network). This expectation is then

$$E_{\widetilde{P}_\pi}[d] = \sum_d \frac{P_\pi(d)d^2}{\langle d \rangle_\pi} = \frac{\langle d^2 \rangle_\pi}{\langle d \rangle_\pi}.$$

So the neighborhoods are expected to expand when $\langle d^2 \rangle_\pi / \langle d \rangle_\pi \geq 2$, and not otherwise.

Thus if we have appropriate expressions for the distribution P_π and some of its moments, then we can find the threshold at which the giant component emerges and determine when diffusion occurs despite having some fraction of immune nodes. Calculate $P_\pi(d)$ as follows. Consider the starting network under P and some node that starts with degree d'. If the node is not immune, then it might lose some of its neighbors when we delete the immune nodes from the network. In particular, the probability that a node starting with degree d' ends up with degree $d \leq d'$ follows a binomial distribution. To see this, note that the probability that some specific set of d of the neighbors is susceptible is $(1 - \pi)^d$, the probability that the rest are immune is $\pi^{d'-d}$, and there are $\binom{d'}{d}$ different subsets of d neighbors that are susceptable, which leads to a probability of

$$\binom{d'}{d}(1 - \pi)^d \pi^{d'-d}.$$

---

6. The analysis here works within the assumptions and structure used to derive (4.8).

Therefore the degree distribution after eliminating immune nodes can be written as

$$P_\pi(d) = \sum_{d' \geq d} P(d') \binom{d'}{d} (1-\pi)^d \pi^{d'-d}. \tag{7.4}$$

From (7.4) it follows that[7]

$$\langle d \rangle_\pi = \langle d \rangle (1-\pi) \tag{7.5}$$

and

$$\langle d^2 \rangle_\pi = \langle d^2 \rangle (1-\pi)^2 + \langle d \rangle \pi (1-\pi). \tag{7.6}$$

The threshold for a giant component of susceptible nodes to emerge is found by rewriting (7.3) using (7.5) and (7.6):

$$\langle d^2 \rangle (1-\pi) = \langle d \rangle (2-\pi) \tag{7.7}$$

or

$$\pi = \frac{\langle d^2 \rangle - 2\langle d \rangle}{\langle d^2 \rangle - \langle d \rangle}. \tag{7.8}$$

For example, if we examine a regular network of degree $\bar{d}$, then (7.8) provides a threshold at $\pi = (\bar{d} - 2)/(\bar{d} - 1)$. Thus, if $\bar{d} = 2$, then the threshold is right where $\pi = 0$. If the degree is 3, then $\pi = 1/2$, so that a giant component emerges if less than half the population is immune.

If we consider a Poisson random network, then since $\langle d^2 \rangle = \langle d \rangle^2 + \langle d \rangle$, and $\langle d \rangle = (n-1)p$, it follows that the threshold in (7.8) corresponds to the threshold $\pi = 1 - \frac{1}{(n-1)p}$ or $p(n-1)(1-\pi) = 1$ (as found in Section 4.3).

If we consider the case of a scale-free network, then we find very different effects. The interesting aspect is that the threshold for contagion is 0 (for $\gamma < 3$), which was first shown by Cohen et al. [161]. To see this, simply note that $\langle d^2 \rangle$ is diverging in n for a degree distribution in which $P(d)$ is proportional to $d^{-\gamma}$ for $\gamma < 3$.[8] For such scale-free degree distributions, (7.8) leads to a limit of $\pi = 1$, and so effectively all nodes need to be immune before the giant component of susceptible nodes will disappear. Under such degree distributions there are enough very-high-degree nodes so that many nodes are connected to high-degree nodes

---

7. To see this note that (7.4) implies that $E_\pi[d^t] = \sum_d \sum_{d' \geq d} d^t P(d') \binom{d'}{d} (1-\pi)^d \pi^{d'-d}$, which is equal to

$$\sum_{d'} P(d') \sum_{d \leq d'} d^t \binom{d'}{d} (1-\pi)^d \pi^{d'-d}$$

or $\sum_{d'} P(d') M_\pi(d^t : d')$, where $M_\pi(d^t : d')$ is the expectation of d^t from a binomial distribution with parameter $(1-\pi)$ and an upper limit of d' draws. Equations (7.5) and (7.6) follow, noting that $M_\pi(d : d') = d'(1-\pi)$ and $M_\pi(d^2 : d') = (d')^2(1-\pi)^2 + (d')\pi(1-\pi)$.

8. For $\gamma < 1$, the mean is also diverging, but the ratio of $\langle d^2 \rangle$ to $\langle d \rangle$ still diverges, so the condition is still satisfied.

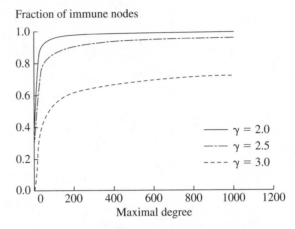

Fraction of immune nodes

FIGURE 7.2 Threshold fraction of nodes that need to be immune in a scale-free network to stop diffusion among susceptible nodes as a function of the maximal degree among nodes in the network.

and the network has a giant component, even when many nodes are eliminated uniformly at random.

It is important to remember that the analysis here is a limiting one, and so the thresholds are at the limit. When we examine a large but finite network, there is a maximal degree in the network; then $\langle d^2 \rangle$ is finite even for a scale-free distribution, and so the threshold in (7.8) is less than 1. We can approximate the threshold using (7.8) and see that it converges to 1 but might differ substantially with small caps on the maximal degree, as pictured in Figure 7.2 for three levels of γ.

Beyond a calculation of the threshold for the emergence of a giant component, we can also make use of the equation that characterizes the size of the giant component (above the threshold where it emerges and below that where the network becomes connected) in the configuration model. Recalling (4.9), the fraction of nodes in the giant component q is characterized by

$$1 - q = \sum_d (1 - q)^d P_\pi(d), \qquad (7.9)$$

where again $P_\pi(d)$ is the appropriate degree distribution associated with the network in which we have deleted a fraction π of immune nodes.

The values of q that solve (7.9) as a function of $1 - \pi$ (recalling (4.13)) are plotted for Poisson random networks for various mean degrees in Figure 7.3.

This analysis shows that the determination of whether a network diffuses an infection (or something else), when there are some immune nodes, depends on the size of the variance in degree relative to the mean degree. Relatively higher variation makes a network more conducive to infection, as it provides enough higher-degree nodes to lead to the formation of giant components. In general, even with Poisson random networks, thresholds are fairly low in terms of expected degree (compared to the degree typically seen in varieties of human interaction), and so diffusion can occur even with high fractions of immune nodes. Of course,

Fraction of susceptible nodes in the giant component
of susceptible nodes

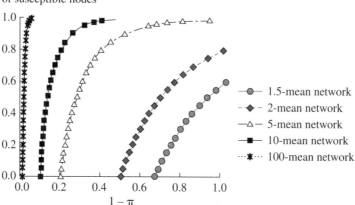

FIGURE 7.3   Fraction in the largest component of the susceptible population as a function of the fraction $1 - \pi$ of the population that is susceptible in a Poisson network.

the extent of the diffusion depends on the proportion of immune nodes and the size of the giant component. As we shall see, the extent to which diffusion occurs is also tempered by other potential impediments.

7.2.2 Breakdown, Attack and Failure of Networks, and Immunization

Another interpretation of the immune nodes that we have been discussing is that these are nodes that have failed and been removed from the system. This interpretation is particularly relevant in computer applications but is also relevant in social settings, including, for example, criminal organizations. If the attacks occur uniformly at random, then the analysis is as discussed above, but there are also many other variations that are of interest, depending on the application. For example, specific nodes or links might be targeted (e.g., high-degree nodes, or links that are important bridges or lie on many paths), links might overload due to excessive use, or there may be a spatial correlation in connection and nodes are targeted geographically.[9]

A flip side to this viewpoint is one of immunization. Suppose that we begin with a network that has a giant component. In such a setting, an initially infected individual can lead to a widespread infection. How many individuals need to be immunized to avoid an epidemic spread? How does this number depend on whether individuals in the population are immunized at random or high-risk (e.g., high-degree) nodes are immunized.

---

9. See Broder et al. [111], Albert et al. [10], Callaway et al. [123], Cohen et al. [162], Watts [656], Rozenfeld et al. [572], Warren et al. [649], Moreno et al. [483], Holme and Kim [335], and Motter and Lai [491].

To see how the issue of targeting can change the analysis, consider a situation in which some percentage of the nodes are removed, except that instead of removing nodes uniformly at random as in the previous analysis, the highest-degree nodes are removed (e.g., Broder et al. [111], Albert, Jeong, and Barabási [10], Callaway et al. [123], and Cohen et al. [162]).

We can work from the methods above, but now using the truncated distributions. So start with a distribution P. We first truncate the distribution removing a percentage π of the highest-degree nodes. The resulting degree distribution of the remaining nodes is denoted $P_{\pi,H}$, where H indicates that the highest-degree nodes have been eliminated. To develop an expression for $P_{\pi,H}$, it helps to break the calculation into two parts. First, eliminating the highest-degree nodes truncates the original distribution P. Second, it also eliminates some of the neighbors of the remaining nodes, thus lowering their degrees. We have to be careful in this second calculation, since the number of links eliminated is not simply proportional to π (as it was with uniformly random elimination of nodes): now it is much higher, as we are eliminating the nodes with the most links.

Consider a case in which π is chosen to cut the distribution nicely right at some degree $\overline{d}$, so that all nodes with degree higher than $\overline{d}$ are removed, and nodes with initial degree of $\overline{d}$ and lower remain. So $\overline{d}(\pi)$ is such that

$$\sum_{d=1}^{\overline{d}(\pi)} P(d) = 1 - \pi.$$

Thus we create a new degree distribution of the remaining nodes, not yet accounting for lost neighbors, which is simply $\frac{P(d)}{1-\pi}$ for degrees up to $\overline{d}(\pi)$ and 0 above that. Next, although we have only removed a fraction π of the nodes, we have removed a much larger fraction of the links, namely

$$f(\pi) = \frac{\sum_{d=\overline{d}(\pi)+1}^{\infty} P(d)d}{\langle d \rangle}.$$

Under the configuration model, a degree d' node then loses each neighbor approximately independently with probability $f(\pi)$. So the new degree distribution is of the form

$$P_{\pi,H}(d) = \sum_{d'=d}^{\overline{d}(\pi)} \frac{P(d')}{1-\pi} \binom{d'}{d} (1 - f(\pi))^d f(\pi)^{d'-d}. \tag{7.10}$$

By the same reasoning used to establish (7.7), it follows from (7.10) that the threshold for there to exist a giant component of the remaining nodes is such that

$$\langle d^2 | d \leq \overline{d}(\pi) \rangle (1 - f(\pi)) = \langle d | d \leq \overline{d}(\pi) \rangle (2 - f(\pi)), \tag{7.11}$$

where $\langle \cdot | d \leq \overline{d}(\pi) \rangle$ indicates that expectations are with respect to the original distribution but are truncated at $\overline{d}(\pi)$.

To see the effect of the truncation most starkly, let us examine (7.11) in the case of a scale-free distribution for which $P(d)$ is proportional to $d^{-\gamma}$ for $d \geq 1$. When $\gamma < 3$, if the maximal degree is high enough then the network exceeds the threshold for a giant component even when some fraction of nodes $1 > \pi > 0$ is removed uniformly at random, as we saw in Section 7.2.1. As an illustration of the process when the highest-degree nodes are removed, let us work with a continuous approximation of P with density $(\gamma - 1)d^{-\gamma}$, which makes the calculations easy. Straightforward calculations[10] imply a threshold equation of

$$(\gamma - 2) \left(1 - \pi^{(\gamma - 3)/(\gamma - 1)}\right) = (\gamma - 3) \left(2 - \pi^{(\gamma - 2)/(\gamma - 1)}\right). \quad (7.12)$$

For example, if we set $\gamma = 2.5$, then (7.12) reduces to $\pi^{1/3} + \pi^{-1/3} = 3$, which is satisfied when $\pi = .056$.[11] This result is quite dramatic: a scale-free network has diffusion that is robust to elimination of a very large fraction of nodes if they are deleted uniformly at random, but it is not robust to deletion of even a small fraction of nodes if the highest-degree nodes are targeted. For instance, note that eliminating only 5 percent of the nodes when $\gamma = 2.5$ eliminates all nodes with degree 4 or higher! Moreover, this targeted deletion eliminates roughly a third of the links in the network. Thus the remaining nodes, which have initial degree 3 or less, lose a nontrivial portion of their links, and so the network that remains can fail to have a giant component. In fact, most of the nodes end up with degree 1 or 0.

Note that this feature of networks has powerful implications for immunization programs. Programs that are aimed at immunizing individuals with the highest connectivity can be much more effective at avoiding an epidemic than a program that immunizes the same number of individuals uniformly at random in a population.

7.2.3 The SIR and SIS Models of Diffusion

One of the canonical models of the spread of disease through a network is what is now known as the *SIS* model, which stands for "susceptible, infected, susceptible" (see Bailey [28]). The idea is that a node can be in one of two states: (1) it is infected, or (2) it is not infected but is susceptible to becoming infected. This model is a variation on the seminal model in the literature, the *SIR* ("susceptible, infected, removed") model, which dates to Kermack and McKendrick [388]. In that

---

10. Note that $\int_1^x (\gamma - 1)d^{-\gamma}\mathrm{d}d = 1 - x^{-\gamma+1}$. It follows that $\bar{d}(\pi) = \pi^{1/(1-\gamma)}$, $f(\pi) = \pi^{(\gamma-2)/(\gamma-1)}$,

$$\langle d^2 | d \leq \bar{d}(\pi) \rangle = \frac{(\gamma - 1)(1 - \bar{d}(\pi)^{-\gamma+3})}{(\gamma - 3)(1 - \pi)},$$

and

$$\langle d | d \leq \bar{d}(\pi) \rangle = \frac{(\gamma - 1)(1 - \bar{d}(\pi)^{-\gamma+2})}{(\gamma - 2)(1 - \pi)}.$$

11. The approximation by a continuous distribution distorts things a bit, since it underweights the lowest-degree nodes, but it gives the right order of magnitude. With a discrete distribution, the results are even lower at 3 percent of the nodes.

model, once infected, a node does not recover but eventually either dies or becomes completely immune to further infection or transmission (as with chickenpox). In contrast, in the SIS model a node can recover from being infected but can then become infected again at a later time (as with certain variations of the common cold).

These models were originally studied in the context of large populations in which any individual could randomly meet any other, with equal probability. It was clear from an early date that the structure of interactions could make a big difference, as was discussed by Rapoport [551], [552]. The modeling has evolved, and models that provided for richer interaction structures were studied by Anderson and May [17] and Sattenspiel and Simon [577]. Network structure was more explicitly brought into play by Kretschmar and Morris [418] and Pastor-Satorras and Vespignani [529]–[531].[12]

7.2.4 The SIR Model

In the SIR model, the diffusion takes place between infected and susceptible nodes. Once a node reaches the "removed" state, it has either recovered and is no longer susceptible or contagious, or it has died. The question of when a small initial infection reaches a nontrivial fraction of nodes is similar to the analysis of diffusion with immune nodes in Section 7.2.1. To see the precise connections, let us examine this question in more detail.

Consider the following scenario. Individuals are connected through a network that is generated under the configuration model and is described by the degree distribution $P(d)$. Suppose that the infection process is such that the probability that an infected node will infect a susceptible neighbor before the infected node is removed can be described by a probability of transmission t. Let the infection operate in an independent fashion across links between susceptible and infected nodes. This model maps into what is known as *bond percolation* in percolation theory, where bonds are the equivalent of social links. The independence assumption is clearly violated in many applications. To see the difficulty with independence, consider the following scenario in line with the SIR interpretation. An infected node infects a neighbor with a random (e.g., Poisson arrival) time with intensity ν. The infected node reaches the recover/removed state also at a random time with intensity γ. The probability t that the node infects a given neighbor before recovering is then $t = 1 - e^{-\beta/\gamma}$.[13] This probability is the same for each neighbor. However, it is not independent across neighbors. For example, suppose that the network is a star and the center is initially infected. If it takes a long time for the center to reach the recover/removed state, then it might infect many neighbors. If it recovers quickly, then it might not infect many neighbors. This scenario introduces correlation in the outcomes of the center's neighbors. Conditional on one neighbor being infected, it is more likely that the others will also be infected, as conditional

---

12. See Morris [486] for a survey of the early literature on the subject.

13. As Newman [502] points out, this reasoning can be extended to much richer ways of generating t, where β or γ could also be random variables, as long as the ex ante expectation of t is the same across links.

on a neighbor being infected it is more likely that the center was infected for a longer time than we would expect unconditionally. Despite this difficulty, the independence assumption still seems to match well with simulations of the actual process with random recovery and infection as reported by Newman [502] (see his Figure 1), at least for large numbers of nodes.

One way to view this process is to start with a network and then pick a node uniformly at random to be the initially infected node. To see the reach of the infection, we can then (in an independent and identical manner) remove links along paths emanating from that node with a probability $1 - t$. We then examine the resulting component size. The analysis becomes analogous to that we have already examined. In particular, generate a degree distribution that represents the remaining network after the links are removed:

$$P_t(d) = \sum_{d' \geq d} P(d') \binom{d'}{d} t^d (1 - t)^{d' - d}. \tag{7.13}$$

Equation (7.13) is analogous to (7.4) with t in place of $(1 - \pi)$.

We then can draw analogous conclusions concerning the threshold for diffusion of the infection to a nontrivial segment of the population, the expected size of the component of infected (and eventually removed) nodes, and so forth.

7.2.5 The SIS Model

In the SIS model, nodes can become infected and then recover in such a way that they become susceptible again, rather than being removed. This model applies to certain diseases but is also useful as a first approximation of models of behavior in which individuals are more likely to undertake a given action as more of their neighbors undertake it, but then can also randomly stop taking the action, with the possibility of taking it up again. The model applies to different settings than does the SIR model, and so is analyzed a bit differently.

Let us start with the following variation on a matching model, which is really a hybrid between a random matching model and a network. Rather than having every member of the population meet any other with equal probability in a certain period, individuals are described by their degree. An individual of degree d_i will have d_i interactions with other individuals from the population in a given period, where probabilities of interaction are governed by the relative probabilities. Let P be the degree distribution in the population. The probability that a meeting of individual i is with an individual who has degree d (e.g., see (4.2)) is governed by

$$\frac{P(d)d}{\langle d \rangle}. \tag{7.14}$$

Thus, an agent is more likely to meet someone who has more meetings, so the probability of meeting someone is proportional to how many meetings she or he has. This model is not the same as a fixed network. Individuals interact randomly, but there are differences across individuals in terms of how many meetings they have. The model is often thought of as an "approximation" to a large network, but that interpretation is a bit of a stretch. It is really a degree-based random meeting model, and the extent to which a large network would have similar properties is not fully understood. Equation (7.14) is also a limiting expression: if we fix a finite set of agents, then the degrees are no longer independently distributed. So, to be careful, consider an infinite number of agents in the matching process, and let $P(d)$ be the measure of agents who have degree d.[14]

Keeping track of individuals' degrees is important because agents with different degrees tend to have different infection rates. In many cases, individuals who have higher degrees will have more interactions and thus be more prone to infection. How infection rates vary across degrees depends on the details of transmission. Suppose that the fraction of individuals of degree d who are infected is currently $\rho(d)$. The chance that a given interaction is with an infected individual, denoted θ, can be calculated using (7.14):

$$\theta = \frac{\sum P(d)\rho(d)d}{\langle d \rangle}. \tag{7.15}$$

Thus θ is the probability that a given meeting is with an infected individual. Note that this value is different from the average infection rate in the population, ρ, which is simply

$$\rho = \sum P(d)\rho(d). \tag{7.16}$$

The difference between (7.15) and (7.16) is that θ is weighted by degrees and represents the chance that a given meeting is with an infected individual, which is important in calculating how infection spreads, while ρ keeps track of the population average, which is important in welfare and policy analyses.

Generally, the likelihood that a susceptible individual becomes infected as a function of θ and the individual's degree can take many different forms. For instance, it might be that the agent is certain to be infected if she or he meets an infected individual. It might also be that it happens by chance. Or perhaps there is some threshold, so that it takes repeated meetings with infected individuals before infection occurs.[15]

Let us consider a simple linear form, which is common in studies of the SIS system (e.g., see Pastor-Satorras and Vespignani [529]), such that the chance that a given susceptible individual who has degree d becomes infected in a given

---

14. This argument can be done with either a countable set of agents or agents indexed by a continuum. The latter makes it easier to describe the matching process (here described for a single matching, but it can be extended to multiple matchings and, in the limit, to continuous time). Take the interval [0, 1] to be the interval of agents' labels, and then let subintervals of length $P(d)$ be those for each degree d. We need to keep track of which subsets of agents in each subinterval are infected at each point, and then who is matched with whom in a given period, so that each agent faces the same distribution of types in each meeting. This tracking can be done in many ways, but one has to be careful to do it in a measurable manner, so that a law of large numbers can be invoked. The calculation takes us beyond the scope of this text, but it is not difficult to do. For details on justifying random matching models with infinite sets of agents, see Aliprantis, Camera, and Puzzello [12], as well as the appendix of Currarini, Jackson, and Pin [182].
15. See Lopez-Pintado [438] for a discussion of the SIS model with various contagion mechanisms.

period when faced with a probability θ that any given meeting is with an infected individual is

$$\nu\theta d, \tag{7.17}$$

where $\nu \in (0, 1)$ is a parameter describing the transmission rate of infection in a given period. Taking $\nu d \leq 1$ for the maximal degree ensures that the probability is well defined. Exercise 7.3 shows that (7.17) is a good approximation when the probability of becoming infected in any given meeting with an infected individual is ν, and ν is relatively small.

In the SIS model, infected individuals can recover and become susceptible again, which can be motivated in various ways. The standard one is that any infected individual recovers in a given period with probability $\delta \in (0, 1)$. Here the recovery is random, is the same across all infected individuals, and does not depend on how long an individual has been infected. This first approximation is tractable, but clearly many diseases involve non-Markovian windows of time for the spread of infection. In this model individuals are either susceptible or infected and alternate between these states, depending on the infection rate in the population.

Thresholds and Steady-State Infection Rates We can now ask a series of questions. First, how high does the infection rate ν have to be relative to the recovery rate δ to have the infection reach some nonzero steady state in the population? Second, can we estimate the long-run steady-state proportion of infected nodes? Third, can we relate the answers to these questions to the network structure or to the degree distribution?

Maintaining a nonzero steady-state infection rate depends on having an infinite set of agents. For a finite set of agents, sooner or later, regardless of the details of the parameters, all individuals will be healthy, and then the infection will die out. Thus in a finite system, the only steady state is one in which all nodes are healthy, unless some random external infection probabilities are introduced, such as the birth of new strains of flu (see Exercise 7.5).

To estimate the steady state, let us work with a mean-field approximation in which $\rho(d)$ is constant over time. The expected change in $\rho(d)$ over time amounts to the measure of susceptible agents who become infected minus the measure of infected agents who recover to become susceptible again. Thus a mean-field approximation to steady state requires that these trends balance:

$$0 = (1 - \rho(d))\nu\theta d - \rho(d)\delta. \tag{7.18}$$

The expression $(1 - \rho(d))\nu\theta d$ in (7.18) represents the fraction of nodes of degree d that were susceptible and become infected, and $\rho(d)\delta$ represents the fraction that recover to become susceptible again. Letting $\lambda = \nu/\delta$, it follows that

$$\rho(d) = \frac{\lambda\theta d}{\lambda\theta d + 1}. \tag{7.19}$$

Equations (7.15) and (7.19) can then be combined to obtain the following steady-state characterization:

$$\theta = \sum_d \frac{P(d)\lambda\theta d^2}{\langle d \rangle (\lambda\theta d + 1)}. \tag{7.20}$$

Equation 7.20 often has more than one solution. There is always a solution $\theta = 0$. That is, if nobody is infected, then the system stays that way. Note that the system does not have a solution $\theta = 1$, unless we let λ become infinite (see Exercise 7.6). This restriction reflects the fact that $\delta > 0$ means that individuals recover over time, so there are always at least some susceptible individuals in the population. So when does the steady state have a solution with $\theta > 0$? And how stable is the solution at 0? That is, if we started at a steady state of 0 and infected a small portion of the population, would the system tend to return to 0 or reach a higher steady state?

Let us first examine the existence of a nonzero steady state in the simplest case, when the degree distribution is regular, so that all individuals have degree $\langle d \rangle$. In that case, (7.20) becomes

$$\theta = \frac{\langle d \rangle \lambda \theta}{\langle d \rangle \lambda \theta + 1}.$$

In addition to the solution of $\theta = 0$, there is another solution:

$$\theta = 1 - \frac{1}{\lambda \langle d \rangle}, \tag{7.21}$$

which is greater than 0 only if $\langle d \rangle > 1/\lambda = \delta/\nu$. This relationship is intuitive: if the number of meetings is sufficiently large compared to the relative recovery/infection rate, then the infection can be sustained. Otherwise, any infection dies out. Thus there is a threshold above which a society can maintain an infection and below which it does not.

There is an interesting difference between such a regular network and other networks. For example, if the degree distribution is scale free, so that a power law holds, then infection can always be sustained, as shown by Pastor-Satorras and Vespignani [529]. For example, if $P(d) = 2d^{-3}$,[16] then the steady-state equation (7.20) becomes

$$\theta = \sum_{d=1}^{\infty} \frac{2\lambda\theta}{\langle d \rangle \left(d^2 \lambda \theta + d \right)}.$$

Approximating the right-hand side by an integral and finding a nonzero solution $\theta \neq 0$ leads to

$$1 = \frac{2\lambda}{\langle d \rangle} \log \left(1 + \frac{1}{\lambda\theta} \right),$$

or (noting that $\langle d \rangle = 2$):

$$\theta = \frac{1}{\lambda \left(e^{\frac{1}{\lambda}} - 1 \right)}.$$

---

16. Note that $\sum_{d=1}^{\infty} 2/d^3$ is twice Apéry's constant (where Apéry's constant is defined to be $\sum_{d=1}^{\infty} 1/d^3$). However, here I use a continuous approximation, and so $\int_1^{\infty} (2/d^3)\mathrm{d}d = 1$, ensuring that the continuous version is a probability. In this case $\langle d \rangle = 2$.

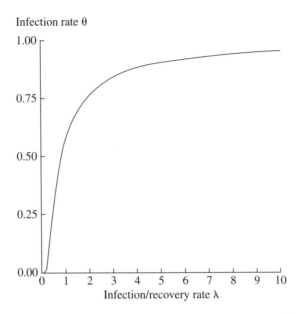

FIGURE 7.4 Infection rate of randomly encountered individual as a function of the infection/recovery rate.

This expression is always positive, regardless of λ, and so there is always a positive steady-state infection rate. The equation is graphed in Figure 7.4.

Nonzero Steady-State Infection Rates While the above makes it clear that it will be difficult to solve for the steady-state infection rates without imposing some structure on the degree distribution, one can still deduce many features across network structures without explicit functional forms. For example, let us examine the following approach taken by Lopez-Pintado [438]. We can ask how the system evolves if we start it with a rate of infection among randomly met individuals at some level θ. The function $H(\theta)$ keeps track of how many people become infected starting from θ:

$$H(\theta) = \sum \frac{P(d)d}{\langle d \rangle} \left(\frac{\lambda d\theta}{\lambda d\theta + 1} \right). \tag{7.22}$$

When $H(\theta) > \theta$ then the new infection rate will be higher than the starting infection rate, while if $H(\theta) < \theta$, the new infection rate is lower than the starting rate. Fixed points of H correspond to steady-state infection levels. Since $H(0) = 0$, 0 is always a fixed point and thus a steady state, as noted before. An important observation is that H is increasing and strictly concave in θ. Thus for H to have another fixed point (steady state) above $\theta = 0$, the slope of H at $\theta = 0$ must be greater than 1; otherwise, increasing θ results in a lower $H(\theta)$, and the

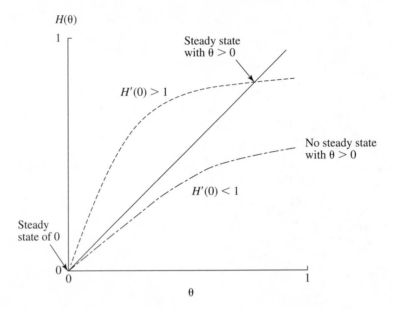

FIGURE 7.5 Existence of a positive steady-state infection rate depends on the derivative of $H(\theta)$ at 0.

infection rate simply declines back to 0. Thus for another steady state to exist, $H'(0) > 1$.[17] This argument is pictured in Figure 7.5.

We can now answer our question concerning the stability of the steady state $\theta = 0$. When $H'(0) > 1$, the steady state of 0 is unstable. That is, for any tiny initial infection rate $\varepsilon > 0$, $H(\varepsilon) > \varepsilon$ so that the infection grows. Indeed, it continues to grow until it hits the unique positive point where $H(\theta) = \theta$. The uniqueness comes from the strict concavity of H. So for $H'(0) > 1$, there exists a unique positive steady-state infection rate and the 0 steady-state is unstable; otherwise, 0 is the only steady-state infection rate. Thus examining $H'(0)$ indicates whether an infection can be sustained in a steady state. Note that

$$H'(\theta) = \sum \frac{P(d)d}{\langle d \rangle} \left(\frac{\lambda d}{(\lambda \theta d + 1)^2} \right),$$

and thus

$$H'(0) = \lambda \frac{\langle d^2 \rangle}{\langle d \rangle}. \tag{7.23}$$

So to have a (unique) positive steady-state infection rate $\theta > 0$ (and a corresponding positive average infection rate $\rho > 0$ across the society), (7.23) implies that

---

17. Noting that H is continuous and increasing in θ, $H(0) = 0$, and $H(1) < 1$ (from (7.22), as this is the expectation of an expression that is always less than 1), it follows that there is a fixed point greater than 0 if and only if $H'(0) > 1$.

$$\lambda > \frac{\langle d \rangle}{\langle d^2 \rangle}. \tag{7.24}$$

For a regular network where $\langle d^2 \rangle = \langle d \rangle^2$, (7.24) corresponds to the threshold $\lambda = 1/\langle d \rangle$ that we found from (7.21). Under a scale-free distribution, $\langle d^2 \rangle$ is divergent (letting the maximal degree grow with the size of the population), and so (7.24) is satisfied for any positive λ. A Poisson degree distribution has $\langle d^2 \rangle = \langle d \rangle^2 + \langle d \rangle$, and so a positive infection requires that

$$\lambda > \frac{1}{\langle d \rangle + 1},$$

which falls between the extremes of the regular and the power distributions.

The intuition behind this result is that individuals with high degrees serve as conduits for infection. Even very low infection rates can lead them to become infected, as they have many meetings. Furthermore, they can infect large numbers of others. In a regular network, every individual has the same degree. If that degree is high enough, then infection occurs. A Poisson distribution with the same average has more of a spread in the distribution and some higher-degree individuals are present. This configuration leads to a lower threshold at which infection can be sustained, because the higher-degree individuals can begin to serve as conduits. As the variation in degree is further increased, as in a scale-free distribution, then there are arbitrarily high-degree individuals who are frequently infected, and infections can be sustained at arbitrarily low net rates of contagion.[18]

Comparisons of Infections across Network Structure   Let us now examine how infection changes as we vary the network structure. There are two different aspects to consider. One is how the threshold for infection varies with interaction structure. The other is how the extent of infection depends on the interaction structure.

Equation (7.24) states that sustaining a positive infection rate depends on whether $\lambda > \langle d \rangle / \langle d^2 \rangle$. From (7.24), it is clear that if the mean $\langle d \rangle$ is held constant and the variance is increased, so that $\langle d^2 \rangle$ increases, then the threshold for infection decreases, which has an intuitive explanation. As before, higher-degree nodes serve as conduits for infection. They have many neighbors and so are easily infected, and they also spread infection to many others. A mean-preserving spread shifts some weight to higher-degree nodes and some to lower-degree nodes. At infection rates near 0, the higher-degree nodes are important to starting an infection, and so such a shift helps lower the threshold for infection.

What happens if we increase the mean of the distribution while keeping the variance constant? For instance, what happens if the distribution is simply shifted by some amount a? A new threshold of $\langle d + a \rangle / \langle (d + a)^2 \rangle$ is established, which is decreasing in a when $\langle d^2 \rangle \leq 2\langle d \rangle$ but is otherwise increasing in a (for small a). Thus such a shift always results in a decrease in the threshold if a is large enough.

---

18. As with the previous models of diffusion, this is a limiting threshold analysis, and so for finite societies there will be thresholds that can be examined through simulations.

So far, we have examined how the possibility of an infection taking root depends on the interaction structure. It can also be very important to know what the extent of an infection is and how the extent varies with the interaction structure. Recall from (7.20) that

$$\theta = \sum_d \frac{P(d)\lambda\theta d^2}{\langle d \rangle \, (\lambda\theta d + 1)}.$$

We saw that (7.20) can be solved in special cases, such as when the network is regular; more generally we need to resort to a numerical estimation of how infection rates behave. Nonetheless, there is still much that we can deduce about how infection varies when the network structure is changed. Jackson and Rogers [354] show that the resulting steady-state infection rates can be ordered when the degree distributions are ordered by stochastic dominance. The basic idea is that the right-hand side of the above equation is an increasing and convex function of d, and it can be ordered when we compare different degree distributions in the sense of stochastic dominance.

The interesting conclusion regarding steady-state infection rates is that they depend on network structure in ways that are very different at low values of the infection rate λ compared to high values. For instance, as already noted, scale-free degree distributions lead to positive infection levels for a wide variety of parameters (at least in the limit when the variance of the distribution becomes large), and in this sense might be more easily infected than networks with regular degrees or links that are formed uniformly at random. While this observation holds for initial thresholds or small infections, the opposite ordering holds for higher rates of infection. Let us examine the ordering in more detail.

Consider two networks with degree distributions P and P', respectively. The following proposition shows that if the distributions are ordered in the sense of strict first-order stochastic dominance or mean-preserving spreads (and hence second-order stochastic dominance, recalling the definitions from Section 4.5.5) as are the corresponding distributions of neighbors' degrees, $\widetilde{P}$ and $\widetilde{P}'$, then we can order the resulting highest steady-state infection levels.

Proposition 7.1 (Jackson and Rogers [354]) *Consider two distributions P' and P, with corresponding highest steady-state average neighbor infection rates $\overline{\theta}'$ and $\overline{\theta}$, and largest steady-state overall average infection rates $\overline{\rho}'$ and $\overline{\rho}$, and suppose that $\overline{\theta} > 0$.*

1. *If P' and $\widetilde{P}'$ strictly first-order stochastically dominate P and $\widetilde{P}$, respectively, then the infection rates are higher under P' than under P (so $\overline{\theta}' > \overline{\theta}$ and $\overline{\rho}' > \overline{\rho}$).*

2. *If P' is a strict mean-preserving spread of P, then the average neighbor infection rate increases: $\overline{\theta}' > \overline{\theta}$. Moreover, there exist bounds on the relative infection to recovery rate, $\underline{\lambda} \leq \overline{\lambda}$, such that:*

   • *If the infection to recovery rate is below the lower bound, so that $\nu/\delta < \underline{\lambda}$, then the steady-state average infection rate is higher under P', so that $\overline{\rho}' > \overline{\rho}$.*

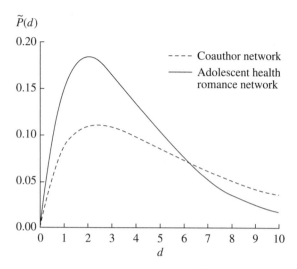

FIGURE 7.6 Distributions of a neighbor's degree $\widetilde{P}(d)$ for the coauthor network of Goyal, Moraga-Gonzalez, and van der Lief [304] and the adolescent health romance network analyzed by Bearman, Moody, and Stovel [51]. The coauthor distribution first-order stochastic dominates the distribution.

- *If the infection to recovery rate is above the upper bound, so that $\frac{\nu}{\delta} > \overline{\lambda}$, then the steady-state average infection rate is higher under P, so that $\overline{\rho}' < \overline{\rho}$.*

Before providing a proof of Proposition 7.1, let us examine the ideas behind it and some of the comparisons needed. First, to get an impression of stochastic dominance relations, Figure 7.6 shows a first-order stochastic dominance relation between two networks, a network of coauthorships in economics and a network of romantic relationships in a U.S. high school. The degree distributions pictured in Figure 7.6 are of neighbors' degrees (of the form $P(d)d/\langle d\rangle$), but the same relationships hold for the degree distributions themselves.[19]

The result that infections increase with first-order stochastic dominance is not surprising, as increasing the density of links in a network increases the possibilities for contagion. The more subtle result is (2), in Proposition 7.1 which concerns mean-preserving spreads and is illustrated in Figure 7.7, which pictures infection rates for three varieties of networks (holding average degree constant). Here the dependence of the rate of infection on network structure is determined by how

---

19. It is possible to find examples of pairs of degree distributions that are ordered by stochastic dominance but whose distributions of neighbors' degrees do not obey the same orderings, or vice versa; but many standard degree distributions display no such reversals. For example, if we consider regular, Poisson random, or scale-free networks and shift the mean of the distribution, the same relative orderings for both the degree distribution and the distribution of neighbors' degrees holds.

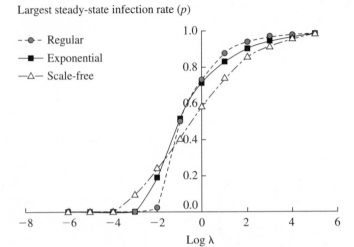

Largest steady-state infection rate (p)

FIGURE 7.7   Average highest steady-state infection in the SIS model as a function of the log infection/recovery rate λ for three network structures with mean degree 4. From Jackson and Rogers [354].

infectious the disease is. The intuition behind this observation is as follows. The change in infection rate comes from competing sources, as a mean-preserving spread leads to relatively more very-high-degree and very-low-degree nodes. As discussed above, high-degree nodes have high infection rates and serve as conduits for infection; thus their presence places upward pressure on average infection rates. Very-low-degree nodes have fewer neighbors to become infected by and so tend to have lower infection rates than other nodes. Thus the impact of a mean-preserving spread depends on the balance between these effects. When infection rates are already high, they tend to increase less than linearly with the degree of a node (note that they cannot increase above 1), which limits the impact of high-degree nodes; the impact of having low-degree nodes can decrease infection. Thus the overall effect of a mean-preserving spread can actually be a decrease in the infection rate. If instead infection rates are low, then the upward pressure effect dominates, as the increase of some nodes' degrees snowballs, leading to further infection.

Proof of Proposition 7.1.   Let $H_{P'}$ denote the function defined in (7.22) with respect to the degree distribution P'. It follows directly that

$$H_{P'}(1) = \sum \frac{P'(d)\lambda(d^2 + x)}{\langle d \rangle_{P'}(1 + \lambda(d + x))} < \sum \frac{P'(d)(d^2 + x)}{\langle d \rangle_{P'}(d + x)} \leq \sum \frac{P(d)d^2}{\langle d \rangle_{P'}d} = 1.$$

Thus, since $\overline{\theta}'$ is the largest θ such that $\theta = H_{P'}(\theta)$, and $1 > H_{P'}(1)$, it follows that $\theta > H_{P'}(\theta)$ for all $\theta \in (\overline{\theta}', 1]$.

Let us then show that under either (1) or (2), $\overline{\theta}' > \overline{\theta}$. Suppose to the contrary that $\overline{\theta}' \leq \overline{\theta}$. Then it follows from the above that

$$\overline{\theta} > H_{P'}(\overline{\theta}). \tag{7.25}$$

Using the expression for $H(\theta)$ from (7.22) and noting that $\lambda d\theta / (1 + \lambda d\theta)$ is strictly increasing when $\theta > 0$, it follows from strict first-order stochastic dominance of $\widetilde{P}'$ over $\widetilde{P}$ that $H_{P'}(\theta) > H_P(\theta)$ for any $\theta > 0$. A similar argument works for a mean-preserving spread, noting that $\lambda d^2 \theta / (1 + \lambda d\theta)$ is strictly convex when $\theta > 0$. Thus, since $\bar{\theta} > 0$, it follows from (7.25) that

$$\bar{\theta} \geq H_{P'}(\bar{\theta}) > H_P(\bar{\theta}),$$

contradicting the fact that $\bar{\theta} = H_P(\bar{\theta})$. We conclude that $\bar{\theta}' > \bar{\theta}$ under the conditions of either (1) or (2).

Next we must show that for (1), $\bar{\rho}' > \bar{\rho}$. Since $\rho(d) = \lambda\theta d / (1 + \lambda\theta d)$ is a strictly increasing function of θ for any positive d, it follows that $\bar{\rho}'(d) > \bar{\rho}(d)$ for any $d > 0$. Thus

$$\bar{\rho}' = \sum_d \bar{\rho}'(d) P'(d) > \sum_d \bar{\rho}(d) P'(d). \tag{7.26}$$

Strict first-order stochastic dominance and the fact that $\rho(d)$ is strictly increasing in d implies that

$$\sum_d \bar{\rho}(d) P'(d) > \sum_d \bar{\rho}(d) P(d) = \bar{\rho}. \tag{7.27}$$

The result follows from (7.26) and (7.27).

Let us now prove the second part of (2). The expression for the steady-state $\rho(d)$ in (7.19) can be rewritten as

$$\rho(d)(\lambda\theta d + 1) = \lambda\theta d. \tag{7.28}$$

Taking expectations with respect to $P(d)$ on each side of (7.28) (and noting that $\sum_d P(d)\rho(d)\lambda\theta d = \lambda\theta^2 \langle d \rangle$ by (7.15)) leads to

$$\lambda\theta^2 \langle d \rangle + \rho = \lambda\theta \langle d \rangle$$

or

$$\rho = \lambda\theta \langle d \rangle (1 - \theta). \tag{7.29}$$

Note that (7.29) implies that ρ is increasing in θ when θ is less than $1/2$ but is decreasing when θ is greater than $1/2$. As we showed earlier, if P' is a strict mean-preserving spread of P, then $\bar{\theta}' > \bar{\theta}$. When $\bar{\theta}'$ and $\bar{\theta}$ are both less than $1/2$, then the corresponding ρ values are similarly ordered, but when the former are both greater than $1/2$, the order is reversed. So to complete the proof, we need only to find bounds on λ that ensure both $\bar{\theta}'$ and $\bar{\theta}$ are below, or above, $1/2$. Note that for sufficiently low λ, $H(1/2) = \frac{\langle d^2 \rangle \lambda / 2}{\langle d \rangle (1 + \lambda / 2)}$ is less than $1/2$, and so the highest steady state must be below $1/2$, for sufficiently high λ the argument is reversed. We can thus find bounds on λ below which both $\bar{\theta}$ and $\bar{\theta}'$ are less than $1/2$, and a corresponding bound on λ above which both $\bar{\theta}$ and $\bar{\theta}'$ are greater than $1/2$. ∎

7.2.6 Remarks on Models of Diffusion

The models of diffusion that we have discussed provide some basic insights about diffusion: mean-preserving spreads in degree distributions (e.g., scale-free networks compared to Poission of the same degree) lead to lower thresholds for infection in terms of a variety of factors, such as infectiousness, the fraction of susceptible nodes, and the probability of transmission along any given link. The full extent of infection can depend on network structure in more complicated ways that relate to how infectious a disease is. Similarly, increases in degree density, such as for first-order stochastic dominance shifts, also lower thresholds and unambiguously increase overall infection rates. Networks with large or diverging variances can sustain infections at very low thresholds, and so scale-free networks can be more easily infected than some other random networks.

While various models of diffusion offer a number of insights, the models that are solved are still relatively particular in terms of the structures that are assumed and the results. There are two obvious shortcomings: analyses are for situations in which degrees of neighbors are independent or at least approximately so, and the analyses generally examine treelike networks with few small cycles or loops. These are both incompatible with many observed networks. Clearly the features omitted in these models can have an impact on diffusion. Correlation in degree can change the basic connectivity patterns, which can increase or decrease the size of components, depending on how it is introduced. Clustering leads to some redundancy, which can change the reach of new paths in a network and also change infection patterns. The basic approach to introducing correlations in degrees is not so difficult a variation on what we have already considered. The main difference is in the calculation of a neighbor's degree, which has been calculated using $\widetilde{P}$. In a model with degree correlation, this function would also depend on the degree of the node whose neighbor we are examining. This generalization changes the model equations into systems of equations. While the modification makes simple closed-forms harder to obtain, approximations can be found using simulations. This generalization has been investigated by Newman [502]; Moreno, Pastor-Satorras, and Vespignani [484]; Boguñá, Pastor-Satorras, and Vespignani [83]; and Vázquez and Weigt [638]. The role of clustering has been examined by Serrano and Boguñá [589] and Newman [504].[20]

Perhaps the most important aspect that is neglected in the above analyses is that the networks often actively react to the ongoing process. For example, in the case of a serious disease outbreak, some individuals react by seeking immunizations and/or avoiding contact with infected individuals. If the diffusion is of valuable information, then individuals might be creating links, or searching through the network specifically to find the information. Such adjustments to the network can either speed up or slow down the diffusion, depending on the application. Some of this dynamic is hinted at in the next section on navigating networks, but the main issue of how agents and the network adjust to diffusion is still largely open.

---

20. See Dorogovtsev and Goltsev [204] for more discussion and references, as well as discussions about component calculations in directed networks and growing random networks.

7.3 ▪ Search and Navigation on Networks

Diffusion concerns how a disease, information, or a behavior moves through a network following the network structure. A related, but quite distinct, issue relates to someone or something actively navigating a network. For instance, consider the problem of finding a node that has a certain attribute. In the classic Milgram [468] experiment, subjects were faced with the task of sending a letter to a particular person. More generally, this task might be the challenge of finding a web page with particular information in it, finding someone who knows how to perform a given task, or locating a file-sharer that has a given file. It can involve following network structure, but with a specific goal in mind and perhaps taking advantage of some knowledge about the network and/or attributes of the target. There are various approaches that could be tried. One could just randomly navigate the network until one bumps into the target node. Taking advantage of the structure of the network to better search is another option. For instance, going to nodes that have more neighbors might save time if such nodes display information about their neighbors. Beyond that, information about the nodes themselves could be used if nodes tend to be connected to other nodes with similar attributes. How do different search methods perform? How does a search depend on the network structure?

Here we examine two different approaches to navigating a network. One makes use of the network structure, for instance, gravitating toward nodes of higher degree but focusing on attributes of nodes that are purely network based (e.g., degree, centrality). The second approach makes use of other information about the nodes, such as profession, or age, which is not directly relevant in the network structure. However, when there is homophily in the network, then navigation that takes such traits into account can dramatically speed up navigation, as we shall see. This result suggests that homophily can provide some social benefits, at least in terms of navigation speed.

7.3.1 Navigating Random Networks

As a benchmark, let us start with the following problem. There is a single target node in a network of n nodes that must be found. We cannot know whether the desired node has been found without examining it.

We can derive an upper bound on the expected number of nodes that must be visited to find a given node. That bound can be found without taking advantage of the network structure at all. We can simply exhaustively visit the nodes one by one, picking the order uniformly at random. In this strategy the network structure is not used in any way, and there is an equal chance that the desired node will be the first node we visit, the second, . . . or the last. The *expected* number of nodes that would have to be visited under this method is the expectation of a uniform distribution, where the probability of it taking k nodes is simply $1/n$. So summing across k yields the expectation of

$$\frac{1}{n} + \frac{2}{n} + \ldots + \frac{n}{n} = \frac{n+1}{2},$$

or roughly half of the nodes. For large numbers of nodes, this method could be very time consuming and inefficient.

Next suppose that we use some aspects of the network structure as follows. Begin by randomly picking a node. If it is not the right node, then randomly pick one of its neighbors, and so forth. This process is similar to the one above.[21] However, let us add a feature to the setting that makes the search easier. When visiting a node, in addition to being able to discern whether it is the target node, we can also tell whether any of its neighbors is the target node. For instance, if we are looking for a person, we can just ask the person we are visiting whether they know the target person. Or, if we are crawling the world wide web, when we visit a given page its links could be labeled in such a way that we can tell whether any of those links point to the target page. In this case, when none of the neighbors of the node currently being visited is the target node, select the next node to be visited uniformly at random from the list of neighbors not yet visited.[22] To get a feel for the improvement in search efficiency, consider a regular network in which each node has degree d. On the first step d nodes are found. On the second step an additional $d - 1$ nodes are found, presuming that the new nodes do not overlap with previously visited nodes. If overlap were never an issue, it would take effectively $n/(d - 1)$ steps to visit the whole network, not counting back-tracking if we hit a dead end. The expected number of steps it would take if there were no overlap is approximated by

$$\sum_{k=1}^{n} \frac{1}{n} \left(\frac{k}{d-1} \right) = \frac{n+1}{2(d-1)}. \tag{7.30}$$

Equation (7.30) just represents the $1/n$ chance that the target is the kth node found and the fact that approximately $k/(d - 1)$ steps are needed to find k nodes if there is no overlap. To see how overlap slows down the search, suppose that for the first half of the nodes searched the rate is only $(d - 1)/2$ new nodes found at each step, so that half of the nodes are ones already visited. Then for the next quarter of the nodes visited the rate is $(d - 1)/4$, and so forth. So if the node happens to be in the first half of the nodes searched, then the expected time to find the node is

$$\frac{\frac{n}{2} + 1}{2 \frac{(d-1)}{2}},$$

or approximately $n/(2d - 1)$. If the node happens to be in the next quarter of nodes visited, then the expected time is

---

21. It could even be less efficient. Suppose that time is measured in terms of the number of links that are crossed to date, and at some point we end up at a node whose neighbors have all been visited before, and we cannot simply leap back to an earlier point. For instance, if the network is a tree and we start at a root node and traverse a branch that does not have the desired node, substantial time could be spent back-tracking.

22. If all neighbors have been visited, then start over by picking an unvisited node from the overall network, with an equal probability on any unvisited node.

$$\frac{\frac{n}{2}}{\frac{(d-1)}{2}} + \frac{\frac{n}{4}+1}{2^{\frac{(d-1)}{4}}},$$

or approximately

$$\frac{n}{(d-1)} + \frac{n}{2(d-1)}.$$

If we continue in this manner, the expected time conditional on the node being in the next eighth is

$$\frac{2n}{(d-1)} + \frac{n}{2(d-1)},$$

and so forth. The overall expected time to finding the node is then approximately

$$\sum_{k=1}^{\infty} \frac{1}{2^k} \left(\frac{(k-1)n}{(d-1)} + \frac{n}{2(d-1)} \right) = \frac{n}{2(d-1)} \sum_{k=1}^{\infty} \frac{k-\frac{1}{2}}{2^k} = \frac{3n}{2(d-1)}.$$

Thus overlaps have tripled the expected time of finding the node. This calculation is not precise, since it presumes that the fraction of new nodes found at a given step is roughly proportional to the fraction of unmet nodes in the network, which might be an over- or underestimate, depending on the architecture of the network. But at least the estimate suggests that while the overlaps slow down the process, it slows down by a factor rather than by a power.

Let us then explore how the search speed changes for other network structures. For a network with degree distribution P, as we randomly pick new nodes the degree of the new node has a distribution described by $\widetilde{P}(d) = d\,P(d)/\langle d \rangle$ (presuming that the degrees of neighboring nodes are independent, as discussed in Section 4.2.1). Then, ignoring overlap, each new node visited through this search process informs us about the expected number of additional nodes given by

$$\sum_d (d-1) \frac{P(d)d}{\langle d \rangle} = \frac{\langle d^2 \rangle}{\langle d \rangle} - 1.$$

Using this expression in place of the $(d-1)$ from the analysis with a regular network, the expected number of steps until the target node is found is roughly a factor times

$$\frac{n\langle d \rangle}{\langle d^2 \rangle - \langle d \rangle}. \tag{7.31}$$

For a Poisson random network it follows that

$$\frac{n\langle d \rangle}{\langle d^2 \rangle - \langle d \rangle} = \frac{n}{\langle d \rangle} = \frac{n}{(n-1)p},$$

which is approximately $1/p$, where p is the probability of a link. Thus searching a Poisson network is quite similar to searching a regular network. Ignoring overlap is a good approximation for many random networks below the threshold at which fixed-sized loops become prevalent, but could lead to underestimation above such thresholds. Providing fully accurate estimates for rich models of networks admitting nontrivial clustering is a difficult problem, and there is little research on that subject.

Next consider a network that has a degree distribution approximated by a power law, so that $P(d)$ is $cd^{-\gamma}$ for some scalar c and $3 > \gamma$, but such that the nodes' degrees are independent (e.g., generated by the configuration model rather than preferential attachment). To calculate $\langle d^2 \rangle$ for such a random network, we have to know the maximal degree in the distribution, denoted $M < n$. Then

$$\langle d^2 \rangle = \sum_{d=1}^{M} cd^2(d^{-\gamma}),$$

which is approximately

$$\langle d^2 \rangle = \int_{d=1}^{M} cd^{2-\gamma} \mathrm{d}d = \frac{c}{3-\gamma} \left(M^{3-\gamma} - 1 \right).$$

A similar calculation leads to

$$\langle d \rangle = \frac{c}{2-\gamma} \left(M^{2-\gamma} - 1 \right).$$

Thus from (7.31) the expected time to finding the desired node in a power distribution truncated at a large maximum degree of M is approximately proportional to

$$\frac{n(3-\gamma)}{2-\gamma} \frac{\left(M^{2-\gamma} - 1 \right)}{\left(M^{3-\gamma} - 1 \right)}. \tag{7.32}$$

For large M and $2 < \gamma < 3$, (7.32) is proportional to

$$\frac{n}{M^{3-\gamma}}. \tag{7.33}$$

Since a change in the truncation or maximum degree M can lead to dramatic changes in the calculation of $\langle d^2 \rangle$ and other moments, it can lead to a significant change in the expected time to finding the desired node. There is no right or wrong way to estimate search time, because each truncation leads to a valid degree distribution that approaches a continuous power distribution as the number of nodes increases. But many finite distributions that approach continuous power distributions in the limit have different features. There are a variety of different specifications that have been explored for the maximum degree (see Newman [503] for some discussion). Cohen et al. [161] suggest setting the maximum on a discrete finite approximate power-law distribution to be $M = n^{\frac{1}{\gamma-1}}$. This cutoff ensures that the expected number of nodes with degree larger than M (under the continuous approximation) on a draw of n nodes does not exceed 1.[23] Thus the

---

23. That is, $n \int_{M}^{\infty} d^{-\gamma} \mathrm{d}d$ is proportional to 1, up to a rescaling to ensure that it is a proper density.

expected number of steps to find the desired node for (7.33) is proportional to

$$\frac{n}{n^{\frac{3-\gamma}{\gamma-1}}} = n^{\frac{2(\gamma-2)}{\gamma-1}}. \tag{7.34}$$

For instance, if $\gamma = 2.5$, then the expected time is $n^{2/3}$, which is much more efficient than the linear-in-n time for the Poisson and regular networks.

Variations on Navigation Techniques Many variations on the navigation procedure have been examined. For example Adamic et al. [6] and Adamic, Lukose, and Huberman [5] analyze situations in which instead of obtaining information about direct neighbors only, a node can also report on second neighbors. Then at each step we learn about a number of new nodes that is proportional to the size of the second neighborhood of a node found by following a random link. Without any overlap, and with independence in neighboring nodes' degrees, the size of the second neighborhood of a node found through such a search is simply

$$\frac{\langle d^2 \rangle}{\langle d \rangle} - 1 + \left(\frac{\langle d^2 \rangle}{\langle d \rangle} - 1 \right)^2 = \frac{\langle d^2 \rangle}{\langle d \rangle} \left(\frac{\langle d^2 \rangle}{\langle d \rangle} - 1 \right).$$

This equation has terms for the direct neighborhood size plus the expected number of distance-two neighbors brought in by the new nodes. The search method roughly squares the number of nodes found at each step. So for the Poisson and regular networks the expected times become proportional to $n/\langle d \rangle^2$; for the power distribution, it becomes proportional to

$$\frac{n}{n^{2\frac{3-\gamma}{\gamma-1}}} = n^{\frac{3\gamma-7}{\gamma-1}}.$$

Networks with larger tails in their distributions lead to much more effective searches. The intuition for this observation is clear. Since we are more likely to find larger degree nodes through following randomly chosen links, as the degree distribution places more weight on higher-degree nodes, more of the network is examined by searching fewer nodes.

Adamic et al. [6] and Adamic, Lukose, and Huberman [5] also consider another variation on a search that takes even greater advantage of the presence of high-degree nodes. When a given node is not the desired node, and neither are its neighbors, then instead of next picking an unvisited neighbor uniformly at random, one chooses the unvisited neighbor with the highest degree. As higher-degree nodes have more neighbors, this choice not only leads to observing more nodes on a given step but also to improved opportunities (through more draws) of finding even-higher-degree nodes. This process quickly results in searches in which most of the nodes being searched are at the high end of the distribution. So, for instance, in the power network after a few steps most of the nodes examined have degree near M, and so a rough approximation for the expected time of search is then n/M, which for $M = n^{\frac{1}{\gamma-1}}$ becomes $n^{\frac{\gamma-2}{\gamma-1}}$. This method is significantly quicker than simply following links chosen uniformly at random at each stage, which was the process that led to (7.34). As Adamic et al. [6] point out, the relative size of improvement depends on γ. If γ is close to 2, then one naturally finds the very largest degree nodes simply by following random links, as higher-degree nodes have more weight in the distribution, whereas when γ is closer to 3, then there is more of an improvement from explicitly following a degree-based search algorithm, as higher-degree nodes are a bit rarer.

7.3.2 Navigating Structured Networks: Taking Advantage of Homophily

The analysis in the previous section suggests that navigating a network should take an amount of time proportional to n in a regular or Poisson network, and some low power of n if the network's degree distribution follows a power law, provided there is no additional structure of the network that can be used. This result seems inconsistent with the Milgram [468] small-world experiments described in Section 3.2.1. In these experiments, people were able to send a letter to a target in a median of 5 steps, of those that were successful. This experiment was conducted in a population on the order of hundreds of millions of people, so that even the square root of n is on the order of 10,000. Thus the individuals in Milgram's experiment were taking advantage of additional structure of the network to choose the recipient of the letter, as even sending the letter to very highly connected individuals is not enough to hit the median number of 5. It is clear that individuals did not just randomly choose a neighbor to send the letter to, but instead tried to send the letter to someone who had something in common with the target, or to someone who might be closer to someone who has something in common with the target. Indeed, Killworth and Bernard [394] show that people in small-world letter experiments are primarily guided by occupation and/or geography in their choices of whom to forward the letter to. To see how this insight might substantially expedite the search, let us examine navigation in a more structured setting.

The main point is that the previous search algorithms were based entirely on network primitives without reference to any other characteristics of the nodes. If the formation of links is actually governed by underlying social structure, there can be much more efficient methods of navigation that use such social information. For instance, consider a society of individuals who form a network described by a hierarchy in the form of a binary tree as follows. Each individual has a label. The first group are of type 0. This group forms the "root" of the tree. Two groups, of types 00 and 01, form the second level of the tree. There are four groups, of types 000, 001, 010, and 011, which form the third level, and so on. A given individual is linked to all other individuals who are of the same type and to all those who are of a type that differs by the addition or deletion of one terminal digit. So, someone of a type 0101 is connected to those with labels 010, 0101, 01010, and 01011. Thus they are connected to the individuals of the same type, as well as the type immediately preceding them in the tree, and the two types that follow them in the tree (unless they are at the last level of the tree). An example is pictured in Figure 7.8. The tree has K levels, so the vectors of types have length at most K.

The types might describe individuals by a list of attributes, which can include all sorts of information, such as ethnicity, gender, profession, education, physical

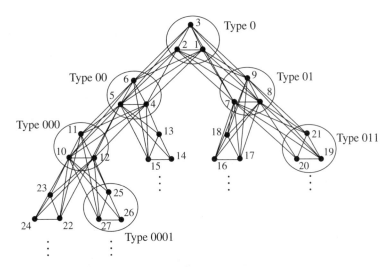

FIGURE 7.8 A network organized by types, with three agents of each type and a tree structure among types.

attributes, hobbies, geographic location, or favorite music; this list is coded as vectors of 0s and 1s.[24]

If there are m individuals of each type, then the society consists of

$$n = m \sum_{k=0}^{K} 2^k = m \left(2^{K+1} - 1\right) \tag{7.35}$$

individuals in total.

Now, let us measure the distance between any two individuals in this network. It takes at most $2(K-1)$ links to get from one individual to another. We can also keep track of how distance relates to the overall population size. Equation (7.35) implies that

$$K = \frac{\log(n+m) - \log(m)}{\log(2)} - 1.$$

Therefore, the maximum distance of $2(K-1)$ is

$$2\frac{\log(n+m) - \log(m)}{\log(2)} - 3.$$

Thus the maximum distance grows proportionally to $\log(n)$ for a fixed m.

Now we find that if the society has hundreds of millions of people, then the maximum distance will be on the order of 10 to 100, and the median distance even

---

24. While this typing may sound restrictive, we can arbitrarily closely approximate any continuous variables by including more entries for a given attribute. It is important to keep in mind how K varies as we code individuals.

less. This value is much more in line with data from the Milgram [468] experiments, in which observed distances had a median of 5 and maximum of 12 steps. To see how an individual might navigate the network, suppose that we pick an individual at random and give him or her a letter and then ask the individual to get the letter to some target agent in the society with a known type. An individual can simply follow the "greedy algorithm" of sending the letter to the neighbor who is closest to the other target, as Milgram [468] discusses in the context of his experiments. This algorithm is clearly optimal and follows a shortest path. The strategy takes the following form for an individual with a type ℓ of length k:

- If the target is a neighbor, then send it directly.
- If the target is not a neighbor, but has a type equal to ℓ plus some additional entries (and so lies further down the tree), then send it to any acquaintance whose type has a $k + 1$st entry that matches that of the target.
- If the target is not a neighbor and has a type that does not match ℓ in the first k entries, then send it to a neighbor who is higher in the tree.

It is worth remarking that implementing the algorithm only requires an individual to have an idea of which neighbor lies closer to the target, rather than having a full appreciation of the network structure. It is also critical, however, that the algorithm make use of the types of the individuals and the underlying social structure that indicates the positions of other individuals in the tree. Moreover, it is clear that the hierarchichal nature of the tree is used in the algorithm, although it could be adapted to other sorts of structures.

While the setting described here is quite stylized, it provides some insight into how the structure of a network can help speed up navigation. The idea of using a hierarchical structure to navigate a network is analyzed in various settings by Kleinberg [402] as well as Watts, Dodds, and Newman [659], who consider random graphs rather than the fully structured trees described above.[25]

To see how the navigation changes in a random network with less structure than the tree mentioned above, let us examine the model of Watts, Dodds, and Newman [659]. Again, let individuals be described by vectors of 0s and 1s, but such that each type is exactly K entries long. This restriction is equivalent to considering only the individuals whose types lie at the leaves of the tree in Figure 7.8. Let us keep track of the social distance between two individuals i and j, and denote it by x_{ij}, defined as follows. If two individuals are of the same type, let their distance be 1. Two individuals who differ only in their last entry are at a distance of 2. Individuals whose first point of difference is their second to last entry or later are at a distance of 4, and whose first point of difference is the third to last entry are at a distance of 6, and so forth. This scoring method keeps track of how many links one would have to travel in the tree in Figure 7.8 to travel from one type at the bottom row to another type at a bottom row. These social distances are a measure of similarity, but they do not yet correspond to actual distances in the network. The random network is then formed as follows. Uniformly at random pick a node i. Next pick a distance $x \in \{1, 2, 4, 6, \ldots, 2(K - 1)\}$, where K is the depth of

---

25. For a precursor to these models, see Killworth and Bernard [395].

the social tree, with probabilities $ce^{-\alpha x}$, where c normalizes the probabilities to sum to 1 and α is a parameter that adjusts how sensitive the link formation process is to similarity. Once x is chosen, then uniformly at random select a node j at that distance x from i and connect those nodes (provided there is not already a connection). Repeat this process until some average number of links per node, d, has been reached. When α is high, then nodes will form most of their links to other nodes that are more similar to themselves, whereas when α is low, then the links are formed more uniformly at random.

Since this network is now random, there might not exist a path between two nodes. Nevertheless, individuals can still follow a greedy algorithm of forwarding the letter to the neighbor who has a minimal social distance x_{ij} from the target, although that strategy might no longer be fully optimal, given that the network structure is now randomly determined. Watts, Dodds, and Newman [659] construct such random networks through simulation and then examine the results of following the algorithm for randomly selected pairs of nodes. They set the population size $n = 10^8$, average degree $d = 300$, and the homophily parameter $\alpha = 1$, and define a tree with 10 branches at each level and 100 individuals in each group at the leaf of a hierarchy.[26] They then posit a probability of .25 that a message is lost during any step, so that it is possible that some messages never reach their targets. Based on this specification, they measure the average distance of messages that eventually reach their targets and find it to be about 6.7, which is quite close to the average of 6.5 from the Milgram experiments.

7.3.3 Social Structure and Navigation Speed*

The Watts, Dodds, and Newman [659] random network based on social types is similar to one analyzed by Kleinberg [402]. Kleinberg proves that there exist parameter values such that as n grows, two nodes picked uniformly at random will be connected at a distance of at most $O(\log(n))$ with a probability of at least $1 - \varepsilon(n)$ for some function $\varepsilon(n) \to 0$.[27] Interestingly, Kleinberg also shows that it is critical for α to be exactly 1 for such a result to hold. As a rough intuition, if α becomes too small, then the network begins to resemble a uniformly random network, which has a longer navigation time, as argued previously. If α is too large, then the network connections are only formed to nearby nodes. Let us explore this dynamic in more detail to sharpen the insights.

Consider a set of n nodes. The primitive distances between nodes are described by a hierarchical tree structure. The tree T has $b \geq 2$ branches at each level and has n leaves, which correspond to the nodes that will be used to form a random directed network. Thus we can write n as b^K, where K is the depth of the tree T. The distance between two nodes i and j, denoted by x_{ij}, is half of the distance

---

26. They also work with two different hierarchies at once. Individuals are assigned to groups in each of the hierarchies uniformly at random, and then the social distance x_{ij} between two individuals i and j is taken to be the minimum distance over the different hierarchies.

27. $O(\log(n))$ is standard "big O" notation, which denotes that something is of an order no larger than $\log(n)$. More generally, $f(n) = O(g(n))$ indicates that $\limsup \frac{|f(n)|}{|g(n)|} < \infty$, or in other words that there exists γ and n' such that $f(n) \leq \gamma g(n)$ for all $n \geq n'$.

in the tree between two nodes i and j. Thus x_{ij} corresponds to the depth of the smallest subtree that contains both i and j. This quantity is not the distance in the random network that is formed based on the tree but just an auxiliary distance that might be thought of as a primitive measure of how dissimilar two nodes are. For each node i, form d directed links, where the node at the other end of a given link is chosen independently at random and the probability of choosing node j is proportional to $b^{-\alpha x_{ij}}$.[28]

An algorithm for searching to find a directed path from node i to node j is *decentralized* if it only uses information about the identities of the neighbors of i and their location as leaves in the tree, and the location of node j as a leaf in the tree.

The search time is *polylogarithmic* if there exists γ for which a starting node and target node picked uniformly at random are connected by a directed path of length at most $O\big([\log(n)]^\gamma\big)$ with a probability of at least $1 - \varepsilon(n)$ for some function $\varepsilon(n) \to 0$. Thus search time is polylogarithmic if two randomly selected nodes are likely to be within a distance that is a power of $\log(n)$. The out-degree d of a sequence of random networks indexed by n is polylogarithmic if there exists γ such that the degree is proportional to $\big(\log(n)\big)^\gamma$.

Theorem 7.1 (Kleinberg [402]) *Consider a sequence, indexed by n, of random networks formed on n nodes based on a group structure as described above.*

1. *If $\alpha = 1$ and $d \geq c \big(\log_b(n)\big)^2$ for some $c > 0$, then there exists a decentralized algorithm for which search time is polylogarithmic (with exponent 1).*

2. *If $\alpha \neq 1$, then there is no polylogarithmic degree for which a decentralized algorithm exists with a search time that is polylogarithmic.*

I sketch a proof of this theorem, as it clarifies the criticality of $\alpha = 1$ and supplies insight into the theorem.

Proof of Theorem 7.1. Recall the procedure used to assign neighbors in the construction of the random network, with neighbors at different distances being assigned different probabilities. The normalizing constant of the distribution of distances over nodes linked to by node i is denoted by $Z(n)$:

$$Z(n) = \sum_{j \neq i} b^{-x_{ij}} = \sum_{k=1}^{\log_b(n)} (b-1)b^{k-1}b^{-k} = \frac{b-1}{b}\log_b(n) \leq \log_b(n).$$

Let us prove (1). Uniformly at random select a starting node i and a target node j. Let them be a social (not network) distance x_{ij} apart, which is the depth of the smallest subtree T' of the tree T that contains them both. Consider the subtree

---

28. Kleinberg is not explicit about how to handle the possibility of duplicate links. Let us proceed as follows. Form the links in an independent and identically distributed manner, and then delete any duplicate links. For any given n there is a probability that fewer than d links per node are formed, but that probability vanishes as n becomes large, if d does not grow too fast as a function of n. In any case, this possibility of duplicate links can be dealt with in the proof of Kleinberg's main result.

T'' of depth $x_{ij} - 1$ that contains j. Since i is at a distance of x_{ij} from each leaf in T'', and there are $b^{x_{ij}-1}$ leaves in T'', the probability that i is not directedly linked to any leaf in T'' is

$$\left(1 - b^{x_{ij}-1}\frac{b^{-x_{ij}}}{Z(n)}\right)^d \leq \left(1 - \frac{1}{\log_b(n)}\right)^{c(\log_b(n))^2} \to e^{-c\log_b(n)} = n^{-c/\log(b)}.$$

Thus the probability that i fails to have a directed link to some node in T'' is at most (a factor times) $n^{-c/\log(b)}$ for large n. If there is a directed link to some node in T'', take one to a node in the smallest subtree possible that contains j, and call this node i'. This new tree has depth no more than $x_{ij} - 1$. Now repeat the argument starting from i', noting that we establish the same upper bound on the probability of failure to find a new directed link to a further subtree, and so forth. Given the maximal depth of the tree T, it takes at most $x_{ij} \leq \log_b(n)$ steps in this manner to reach the target j from the starting node i, and thus the search time is polyalgorithmic with exponent 1. Moreover, the probability of a failure at any step is at most $n^{-c/\log(b)}$. So the overall probability of failing to find a directed path is at most $\log_b(n)n^{-c/\log(b)}$, which converges to 0 as n grows, establishing (1).

To see (2), consider two separate cases. When $\alpha < 1$, the normalizing constant $Z(n)$ is such that for large n,

$$Z(n) = \sum_{j\neq i} b^{-\alpha x_{ij}} = \sum_{k=1}^{\log_b(n)} (b-1)b^{k-1}b^{-\alpha k} \geq \frac{n^{1-\alpha}}{b}.$$

Consider a subtree T' containing a target node j and having between n^γ and bn^γ leaves, where $0 < \gamma < 1 - \alpha$. For any node i' not in T', the probability that i' has any link into T' is at most

$$\frac{dbn^\gamma}{n^{1-\alpha}/b} = dbn^{\gamma+\alpha-1}.$$

For any polylogarithmic d the term $n^{-1+\gamma+\alpha}$ still dominates the expression, and so with a high probability it takes more than a polylogarithmic number of draws of nodes before finding any one with a link into T'. It follows easily that in any decentralized algorithm with high probability, it takes more than a polylogarithmic number of steps from a starting node i outside T' before any link into T' is found. The claim then follows.

If $\alpha > 1$, then $Z(n)$ is larger than some constant Z. Consider a node i such that the distance of i to the target j is $\log_b(n)$, so that the smallest subtree containing i and j is T. Note that such a starting node and target are selected with a nonvanishing probability (in fact, one of just more than $(b-1)/b$). Let T' be the tree of depth $\log_b(n) - 1$ that contains j. Each of i's directed out-links go to any given node in T' with a probability of no more than $b^{-\alpha \log_b(n)}/Z = n^{-\alpha}/Z$. Thus the probability that any of i's directed links goes to a node in T' is at most $dn^{1-\alpha}/Z$. Again, it follows easily that in any decentralized algorithm with high probability, it will take more than a polylogarithmic number of steps from starting node i outside T' before any link into T' is found, and so (2) is established. ▪

The remarkable thing about Kleinberg's result is that the distribution over links in the random network needs to be very specific to obtain the small-world conclusions that were observed in Milgram's letter experiments and that have emerged in many follow-ups.[29] To better understand this result, note that the critical aspect of the proof of (1) is that we have a similar probability of a link going into any subtree. The size of the subtree in terms of nodes is balanced by the probability that a link goes to that subtree. That is, there are few nearby nodes in terms of social distance, but they have a proportionally higher probability of being linked to; and there are more nodes that are further away in terms of social distance, but they have a proportionally lower probability of being linked to. When $\alpha = 1$ the balance is just right, so that a sort of uniformity develops in the distribution over the social distances that different links span. When α differs from 1 this proportionality is upset and either there is a limiting probability that almost all links span socially dissimilar nodes, which makes it difficult to approach a node; or else almost all links span socially similar nodes, which makes it difficult to reach distant nodes.

This line of argument is not specific to the hierarchical tree structure. In fact, Kleinberg [402] shows that the results extend to the following more general random network process that is governed by a social structure.[30] Consider a set of n nodes. The primitive distances between nodes are described by what Kleinberg refers to as a group structure. There are subsets of the n nodes, denoted by a generic element R. The collection of groups $\mathcal{R}$ satisfies the following criteria for some $\lambda < 1$ and $\beta > 1$:

- The set $\{1, \ldots n\}$ is a group.
- If $R \in \mathcal{R}$ is a group with at least two members and $i \in R$, then there exists another group R' that is a strict subset of R, includes i, and has a size of at least $\min(\lambda |R|, |R| - 1)$.
- If i is in each of groups $R_1, \ldots R_k$, and each of these groups has fewer than q members, then the union of the groups has fewer than βq members.

As Kleinberg points out, the last condition is a sort of bounded growth condition so that as the network grows an individual does not belong to an increasing number of disparate categories.

Let the distance between i and j, again denoted by x_{ij}, be defined as the size of the smallest group that contains both i and j: $x_{ij} = \min_{R:\{i,j\}\subset R} |R|$. Given a group structure, form a random network as follows. For each node i, form d links, where each link is formed to a node that is chosen independently at random, and the probability of choosing node j is proportional to $x_{ij}^{-\alpha}$.

As Kleinberg [402] shows, conclusions (1), and (2) for the case $\alpha < 1$, of Theorem 7.1 extend to such network structures. The case $\alpha > 1$ requires some additional structure so that nodes are not too close in underlying the group structure.

---

29. Polylogarithmic distance may not be very short, depending on the exponent. Nevertheless, in a large society it will generally be orders of magnitude shorter than distances that are not polylogarithmic. Thus, it still offers substantial insight into shorter versus longer distances and what is needed to obtain shorter distances.
30. Kleinberg [402] also shows that the results extend to situations in which nodes have sufficiently large but constant out-degree as a function of n.

We are left with the following conundrum. Theorem 7.1 seems to suggest that a very special balance in network formation based on social distance is necessary for the network to be easily navigable. Yet we observe such quick navigation in quite large societies. It is unlikely that societies just happen to hit the right balance. More likely there is something missing from the models, and it is clear that the network-formation process underlying many social networks is much more complex than in these models. Nevertheless, the models do provide a basic insight: it is necessary to have some balance of long-distance as well as short-distance links to make a network easily navigable. Such a balance can arise for a variety of reasons; and moreover it can be seen not only in terms of some random link formation process, but also in terms of the incentives that individuals have to form such links. Generally, we expect links at shorter social distances to be easier or less costly to maintain. This follows since individuals who are quite similar have more in common and may find it easier to communicate, trust each other, and so forth. While individuals who are at greater distance might find it more costly to maintain a relationship, relationships that span greater social distance can be valuable because they provide more access to other distant individuals, as well as to information, ideas, and insights that are less likely to duplicate those already obtained from individuals who are similar in type. The most important point is that such longer distance relationships become more valuable to the extent that there are fewer of them. Thus the balance can naturally arise *precisely because* people wish to navigate the network, or are interested in obtaining varied information, or the like. This idea behind small worlds is discussed in Section 6.5, but the analysis above also provides a basis for understanding the conclusions there in terms of the concept of navigation.

7.4 ▪ Exercises

7.1 *Immunity Thresholds and Mean-Preserving Spreads* Consider diffusion in the presence of immune nodes as discussed in Section 7.2.1. Consider a degree distribution for which the threshold π for the emergence of a giant component of susceptible nodes is between 0 and 1. Describe how the threshold in (7.8) changes with the introduction of a mean-preserving spread of the degree distribution.

7.2 *Immunity in the SIR Model* Consider the SIR model in a situation in which a proportion π of the nodes (selected uniformly at random) are immune from the outset. Develop an estimate of the threshold for infection analogous to (7.8), relating π and the probability of transmission t to statistics from the degree distribution. Plot combinations of π and t at the threshold for several values of average degree in the Poisson random-network model.

7.3 *Linear Infection Approximations in the SIS Model* Consider a case in which the probability of becoming infected in any given meeting with an infected individual is v, while one cannot become infected from a susceptible (noninfected) individual. Then the probability of becoming infected in d random meetings with individuals who are (independently) infected with a probability θ is

$$\sum_{x=1}^{d} \left(1 - (1 - v)^x\right) \binom{d}{x} \theta^x (1 - \theta)^{d-x}.$$

Here the sum is over the number of infected neighbors, x, that an individual with d meetings is likely to have, where $\binom{d}{x}\theta^x(1-\theta)^{d-x}$ is the probability of having x infected meetings. The expression $(1-(1-\nu)^x)$ is then the probability of becoming infected in at least one of the meetings, which is just 1 minus the probability of not becoming infected in any of the meetings with infected individuals.

Show that if ν is small relative to d (so that $(1-\nu)^x$ is approximately $1-\nu x$ for any $x \leq d$), then this expression approaches (7.17).

7.4 *Average Neighbor Behavior in the SIS Model*   Consider a variation on the SIS model: instead of a probability of $\nu\theta d$ of becoming infected when being of degree d and having a neighbor infection rate of θ, the probability is $\nu\theta$. The latter would be the case if, for instance, infection transmission depends on the fraction of neighbors being infected rather than on the absolute number. Show that in this case, $\rho(d)$ is independent of d and that $\rho = \theta = \frac{\lambda-1}{\lambda}$ if $\lambda > 1$ and is 0 otherwise.

7.5 *Mutation in the SIS Model*   Consider the SIS model in which the infection rate (7.17) is modified to be

$$\nu\rho d + \varepsilon,$$

where $\varepsilon > 0$ is a rate at which a node "mutates" to become infected, regardless of contact with infected nodes. Develop an expression for the steady-state θ as a function of the degree distribution, and provide a solution for regular degree distributions.

7.6 *Less Than Full Infection in the SIS Model*   Show that a solution θ to (7.20) is less than 1 whenever λ and the support of P are finite.

7.7 *Navigating a Structured Network on a Hypercube*   As in Section 7.3.2, code individuals in a binary manner as vectors of entries of 0s and 1s, but now have each individual have a vector of length K. An individual with a given vector of attributes is linked to all other individuals with exactly the same attributes, and also to those who have vectors of attributes that differ by one entry. Thus the network can be thought of as a hypercube or a regular lattice. So, for instance, if $K = 4$, then an individual with attributes $(0, 1, 0, 0)$ is connected to individuals of the same type as well as individuals of types $(0, 0, 0, 0)$, $(1, 1, 0, 0)$, $(0, 1, 1, 0)$, and $(0, 1, 0, 1)$. Again, there are m individuals with each label. Calculate the average distance in path length between two nodes picked uniformly at random (allowing for the second node to be identical to the first). How does this distance vary with the number of individuals in the society n? How does average degree grow in this network?

7.8 *Navigating Random Networks Based on Groups**   Consider the extension of Theorem 7.1 to the situation in which the random network is governed by group structures, as described in Section 7.3.3. Show that (1) of Theorem 7.1 extends to this setting. Show that (2) holds when $\alpha < 1$.

CHAPTER 8

Learning and Networks

Social networks play a central role in the sharing of information and the formation of opinions. This is true in the context of advising friends on which movies to see, relaying information about the abilities and fit of a potential new employee in a firm, debating the relative merits of politicians, or even simply providing information about scientific research and results. Given the role of social networks in the formation of opinions and beliefs, and the subsequent shaping of behaviors, it is critical that we have a thorough understanding of how the structure of such networks affects learning and the diffusion of information. Some fundamental questions concern how social networks influence

- whether individuals in a society come to hold a common belief or remain divided in opinions,
- which individuals have the most influence over the beliefs in a society,
- how quickly individuals learn, and
- whether initially diverse information scattered throughout the society can be aggregated in an accurate manner.

Various answers have been given to these questions, and in this chapter we explore some of the basic models. I start with some background on a few of the classic sociological studies that sparked theories of opinion leaders, and provide basic evidence of the role of social networks in the formation of beliefs and opinons. This material provides initial answers to the second question above, as certain studies have focused on questions about which individuals tend to become opinon leaders. From there, I discuss two different types of models. The first is a Bayesian learning model, in which individuals observe actions and results experienced by their neighbors and process the information in a sophisticated manner. While Bayesian updating has firm normative foundations, it is a bit cumbersome to use in such complex settings. Nevertheless, the model does provide insight into the first question above, as it provides conditions under which individuals come to act similarly over time. The second model is based on a much more naive, but still natural, form of updating in which individuals exchange information with their neighbors over time and then update by taking a weighted average of what they hear. This class

223

of models is quite tractable, and allows us to incorporate rich network structures and provide explicit answers to each of the above questions.

8.1 • Early Theory and Opinion Leaders

Early theory on information transmission in social networks includes the seminal work on the role of opinion leaders of Lazarsfeld, Berelson, and Gaudet [426][1] and the subsequent fuller development by Katz and Lazarsfeld [380]. These studies examined the formation of opinions in voting and various other household decisions. The study by Lazarsfeld, Berelson, and Gaudet [426] provides a basis for the identification of opinion leaders through observations of how individuals reached voting decisions in an Ohio town in the 1940 U.S. presidential campaign. These researchers define opinion leaders to be individuals in a society who become informed through various media and other interactions; the opinion leaders convey their opinions and information to and influence other individuals who are less directly informed. This theory of opinion leaders was developed in more detail and structure by Katz and Lazarsfeld [380]. They also conducted a very influential and extensive study of opinion formation, this time in Decatur, Illinois, in the early 1950s. They conducted two sets of interviews with women over the age of 16 and asked the subjects about things beyond political opinions. In particular, Katz and Lazarsfeld asked the women about their opinions regarding household goods, fashion, movies, and local public affairs (including politics). The study was cleverly designed to interview the same women multiple times, a couple of months apart, so that changes in opinions could be identified. When finding a change in opinion, Katz and Lazarsfeld then asked questions that helped them trace the sources that influenced the change. They asked the women whose opinions had changed to identify who had influenced their decisions and whether those individuals had influenced others' decisions. Thus Katz and Lazarsfeld were able to identify individuals who played a role in multiple changes of opinion, and they dubbed these individuals the "opinion leaders." Katz and Lazarsfeld found evidence that while sometimes opinion leaders held higher social status, often opinion leaders were of the same social status as those whom they influenced, especially when it came to various household decisions. Opinion leaders were often distinguished by their gregariousness and the size of their families (which is then correlated with their age and experience).[2]

Beyond the applications mentioned above, which provide just a glimpse of some of the early research on opinion leaders, there are important influences on opinions from families, education, religion, and various organizations. The role that opinion leaders play in the dissemination of information and their influence on opinions and decisions has been of primary importance to those interested in marketing, social

---

1. There are earlier studies that provide roots for theories on opinion leaders, such as Merton [465]. See the background chapter in Katz and Lazarsfeld [380] for more details on the birth of the theory.
2. There is also some evidence from Coleman, Katz, and Menzel [165], in another influential study about the adoption of a drug (see Section 3.2.10), that early adopters tend to be more highly connected, and that adoption later spreads to less socially connected individuals, which often correlates with age and experience.

programs, education, campaigning, and more general diffusion properties (e.g., see Rogers [564]). I do not wish to survey the large empirical literature on belief and opinion formation; but it is important to mention the work above, because it was critical in showing how social connections play a role in learning and the formation of opinions, and that different agents in a society have different influences. To more fully understand the process, we need to examine models that explicitly account for social network structure in the patterns of information dissemination. Some of these models allow us to operationalize social influence and opinion leadership in the context of explicit social networks.

8.2 ▪ Bayesian and Observational Learning

I take the models out of historical sequence, starting with later models on observational learning and then returning to the classic models of social influence, consensus formation, and opinion formation. This approach allows us first to digest a fairly basic insight on consensus, at least in terms of actions, which provides a benchmark result based on strong assumptions. I then turn to the richer modeling of social influence.

A central conclusion in the observational learning setting is that if agents can observe one another's actions and outcomes over time, and all agents have the same preferences and face the same form of uncertainty, then they develop similar payoffs over time. The idea is that an agent who is doing significantly worse than a neighbor must realize this over time and will eventually change actions and do as well as the neighbor. Thus all connected agents must end up with the same limiting payoffs. It does not imply that they all learn to take the best possible action—it could be that they all have suboptimal payoffs. However, if in addition, agents start with sufficient diversity in opinions so that they have incentives to experiment with different actions, then they will have a high probability of converging to the optimal action. Of course, this conclusion relies on a static environment and similarity in preferences and situations across agents but it provides a benchmark.

The following learning environment is a variation on that studied by Bala and Goyal [29]. There are n agents connected in an undirected social network. In each period $t \in \{1, 2, \ldots\}$, the agents simultaneously choose among a finite set of actions. The payoffs to the actions are random, and their distribution depends on an unknown state of nature. The agents are all faced with the same set of possible actions and the same unknown state of nature. They have identical tastes and face the same uncertainty about the actions. At each date, in addition to observing his or her own outcome, an agent also observes his or her neighbors' choices and outcomes.

The main results and ideas can be most easily seen for two choices of action, A or B, which can readily be generalized to any finite number of actions. For the purpose of exposition, suppose that A results in a payoff of 1 per period with probability 1, while B pays 2 with probability p and 0 with probability $1 - p$. An agent would like to maximize the expected sum of discounted payoffs,

$$E\left[\sum_t \delta^t \pi_{it}\right],$$

where $\delta \in (0, 1)$ is a discount parameter, and π_{it} is the payoff that i receives at time t. If $p > 1/2$ then every agent would prefer to choose B, while if $p < 1/2$ then every agent would prefer to choose action A. However, p is unknown to the agents and can take on a finite set of values $p \in \{p_1, \ldots, p_K\}$ with each $p_k \neq 1/2$.[3] Let agent i begin with a prior μ_i over this set, such that $\mu_i(p_k) > 0$ is the probability that i initially assigns to p_k being the probability that action B pays 2. This scenario is a standard multi-armed bandit problem, except that an individual can observe the actions and outcomes chosen by his or her neighbors.

The learning in such an environment can be quite complicated. For instance, seeing that a neighbor chooses an action B might indicate that the individual's neighbors have had good outcomes from B in the past. Thus, beyond simply seeing actions and outcomes, an individual can make inferences about outcomes of indirect neighbors by observing the action choices of neighbors. Such full Bayesian learning is explored in the context of three-link networks by Gale and Kariv [266] (see also Choi, Gale, and Kariv [151] and Celen, Kariv, and Schotter [144]), but it quickly becomes intractable in larger networks. Instead, Bala and Goyal [29] examine a limited form of Bayesian updating, in which agents only process the information from actions and outcomes and ignore any indirect information that might be gleaned from the action sequences of neighbors.

In this version of the model, observing a 0 or 2 must obviously indicate that an agent took action B, and a payoff of 1 indicates that the agent took action A. Thus we can keep track of an agent i's beliefs at time t simply by knowing the initial beliefs μ_i and the history of the payoffs of i and each of i's neighbors, denoted by h_{it}. So, let $\mu_i(h_{it})$ denote i's belief in the probability that action B will result in a payoff of 2, updated according to Bayes's rule conditional on observing history h_{it}.

The following proposition is a slight strengthening of the main conclusion from Bala and Goyal [29], as it concludes that players play the same strategy after some time rather than just getting the same limiting payoff. This stronger conclusion follows easily from having different payoffs for different actions.

Proposition 8.1 (Bala and Goyal [29]) *With probability 1, there exists a time such that all agents in a component settle down to play the same action from that time onward.*

The proof is straightforward, and the main points are as follows. It suffices to consider a given component, so let the n agents be in a single component. If action B is not played infinitely often, then all agents must play A after some time and the conclusion follows. So consider the case in which action B is played infinitely often. Then some agent, say i, plays B infinitely often. By the strong law of large numbers (see Theorem 4.3), with probability 1, $\mu_i(h_{it})$ converges to the true p_k.[4]

---

3. The finiteness of values is not critical to the result. However, the conclusion that all agents take the same action depends on the two actions providing different average payoffs. If the two actions lead to the same average payoff, then it is still possible that agents end up with the same long-run utility.

4. Here, one does not need to invoke the martingale convergence theorem, as the convergence of μ_i is easily established directly from the strong law of large numbers and Bayes's rule. A detail

Thus for agent i to play B infinitely often it must be that $p_k > 1/2$. Since $\mu_i(h_{it})$ converges to p_k almost surely, there is a (random) time after which i will play B exclusively. Each neighbor sees i's play and, again with probability 1, will have the same limiting beliefs. Thus each neighbor must have limiting beliefs that B offers a higher payoff, and so each of i's neighbors will play B exlusively after some random time. Iterating on this logic leads to the claim.

The basic reasoning behind this result extends to other behavioral assumptions. For example, a similar conclusion holds if each agent chooses the action chosen last period by the agent who has the highest payoff to date in his or her neighborhood (including the agent), although the proof for that case is different. The critical point is that players observe the streams of payoffs of their neighbors and can see whether there is an action that would do better than the one they are currently choosing.

The fact that all agents end up choosing the same action does not imply that they have the same limiting beliefs, nor does it imply that they choose the "right" action. For instance, they might choose A, while $p > 1/2$ and B would offer a higher payoff, because each agent starts with low μ_i values. Moreover, the beliefs can remain different from one another. In terms of picking the right action, it is impossible (i.e., it happens with probability 0) that agents choose B infinitely often when A is the right action, as they will learn the payoff to B almost surely. So in this version of the model, for them to fail to pick the right action it must be that they end up picking action A, while B would have been better. If there is enough optimism about the payoff of B by at least one agent in the society, then that agent will play B for a long time regardless of the observed history, and so the society will have a high probability of learning the higher payoff action. This logic is the content of the following claim, which is a variation on a proposition in Bala and Goyal [29].[5]

Claim 8.1 *For any ε, there exists a $\mu < 1$ such that if there exists at least one agent i who has an initial belief that action B will pay 2 with probability greater than μ, then the probability that all agents in i's component eventually converge to choosing the correct action (i.e., that with the higher true expected payoff) is at least $1 - \varepsilon$.*

The idea is that if at least one agent is sufficiently optimistic, then regardless of the initial string of outcomes, that agent will play B enough times so that there is a probability of at least $1 - \varepsilon$ that the component will learn whether B has a different payoff than that of A. Bala and Goyal [29] also investigate other conditions ensuring convergence to the correct action, which are conditions on having enough agents with independent neighborhoods, so that some agents try the better action often enough to learn of its superior payoff.

It is easy to see that the above results readily extend to situations in which A's payoff is also uncertain or there are more than two actions. Variations on such a

---

to consider is that i also sees some finite set of neighbors' actions and payoffs. However, that can only lead to more observations of B, and the beliefs still converge almost surely to knowing the true probability of B.

5. The proposition of Bala and Goyal (their Proposition 4.1) has an additional condition bounding the number of neighbors that an agent has. That is not needed here, since I consider a finite society.

result can also be established in a fully Bayesian rational setting, as by Gale and Kariv [266]. Convergence to a consensus action does depend on various aspects of the formulation, and the limits of such results with respect to variations on who observes what and when, and other aspects of the formulation, are explored by Rosenberg, Solan, and Vieille [567].

While the above results show the potential for reaching an eventual consensus, at least in terms of beliefs about best actions, they do not give us much of an impression of shorter run behavior, which might be relevant, especially if the world is not stationary. More importantly, the network structure does not enter the above analysis in a substantial way, as connectedness and optimisim (or some large number of agents with independent neighborhoods) are the only attributes used to derive these results. Let us now turn to a set of models that bring network structure explicitly into the analysis.

8.3 ▪ Imitation and Social Influence Models: The DeGroot Model

The seminal network interaction model of information transmission, opinion formation, and consensus formation is due to DeGroot [189]. It is a very simple and quite natural starting point for a theory that allows us to more fully understand how the structure of a network influences the spread of information and opinion formation.

Individuals in a society start with initial opinions on a subject. Let these be represented by an n-dimensional vector of probabilities, $p(0) = \big(p_1(0), \ldots, p_n(0)\big)$. Each $p_i(0)$ lies in the interval $[0,1]$, and might be thought of as the probability that a given statement is true, or the quality of a given product, or the likelihood that the individual might engage in a given activity, or the like. The interaction patterns are captured through a possibly weighted and directed $n \times n$ nonnegative matrix T. In particular, let T be a (row) stochastic matrix, so that its entries across each row sum to 1. The interpretation of T_{ij} is that it represents the weight or trust that agent i places on the current belief of agent j in forming i's belief for the next period. Beliefs are updated over time so that

$$p(t) = Tp(t-1) = T^t p(0). \tag{8.1}$$

This process is illustrated in the following example.

Example 8.1 (Updating in the DeGroot Model)   *There are three individuals and an updating matrix described by*

$$T = \begin{pmatrix} 1/3 & 1/3 & 1/3 \\ 1/2 & 1/2 & 0 \\ 0 & 1/4 & 3/4 \end{pmatrix}.$$

This updating process is pictured in Figure 8.1.

Agent 1 weights all beliefs equally, while agent 2 weights agents 1 and 2 equally but ignores 3, and agent 3 weights 2 and 3 but ignores 1; and agent 3 places more

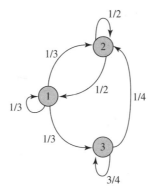

FIGURE 8.1   An updating process.

weight on his or her own belief. Suppose that initially the vector of beliefs is

$$p(0) = \begin{pmatrix} 1 \\ 0 \\ 0 \end{pmatrix}.$$

So agent 1 starts with a belief of 1 (of the probability of some event), while agents 2 and 3 start with a belief of 0. Here

$$p(1) = Tp(0) = \begin{pmatrix} 1/3 & 1/3 & 1/3 \\ 1/2 & 1/2 & 0 \\ 0 & 1/4 & 3/4 \end{pmatrix} \begin{pmatrix} 1 \\ 0 \\ 0 \end{pmatrix} = \begin{pmatrix} 1/3 \\ 1/2 \\ 0 \end{pmatrix}.$$

As agents update again, beliefs become

$$p(2) = Tp(1) = \begin{pmatrix} 1/3 & 1/3 & 1/3 \\ 1/2 & 1/2 & 0 \\ 0 & 1/4 & 3/4 \end{pmatrix} \begin{pmatrix} 1/3 \\ 1/2 \\ 0 \end{pmatrix} = \begin{pmatrix} 5/18 \\ 5/12 \\ 1/8 \end{pmatrix}.$$

Iterating this process leads to beliefs that converge:

$$p(t) = Tp(t-1) = T^t p(0) \rightarrow \begin{pmatrix} 3/11 \\ 3/11 \\ 3/11 \end{pmatrix}.$$

A method for calculating the limit belief is discussed in detail below.

This process has the following motivation, discussed by DeMarzo, Vayanos, and Zwiebel [193]. Agents are connected by a (possibly directed) network indicating whose information they are able to observe over time. At time $t = 0$, each agent sees a noisy signal $p_i(0) = \mu + e_i$, where $e_i \in \mathbb{R}$ is a noise term. An agent i then hears the opinions of his or her neighbors and assigns precision π_{ij} to the signal of agent j. If agents are Bayesian and the noise terms are normally distributed with 0 mean, then agent i updates according to (8.1), setting $T_{ij} = \pi_{ij} / \sum_k \pi_{ik}$, where $\pi_{ik} = 0$ if i does not have a directed link to k.[6] At time $t = 2$, agents realize that their neighbors have new information (collected by the neighbors at time $t = 1$), and so it is worthwhile listening to their neighbors again to collect this indirect information. The fully optimal processing of the neighbors' new beliefs within this second stage is a bit more complicated, as one has to account for how much new information is in the signal and what the precision is at that stage. With each iteration, the inference problem becomes more complicated. The DeGroot model can be thought of as a boundedly rational version of this process, where the agents do not adjust their weightings over time. Nevertheless, iterating on the updating process allows agents to incorporate more distant information and possibly to reach a consensus. Moreover, as we shall see, there are situations in which updating according to this simple process still leads agents to converge to a fully accurate belief in the limit.

8.3.1 Incorporating Media and Opinion Leaders

In view of the discussion of media and opinion leaders in Section 8.1, it is also of interest to understand how external sources of information can influence a society and how opinion leaders shape the opinions of others. This model easily incorporates various forms of external information providers, who are not influenced by the opinions of the members of the society, but who are listened to. Such fixed sources of information can simply be viewed as is with $T_{ii} = 1$ and $T_{ij} = 0$ for all $j \neq i$, but for whom $T_{ji} > 0$ for some js. Thus, an external source of information is modeled as an agent i whose opinion stays fixed at $p_i(0)$ to whom other nodes pay attention.

Opinion leaders arise naturally in the model as individuals who are listened to by others and who have nonnegligible influence on the opinions of at least some other agents. That is, the influence of an agent j over the final beliefs in the society depends on how much weight other individuals place on the agent, as captured through T_{kj}s. This model is explored more explicitly below.

8.3.2 Convergence

Under what conditions does the updating process in (8.1) converge to a well-defined limit? What limit does the process converge to? Let us examine these questions in sequence.

A social influence matrix T is *convergent* if $\lim_t T^t p$ exists for all initial vectors of beliefs p. This concept is illustrated in the following example.

Example 8.2 (Convergence) *Suppose that there are three individuals and an updating matrix described by*

$$T = \begin{pmatrix} 0 & 1/2 & 1/2 \\ 1 & 0 & 0 \\ 0 & 1 & 0 \end{pmatrix}.$$

6. See also DeGroot and Schervish [190] (their Section 6.3) for background on such updating.

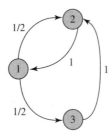

FIGURE 8.2   A society with a convergent updating process.

The updating process is pictured in Figure 8.2.
   Here, one can check that

$$T^2 = \begin{pmatrix} 1/2 & 1/2 & 0 \\ 0 & 1/2 & 1/2 \\ 1 & 0 & 0 \end{pmatrix}, \quad T^3 = \begin{pmatrix} 1/2 & 1/4 & 1/4 \\ 1/2 & 1/2 & 0 \\ 0 & 1/2 & 1/2 \end{pmatrix},$$

$$T^4 = \begin{pmatrix} 1/4 & 1/2 & 1/4 \\ 1/2 & 1/4 & 1/4 \\ 1/2 & 1/2 & 0 \end{pmatrix}, \ldots,$$

and

$$T^t \rightarrow \begin{pmatrix} 2/5 & 2/5 & 1/5 \\ 2/5 & 2/5 & 1/5 \\ 2/5 & 2/5 & 1/5 \end{pmatrix}.$$

Thus, no matter what beliefs $p(0)$ the agents start with, they all end up with limiting beliefs corresponding to the entries of $p(\infty) = \lim_t T^t p(0)$ where

$$p_1(\infty) = p_2(\infty) = p_3(\infty) = \tfrac{2}{5}p_1(0) + \tfrac{2}{5}p_2(0) + \tfrac{1}{5}p_3(0).$$

Example 8.2 not only shows that beliefs converge over time, but it also illustrates that the agents reach a consensus, and that agents 1 and 2 have twice as much influence over the limiting beliefs as agent 3 does. This sort of result is obtained under very natural conditions, and we shall see how to characterize the limiting influence below. It is also possible for an updating process to fail to converge, as illustrated in the following example.

Example 8.3 (Nonconvergence)   *This case is a slight variation on Example 8.2: we change only the third person's weights so that she places all of her weight on agent 1:*

$$T = \begin{pmatrix} 0 & 1/2 & 1/2 \\ 1 & 0 & 0 \\ 1 & 0 & 0 \end{pmatrix}.$$

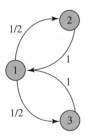

FIGURE 8.3 A society with a nonconvergent updating process.

The updating process is pictured in Figure 8.3. In this case,

$$T^2 = \begin{pmatrix} 1 & 0 & 0 \\ 0 & 1/2 & 1/2 \\ 0 & 1/2 & 1/2 \end{pmatrix}, \; T^3 = \begin{pmatrix} 1/2 & 1/2 & 0 \\ 1 & 0 & 0 \\ 1 & 0 & 0 \end{pmatrix}, \; T^4 = \begin{pmatrix} 1 & 0 & 0 \\ 0 & 1/2 & 1/2 \\ 0 & 1/2 & 1/2 \end{pmatrix} \dots,$$

and so the matrix simply oscilates and there is no convergence. For example, if we start with an initial set of beliefs with $p_1(0) = 1$ and $p_2(0) = p_3(0) = 0$, then since 1 updates based on 2 and 3's beliefs, and 2 and 3 update based on 1's belief, the agents simply swap their beliefs over time.

The key to the failure of convergence in Example 8.3 is that there is a directed cycle in the network pictured in Figure 8.3 and all cycles are of lengths that are multiples of 2. The updating/interaction matrix T is said to be *periodic*, and this allows the process to cycle without converging.

To be more specific, let us say that T is *aperiodic* if the greatest common divisor of all directed cycle lengths is 1, where the directed cycles are defined relative to a directed network in which a directed link exists from i to j if and only if $T_{ij} > 0$. This condition is satisfied in Example 8.2, which has directed cycles of lengths 2 and 3, so that the greatest common divisor of the directed cycle lengths is 1; it fails in Example 8.3 whose directed cycles are all of lengths that are multiples of 2.

Standard results in Markov chain theory[7] are easily adapted to this model to conclude that if T is strongly connected (so that there is a directed path from any node to any other node, also referred to as being *irreducible*) and aperiodic, then it is convergent. Much of the literature simply presumes that T is strongly connected and that $T_{ii} > 0$ for some or all i, which implies that the matrix is aperiodic (since it has at least one cycle of length 1) and therefore convergent. However, it is not necessary to have $T_{ii} > 0$ for even a single i to ensure convergence, as we see from Example 8.2. The full necessary and sufficient condition for convergence follows from standard results, as pointed out in Golub and Jackson [294], and is given below in Theorem 8.1.[8]

7. See Meyer [466], for example. Section 4.5.8 provides some basic definitions concerning Markov chains.

8. There are also other necessary and sufficient conditions. For instance, it is necessary and sufficient that the submatrix of T restricted to any strongly connected and closed group of nodes be a primitive matrix (e.g., see Theorem 1 in the appendix of Hegselmann and Krause [325]). This condition can be shown to be implied by the aperiodicity of T on these nodes.

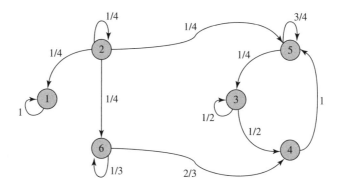

FIGURE 8.4 Closed sets of nodes.

Before stating the convergence result, the following definition is needed. A *closed* set of agents is a $C \subset \{1, \ldots, n\}$ such that there is no directed link from an agent in C to an agent outside C; that is, there is no pair $i \in C$ and $j \notin C$ such that $T_{ij} > 0$.

Example 8.4 (Closed Sets of Nodes) *Consider a society of $n = 6$ agents with an updating process*

$$T = \begin{pmatrix} 1 & 0 & 0 & 0 & 0 & 0 \\ 1/4 & 1/4 & 0 & 0 & 1/4 & 1/4 \\ 0 & 0 & 1/2 & 1/2 & 0 & 0 \\ 0 & 0 & 0 & 0 & 1 & 0 \\ 0 & 0 & 1/4 & 0 & 3/4 & 0 \\ 0 & 0 & 0 & 1/3 & 0 & 2/3 \end{pmatrix}.$$

as pictured in Figure 8.4.

There are many closed sets of nodes: for instance, $\{1, 2, 3, 4, 5, 6\}$ is closed. Also, $\{1\}$ is closed, as are $\{3, 4, 5\}$, $\{1, 3, 4, 5\}$, $\{3, 4, 5, 6\}$, and $\{1, 3, 4, 5, 6\}$. The only strongly connected and closed sets of nodes are $\{1\}$ and $\{3, 4, 5\}$, as pictured in Figure 8.5

Convergence for the overall society holds if and only if each closed and strongly connected set of nodes converges, which happens if and only if each such set is aperiodic (as in Figure 8.5).

Theorem 8.1 *T is convergent if and only if every set of nodes that is strongly connected and closed is aperiodic.*

The sufficiency of aperiodicity for convergence follows fairly easily by adapting theorems on steady-state distributions of Markov chains. The necessity can be shown by providing an algorithm for constructing a nonconvergent $p(0)$ when there is a strongly connected and closed group such that all directed cycles have a common divisor greater than 1. A proof is provided in Golub and Jackson [294].

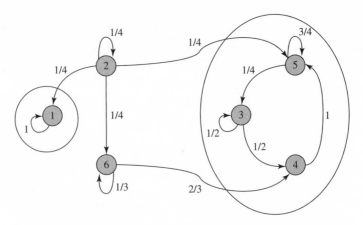

FIGURE 8.5 The only closed and strongly connected sets of nodes.

8.3.3 Consensus in Beliefs

Beyond knowing whether beliefs converge, we are also interested in characterizing what beliefs converge to when they do converge, which agents have substantial influence in the society, and when it is that a concensus is reached. These correspond to several of the questions stated at the start of the chapter.

Let us begin with a couple of simple observations. First, it is straightforward to see that if beliefs converge, then a strongly connected and closed group of agents will reach a concensus. A group of agents $C \subset \{1, \ldots, n\}$ *reaches a consensus* under T for an initial vector of beliefs $p(0)$ if $\lim_t p_i(t) = \lim_t p_j(t)$ for each i and j in C.

Proposition 8.2 *Under T, any strongly connected and closed group of individuals reaches a consensus for every initial vector of beliefs if and only if it is aperiodic.*

Proof of Proposition 8.2. We know that such a group will have convergent beliefs if and only if it is aperiodic. Since a consensus cannot be reached if beliefs do not converge, to conclude the claim it suffices to show that aperiodicity of a strongly connected and closed group implies consensus. Suppose to the contrary that for some $p(0)$ the belief of the agents in C converge, but to some p' such that $p_i' \neq p_j'$ for some i and j in C. Without loss of generality, we may ignore agents outside C, so consider C to be the full set of agents $\{1, \ldots, n\}$ and relabel the agents so that $p_1' \geq p_2' \geq \cdots \geq p_n'$. Find the minimal i such that $p_i' > p_{i+1}'$. Given that agents are strongly connected, some agent $k \leq i$ has $T_{kh} > 0$ for some $h \geq i + 1$. Convergence then implies that for any ε, we can find a large enough t' such that for all $t > t'$:

$$p_k(t) = \sum_j T_{kj} p_j(t - 1) \leq \sum_j T_{kj} p_j' + \varepsilon \leq (1 - T_{kh}) p_1' + T_{kh} p_h' + \varepsilon.$$

But since the right-hand expression is less than p_k' for a small enough ε, we reach a contradiction. ∎

Thus we have a complete characterization of consensus for strongly connected and closed groups, which is that they reach a consensus when they converge. The logic is evident from the proof, which is basically that a given individual cannot reach a higher limit than each of his or her neighbors, since his or her belief involves a weighted average of those beliefs.

So let us partition the society into strongly connected and closed groups of individuals, and then the remaining individuals. If there is more than one strongly connected and closed group, then clearly the society does not always reach a consensus except in the rare cases when the initial beliefs are such that separate closed and strongly connected groups happen to reach the same limit. For instance, if one closed group starts with common beliefs at 1 and a separate closed group starts with common beliefs at 0, they forever stay apart. To reach consensus when the initial beliefs are not chosen exceptionally, there must be exactly one closed and strongly connected group in the society. It is easy to see that there always exists at least one such group (see Exercise 8.7), and that a society can be partitioned into some number of strongly connected and closed groups and then a remaining set of agents who each have at least one directed path to an agent in a strongly and connected group.

Corollary 8.1 *A consensus is reached in the DeGroot model if and only if there is exactly one strongly connected and closed group of agents and T is aperiodic on that group.*

This corollary leads to another characterization of consensus due to Berger [56].

Corollary 8.2 (Berger [56]) *A consensus is reached in the DeGroot model if and only if there exists t such that some column of T^t has all positive entries.*

We can think of an entry $[T^t]_{ij}$ as keeping track of the indirect weight that an agent i places on agent j, through paths of length t. Thus, if some column of T^t has all positive entries, then every agent is putting some indirect weight on the agent corresponding to that column. Thus all agents must have an indirect path to some agent, and so there must be exactly one strongly connected and closed group of agents. Once a column is all positive, it stays that way, which guarantees aperiodicity (a not too difficult proof, but one that takes some thought).

8.3.4 Consensus and Nonconstant Updating Rules

The consensus result does not rely on T being a stationary matrix. It holds in a variety of models in which T varies with time or with the beliefs themselves. The following are some generalizations of the DeGroot model such that the updating can vary with time and circumstances.

Example 8.5 (Time-Varying Weight on Own Beliefs) *DeMarzo, Vayanos, and Zwiebel [193] examine a variation on the DeGroot model (related to a variation on DeGroot's model of Chatterjee and Seneta [148]) in which the updating rule is*

$$p(t) = \left[(1 - \lambda_t)I + \lambda_t \widehat{T} \right] p(t-1),$$

where I is the identity matrix, $\lambda_t \in (0, 1]$ is an adjustment factor, and $\widehat{T}$ is a stochastic matrix. If λ_t is constant over time, then this corresponds to the DeGroot model; otherwise, it allows updating to vary over time, so that an agent might place more (or less) weight on his or her own belief over time.

Example 8.6 (Only Weighting Those with Similar Beliefs) *The following model of Krause [417] (see also Hegselmann and Krause [325]) allows an agent to pay attention only to other agents whose beliefs do not differ much from his or her own. Thus the agent has a distrust of information that diverges too much from his or her own:*

$$T(p(t), t)_{ij} =$$

$$
\begin{cases}
\frac{1}{n_i(p(t))}, & \text{if } |p_i(t) - p_j(t)| < d, \text{ and } n_i(p(t)) = \#\{k : |p_i(t) - p_k(t)| < d\} \\
0, & \text{otherwise.}
\end{cases}
$$

An agent places equal weight on all opinions that are within some distance d of his or her own current opinion. This weighting substantially complicates the process, as the updating depends on the specifics of the opinions rather than just on time. This model is also closely related to that of Deffuant et al. [188] and Weisbuch et al. [661], in which at each time two agents are randomly matched and then update their beliefs only if the beliefs are sufficiently close.

The reaching of a consensus in the above generalizations of the DeGroot model is covered under the following result of Lorenz [440].

Theorem 8.2 (Lorenz [440]) *Suppose that $T(p(t), t)$ satisfies the following conditions:*

- *There exists $\delta > 0$ such that $T(p(t), t)_{ij} > 0$ if and only if $T(p(t), t)_{ij} > \delta$, for all t, i, j, and $p(t)$.*
- *$T(p(t), t)_{ii} > 0$ for all i, t, and $p(t)$.*
- *$T(p(t), t)_{ij} > 0$ if and only if $T(p(t), t)_{ji} > 0$ for all t, i, j, and $p(t)$.*

Then the society can be partitioned into sets of agents such that each group of agents reaches a consensus, and any two agents who place weight on each other infinitely often are in the same group.

The proof is carried out by bounding below the weights that agents must place on anyone with whom they communicate over time and then tying these weights back to the initial beliefs and establishing convergence. Effectively, with weights bounded below, a belief cannot stay too much above or below those of its neighbors over time, and so convergence to a consensus is still guaranteed. The reader is referred to Lorenz [440] for the details. The assumptions of each agent updating based on his or her own beliefs at every time, agents paying mutual attention to each other, and having a lower bound on attention (which then also puts an upper bound on how many agents a given agent can pay attention to) are stronger than necessary. But they simplify the analysis and allow the theorem to capture quite nonstationary updating processes in a class that significantly generalizes the basic DeGroot model in some directions. The requirement of symmetric attention implies that any path-connected agents are strongly connected, which is a way of guaranteeing that beliefs reach an overall consensus. If, for instance, this condition is violated, and there are two separate strongly connected groups and an agent who pays attention to both of these groups, then the two groups might reach different consensus beliefs, and the agent paying attention to both might have an intermediate limiting belief. The assumption that $T_{ii} > 0$ is not essential to the result, but it is an easy way to guarantee that the setting is aperiodic no matter how T varies with time and circumstances, so that the beliefs converge nicely. The bounding below of positive weights by some $\delta > 0$ is not a necessary condition, but some bound is needed. To see this, simply consider an example in which two agents pay attention to each other, but the weight each one places on the other's belief goes to 0 at a sufficiently high rate so that their beliefs never converge (see Exercise 8.9).

It is clear that Theorem 8.2 must allow for various sets of agents to have different limiting beliefs, since in the Krause [417] model no updating occurs between agents whose beliefs start out at a distance of more than d. For example, if $d < 1/2$ it is easy to find examples with multiple limiting sets of beliefs.

The Krause model has a discontinuity in it, as an agent will pay attention to one belief at a certain distance, but not another one that is a distance of more than ε away. That discontinuity is important in determining whether beliefs differ across different subgroups. For instance, consider the following continuous variation of that model.

Example 8.7 (Continuous Updating with "Close" Beliefs) *Suppose that agents place positive weight on all others, but place higher weight on opinions closer to their own:*

$$T_{ij}(p(t), t) = \frac{e^{-\gamma_{ij}|p_i(t) - p_j(t)|}}{\sum_k e^{-\gamma_{ik}|p_i(t) - p_k(t)|}},$$

where $\gamma_{ij} > 0$ for all ij. There is a lower bound on weights and all are positive, so a consensus is reached among all agents from any starting belief.

The following generalization of the DeGroot model shows that it is possible that a consensus is never reached, even when all agents are strongly connected, if agents persist in weighting their initial beliefs when updating.

Example 8.8 (Time-Varying Weight on Own Beliefs) *Friedkin and Johnsen [259] examine a model in which updating always mixes in some weight on an individual's initial beliefs. Let D be an $n \times n$ matrix whose entries are positive only along the diagonal, and $D_{ii} \in (0, 1)$ indicates the extent to which agent i pays attention to others' attitudes. The evolution is described by*

$$p(t) = D\widehat{T} p(t-1) + (I - D) p(0).$$

It is easy to see that consensus may never be reached. For example, simply set $n = 2$, $D_{ii} = 1/2$ and $\widehat{T}_{12} = 1 = \widehat{T}_{21}$. Then an agent is always averaging his original belief with the latest belief of the other agent:

$$p(t) = \begin{pmatrix} 1/2 & 0 \\ 0 & 1/2 \end{pmatrix} \begin{pmatrix} p_1(t-1) \\ p_2(t-1) \end{pmatrix} + \begin{pmatrix} 1/2 & 0 \\ 0 & 1/2 \end{pmatrix} \begin{pmatrix} p_1(0) \\ p_2(0) \end{pmatrix},$$

so that

$$p_i(t) = \frac{p_j(t-1)}{2} + \frac{p_1(0)}{2}.$$

Starting agent 1 with belief 1 and agent 2 with belief 0, it is straightforward to check that beliefs converge to 2/3 for agent 1 and 1/3 for agent 2.

These examples demonstrate that there is a wide set of circumstances in which a consensus is eventually reached, provided that agents do not (discontinuously) select whom they pay attention to based on agreement with their current beliefs, and as long as they do not cling too directly to the past. But the examples do not address some important questions, as we still are interested in knowing what beliefs converge to when they do converge and how quickly beliefs change. If convergence takes many iterations and updating is infrequent, then eventual convergence may not be so relevant. These are the issues that I discuss next.

8.3.5 Social Influence

To ascertain how each agent in the social network influences the limiting belief, let us return to the DeGroot model. To start, consider a closed and strongly connected group of agents, and for now let them be $\{1, \ldots, n\}$. Suppose that T is aperiodic so that by Proposition 8.2 all beliefs converge, and a consensus is reached. Let $p(0)$ be an arbitrary starting belief vector and $p(\infty) = (p^\infty, \ldots, p^\infty)$ be the vector of limiting consensus beliefs. To keep track of the limiting influence that each agent has, we seek a vector $s \in [0, 1]^n$ such that $\sum_i s_i = 1$ and

$$p^\infty = s \cdot p(0) = \sum_i s_i p_i(0).$$

If such an s exists, then the limiting beliefs would be weighted averages of the initial beliefs, and the relative weights would be the influences that the various agents have on the final consensus beliefs.

To get a feel for the origins of such weights, suppose that an influence vector exists that keeps track of the influence of each agent regardless of the initial beliefs. Since starting with $p(0)$ or with $p(1) = Tp(0)$ yields the same limit, it must be that $s \cdot p(1) = s \cdot p(0)$. Therefore

$$s \cdot (Tp(0)) = s \cdot p(0).$$

Since this equality has to hold for every $p(0)$, it follows that

$$sT = s.$$

Thus s is a left-hand unit eigenvector of T.[9] When T is strongly connected, aperiodic, and row stochastic, there is a unique such unit eigenvector (eigenvector with eigenvalue 1) that has nonnegative values, and in fact it has all positive values.[10]

---

9. See Section 2.4 for definitions and discussion of eigenvectors and eigenvalues.

10. This fact follows from variations on the Perron-Frobenius theorem and from results in Markov chain theory. For details, see Lemma 5 in Golub and Jackson [294].

Indeed, there is an easy way to calculate this eigenvector, since $s \cdot p(0)$ must lead to the same belief as any entry of $p(\infty) = (p^\infty, \ldots, p^\infty) = T^\infty p(0)$. Thus each row of T^∞ must converge to s as seen in Example 8.2.

Example 8.9 (Social Influence in Example 8.2) *Recall the updating matrix described by*

$$T = \begin{pmatrix} 0 & 1/2 & 1/2 \\ 1 & 0 & 0 \\ 0 & 1 & 0 \end{pmatrix},$$

where

$$T^t \to \begin{pmatrix} 2/5 & 2/5 & 1/5 \\ 2/5 & 2/5 & 1/5 \\ 2/5 & 2/5 & 1/5 \end{pmatrix}.$$

Note that $s = (2/5, 2/5, 1/5)$ is a unit eigenvector of T, that is,

$$sT = (2/5, 2/5, 1/5) \begin{pmatrix} 0 & 1/2 & 1/2 \\ 1 & 0 & 0 \\ 0 & 1 & 0 \end{pmatrix} = (2/5, 2/5, 1/5) = s.$$

In general, an easy way to calculate or at least approximite the left-hand unit eigenvector of a stochastic matrix T is simply to iterate on T^t and find its limits. One can also solve $sT = s$ directly if n is not too large.

It is worth noting the relationship between this measure of social influence and the eigenvector-based centrality measures discussed in Section 2.2.4. Indeed, this model can be thought of as providing an explicit basis for some of the eigenvector-based measures of centrality and influence.

If there is only one closed strongly connected group, then the above reasoning tells us what its beliefs converge to and the relative social influences that each of its members has. The remaining agents must then each have directed paths leading to the strongly connected group and must reach the same consensus belief (Proposition 8.2). Thus the other agents exert no social influence on the limiting belief, and their initial beliefs are irrelevant in determining the limiting belief.

When there are several closed strongly connected groups, each reaches its own consensus, with its own social influence weights, and then the remaining agents who are path-connected to the strongly connected groups acquire some weighted average of the limit beliefs of the strongly connected groups. This is stated by DeMarzo, Vayanos, and Zwiebel [193] as follows.[11]

Theorem 8.3 (DeMarzo, Vayanos, and Zwiebel [193]) *Given T, partition the set of agents into closed and strongly connected groups $B_1, \ldots, B_K$; and R, the remaining agents who are not in any closed and strongly connected group. A stochastic matrix T is convergent if and only if there is a nonnegative row vector $s \in \mathbb{R}^n$ such that*

---

11. Their result assumes that $T_{ii} > 0$ for each i, which is not necessary. Theorem 8.3 is the version stated and proven in Golub and Jackson [294].

- $\sum_{i \in B_k} s_i = 1$ for any closed and strongly connected group of agents B_k;
- $s_i > 0$ if i is in a closed and strongly connected group, and $s_i = 0$ otherwise;
- s_{B_k} is the left-hand nonnegative unit eigenvector of T restricted to B_k; and
- for any vector p and B_k, $\left(\lim_{t \to \infty} T_{B_k}^t p\right)_{B_k} = s_{B_k} p_{B_k}$,

and for each agent $j \in R$ not in any closed strongly connected group, there exists a $w_{B_k}^j \geq 0$ for each B_k and such that $\sum_k w_{B_k}^j = 1$ and such that $(\lim_{t \to \infty} T^t p)_j = \sum_k w_{B_k}^j s_{B_k} p_{B_k}$.

Theorem 8.3 states that (provided a society converges) each closed and strongly connected set of nodes converges to a consensus belief that is determined by the social influence vector for that group times the group's initial beliefs. Agents outside the closed and strongly connected sets then converge to some weighted average of the closed and strongly connected groups' limiting beliefs. This theorem is illustrated below in the context of Example 8.4.

Example 8.10 (Social Influence in Example 8.4) *Given the updating in the six-person society from Example 8.4 we can determine the social influence weights and other weights as follows. It is clear that agent 1 will simply adhere to his or her initial beliefs, and so $s_1 = 1$. The only other closed and strongly connected set of agents is $\{3, 4, 5\}$. These agents only pay attention to one another when updating and reach a consensus as if they were an isolated society. If we restrict T to these three agents, it becomes*

$$T_{\{3,4,5\}} = \begin{pmatrix} 1/2 & 1/2 & 0 \\ 0 & 0 & 1 \\ 1/4 & 0 & 3/4 \end{pmatrix},$$

which has a unit eigenvector of $(2/7, 1/7, 4/7)$, and so these are the corresponding entries of s. Thus these three agents converge to a belief of $\frac{2}{7} p_3(0) + \frac{1}{7} p_4(0) + \frac{4}{7} p_5(0)$, and the overall influence vector is

$$s = (1, 0, 2/7, 1/7, 4/7, 0).$$

Note that this vector sums to 2, the number of closed strongly connected groups, and each group converges to its own consensus. As for the remaining agents, 6 only pays attention to 6 and 4. Given that 4 eventually converges, 6's belief will converge to 4's belief regardless of 6's initial belief. Therefore 6 has a weight of 1 on $\{3, 4, 5\}$, and so $w_{\{1\}}^6 = 0$ while $w_{\{3,4,5\}}^6 = 1$. Next consider agent 2's limiting belief. Agent 2 is paying equal attention to 1, 2, 5, and 6. As the beliefs of 1, 5, and 6 converge to various limits, 2's initial belief does not matter. Given that 6 converges to the same belief as 5, then effectively 2 has twice as much weight on the limiting belief of 5 (2/4) compared to that of 1 (1/4). We can also arrive at this result by noting that 2's limiting beliefs have to satisfy

$$p_2(\infty) = \tfrac{1}{4} p_1(\infty) + \tfrac{1}{4} p_2(\infty) + \tfrac{1}{4} p_5(\infty) + \tfrac{1}{4} p_6(\infty),$$

and so given that $p_6(\infty) = p_5(\infty)$ we have that

$$\tfrac{3}{4} p_2(\infty) = \tfrac{1}{4} p_1(\infty) + \tfrac{1}{2} p_5(\infty),$$

or

$$p_2(\infty) = \tfrac{1}{3} p_1(\infty) + \tfrac{2}{3} p_5(\infty).$$

Thus $w^2_{\{1\}} = 1/3$ and $w^2_{\{3,4,5\}} = 2/3$, so 2's limiting beliefs are obtained by giving weight $1/3$ to 1's belief and $2/3$ to the consensus limit of 3, 4, and 5.

These calculations provide a general approach to solving for w. We know that for any i not in a closed and strongly connected group, the limiting beliefs of i have to satisfy the following equation to converge:

$$p_i(\infty) = T_{ii} p_i(\infty) + \sum T_{iB_k} p_{B_k}(\infty) + \sum_{j \in R, j \neq i} T_{ij} \sum_k w^j_k p_{B_k}(\infty)$$

for any limiting beliefs $p_{B_k}(\infty)$ of the closed strongly connected groups B_k, where $T_{iB_k} = \sum_{\ell \in B_k} T_{i\ell}$. Thus

$$w^i_{B_k} = \frac{T_{iB_k} + \sum_{j \in R, j \neq i} T_{ij} w^j_k}{1 - T_{ii}}.$$

This system is generally easily solved.[12]

We can deduce a few general insights about influence. Note that since s corresponds to a (left-hand unit) eigenvector of T, it follows that

$$s_j = \sum_i T_{ij} s_i \tag{8.2}$$

for all j. Equation 8.2 implies that an agent acquires influence from being headed by agents who are themselves influential.

From this observation we can derive some easy conclusions about social influence. If one agent, say, j, receives systematically more weight than another, k, so that $T_{ij} \geq T_{ik}$ for all i, then j has more influence than k. This definition gives us an obvious and natural notion of opinion leaders—they are individuals who dominate others in the weights assigned to them in the communication and updating process. It is also easy to see that if two individuals receive similar weights, then the one who receives weight from agents who themselves have higher social influence will have more influence. There are also some benchmark cases: If the society is reciprocal, so that $T_{ij} = T_{ji}$ for all i and j, then all agents have equal weight, regardless of the actual distribution of weights. This conclusion follows from the result that if $\sum_i T_{ij} = 1$ for all j, then each agent has the same influence (see Exercise 8.10).

To get a feel for how the social influence vector depends on the social structure, let us examine a particular application.

---

12. The equations always have a unique solution, by the following argument pointed out to me by Ben Golub. The equations can be written as $(I - T_R) w_k = c$ for $k \in \{1, \ldots, K\}$, where T_R is the restriction of T to R, I is the $|R| \times |R|$-identity matrix, w_k is an $|R| \times 1$ vector that has jth entry $w^j_{B_k}$, and c is a vector that does not involve w_k. There is a unique solution if $I - T_R$ is invertible. That follows from observing that $T^t_R \to 0$, since the weights in T_R sum to less than 1 for at least one row, which then implies that T_R does not have a unit eigenvalue.

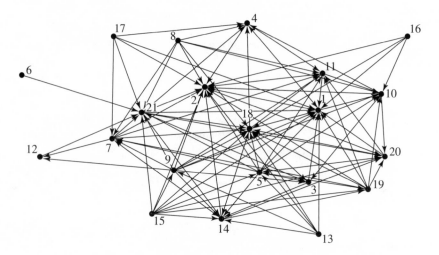

FIGURE 8.6   Krackhardt's network of advice among managers.

Example 8.11 (An Application: Influence in Krackhardt's Advice Network)   *Krackhardt [411] collected data regarding a small manufacturing firm on the west coast of the United States. The firm had about 100 employees and 21 managers. Krackhardt collected information from the managers about who sought advice from whom. The resulting directed network is pictured in Figure 8.6.*[13]

Based on these data, we can develop a T matrix by normalizing each row of the advice matrix to sum to 1, so that the row reflects how a given agent weights others' opinions when forming his or her own. So if i seeks advice from seven different agents, including agent j, then $T_{ij} = 1/7$. This normalization might not correspond to the actual weights that different agents place on one another, but the data do not contain any weighted information. The process also does not include self-weighting, which could be added, but again the data do not provide any direct insight into how much agents self-weight. Given this matrix T, we can calculate the vector s directly as the left-hand unit eigenvector.[14] *The resulting influence weights s are reported in Table 8.1.*

There are some interesting patterns regarding the influence weights s. First, there are four agents (6, 13, 16, and 17) from whom no other agents seek advice. Each of these agents is then outside the single closed strongly connected group, which contains all other agents. Next, the influence can be much different from in-degree. For example, agent 21 has the highest influence, even though he or she is the advisor of 8 other agents, while agent 18 advises 12 other agents. This disparity occurs for at least two reasons. First, influence is higher when an agent is heeded

---

13. Krackhardt asked each manager to fill out a questionnaire indicating the full advice network. The data used here have a directed relation from i to j if both i and j responded that i seeks advice from j (these are the data in the "LAS" matrix from page 129 of Krackhardt [411]).

14. This calculation was done using the program Matlab, which reports all 21 eigenvectors and eigenvalues for the 21×21 T matrix.

TABLE 8.1
Influence in Krackhardt's network of advice among managers

Agent number	s	Level	Department	Age (years)	Tenure (years)
1	0.048	3	4	33	9.3
2	0.132	2	4	42	19.6
3	0.039	3	2	40	12.8
4	0.052	3	4	33	7.5
5	0.002	3	2	32	3.3
6	0.000	3	1	59	28
7	0.143	1	0	55	30
8	0.007	3	1	34	11.3
9	0.015	3	2	62	5.4
10	0.024	3	3	37	9.3
11	0.053	3	3	46	27
12	0.051	3	1	34	8.9
13	0.000	3	2	48	0.3
14	0.071	2	2	43	10.4
15	0.015	3	2	40	8.4
16	0.000	3	4	27	4.7
17	0.000	3	1	30	12.4
18	0.106	2	3	33	9.1
19	0.002	3	2	32	4.8
20	0.041	3	2	38	11.7
21	0.201	2	1	36	12.5

Note: Level indicates the level of hierarchy in the firm (e.g., 1 denotes the highest, 3 the lowest level).

by other agents who are themselves more influenced; and second, one gets more influence when advising agents who seek advice from relatively fewer agents. For instance, agent 7 (the head of the firm) has substantial influence even though he or she only advised six other managers. What is crucial is that agent 7 advises 2, 18, and 21, who all have substantial influence themselves (and are at the second level of the hierarchy in the firm).

As mentioned above, social influence as defined in the context of the DeGroot model provides a foundation for eigenvector-based centrality measures. It also provides a basis for understanding other related systems. In particular, the structure of Google's PageRank system[15] is analogous to the influence vectors here, where the T matrix is derived by normalizing the directed matrix of links between web

---

15. Some other citation and ranking measures are also based on eigenvectors, for instance, the measure of Palacios-Huerta and Volij [520].

pages (so that $T_{ij} = 1/d_i$ if page i has a link to page j, and d_i is the number of directed out-links that page i has to other pages).

8.3.6 Segregation and Time to Consensus*

While eventual convergence of beliefs and the reaching of a consensus is important, it is also important to know how quickly beliefs reach their limits. In many applications there might be only a few rounds of updating, or new information might enter the system over time, or updating might occur over long time horizons. If the convergence is slow, then we might observe heterogeneous beliefs in a society even though it tends toward consensus. Let us examine how quickly beliefs reach a consensus, and whether we can quantify this rate.

Two Agents To develop the basic intuition and begin to quantify the speed of convergence, it is useful to start with the case of two individuals, as that is particularly transparent and provides the basis for a general analysis.

Before developing the formal analysis, it is useful to discuss the basic ideas. If the two individuals have very similar weightings, so that the weight that 1 places on 1, T_{11}, is similar to the weight that 2 places on 1, T_{21} (which happens if and only if T_{12} is close to T_{22}), then they will clearly quickly develop similar beliefs. In fact the end social influence of 1 will be close to T_{11} and the influence of 2 will be close to T_{22}. Thus, similar weightings (similar rows of T) produce fast convergence. Conversely, if the weightings of the two agents are very different, then their beliefs can differ substantially for longer periods, and convergence is relatively slower. For instance, if each agent weights his or her own opinion very heavily and pays little attention to the other agent, then the beliefs are slow to converge to reach a consensus.[16] So heuristically, we should expect that the rate of convergence should be related to some measure of how much T_{11} differs from T_{21}.

In what follows, let us focus on the only nontrivial case, which is that of the strongly connected aperiodic setting (so the off-diagonals of T are positive) in which the agents do not place identical weights on each other. Thus opinions converge, each agent has some limiting influence, and convergence is not instantaneous.

To develop these ideas more formally, we need to know how beliefs at time t, $p(t) = T^t p(0)$, differ from the limiting beliefs $p(\infty)$. One way to keep track of this difference is to track the difference between T^t and its limit T^∞ (which, recall, has as its rows the influence vector s, which is the left-hand unit eigenvector). To see how T^t behaves as we increase t, it is useful to rewrite T using what is known as its diagonal decomposition. In particular, let u be the matrix of left-hand eigenvectors of T. Then u has the form

$$u = \begin{pmatrix} s_1 & s_2 \\ 1 & -1 \end{pmatrix}, \tag{8.3}$$

---

16. Note that slow convergence can also occur when each agent weights the other agent heavily and himself or herself only slightly, in which case beliefs can oscillate over time.

where (s_1, s_2) is the unit eigenvector corresponding to the social influence weights, and $(1, -1)$ is the other eigenvector. It is easy to check that

$$s_1 = \frac{T_{21}}{1 + T_{21} - T_{11}} \text{ and } s_2 = \frac{1 - T_{11}}{1 + T_{21} - T_{11}}.$$

Also, in the case of a row-stochastic T with $n = 2$, $(1, -1)$ is always the second eigenvector, since

$$(1, -1)T = (T_{11} - T_{21}, T_{12} - T_{22}) = (T_{11} - T_{21}, -(T_{11} - T_{21}))$$

$$= (T_{11} - T_{21})(1, -1).$$

Moreover, not only is $(1, -1)$ the second eigenvector, but its associated eigenvalue is $T_{11} - T_{21}$.

Since u is the matrix with its rows being the eigenvectors, we know that

$$uT = \Lambda u, \tag{8.4}$$

where Λ is the matrix with the first and second eigenvalues λ_1 and λ_2 (ranked in terms of absolute values) on its diagonal:

$$\Lambda = \begin{pmatrix} \lambda_1 & 0 \\ 0 & \lambda_2 \end{pmatrix} = \begin{pmatrix} 1 & 0 \\ 0 & T_{11} - T_{21} \end{pmatrix}.$$

From (8.3) and given that $s_1 > 0$ and $s_2 > 0$, we know that u has an inverse u^{-1}, which is easily seen to be

$$u^{-1} = \begin{pmatrix} 1 & s_2 \\ 1 & -s_1 \end{pmatrix}.$$

From (8.4) it follows that

$$T = u^{-1}\Lambda u. \tag{8.5}$$

Equation 8.5 is the *diagonal decomposition* of T.[17] From (8.5) it follows that

$$T^2 = u^{-1}\Lambda u u^{-1}\Lambda u = u^{-1}\Lambda^2 u,$$

and more generally that

$$T^t = u^{-1}\Lambda^t u. \tag{8.6}$$

---

17. It is sometimes useful to note that u^{-1} is the matrix of right-hand eigenvectors of T, and that they have the same matrix of eigenvalues as u. To see this, note that from (8.4) it follows that $uTu^{-1} = \Lambda uu^{-1} = \Lambda$. Thus $u^{-1}uTu^{-1} = u^{-1}\Lambda$, and so $Tu^{-1} = u^{-1}\Lambda$, and u^{-1} is the vector of right-hand eigenvectors. The left-hand and right-hand eigenvectors are also called the *row* and *column eigenvectors,* respectively. The diagonal decomposition can then be stated in terms of u and its inverse, or equivalently, in terms of the two matrices of column and row eigenvectors.

So the convergence of T^t is directly related to the convergence of Λ^t:

$$\Lambda^t = \begin{pmatrix} \lambda_1^t & 0 \\ 0 & \lambda_2^t \end{pmatrix} = \begin{pmatrix} 1 & 0 \\ 0 & (T_{11} - T_{21})^t \end{pmatrix}.$$

Given that $\lambda_2 = (T_{11} - T_{21}) < 1$, it follows that λ_2^t converges to 0, and so the distance of T^t from T^∞ is simply a factor times the second eigenvalue raised to the tth power, or $(T_{11} - T_{21})^t$. Thus the heuristic relationship that we started with is confirmed: the convergence of T is tied to the level of differences in the weights of the agents.

To get the full expression for T^t, note that by (8.6):

$$T^t = \begin{pmatrix} 1 & s_2 \\ 1 & -s_1 \end{pmatrix} \begin{pmatrix} \lambda_1^t & 0 \\ 0 & \lambda_2^t \end{pmatrix} \begin{pmatrix} s_1 & s_2 \\ 1 & -1 \end{pmatrix} = \begin{pmatrix} s_1 + \lambda_2^t s_2 & s_2 - \lambda_2^t s_2 \\ s_1 - \lambda_2^t s_1 & s_2 + \lambda_2^t s_1 \end{pmatrix}.$$

Therefore

$$p_1(t) = \left(s_1 + \lambda_2^t s_2\right) p_1(0) + \left(s_2 - \lambda_2^t s_2\right) p_2(0)$$

$$= p_1(\infty) + \left(T_{11} - T_{21}\right)^t s_2 \left(p_1(0) - p_2(0)\right), \qquad (8.7)$$

or

$$|p_1(t) - p_1(\infty)| = \left|T_{11} - T_{21}\right|^t s_2 \left|p_1(0) - p_2(0)\right|,$$

or

$$|p_1(t) - p_1(\infty)| = \frac{\left|T_{11} - T_{21}\right|^t (1 - T_{11})}{1 - T_{11} + T_{21}} \left|p_1(0) - p_2(0)\right|.$$

Thus the difference between the beliefs of agent 1 at time t and the limit varies linearly with the difference in the agents' starting beliefs and with the influence of the other agent. It declines exponentially in time: proportionally to the tth power of the difference in weights that the agents have.

Convergence that is exponential in t is generally considered to be fast, although the speed depends on how frequently updating occurs and how different the weights are. If agents update infrequently and they have very different weights, then they can maintain quite different beliefs. Having a direct expression for T^t allows one to see how close beliefs are to their limit (and also to consensus) at any date.

Many Agents   We relied on $n = 2$ in the above derivation for the precise expression of the second eigenvalue and eigenvectors. The facts that T has a diagonal decomposition $T = u^{-1}\Lambda u$ and

$$T^t = u^{-1}\Lambda^t u \qquad (8.8)$$

extend readily to $n > 2$, where Λ is the diagonal matrix with entries that are the eigenvalues of T and u is the matrix of corresponding left-hand (row) eigenvectors.[18] So, for the many-agent case the convergence of T^t to the limit T^∞ depends on how quickly λ_k^t goes to 0 for each k, where λ_k is the kth-largest eigenvalue of T. This rate will generally be governed by the second-largest eigenvalue, as the others will converge more quickly.[19] By (8.8) (noting that u^{-1} has a first column of all 1s[20]) it follows that

$$[T^t]_{ij} = s_j + \sum_{k \geq 2} \lambda_k^t u_{ik}^{-1} u_{kj}.$$

Thus the exact expression for the difference between the beliefs of agent i at time t and the limiting beliefs is given by

$$p_i(t) - p_i(\infty) = \sum_j p_j(0) \sum_{k \geq 2} \lambda_k^t u_{ik}^{-1} u_{kj}. \tag{8.9}$$

The expression (8.9) provides the basis for the following theorem. Many such convergence results exist in the theory of Markov chains, as well as in the specific context of social influence models (e.g., see Seneta [587] or DeMarzo, Vayanos, and Zwiebel [193]).

Theorem 8.4 *Let T be strongly connected and aperiodic, and let λ_2 be the second-largest eigenvalue of T. Moreover, suppose that the matrix of left-hand eigenvectors u is nonsingular and thus invertible, so that T is diagonalizable.[21] Then $|\lambda_2(T)| < 1$ and there exists $C > 0$ such that for each $i \in \{1, \ldots, n\}$,*

$$|p_i(\infty) - p_i(t)| \leq C \left| \lambda_2^t \right|. \tag{8.10}$$

Moreover, if T is nonsingular, then there exists some i, $p(0)$, and $c > 0$ such that for all large enough t,

$$|p_i(\infty) - p_i(t)| \geq c \left| \lambda_2^t \right|. \tag{8.11}$$

The inequality (8.10) follows direcly from (8.9), noting that $|\lambda_2^t|$ is larger than any other $|\lambda_k^t|$ for any k, and that the other parts of the expression are independent of t. The inequality (8.11) also follows from (8.9), although a full proof requires showing that the weighting term on the second eigenvalue (plus any further eigenvalues that take on the same value as the second) on the right-hand

---

18. This holds when u is invertible, and thus when u is nonsingular, so for generic T. When $n > 2$, some eigenvalues can be complex-valued but the decomposition is still valid.

19. The Perron-Frobenius theorem (e.g., see Meyer [466]) states that if T is stochastic, strongly connected, and aperiodic, then the second eigenvalue is less than 1. The Perron-Frobenius theorem applies to primitive matrices (those with all positive entries), but T^t has all positive entries for large enough t, which can be shown by variations on the Perron-Frobenius theorem, or more directly, given the strong connection and aperiodicity of T.

20. It is easy to check that the unit column eigenvector is a vector of all 1s.

21. The case of nondiagonalizable T is nongeneric, but that does not mean that such matrices might not pop up in practice, especially if T is derived from normalizing some adjacency matrix that has 0 or 1 for entries. The case of nondiagonalizable T has a similar result with an adjustment to the expression in (8.10); see Seneta [587].

side of (8.9) (i.e., $u_{i2}^{-1}u_{2j}$) must be nonzero for some i and j, then setting $p_j(0) = 1$ and $p_k(0) = 0$, $k \neq j$. For a full proof of the second claim, see Karlin and Taylor [379].

Example 8.12 (Convergence Speed in Krackhardt's Advice Network) *The second-largest eigenvalue in magnitude corresponding to Krackhardt's advice network discussed in Example 8.11 is* $.4825$.[22] *From (8.9) we can obtain a rough upper bound on C in (8.10) simply by noting that the sum is less than* $\lambda_2^t 20$, *given that* $n = 21$. *Thus a crude upper bound on the distance from a consensus at time t is* $20 \times (.5)^t$.

While Theorem 8.4 offers detailed information about the rate of convergence of beliefs, the second-largest eigenvalue is not directly intuitive. In the case of two agents, we were able to relate the second eigenvalue to the difference in one of the agent's self-weight and the incoming weight from the other agent, and so it had a natural interpretation. The case of more than two agents, however, does not have quite as direct an interpretation. Nonetheless, the second eigenvalue still has a relationship to the level of difference in the weights that different agents place on one another. A result developed by Hartfiel and Meyer [323] relates the second eigenvalue of a stochastic matrix to another measure that has a more intuitive feel and is related to the difference of weights in the two-agent case.

Given some $A \subset N$, let

$$T_{A,A^c} = \sum_{i \in A, j \notin A} T_{ij}.$$

Thus, this sum is the total weight that all agents in A place on agents outside A. Let the *coupling of T*, denoted $\sigma(T)$, be defined by

$$\sigma(T) = \min_{B \neq \emptyset, C \neq \emptyset: B \cap C = \emptyset} \left(T_{B,B^c} + T_{C,C^c} \right). \tag{8.12}$$

Equation 8.12 reflects how strongly interconnected the different subgroups of a society are. If the coupling measure is low, then there are two disjoint groups who pay little attention to anyone outside their respective groups. In such a situation, those two groups can maintain different beliefs for a long time, convergence is relatively slow, and it will take a long time to approach a consensus. To understand why the uncoupling measure looks at two groups and not just one, note that if there is just one group that is introspective, then the rest of the society must be paying some attention to that group so that convergence can still be fast. For instance, if everyone is paying substantial attention to a single agent, say i, then the convergence can still be fast. Thus the coupling measure needs to reflect that two disjoint groups with little communication have slow convergence.

Theorem 8.5 (Hartfiel and Meyer [323]) *For any* $\varepsilon > 0$, *there exists a* $\delta > 0$ *such that if T is a strongly connected stochastic matrix and* $\sigma(T) < \delta$, *then*

---

22. Again, this was calculated using the program Matlab from the stochastic matrix described in Example 8.11.

$|\lambda_2(T)| > 1 - \varepsilon$. *Conversely, for any* $\delta > 0$ *there exists a* $\varepsilon > 0$ *such that if* T *is a strongly connected stochastic matrix and* $|\lambda_2(T)| > 1 - \varepsilon$, *then* $\sigma(T) < \delta$.

Using Theorems 8.4 and 8.5 together, we can conclude that for any $\varepsilon > 0$, there exists a $\delta > 0$ such that if T is strongly connected and aperiodic with coupling less than δ, then we can find initial beliefs $p(0) \in \mathbb{R}^n$, an agent $i \in A$, and a $C > 0$ such that for all large enough t,

$$|p_i(\infty) - p_i(t)| \geq C(1 - \varepsilon)^t.$$

Conditions that ensure that the second eigenvalue is close to 0, and so convergence is fast, are not as readily available. The above analysis suggests that the coupling measure is closely related to the second eigenvalue, so that if coupling is close to 1 then the second eigenvalue is close to 0. When $n = 2$, this is true, and in fact the coupling measure is precisely 1 minus the absolute value of the second eigenvalue (see Exercise 8.13). When $n > 2$ such a relationship does not hold in general, although there are some special cases for which such results can be found.

Consider the special class of Ts known as *expander graphs* (e.g., see Hoory, Linial, and Widgerson [338]). These are matrices that are d-regular and symmetric, so that there exists a degree $d \geq 1$ such that each i has $T_{ij} = 1/d$ for d agents $j \neq i$, and $T_{ij} = T_{ji}$. So, this class can be thought of as representing situations in which communication is mutual and agents split their time evenly between a set number of neighbors. It is clear that the influences of the agents are equal (and are $s_i = 1/n$ for each i, as in Exercise 8.10), so the limit belief is the average of initial beliefs. Even though the influences are equal, the rate of convergence can vary, depending on how the agents are arranged.

Various studies have examined how quickly neighborhoods grow in expansion graphs. If the expansion is slow, then some neighborhoods of agents are talking mainly among themselves and not so much to the remaining agents, and we would expect slow convergence. For situations in which the expansion is high, so that all neighborhoods of agents are placing great weight on outsiders, the expansion is faster and the convergence should be as well. A measure of this expansion rate is the *expansion ratio* of a symmetric and d-regular T, defined by

$$h(T) = \min_{B:|B| \leq n/2} \frac{T_{B,B^C}}{|B|}. \tag{8.13}$$

Given the normalization by the size of B, it is clear that h lies between 0 and 1. So (8.13) is keeping track of the fraction of the weight that the most inward-weighting group B places on outside groups. This measure has a similar structure to that for coupling, except that it only looks at one B rather than two disjoint groups. Given the symmetry of the network, if B places low weight on the outside world, then the outside world must also place a low weight on B, and similarly for high weight. Thus it is sufficient to look across single groups.

Several results on expander graphs relate the second eigenvalue to the expansion ratio. For example, it is known that $\lambda_2(T) \leq 1 - (h(T))^2/2$ (such results date to Cheeger [149]; e.g., see Theorem 2.4 in Hoory, Linial, and Widgerson [338]). Combining this with Theorem 8.4 leads to the following proposition, as noted by Golub and Jackson [294].

Proposition 8.3 *If T is strongly connected, d-regular, and symmetric, then there exists C > 0 such that for each i,*

$$|p_i(\infty) - p_i(t)| \le C \left(1 - \frac{(h(T))^2}{2}\right)^t.$$

With some feel for the speed of convergence, let us now turn to the question of the conditions for a society to reach an accurate consensus.

8.3.7 When a Consensus Is Correct: Wise Crowds

While we have seen that strongly connected and closed groups' agents following variations of repeated updating rules will reach consensus beliefs, provided that the updating rules they follow are convergent, we do not know whether the consensus beliefs are "correct." The information that individuals are sharing might concern some objectively measurable event. For instance, they might be estimating the reliability of a product, or the probability that something will occur. In such cases, as they update their beliefs and those beliefs converge, we can ask whether those beliefs converge to the right probability (expectation, or the like).

This question was analyzed in the context of the DeGroot model by Golub and Jackson [294]. They consider a sequence of societies, indexed by n, where n grows. Each society is strongly connected and convergent[23] and described by the updating matrix T^n. There is a true state of nature described by μ, and each agent i in network n observes a signal so that the initial belief of i in society n is a random variable $p_i^n(0)$ that is distributed with mean μ, a finite variance of at least $\sigma^2 > 0$, and support that is a subset of a compact set $[-M, M]$. The signals are independently but not necessarily identically distributed.

In each social network of the sequence, the belief of each agent i in network n converges to the consensus limit belief $p^n(\infty) = \sum_i s_i^n p_i^n(0)$, where the social influences s_i^n are defined in Theorem 8.3. The sequence of networks is *wise* if $p^n(\infty)$ converges in probability to the true state μ as n grows. That is, the sequence $(T_n)_{n=1}^{\infty}$ is *wise* if for every $\varepsilon > 0$

$$\lim_n \Pr\left[\left|p^n(\infty) - \mu\right| \ge \varepsilon\right] = 0.$$

We know from the law of large numbers (Theorems 4.2 and 4.3) that if we average the signals, then the limiting average will converge to be accurate with probability 1. Thus if the agents in the sequence of societies have equal influence, then they converge to accurate beliefs, and so such a sequence of societies is wise. So it is sufficient to have, for instance, $\sum_i T_{ij}^n = 1$ for each j, or to have reciprocal weights. But these are clearly not necessary conditions. What is necessary and sufficient is that no agent retain too much influence. Thus the necessary and sufficient condition for a wise sequence of societies is that $\max_i s_i^n$ tends to 0 (see

---

23. Clearly, this condition is without loss of generality, as it is necessary for the society to reach a concensus.

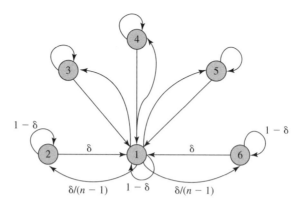

FIGURE 8.7 Agent 1 retains social influence 1/2.

Exercise 8.14). If some individual retains a nonvanishing weight, then the consensus involves a nontrivial weight on this single signal and retains a nonvanishing variance; it then cannot be wise. So characterizing wise sequences of societies amounts to understanding when it is that no agent wields too much influence as the society grows.

It is clear that if there is some $\delta > 0$ and j such that all agents have weight at least δ on j (so $T_{ij}^{n} > \delta$ for all i and n), then j has a weight of at least δ in each society, so that $s_j^n \geq \delta$ for all n (which follows directly from (8.2)). Thus too strong an opinion leader leads the society to a limiting belief that will not be accurate, as it retains too much weight on the single opinion. If that opinion is infinitely more accurate than others, its influence could be useful, but when all agents have some error in their initial beliefs, this overweighted belief hinders the accuracy of the consensus. Is it possible to have some opinion leader j, but such that the weights of all agents $i \neq j$ on j go to 0 as n grows? The following example from Golub and Jackson [294] shows that this case can still lead to problems.

Example 8.13 (An Influential Agent with High Relative Weight) *Consider the following updating rule. Each agent places weight δ on himself or herself. Agents other than 1 place weight δ on agent 1, while agent 1 places weight δ on each $j > 1$, as in Figure 8.7. Regardless of the level of δ, even if it goes to 0 much more rapidly than $1/n$, agent 1 maintains social influence 1/2, and the limiting consensus is half agent 1's signal and half an average of all the other agents' signals (see Exercise 8.15). Agent 1 thus receives much more weight relative to any other agent, even though that could be a small amount of weight.*

One might expect that bounding the relative weight placed on any agent relative to the weight accorded to others would be enough to overcome this problem. However, the following example shows that this is not the case: indirect weight matters, too.

Example 8.14 (An Influential Agent with High Indirect Weight) *Each agent places weight δ on the agent with the next lower label and $1 - \delta$ on the agent with the next higher label, except agent 1 who places weight δ on himself or herself,*

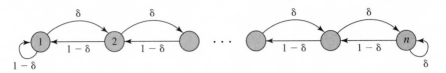

FIGURE 8.8 A lack of balance in and out—many agents maintain nonvanishing influence.

and agent 2 with self-weight $1 - \delta$. This scenario is pictured in Figure 8.8. One can verify from (8.2) (see Exercise 8.16) that

$$s_i^n = \left(\frac{\delta}{1-\delta}\right)^{i-1} \cdot \frac{1 - \left(\frac{\delta}{1-\delta}\right)}{1 - \left(\frac{\delta}{1-\delta}\right)^{n+1}}.$$

For instance, if δ is small, then s_1^n becomes large (close to 1) as n grows. This example shows the difficulty in providing conditions to guarantee the wisdom of crowds, as indirect trust is critical in building an agent's social influence.

Golub and Jackson [294] provide the following sufficient conditions for wise crowds.

Let $T_{AB} = \sum_{i \in A, j \in B} T_{ij}$ be the total weight placed by agents in A on agents in B.

The first condition is as follows. A sequence T^n is *balanced* if there exists a sequence $j_n \to \infty$ such that for every sequence of sets of agents B_n such that $|B_n| \leq j_n$ for each n,

$$\sup_n \frac{T^n_{B_n^C, B_n}}{T^n_{B_n, B_n^C}} < \infty,$$

where B_n^C is the complement of B_n. This condition requires that a smaller group cannot receive infinitely more weight from a larger group than it gives back. This condition rules out not only cases like Example 8.13, where a single agent receives an unbalanced amount of weight, but also those in which some group of agents obtains an unbalanced weight, as otherwise some subset of them could retain nonvanishing influence.

The next condition is not relative, but one based on absolute weights. A sequence T^n satisfies *minimal dispersion* if there is a $q \in \{1, 2, \ldots\}$ and $r > 0$ with for any sequence of B_n and C_n with $|B_n| \geq q$ and $|C_n|/n \to 1$, we have $T^n_{B_n, C_n} > r$ for all large enough n. This condition requires groups that have some minimal number of agents in them to give at least a specific weight to groups that contain almost all of the agents.

Theorem 8.6 (Golub and Jackson [294]) *If $(T_n)_{n=1}^\infty$ is a sequence of convergent stochastic matrices satisfying balance and minimal dispersion, then it is wise.*

The reader is referred to Golub and Jackson [294] for the proof and other sufficient conditions. The basic outline of the proof is as follows. If the wise crowd condition fails, then there is some group of agents who retain nontrivial influence as the society grows. The number of agents who can retain influence, and the amount of influence that they can retain, is limited by the fact that the influences sum to 1 in any society. The balance condition implies that the influential group cannot receive its weight just from the noninfluential agents, because it would need a large amount of weight from them as their influence tends to 0. But then the dispersion condition implies that there is not enough weight coming solely from the influential group, as they must accord some of their weight to the outside.

Interestingly, the conditions for the wisdom of a society are not closely related to the conditions that affect the speed of convergence. It is possible for a society to be wise and to converge quickly, as for example when every agent weights all agents equally, so that convergence occurs in one period. Or the society could be wise and yet converge quite slowly, as for example when the agents are connected in a large circle, each paying most of his or her attention to himself or herself and a small amount to his or her two closest neighbors. It is also possible for a society to be unwise and converge either quickly or slowly. For instance, if all agents pay substantial attention to a single agent, then convergence can be quite fast but unduly influenced by the initial beliefs of that agent. Or if society is split into two groups who pay most of their attention to two different agents, but with a small amount of communication between groups, then those two agents will have substantial influence and convergence can be quite slow.

The DeGroot model and its variants are tractable and powerful tools for studying a variety of issues associated with the diffusion of information and learning. As the T matrix is something that can be examined empirically, it holds substantial promise as a tool for empirical research. In closing this chapter, let me offer a couple of thoughts on important directions for further development of the theory. In many settings, it is clear that opinions and beliefs do not converge to a consensus, even though societies are strongly connected, and yet in other contexts we do see a consensus develop. Persistent differences can be due to a number of factors, including updating that occurs infrequently or for only a few iterations, weights that change over time, agents who weight events heterogeneously and update differently, or nonstationarities in the underlying environment so that new information is constantly coming in. The DeGroot model and some of its variations provide powerful starting points for analyzing the evolution of beliefs, but there is still much for us to learn about the circumstances in which these models provide useful predictions and the further development of the theory.

8.4 ▪ Exercises

8.1 *Extension of Proposition 8.1*   Show that Proposition 8.1 extends to situations in which both A and B have uncertain payoffs, provided there are a finite number of possible distributions of payoffs for each and that each distribution over payoffs for A has a different mean expected payoff from each distribution of B. Argue that it then also holds with some finite number of action choices, under the same assumption of distinguished distributions.

8.2 *Failure of Action Consensus*   Argue that Proposition 8.1 fails to be true if there is some p_k for which $p_k = 1/2$. Argue that even in this case all agents must still have the same limiting payoff.

8.3 *Failure of Payoff Consensus*   (a) Argue that Proposition 8.1 fails to be true if different agents face different realizations of p.

(b) Argue that Proposition 8.1 fails to be true if agents face the same p as their neighbors at each point in time, but there is an (arbitrarily) small probability $\varepsilon > 0$ in each period that p changes and the p stays fixed until the next time it is randomly changed.

8.4 *Observational Learning in a Directed Network*   Argue that Proposition 8.1 holds if g is strongly connected and possibly directed.

8.5 *Observational Learning in a Directed Network*   Consider the setting of Proposition 8.1, with the following variation. There are two agents. Agent 1 observes agent 2's actions and payoffs over time, but not her own, and agent 2 observes agent 1's actions and payoffs over time, but not his own. Suppose that an agent chooses the action that maximizes his or her expected current period payoff given his or her updated beliefs conditional on his or her information. Show that the two agents eventually choose the same action, almost surely.

8.6 *Herding, Cascades, and Observational Learning*   Consider the following variation of the observational learning model, which is due to Banerjee [36] and Bikhchandani, Hirshleifer, and Welch [65]. Agents choose an action only once. Action A pays 1, while action B pays either 0 or 2 with equal probability. Agents choose sequentially and cannot communicate other than to observe one another's actions (but not payoffs). Agents see all of the previous actions before making a choice themselves. So, agent 1 makes a choice and gets a payoff. Then agent 2 sees agent 1's choice (but not payoff) and chooses an action and gets a payoff. Then agent 3, having seen the actions of 1 and 2, but not their payoffs, chooses an action and gets a payoff, and so forth. Action B is either "good" and pays 2 for all of the agents who choose it, or "bad" and pays 0 for all of the agents who choose it. Agents know that the chances that B is bad or good are equal. In addition, they each observe a private signal about the state of nature ("good" or "bad"). The signal of agent i, denoted by s_i, takes on a value "good" or "bad." If the true state is "good," then the signal is "good" with a probability p, where $1 > p > 1/2$, and similarly if the state is "bad," then the signal is "bad" with probability p. The signals are independent across the agents, conditional on the state. All agents have had a course in basic probability, and choose the action they think leads to the highest expected payoff conditional on all the information they have. If they think there is an equal chance of good or bad, then they flip a coin.

How does this process evolve? The first agent has only her own signal. Thus if she sees "good," then she thinks there is a p probability that the state is "good" and so chooses action B. If she sees "bad," then she thinks that there is a p probability that the state is "bad," and so chooses action A. Based on the actions of the first agent, all subsequent agents can deduce her signal. So consider the case in which the first agent chose B (the other case is analogous). If the second agent sees "good," then he has effectively seen two good signals and so chooses A. If the second agent sees "bad," then he has effectively seen one good and one bad signal and so chooses with equal probability. Note that the third agent cannot always be sure of the second agent's signal. If the third agent sees an action sequence B, A, then she can be sure that the signals were "good," "bad." But if the third agent sees B, B, she cannot be sure of what the second agent's signal was. Show that the third agent, conditional on seeing B, B, will ignore her own signal and choose B regardless. Show that this is true for all subsequent agents.

8.7 *Existence of a Strongly Connected Closed Group*   (a) Show that for any row-stochastic T, there always exists at least one closed and strongly connected set of nodes.

(b) Show that the society can be partitioned into some number of strongly connected and closed groups, and then a remaining set of agents who each have at least one directed path to an agent in a strongly connected and closed group.

8.8 *Convergence with Asynchronous Updating and Periodicity**   Consider the following variation on the DeGroot model. Each period is divided into n subperiods. In period t agents start with a vector of beliefs $p(t-1)$, and updating proceeds as follows. Agents update one at a time. Set $p(0, t) = p(t-1)$. Agent 1 updates first, and so a vector $p(1, t)$ is defined that only 1 has updated from $p(0, t)$. Next, agent 2 updates from $p(1, t)$ to form $p(2, t)$, and so on, so that i updates from $p(i-1, t)$ to form $p(i, t)$, and then $p(t) = p(n, t)$. Argue that beliefs converge to a consensus in any closed strongly connected component, regardless of whether it is aperiodic. (Hint: consider the highest and lowest beliefs at any time, and show that they must converge over time.)

8.9 *Nonconvergence with Time/Distance Varying Weights*   Provide an example with $n = 2$ of the model in Example 8.5 where $p(0) = (0, 1)$, and all entries of $T(p(t), t)$ are positive at all times, and beliefs converge, but such that a consensus is not reached. Why does this example fail to contradict Theorem 8.2?

8.10 *Equal Influence in an Updating Network*   Use $sT = s$ (and the result on the uniqueness of the influence vector in Section 8.3.5) to show that if $T_{ji} = T_{ij}$ for all i in a strongly connected and aperiodic society, then every agent has the same influence. Show this result is a consequence of the stronger claim (from Golub and Jackson [294]) that if $\sum_j T_{ji} = 1$ for all i, then every agent has the same influence.

8.11 *A Stubborn Agent in an Updating Network**   Start with a strongly connected T in the DeGroot model. Pick some agent i, and generate $T(\varepsilon)$ as follows. For any $j \neq i$ set $T_{jk}(\varepsilon) = T_{jk}$. For i, set $T_{ik}(\varepsilon) = \varepsilon T_{ik}$ for $k \neq i$ and $T_{ii}(\varepsilon) = 1 - \sum_{k \neq i} T_{ik}(\varepsilon)$. Thus as ε becomes small, the weight that i places on himself or herself increases, and the weight that i places on other agents is scaled down, while the weights of any agent other than i are left unchanged. Let $s_i(\varepsilon)$ be i's social influence. Show that as $\varepsilon \to 0$, i's social influence goes to 1. That is, show that $s_i(\varepsilon) \to 1$ and so the consensus beliefs in the society converge to i's initial belief as ε becomes small.

8.12 *Convergence Speed with Two Agents*   Consider the DeGroot model with two agents and

$$T = \begin{pmatrix} 1/2 & 1/2 \\ 1/4 & 3/4 \end{pmatrix}. \tag{8.14}$$

Find the social influence vector and the second eigenvalue. Using (8.14) and (8.7) find the exact expression for $p_1(t)$ when $p_1(0) = 1$ and $p_2(0) = 0$.

8.13 *Convergence Speed and Integration with Two Agents* Show that in the DeGroot model, when $n = 2$ and T is strongly connected, the coupling measure is exactly 1 minus the absolute value of the second eigenvalue.

8.14 *Necessary and Sufficient Conditions for Wise Crowds* Prove the following Lemma from Golub and Jackson [294].

Lemma 8.1 *If $(T_n)_{n=1}^{\infty}$ is a sequence of strongly connected and aperiodic updating matrices, then*

$$\lim \Pr\left[\left|\sum_i s_i^n p_i^n(0) - \mu\right| \geq \varepsilon\right] = 0$$

for all $\varepsilon > 0$ if and only if $\max_i s_i^n \to 0$.

For the "if" part, build from the proof of Theorem 4.2. Argue the "only if" part directly.

8.15 *Unwise Crowds: High Relative Weight* Show that agent 1's social influence is 1/2 in Example 8.13 for all n.

8.16 *Unwise Crowds: Indirect Weight* Show that agent 1's social influence is

$$s_i^n = \left(\frac{\delta}{1-\delta}\right)^{i-1} \cdot \frac{1 - \left(\frac{\delta}{1-\delta}\right)}{1 - \left(\frac{\delta}{1-\delta}\right)^{n+1}}$$

in Example 8.14.

8.17 *Wise Crowds* Consider the updating rules from Examples 8.13 and 8.14, and show whether each satisfies or violates the balance condition and the dispersion condition.

CHAPTER 9

Decisions, Behavior, and Games on Networks

Peers exert enormous influence on human behavior. It is easy to cite examples, ranging from which products we buy, whether we engage in criminal activities, how much education we pursue, to which profession we choose. There is a vast literature on the subject, including research by social psychologists, sociologists, researchers in education, and economists. We have already seen examples of studies that touch on these issues, and there is also a good bit known about statistical issues of identifying peer effects (see Chapter 13). The purpose of this chapter is to provide the foundations for understanding how the structure of social networks influences behavior. For example, if we change the network of social interactions, how does behavior change? This issue has a rich history in sociology and has more recently emerged in economics and computer science. Some aspects of this were touched on in Chapters 7 and 8 on diffusion and learning, but there are many situations in which social influences involve human decision making rather than pure contagion or updating. The focus in this chapter complements those earlier chapters by expanding the analysis to explicitly account for how individuals' decisions and strategies are influenced by those of their neighbors.

The main complication in the analysis of this chapter compared to the study of diffusion and learning is that behavior depends in more complicated ways on what neighbors are doing. For example, if an individual is choosing a piece of software or some other product and wants it to be compatible with a majority of neighbors, then this is a coordination game, and behavior can change abruptly, depending on how many neighbors are taking a certain action. It might also be that an individual only wants to buy a product or make an investment when his or her neighbors do not. These interactive considerations require game-theoretic reasoning, adapted and extended to a network setting.

The chapter begins by showing how a variation on the ideas encountered in the learning and diffusion analyses can be used to study behaviors. In particular, I start with a model in which people react to their neighbors in a way that can be captured probabilistically, predicting their actions as a function of the distribution of their neighbors' play. I then introduce game-theoretic settings, where richer behaviors are studied. There are two main types of situations considered that capture many of the applications of interest. One type deals with strategic complementarities, as in choosing compatible technologies or pursuing education, and players' incentives to take a given action increase as more neighbors take that action. This type leads to nice properties of equilibria, and we can deduce quite a bit about how players' strategies vary with their position in a network and how overall behavior in the society responds to network structural changes. The second type of strategic interaction considered has the opposite incentive structure: that of strategic substitutes. In it, players can "free ride" on the actions of their neighbors, such as in gathering information or providing certain services, and a player's incentive to take a given action decreases as more neighbors take that action. This dynamic leads to quite different conclusions about player behavior and equilibrium structure. I also discuss models that are designed to capture the dynamics of behavior.

9.1 • Decisions and Social Interaction

Let me begin by discussing an approach to modeling interactive behavior that builds on the tools from Chapters 7 and 8 on diffusion and learning. To model decisions in the face of social interaction, one needs to characterize how a given individual behaves as a function of the actions of his or her neighbors. For instance, will the individual want to buy a new product if some particular subset of his or her neighbors are buying the product? Generally, the answer to such a question depends on a series of characteristics of the individual, the product, the alternative products, and on the set of neighbors buying the product.[1] As a useful starting point, we can think of this process as being stochastic, with a probability of the individual taking a given action depending on the actions chosen by the neighbors.

9.1.1 A Markov Chain

To fix ideas and provide a benchmark, consider a setting in which interaction is symmetric in the sense that any individual can be influenced by any (or every) other, and the particular identities of neighbors are not important, just the relative numbers of agents taking various actions. Each individual chooses one of two actions. This choice might be between smoking or not, adopting one technology or another, going to one park or another, voting or not, voting for one candidate or another, and so on.

The two actions are labeled as 0 and 1, and time evolves in discrete periods, $t \in \{1, 2, \ldots\}$. The state of the system is described by the number of individuals who are taking action 1, denoted by s_t, at the end of period t.

---

1. The implications of interdependencies in consumer choices have been studied in a variety of contexts, including the implications for how firms might price goods (e.g., see Katz and Shapiro [382]) and implications for increasing returns (e.g., see Arthur [21] and Romer [566]). For more background on consumers' behavior and network externalities, see the survey by Economides [216].

In some applications, if the state of the system is $s_t = s$ at time t, then there is a well-defined probability that the system will be in state $s_{t+1} = s'$ at time $t + 1$. That is, if we know how many people are taking action 1 at date t, then there is a well-defined distribution over the number of people taking action 1 at date $t + 1$. This might be deterministic or random, depending on what is assumed about how people react to others. For example, each person might look at the state at the end of time t and then choose the action that leads to the greatest benefit for himself, presuming that others will act as at time t; for instance if we pick one person at random from the society and ask him to update his choice. All individuals may respond, or each one may respond probabilistically. What is critical is that there are well-defined probabilities of being in each state tomorrow as a function of the state today. Let Π be the $n \times n$ matrix describing these *transition probabilities* with the entry in row s and column s' being

$$\Pr(s_{t+1} = s' \mid s_t = s).$$

This results in a (finite-state) Markov chain (recalling definitions from Section 4.5.8). If the Markov chain is irreducible and aperiodic, then it has a steady-state distribution described by the vector $\mu = (\mu_0, \ldots, \mu_n)$, where μ_s is the probability of state s, or that s players choose action 1. Thus when behavior can be described by a Markov chain, we have sharp predictions about behavior over the (very) long run. Let us now examine some applications of this reasoning.

9.1.2 Individual-by-Individual Updating

Consider a setting in which at the beginning of a new period, one individual is picked uniformly at random[2] and updates her action based on the current number of people in the society taking action 1 or 0.[3] How an individual updates her action depends on the state of the system. In particular, let p_s denote the probability that the individual chooses action 1 conditional on s out of the other $n - 1$ agents choosing action 1. This form of updating could be a form of best-reply behavior, in which an individual (myopically) chooses an action that gives the highest payoff given the current actions of other individuals. So, for instance, the individual decides on whether to go / to stay at the beach based on how many others are there. It could also be something similar to the contagion of a disease (see Section 7.2), but now the transition between infection and susceptible is a richer function of the state of the system.

The transition probabilities from one state to another are completely determined by the vector $p = (p_0, \ldots, p_{n-1})$. For example the probability of going from s to $s + 1$ is the probability that one of the $n - s$ players choosing 0 out of the n

---

2. The method is not important, as one can use any method of picking the individual as long as it is Markovian. It could weight different individuals differently and be dependent on the state.

3. Note that the interpretation of dates here is flexible. Effectively, time simply keeps track of the moments at which some agent makes a decision and need not correspond to any sort of calendar time. The arrival process of when decisions are made can be quite general, with the main feature here being that only one agent is updating at any given moment.

players in total is selected to update, and then that player has a probability of p_s of selecting action 1, and so the probability is $\frac{n-s}{n} p_s$. The full list of transition probabilities is

$$\Pr\left(s_{t+1} = s + 1 \mid s_t = s\right) = \frac{n-s}{n} p_s, \text{ for } 0 \leq s \leq n - 1,$$

$$\Pr\left(s_{t+1} = s - 1 \mid s_t = s\right) = \frac{s}{n}\left(1 - p_{s-1}\right), \text{ for } 1 \leq s \leq n,$$

$$\Pr\left(s_{t+1} = s \mid s_t = s\right) = \frac{n-s}{n}\left(1 - p_s\right) + \frac{s}{n} p_{s-1}, \text{ for } 0 \leq s \leq n,$$

$$\Pr\left(s_{t+1} = s' \mid s_t = s\right) = 0 \text{ if } s' \notin \{s - 1, s, s + 1\}.$$

This Markov chain has several nice properties. Provided that $1 > p_s > 0$ for each s, any state is eventually reachable from any other state, and so the Markov chain is irreducible. Moreover, there is a chance of staying in any state, and so the Markov chain is aperiodic. Thus it has a unique steady-state distribution over states.

In this setting, the Markov chain takes a particularly simple form as only one individual is changing actions at a time. At steady state, the probability of ending in state s is simply the probability of being in an adjacent state $s - 1$ or $s + 1$ and then getting one more or less individual to choose action 1, or else starting at s and staying there. Thus

$$\mu_0 = \mu_0(1 - p_0) + \mu_1\left(\frac{1}{n}\right)(1 - p_0),$$

$$\mu_s = \mu_{s-1}\left(\frac{n - (s - 1)}{n}\right) p_{s-1} + \mu_{s+1}\left(\frac{s + 1}{n}\right)(1 - p_s)$$

$$+ \mu_s\left(\frac{n - s}{n}\right)(1 - p_s) + \mu_s\left(\frac{s}{n}\right) p_{s-1},$$

$$\mu_n = \mu_{n-1}\left(\frac{1}{n}\right) p_{n-1} + \mu_n p_{n-1}.$$

Solving this system leads to

$$\frac{\mu_{s+1}}{\mu_s} = \left(\frac{n - s}{s + 1}\right)\left(\frac{p_s}{1 - p_s}\right) \tag{9.1}$$

for all $0 \leq s \leq n - 1$; which, together with $\sum_s \mu_s = 1$, completely determines the solution.

Ants, Investment, and Imitation   To see an application of such a Markov model of behavior, consider a special case due to Kirman [397] in which individuals imitate one another. In particular, Kirman's work was motivated by the observation that ants tend to herd on the food sources that they exploit, even when faced with equally useful sources. Moreover, ants also switch which source they exploit more intensively. Kirman discusses similar patterns of behavior in human

investment and other forms of imitation. He models this behavior by considering a
dynamic such that at each time interval an individual is selected uniformly at random from the population. With a probability ε the individual flips a coin to choose
action 0 or 1, and with a probability $1 - \varepsilon$ the individual selects another individual
uniformly at random and mimics his or her action.

That model is a special case of the framework above, in which[4]

$$p_s = \frac{\varepsilon}{2} + (1 - \varepsilon) \, \frac{s}{n - 1}.$$

Then from (9.1) it follows that

$$\frac{\mu_{s+1}}{\mu_s} = \frac{n - s}{s + 1} \left(\frac{(n - 1)\varepsilon + 2s \, (1 - \varepsilon)}{2(n - 1) - (n - 1)\varepsilon - 2s \, (1 - \varepsilon)} \right). \qquad (9.2)$$

Given the symmetry of this setting, the long-run probability (starting from a
random draw of actions) of any individual choosing either action is 1/2. If $\varepsilon = 0$,
then this system is one of pure imitation, and eventually the system is absorbed
into the state 0 or n and then stays there forever. If $\varepsilon = 1$ then the individuals simply
flip coins to choose their actions irrespective of the rest of society. Then

$$\frac{\mu_{s+1}}{\mu_s} = \frac{n - s}{s + 1}$$

and the system has a pure binomial distribution with a parameter of $1/2$, so that the
probability of having exactly s individuals choosing action 1 at a given date in the
long run is $\mu_s = \binom{n}{s} \left(\frac{1}{2} \right)^n$. In that case, the extreme states are less likely, and the
most likely state is that half the society is choosing action 1 and half is choosing
action 0.[5]

To obtain a uniform distribution across social states it must be that $\mu_{s+1} = \mu_s$ for
each s. It is easily checked from (9.2) that this holds when $\varepsilon = 2/(n + 1)$.[6] To model
the behavior of the ants with tendencies to herd on action choice, the high and low
states need to be more likely than the middle states. This requires that individuals
pay more attention to others than in the situation just examined, in which the
distribution across states is uniform. Thus the probability that an individual ignores
society and flips a coin must be $\varepsilon < 2/(n + 1)$. Then behavior is sensitive enough
to the state of the society so that the more extreme states, in which all individuals
follow one action or the other ($s = 0$ and $s = n$), are more heavily weighted. This
distribution is pictured in Figure 9.1 for several different values of ε. Thus, to

---

4. Kirman also allows for the probability that an individual does not change his or her action, but
that is equivalent in terms of long-run distributions, as it simply slows the system down and can
be thought of as changing the length of the time period.

5. Note that each configuration of choices across individuals is equally likely, but there are many
more configurations (keeping track of players' labels) in which half the society is choosing 1 and
half 0 than there are configurations where all are choosing 0.

6. This is slightly different from Kirman's expression since ε here is the chance that an agent
flips a coin, whereas it is the chance that an agent changes actions in Kirman's labeling.

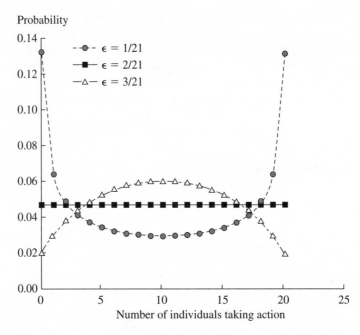

FIGURE 9.1   Kirman's [396] ant-imitation model for three levels of random behavior.

model herding, one needs to have a sufficiently high probability of imitating other agents so that the natural tendency toward even mixing is overturned.

Sensitivity to Societal Action and Herding   As I now show, Kirman's result about the relationship between social sensitivity and herding can be formalized and is true for more general processes than the pure imitation process described above. To see this, consider a class of processes that treat actions 0 and 1 symmetrically. That is, suppose that $p_s = 1 - p_{n-1-s}$, so that the chance of choosing action 1 conditional on s out of the other $n - 1$ agents choosing action 1 is the same as the chance of choosing action 0 conditional on s out of the other $n - 1$ agents choosing action 0.

Say that a social system $p' = (p'_0, \ldots, p'_n)$ is *more socially sensitive* than another social system $p = (p_0, \ldots, p_n)$ if $p'_s \geq p_s$ when $s > (n - 1)/2$ and $p'_s \leq p_s$ when $s < (n - 1)/2$, with at least one strict inequality. Thus an individual in a society described by p' is more likely to choose the same action as the majority of the population than an individual in a society described by p. Note that this definition does not require that an individual want to match the majority of a society. It admits processes in which $p'_s < 1/2$ when $s > n/2$ so that an individual is actually choosing against the current of the society. The social sensitivity comparison is a relative comparison between two societies.

Proposition 9.1   *If two different societies, described by p and p', each treat actions 0 and 1 symmetrically and the process p' is more socially sensitive than p, then the steady-state distribution over numbers of agents taking action 1,*

$(\mu'_0, \ldots, \mu'_n)$ *corresponding to p', is a mean-preserving spread of the steady-state distribution* $(\mu_0, \ldots, \mu_n)$ *corresponding to p. In fact, there exists* $\bar{s} > n/2$ *such that* $\mu'_s > \mu_s$ *if and only if* $s > \bar{s}$ *or* $s < n - \bar{s}$.

The proof of Proposition 9.1 can be deduced from (9.1) and is left as Exercise 9.2. This result is not surprising, as it states that increasing the extent to which an individual's choice of action matches that of a majority of the society then increases the extent to which the society tends to extremes, in terms of spending more time in states in which higher concentrations of individuals choose the same action. Nevertheless, such a result provides insight as to the type of behavior that leads to herding (or lack thereof). This is a different sort of herding than the consensus formation we saw in our discussion of learning in Chapter 8,[7] as the society here oscillates between the actions over time, but with specific patterns of either herding to one action or the other, or else splitting among the two.[8]

Calvó-Armengol and Jackson [130] use such a model to study social mobility. Action 1 can be interpreted as pursuing higher education, and the n individuals as being the families in some community. When a family is randomly selected to make a new choice, it is interpreted as having a child in the family replacing the parent and making a choice. Social mobility patterns are determined by how likely it is that a child makes the same choice that his or her parent did. Provided p_s is nondecreasing in s, a parent's and child's decisions are positively correlated. The idea is that the child is most likely to choose action 1 when there are many others in the community who have chosen action 1, which is relatively more likely to happen if the parent also chose action 1. Thus parent and child actions are correlated even though there is no direct link between the two, only the fact that the surroundings that influence their decisions overlap. Thus this explanation of social mobility is complementary (no pun intended) to that of direct parent-child interaction.

For example, the probability that the parent and child both choose action 1 is $\sum_s \mu_s \frac{s}{n} p_{s-1}$. This comes from summing across states the probability of state s, μ_s, times the probability that the parent is a 1, $\frac{s}{n}$, times the probability that the child also chooses action 1, p_{s-1} (as there are $s - 1$ others choosing action 1, given that the parent was choosing 1).

As a quick demonstration that such a model can easily generate outcomes consistent with observed patterns, Calvó-Armengol and Jackson [131] restrict attention to a special case for which there are just two parameters that govern the Markov chain. For this case $p_s = q$ for some $1 > q > 0$ when $s \geq \tau$, and $p_s = 1 - q$ when $s < \tau$, where $\tau \in \{0, \ldots, n - 1\}$ is a threshold. Thus individuals choose action 1 with probability q if at least τ others have, and with probability $1 - q$ otherwise (see Exercise 9.3).

Table 9.1 shows a few representative observations of father-daughter education decisions from the Calvó-Armengol and Jackson [131] fitting exercise (see their

---

7. This is also distinct from the herding literature of Banerjee [36]; Bikhchandani, Hirshleifer, and Welch [65]; and others, who examine how situations with uncertainty and private information about the benefits of actions can lead individuals to follow a herd of others in choosing an action (see Exercise 8.6).

8. For example, if people wish to avoid congestion, so that they are likely to choose the action chosen by the minority, then the most likely state becomes that of $s = n/2$.

TABLE 9.1
Father/daughter education choices for selected
European countries

	Data			Estimation		
AU	0	1			0	1
0	.903	.033		0	.902	.048
1	.054	.008		1	.048	.003
				$q = .95, \tau = 14$		
GR	0	1			0	1
0	.646	.192		0	.648	.157
1	.124	.038		1	.157	.038
				$q = .81, \tau = 15$		
UK	0	1			0	1
0	.246	.223		0	.230	.250
1	.254	.278		1	.250	.270
				$q = .52, \tau = 25$		

Note: AU, Austria; GR, Greece; UK, United
Kingdom.

supplementary material). The data are the relative frequencies of observations of
father and daughter education choices based on wave 5 of the European Community Household Panel data set. A value of 0 represents that the individual at most
graduated from high school; while 1 represents pursuing some education beyond
high school. The row choice of 0 or 1 is the father's choice, while the column is
the daugher's.

The second column of the table displays the probabilities from the simple
threshold model when the community size is 25 families, with the best fit q and τ
(from a search on a grid of q to hundredths, and across each τ). The fitted values
do not match exactly; for instance in the fitted matrices the probability of 0,1 is
the same as 1,0, which is a function of the simplified threshold model. Allowing
for richer choices of the p_s values would provide a better fit. This example shows
that such simple models can lead to patterns that are quite close to observed ones,
and that social interaction is a viable part of explaining parent-child correlations,
among other things.

9.1.3 An Interaction Model with Network Structure

While the simple Markov model of social interactions discussed in the previous
section provides insight into broad patterns of social behavior, it does not incorporate the micro-details of who interacts with whom. Such network relations can have
a profound effect on the process. To incorporate networked interactions, we need

a richer structure. Consider the following process, which allows us to incorporate a (possibly weighted and directed) network.

As before, individuals choose between two actions 0 and 1; now the social state needs to keep track of which agents are taking which actions. The social state is thus an n-dimensional vector $x(t)$, where $x_i(t)$ for $i \in \{1, \ldots, n\}$ is the action that agent i took at period t. Interaction is described by w, which is an $n \times n$-dimensional matrix, where entry $w_{ij} \in [0, 1]$ is a weight that describes the probability that individual i's choice in period $t + 1$ is the action that j took in period t. This matrix is row stochastic.

In addition, let us allow for a probability $\varepsilon_i(1)$, that i chooses action 1 independently of the state of the system and a probability $\varepsilon_i(0)$, that i chooses action 1 independently of the state of the system. Letting $\Pr(x_i(t + 1) = 1 | x(t))$ denote the probability that $x_i(t + 1) = 1$ given the state at time t is described by the vector $x(t)$, it follows that

$$\Pr\left(x_i(t + 1) = 1 | x(t)\right) = \varepsilon_i(1) + (1 - \varepsilon_i(1) - \varepsilon_i(0)) \sum_j w_{ij} x_j(t).$$

Allowing for $w_{ii} > 0$, we can encode the possibility that the agent does not update his or her action at all. The Kirman ants model considered in Section 9.1.2 is the special case for which $w_{ij} = 1/(n - 1)$ for $j \neq i$ and $j \leq n$, and $\varepsilon_i(1) = \varepsilon_i(0) = \varepsilon/2$.

Much richer models than the ants model can now be encoded. To get a feel for the variables, consider the following example. With probability $1/4$, individual i sticks with his or her previous action; with probability $1/4$, i follows the action of agent $i - 1$; with a probability $1/4$, i follows the action of agent $i + 1$, and with probability $1/4$, i randomizes between 0 and 1. This strategy corresponds to $\varepsilon_i(1) = \varepsilon_i(0) = 1/8$ and then having equal $(1/3)$ weights on i, $i + 1$, and $i - 1$ (mod n):

$$w = \begin{pmatrix} 1/3 & 1/3 & 0 & 0 & \ldots & 0 & 0 & 1/3 \\ 1/3 & 1/3 & 1/3 & 0 & \ldots & 0 & 0 & 0 \\ 0 & 1/3 & 1/3 & 1/3 & \ldots & 0 & 0 & 0 \\ & & \vdots & & \ddots & & \vdots & \\ 0 & 0 & 0 & 0 & \ldots & 1/3 & 1/3 & 1/3 \\ 1/3 & 0 & 0 & 0 & \ldots & 0 & 1/3 & 1/3 \end{pmatrix}.$$

This framework allows an individual's choice of actions to depend on arbitrary neighborhoods of others, placing varying weights on different agents. Correspondingly, the process becomes more cumbersome to deal with, as we now need to keep track of each agent's social state in each period, rather than just an aggregate statistic. Nonetheless, this is still a well-defined finite-state Markov chain, which allows us to deduce quite a bit about actions over time. First, in situations in which the system is irreducible and aperiodic, a steady-state distribution exists (see Section 4.5.8). Second, although the steady-state distribution is a potentially complex joint distribution on the full vector of agents' actions, it is easy to deduce the steady-state probability that any given agent takes action 1.

We determine this probability by using the following trick to encode the ε_i-based choices of the individuals. Expand the society to include two fictitious extra individuals, labeled $n + 1$ and $n + 2$. Agent $n + 1$ always takes action 0, and agent $n + 2$ always takes action 1. Now, we can encode an individual's actions entirely in terms of an $(n + 2) \times (n + 2)$ matrix W. This is done by setting $W_{i,n+1} = \varepsilon_i(0)$, $W_{i,n+2} = \varepsilon_i(1)$, and $W_{ij} = (1 - \varepsilon_i(0) - \varepsilon_i(1))w_{ij}$ for $j \leq n$, and weights 1 on themselves for the extra individuals.

Then let $X(t) = (x_1(t), \ldots, x_n(t), 0, 1)$ be the $(n + 2) \times 1$ vector representing the larger state space including the extra "constant" individuals' actions. Then

$$\Pr(X_i(t + 1) = 1 | X(t)) = [W X(t)]_i .$$

More importantly, for any $t' > t$, it follows that

$$\Pr(X_i(t') = 1 | X(t)) = \left[W^{t'-t} X(t) \right]_i , \qquad (9.3)$$

where $W^{t'-t}$ is the matrix W raised to the $t' - t$ power. In many cases, the vector of probabilities in (9.3) converges to a steady-state probability from any starting state. For example, if there is a directed path from each i to at least one of $n + 1$ and $n + 2$, then the limit is unique and well behaved.

Proposition 9.2 *If for each individual i either $\varepsilon_i(0) + \varepsilon_i(1) > 0$ or there is a directed path[9] from i to some j for whom $\varepsilon_j(0) + \varepsilon_j(1) > 0$, then there is a unique limiting probability that i chooses action 1, and this limiting probability is independent of the starting state. Moreover, the vector of limit probabilities is the unique (right-hand) unit eigenvector of W such that the last two entries ($n + 1$ and $n + 2$) are 0 and 1.*

Proof of Proposition 9.2. The convergence of W^t to a unique limit follows from Theorem 2 in Golub and Jackson [294]. Given that there is a directed path from each i to at least one of $n + 1$ and $n + 2$, and each of these places weight 1 on itself, the only minimal closed sets (minimal directed components with no directed links out) are the nodes $n + 1$ and $n + 2$ viewed as separate components. It then follows from Theorem 3 in Golub and Jackson [294] that $\lim_t W^t X(0)$ is of the form $W^\infty X(0)$, where

$$W^\infty = \begin{pmatrix} 0 & 0 & 0 \ldots & 0 & \gamma_1 & \pi_1 \\ 0 & 0 & 0 \ldots & 0 & \gamma_2 & \pi_2 \\ & \vdots & \ddots & & \vdots & \\ 0 & 0 & 0 \ldots & 0 & \gamma_i & \pi_i \\ & \vdots & \ddots & & \vdots & \\ 0 & 0 & 0 \ldots & 0 & \gamma_n & \pi_n \\ 0 & 0 & 0 \ldots & 0 & 1 & 0 \\ 0 & 0 & 0 \ldots & 0 & 0 & 1 \end{pmatrix} .$$

---

9. This path refers to a directed path in the directed graph on the original n individuals, where a link ij is present if $w_{ij} > 0$.

It also must be that $WW^\infty = W$. Thus $W\pi = \pi$, where $\pi = (\pi_1, \ldots, \pi_n, 0, 1)$, which implies that π is a right-hand unit eigenvector of w. ▪

To get a feel for Proposition 9.2, let us consider an application.

Example 9.1 (An Application: Probability of Action in Krackhardt's Advice Network) *We revisit Example 8.11, which concerned Krackhardt's [411] data on an advice network among managers in a small manufacturing firm on the West Coast of the United States. Suppose that the action is whether to meet to go to a bar after work. This happens repeatedly, and each day the managers find out who went the previous night. Consider the following action matrix based on Krackhardt's data. A manager with out-degree d chooses to go to the bar with probability $1/(d+2)$, not to go with probability $1/(d+2)$, and with the remaining probability of $d/(d+2)$ uniformly at random picks one of his or her neighbors and then goes to the bar if that neighbor did on the previous day. This strategy is used by all managers except the top-level ones (labeled 2, 7, 14, 18, and 21 in Example 8.11) who are biased toward going to the bar. The top-level managers have a similar rule except that they use weights $1/(d+1)$ and do not place any weight on action 0. We can then calculate the frequency with which each manager goes to the bar in the long run, as listed in Table 9.2. The probabilities were calculated via Matlab by finding the (right-hand) unit eigenvector of W.*

It is important to emphasize that Table 9.2 does not give us the joint distribution over people going to the bar. It is easy to see that there will be correlation and long periods of time when many people go to the bar, and then long periods of time where few people go. Moreover, even this myopic sort of behavior can lead individuals to coordinate. The technique of calculating the marginals, however, only allows us to see the individual probabilities and not the joint distribution. The joint distribution is over 2^{21} (more than 2 million) different states, and so it is a bit difficult to keep track of.

Beyond the difficulties in tracking the full joint distribution, there are two other limitations to this analysis, which motivates a game-theoretic analysis. The first is that the individuals are backward looking. That is, they look at what their neighbors did yesterday in deciding whether to go to the bar, rather than coordinating with their neighbors on whether they plan to go to the bar today. Second, agent i weights the actions of the others in a separable way. Alternatively, an individual might prefer to go when a larger group is going (or might, in contrast, want to avoid congestion and stay away on crowded nights). This sort of decision is precluded by the separability in the way that an individual i treats the actions of the others. In particular, it is that aspect of the framework above that allows us to solve for the probabilities of action. To see this explicitly, consider an example with $n = 3$ people. Suppose that individual 1 has $w_{12} = w_{13} = 1/2$ and $\varepsilon_1(0) = \varepsilon_1(1) = 0$. Consider how individual 1 will play in period 1, as a function of how we choose the starting state. In one case, pick the initial choices of individuals 2 and 3 in an independent manner, selecting 1 with probability p. In that case, individual 1 will choose 1 with probability p. But pick the initial actions of individuals 2 and 3 to be the same, so with probability p they are both 1, and with probability $1 - p$ they are both 0. This choice does not make any difference in calculating the probability that individual 1 chooses action 1 in the first period. It is still p. The joint distribution

TABLE 9.2
Probability of action in Krackhardt's network of advice among managers

Agent	Probability of taking action 1	Level	Department	Age (years)	Tenure (years)
1	0.667	3	4	33	9.3
2	0.842	2	4	42	19.6
3	0.690	3	2	40	12.8
4	0.666	3	4	33	7.5
5	0.690	3	2	32	3.3
6	0.585	3	1	59	28
7	0.771	1	0	55	30
8	0.676	3	1	34	11.3
9	0.681	3	2	62	5.4
10	0.660	3	3	37	9.3
11	0.656	3	3	46	27
12	0.585	3	1	34	8.9
13	0.680	3	2	48	0.3
14	0.821	2	2	43	10.4
15	0.687	3	2	40	8.4
16	0.651	3	4	27	4.7
17	0.671	3	1	30	12.4
18	0.737	2	3	33	9.1
19	0.685	3	2	32	4.8
20	0.685	3	2	38	11.7
21	0.755	2	1	36	12.5

Note: Level indicates the level of hierarchy in the firm (e.g., 1 denotes the highest, 3 the lowest level).

over all players' actions will change, but the probability that any given individual chooses action 1 is unaffected.

This special property that leads to such power in calculating the steady-state probabilities of actions is not always satisfied. In fact, many applications of interest have a more complex structure to the incentives. For example, suppose that individual 1 would like to choose 1 if both of the others choose action 1, but not otherwise. These incentives would arise if action 1 has a cost associated with it (slightly higher than the cost of taking action 0), but agent 1 wants to choose an action that is compatible with as many other agents as possible. Then if we pick 2's and 3's actions independently, there is only a p^2 chance that both 2 and 3 will choose action 1 and that individual 1 will choose action 1 in period 1. If instead we pick 2's and 3's actions to be the same, then there is a p chance that they will both choose action 1 and that individual 1 will then choose action 1 in period 1.

Although we can still write the system as a Markov chain, it is not quite as powerful a tool now, as the transition probabilities and evolution of behavior are more complicated. To get a better handle on such more complicated interaction structures, let us turn to game-theoretic reasoning.

9.2 ▪ Graphical Games

Individual decisions often depend on the relative proportions of neighbors taking actions, as in deciding on whether to buy a product, change technologies, learn a language, smoke, engage in criminal behavior, and so forth. This can result in multiple equilibrium points: for instance, some people may be willing to adopt a new technology only if others do, and so it would be possible for nobody to adopt it, or for some nontrivial fraction to adopt it. A way of introducing such reactive or strategic behavior into the analysis of social interaction is to model the interaction as a game.

A useful class of such interactions was introduced by Kearns, Littman, and Singh [385] as what they called *graphical games*.[10]

More formally, there is a set N of players, with cardinality n, who are connected by a network (N, g). Each player $i \in N$ takes an action in $\{0, 1\}$. The payoff of player i when the profile of actions is $x = (x_1, \ldots, x_n)$ is given by:

$$u_i(x_i, x_{N_i(g)}),$$

where $x_{N_i(g)}$ is the profile of actions taken by the neighbors of i in the network g.

There is nothing about this definition that precludes the network from being directed. For instance, player i may care about how player j acts, but not the reverse. Most of the definitions that follow work equally well for directed and undirected cases. I note points at which the analysis is special in the directed case. Most of the examples examine the undirected case for ease of exposition.

In a graphical game, players' payoffs depend on the actions taken by their neighbors in the network. Nevertheless, a player's behavior is related to that of indirect neighbors, since a player's neighbors' behavior is influenced by their neighbors, and so forth, and equilibrium conditions tie together all the behaviors in the network. Note that we can view this formulation as being without loss of generality, because we can define the network to include links to all of the players that affect a given player's payoff. As an extreme case, if a player cares about everyone's behavior, we have the complete network, which is a standard game with n players.

---

10. These can also be viewed as a special case of multi-agent influence diagrams, often referred to as MAIDs, which are discussed by Koller and Milch [408]. Earlier discussion of the possibility of using MAIDs to model strategic interactions in which players only respond to other subsets of agents' actions dates to Shachter [590], but the first fuller analysis appears in Koller and Milch [408]. Even though the general MAID approach allows for the encoding of multiple types of problems, information structures, and complex interaction structures, and hence includes graphical games as a special case, most of the analysis examines special cases that preclude the graphical games structure discussed here.

9.2.1 Examples of Graphical Games

To fix ideas, consider a couple of examples.

Example 9.2 (Threshold Games of Complements) *Many of the applications mentioned so far involve strategic complements, such that a player has an increasing incentive to take a given action as more neighbors take the action. In particular, consider situations in which the benefit to a player from taking action 1 compared to action 0 (weakly) increases with the number of neighbors who choose action 1, so that*

$$u_i(1, x_{N_i(g)}) \geq u_i(0, x_{N_i(g)}) \text{ if and only if } \sum_{j \in N_i(g)} x_j \geq t_i,$$

where t_i is a threshold. In particular, if more than t_i neighbors choose action 1, then player i should choose 1, and if fewer than t_i neighbors choose action 1 then it is better for player i to choose action 0.

A special case occurs if action 1 is costly (e.g., investing in a new technology or product) but the benefit of that action increases as more neighbors undertake the action:

$$u_i(1, x_{N_i(g)}) = a_i \left(\sum_{j \in N_i(g)} x_j \right) - c_i,$$
$$u_i(0, x_{N_i(g)}) = 0,$$

for some $a_i > 0$ and $c_i > 0$. Here the threshold is such that if at least $t_i = \frac{c_i}{a_i}$ neighbors choose action 1, then it is better for player i to choose 1, and otherwise player i should choose action 0.

Example 9.3 (A "Best-Shot" Public Goods Game) *Another case of interest has the opposite incentive structure. For example, if a player or any of the player's neighbors take action 1, then the player obtains a benefit of 1. For instance, the action might be learning how to do something, where that information is readily communicated; or buying a book or other product that is easily lent from one player to another.[11] Taking action 1 is costly, and a player would prefer that a neighbor take the action rather than having to do it himself or herself; but taking the action and paying the cost is better than having nobody take the action. This scenario is known as a best-shot public goods game (e.g., see Hirshleifer [329]), where*

$$u_i(1, x_{N_i(g)}) = 1 - c.$$
$$u_i(0, x_{N_i(g)}) = 1 \quad \text{if } x_j = 1 \text{ for some } j \in N_i(g), \text{ and}$$
$$u_i(0, x_{N_i(g)}) = 0, \quad \text{if } x_j = 0 \text{ for all } j \in N_i(g),$$

where $1 > c > 0$.

---

11. The distinction between private and public goods is a standard one in economics. The term *public good* refers to the fact that one player might acquire the good, information or the like, that can be consumed by others and hence is not private to that player but is publicly available. The term *local* refers to the benefits of a given player's action being public only in the player's neighborhood.

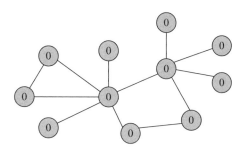

FIGURE 9.2   An equilibrium in a game of complements with threshold 2.

9.2.2 Equilibrium

Given the graphical game structure, we can use game theory to make predictions about players' behavior and how it depends on the network structure. I strongly recommend that those not familiar with game theory read Section 9.9, which provides a quick and basic tutorial in game theory, before proceeding with the remainder of this chapter.

In a graphical game, a *pure strategy Nash equilibrium* is a profile of strategies $x = (x_1, \ldots, x_n)$ such that

$$u_i(1, x_{N_i(g)}) \geq u_i(0, x_{N_i(g)}) \quad \text{if } x_i = 1, \text{ and}$$
$$u_i(0, x_{N_i(g)}) \geq u_i(1, x_{N_i(g)}) \quad \text{if } x_i = 0.$$

So the equilibrium condition requires that each player choose the action that offers the highest payoff in response to the actions of his or her neighbors: no player should regret the choice that he or she has made given the actions taken by other players.

Figures 9.2–9.4 show some pure strategy equilibria for the threshold game of complements outlined in Example 9.2 when the threshold is 2 for all players and the network is undirected. That is, a player prefers to buy a product if at least two of his or her neighbors do, but prefers not to otherwise. Note that the case pictured in Figure 9.2, in which all players take action 0, is an equilibrium for any game in which all players have a threshold of at least 1 for taking action 1.

There is generally a multiplicity of equilibria in such threshold games. For instance, the configuration pictured in Figure 9.3 is also an equilibrium. An equilibrium also occurs when a maximal configuration of players takes action 1. The configuration pictured in Figure 9.4 shows each player taking the maximal action that he or she can in any equilibrium. While these figures show that multiple equilibria can exist in graphical games, it is also possible for none to exist. This is illustrated in Example 9.4.

Example 9.4 (A Fashion Game and Nonexistence of Pure Strategy Equilibria)   *Consider a graphical game in which there are two types of players. One type consists of "conformists" and the others are "rebels." Conformists wish to take an action that matches the majority of their neighbors, while rebels prefer to take an action that matches the minority of their neighbors. This game is a variation on a classic one called "matching pennies," in which one player wishes to choose*

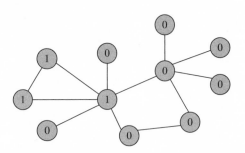

FIGURE 9.3   Another equilibrium in a game of complements with threshold 2.

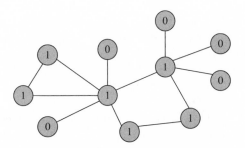

FIGURE 9.4   Maximal equilibrium in a game of complements with threshold 2.

the same action as the other, while the second player wishes to mismatch (see Section 9.4). It is easy to check that pairing one conformist and one rebel in a dyad results in no pure strategy equilibrium, as in the game of matching pennies. More generally, some graphical game structures have no pure strategy equilibrium. One class of networks with pure strategy equilibria is that in which all players have more than half of their neighbors being conformists: there is a pure strategy equilibrium where all the conformists take one action, and all the rebels take the other.

In light of Example 9.4, it is useful to define mixed strategy equilibria for a graphical game. Denote a mixed strategy of i by σ_i, where $\sigma_i \in [0, 1]$ is the probability that player i chooses $x_i = 1$ and $1 - \sigma_i$ is the probability that the player chooses $x_i = 0$. Let σ_{-i} denote a profile of mixed strategies of the players other than i, and let $u_i(x_i, \sigma_{N_i(g)})$ denote the expected utility of player i who plays x_i and whose neighbors play $\sigma_{N_i(g)}$. Let $u_i(\sigma_i, \sigma_{N_i(g)})$ be the corresponding expected utility when i plays a mixture σ_i.[12] Then a mixed strategy equilibrium is a profile

---

12. Thus

$$u_i(x_i, \sigma_{N_i(g)}) = \sum_{x_{N_i(g)} \in \{0,1\}^{d_i(g)}} u_i(x_i, x_{N_i(g)}) \Pr(x_{N_i(g)} | \sigma_{N_i(g)}),$$

and

$$u_i(\sigma_i, \sigma_{N_i(g)}) = \sigma_i u_i(1, \sigma_{N_i(g)}) + (1 - \sigma_i) u_i(0, \sigma_{N_i(g)}).$$

$\sigma = (\sigma_1, \ldots, \sigma_n)$ of mixed strategies such that for every i,

$$u_i(\sigma_i, \sigma_{N_i(g)}) = \max \left[u_i(1, \sigma_{N_i(g)}), u_i(0, \sigma_{N_i(g)}) \right].$$

As graphical games have a finite set of strategies and players, they always have at least one equilibrium, which may be in mixed strategies (see Section 9.9.4). Many graphical game settings always have pure strategy equilibria, including the examples of best-shot and threshold games mentioned above. But we have also seen some natural situations that will only have mixed strategy equilibria, such as the fashion game (Example 9.4).

Let us turn to a class of graphical games that covers many applications of interest and also is nicely behaved, allowing for a tractable analysis of how social structure relates to behavior.

9.3 · Semi-Anonymous Graphical Games

There are many situations in which a player's choice is influenced mainly by the relative popularity of a given action among his or her neighbors and is not dependent on the specific identities of the neighbors who take the action. While an approximation, this type of game can be very useful and has many natural applications. A class of such graphical games is examined by Galeotti et al. [274], and I refer to these as semi-anonymous graphical games.

These games are not quite anonymous (e.g., see Kalai [375]), in which a player is affected by the actions of all other players in a symmetric way; since in a graphical game a player cares only about a subset of the other players' actions. But it is anonymous in the way that a player is influenced by his or her neighbors. That is, the player cares only about how many of the neighbors take action 0 versus 1, but not precisely which of the neighbors take action 1 versus action 0. So semi-anonymity refers to this anonymity on a neighborly level. In addition, another aspect of anonymity is invoked here: players have similar payoff functions, so that differences between players arise from the network structure and not some other innate characteristics. A player's utility function depends on his or her degree and not on his or her label.

9.3.1 Payoffs and Examples

Formally, a *semi-anonymous graphical game* is a graphical game such that the payoff to player i with a degree d_i who chooses action x_i is described by a function $u_{d_i}(x_i, m)$, where m is the number of players in $N_i(g)$ taking action 1.

Thus the payoff function is dependent on the player's degree, the player's own action, and the number of neighbors who take each action. Note that since the function depends on both the degree d_i and the number of neighbors choosing action 1, m, we could equivalently have defined it to be a function of the degree and the number of players taking action 0 (which is simply $d_i - m$) or of the degree and the fraction of players taking action 1 (or 0). Note also that the best-shot game of Example 9.3 is a semi-anonymous graphical game, as is the fashion game in Example 9.4. The threshold games in Example 9.2 are semi-anonymous when each player's threshold depends only on his or her degree. Here are some other examples.

Example 9.5 (A Local Public Goods Game) *Consider a game in which each player's action contributes to some local public good: that is, an action by a given player provides some local benefits to all neighbors. This example generalizes the best-shot public goods graphical game to allow for situations in which having multiple players take action 1 is better than having just one player take the action. For example, having each player study a given candidate's record and then share that information with his or her neighbors could lead players to be more informed about how to vote in an election than having just one player study a candidate's record. In this case, a player who has m neighbors take action 1 gets a payoff of*

$$u_{d_i}(x_i, m) = f(x_i + \lambda m) - cx_i,$$

where f is a nondecreasing function, and $\lambda > 0$ and $c > 0$ are scalars. The case where $\lambda = 1$, $f(k) = 1$ for all $k \geq 1$, and $f(0) = 0$ is the best-shot public good graphical game.

Example 9.6 (A "Couples" Game) *Imagine learning a skill that is most easily enjoyed when there is at least one friend to practice it with, such as playing tennis, some video games, or gin rummy. In this situation, a player prefers to take action 1 if at least one neighbor takes action 1, but prefers to take action 0 otherwise. We can think of this as having a cost of investing in the skill of c, along with a benefit of 1 if there is a partner to participate with. Here*

$$\begin{aligned} u_{d_i}(1, m) &= 1 - c \quad \text{if } m \geq 1, \\ u_{d_i}(1, 0) &= -c, \quad \text{and} \\ u_{d_i}(0, m) &= 0. \end{aligned}$$

This game is a special case of a threshold game of complements with a threshold of 1.

Example 9.7 (A Coordination Game) *Consider a situation in which a player prefers to coordinate his or her action with other players, and his or her payoff is related to the fraction of neighbors who play the same strategy:*

$$\begin{aligned} u_{d_i}(1, m) &= a\frac{m}{d_i}, \quad \text{and} \\ u_{d_i}(0, m) &= b\frac{d_i - m}{d_i}. \end{aligned}$$

This is a special case of a threshold game of complements, with the threshold for player i of degree d_i being $t_i = d_i \frac{b}{a+b}$.

9.3.2 Complements and Substitutes

In the examples, as in many applications, the incentives of a player to take an action either increase or decrease as other players take the action. Distinguishing between these cases is important, since they result in quite different behaviors. These two broad cases are captured by the following definitions.

A semi-anonymous graphical game[13] exhibits *strategic complements* if it satisfies the property of increasing differences; that is, for all d and $m \geq m'$:

$$u_d(1, m) - u_d(0, m) \geq u_d(1, m') - u_d(0, m').$$

A semi-anonymous graphical game exhibits *strategic substitutes* if it satisfies decreasing differences; that is, for all d and $m \geq m'$:

$$u_d(1, m) - u_d(0, m) \leq u_d(1, m') - u_d(0, m').$$

These notions are said to apply strictly if the inequalities above are strict when $m > m'$.

The best-shot public goods game is one of strategic substitutes, as is the local public goods game when f is concave. In that case, higher levels of actions by neighbors (i.e., higher m) lead to an incentive for a given player to take the lower action or to free-ride. Other examples are items that can be shared, such as products and information, as well as other situations with externalities, such as pollution reduction and defense systems (where the network could be one of treaties and the players are countries).

The couples game is one of strategic complements. Local public goods games when f is convex, so there are increasing returns to action 1, also exhibit strategic complementarities. Other examples of strategic complementarities include situations in which peer effects are important. It is important to emphasize that complementarities can exist for many different reasons. For instance, in decisions of whether to pursue higher education, neighbors may serve as role models for a given individual. They may also provide information about the potential benefits of higher education, or serve as future contacts in relaying job information. An individual may have an incentive to conform to the patterns of behavior of his or her peers. The critical common feature is that increased levels of activity among a given player's neighbors increase the incentives or pressures for that player to undertake the activity.

9.3.3 Equilibria and Thresholds

The nice aspect of semi-anonymous graphical games is that the behavior of a given individual can be succinctly captured by a threshold. In the case of strategic complements, there is a threshold $t(d)$, which can depend on a player's degree d such that if more than $t(d)$ neighbors choose action 1, then the player prefers action 1, while if fewer than $t(d)$ neighbors choose 1, then the player prefers 0. It is possible to have situations in which an individual is completely indifferent at the threshold. For the case of strategic substitutes, there is also a threshold, but the best response of the player is reversed, so that he or she prefers to take action 0 if more than $t(d)$ neighbors take action 1, and prefers action 1 if fewer than $t(d)$

---

13. The definitions extend readily to settings beyond the semi-anonymous case, by working with set inclusion. That is, complements are such that if the actions of all of a player's neighbors do not decrease and some increase, then the player's gain in payoffs from taking action 1 compared to action 0 increases; substitutes are the reverse case. See Exercise 9.8.

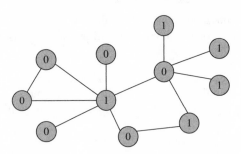

FIGURE 9.5   An equilibrium in a best-shot public goods graphical game.

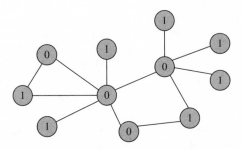

FIGURE 9.6   Another equilibrium in a best-shot public goods graphical game.

neighbors take action 1. The best-shot public goods game and the couples game are games of strategic substitutes and strategic complements, respectively, where the threshold is 1 (irrespective of degree).

As discussed in more detail in Section 9.8, semi-anonymous graphical games with strategic complementarities always have a pure strategy equilibria. In fact, the set of pure strategy equilibria has a nice structure and is what is known as a *complete lattice,* as outlined in Exercise 9.5. This structure implies that there exists a maximum equilibrium such that each player's action is at least as high as in every other equilibrium, and similarly a minimum equilibrium where actions take their lowest values out of all equilibria. For the network pictured in Figures 9.2–9.4, the minimum equilibrium is in Figure 9.2 and the maximum is in Figure 9.4. (There is one other pure strategy equilibrium for this network not pictured in Figures 9.2–9.4, which is the topic of Exercise 9.4.)

Semi-anonymous graphical games of strategic substitutes do not always have a pure strategy equilibrium, but they always have at least one equilibrium in mixed strategies. There are games of strategic substitutes that have pure strategy equilibria, as we have seen in the case of the best-shot public goods game, and they can have multiple equilibria. Figures 9.5 and 9.6 exhibit two different pure strategy equilibria in a best-shot public goods game for the same network as in Figures 9.2–9.4.

There are other equilibria for this network (see Exercise 9.4), but the structure of equilibria in the best-shot public goods game does not exhibit a lattice structure. Nevertheless, there still always exists at least one pure strategy equilibrium for a best-shot public goods graphical game when the network is undirected. Exercise 9.6 shows that an equilibrium may not exist in directed networks.

9.3.4 Comparing Behavior as the Network Is Varied

With some examples and definitions in hand, let us examine a few basic properties of how behavior changes as we vary the structure of a network. Such comparisons show how social structure influences behavior.

First, it is easy to see that in games of complements in which the threshold for taking action 1 is nonincreasing in degree, adding links will lead to (weakly) higher actions, as players will have more neighbors taking action 1.

Proposition 9.3 *Consider a semi-anonymous graphical game of strategic complements on a network (N, g) such that the threshold for taking action 1 is nonincreasing as a function of degree, so that $t(d + 1) \leq t(d)$ for each d. If we add links to the network to obtain a network g' (so that $g \subset g'$), then for any pure strategy equilibrium x under g, there exists an equilibrium x' under g' such that all players play at least as high an action under x' as under x.*

Proposition 9.3 notes that if incentives to take an action increase as more neighbors take an action, then denser networks lead to higher numbers of players choosing the action. This result is not dependent on the 0-1 action space we have been considering, but extends to more general action spaces as outlined in Exercise 9.8. This conclusion requires that players react to the absolute level of activity by neighbors rather than to a proportion.

The more subtle case is that of strategic substitutes. Adding links can change the structure of payoffs in unpredictable ways, as illustrated in the following example of adding a link in a best-shot graphical game, which is drawn from an insight of Bramoullé and Kranton [103]. One might expect, reversing the intuition from the complements case, that adding links leads to new equilibria at which all agents take less action than they did before. However this is not quite right, as decreasing actions for some agents can lead to increasing actions for others in the case of strategic substitutes, and so changing network structure leads to more complex changes in behavior, as pictured in Figure 9.7.

The top panel of Figure 9.7 shows an equilibrium in which both players who are the centers of their respective stars provide the public good and other players free-ride. This configuration is the overall cheapest way of providing the public good to all players and so has a strong efficiency property. However, when we add a link between these two center agents, it is no longer an equilibrium for both of them to provide the public good. Thus adding a link can change the structure of equilibria in complicated ways.

Nevertheless, there is still a well-defined way in which the equilibrium adjusts so that, despite Figure 9.7, adding links still decreases actions if tracking how all equilibria change.

Proposition 9.4 (Galeotti et al. [274]) *Consider a best-shot graphical game on a network (N, g) and any pure strategy equilibrium x of $g + ij$. Either x is also an equilibrium of g, or there exists an equilibrium under g in which a strict superset of players chooses 1.*

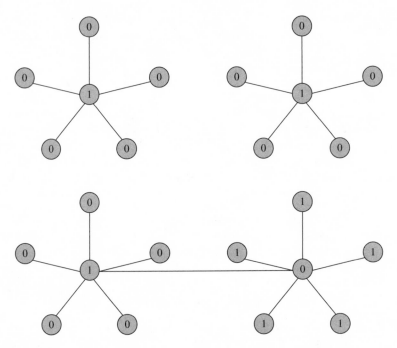

FIGURE 9.7 Adding a link changes the equilibrium structure in a best-shot graphical game.

Thus, although there are generally multiple equilibria and a particular equilibrium might not be an equilibrium when a link is added, any new equilibrium in the network with a new link has a subset of players who take action 1 compared to some equilibrium of the old network. Indeed, in Figure 9.7 the equilibrium in the bottom half of the figure is also an equilibrium without the link being present. It is just that the equilibrium in the top panel no longer survives. Thus in some sense adding links implies fewer players providing the public good. The proof of the proposition follows easily from the structure of maximal independent sets (e.g., see Observation 2.1).

The proposition is illustrated by examining all pure strategy equilibria on the networks of three individuals, as pictured in Figure 9.8. In Figure 9.8, we see how the equilibria vary as links are added or deleted, and although a particular equilibrium might not survive when links are deleted, it can be compared to some other equilibrium of the resulting network.

Making changes beyond the addition of links (e.g., moving links or changing the degree distribution but keeping the mean degree constant) can alter the landscape of equilibria in more complicated ways. The sensitivity of behavior to network changes leaves the graphical games model without sharp comparisons of behavior resulting from changes in network structure. However, there is a variation on graphical games in which behavior varies in predictable ways in response to general changes in network structure. This is not to say that one or the other is a better model, as they fit different situations, and the difference in the sharpness of their predictions is reflective of the differences across those settings.

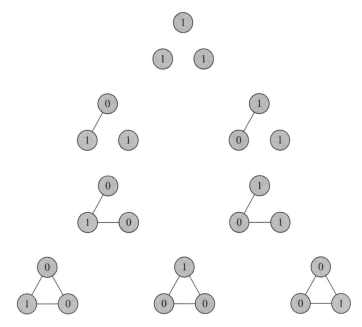

FIGURE 9.8   Equilibria in various three-player best-shot graphical games.

9.4 ▪ Randomly Chosen Neighbors and Network Games

While graphical games nicely model a number of networked interactions in which players have a good idea of their neighbors' actions when choosing their own (or they can adjust their behavior), there are also many situations in which players choose actions in at least partial ignorance of what their neighbors will do, or even in ignorance of who their neighbors will be. This limited knowledge applies when players are learning a skill or making some investment and are unsure of their future interactions. For example, they might be choosing majors in college, which will eventually be very important in interactions with their employers, their colleagues, and so forth. In choosing what major to undertake, they might know something about the number of other people choosing that action and the job market for different majors, without knowing with whom they will be interacting.[14] These ideas are formalized in a setting from Galeotti et al. [274], Jackson and Yariv [362], [364], and Sundararajan [618]. For expositional purposes, I use a setting with just two actions, but the analysis extends to richer settings.

A player knows his or her own degree as well as the distribution over the likely degrees of his or her neighbors, but nothing more about the network structure when choosing an action. Degree can be thought of as the number of interactions that a player is likely to have in the future.

_____

14. See Pasini, Pin, and Weidenholzer [528] for some discussions of another application (to buyer-seller markets) where this network games formulation is appropriate.

Define a strategy to indicate which action is chosen as a function of a player's degree. In particular, let $\sigma(d) \in [0, 1]$ be the probability that a player of degree d chooses action 1. For most degrees (d) will be either 0 or 1, but in some cases there might be some mixing. This definition implicitly builds in a symmetry such that players of the same degree follow the same strategy.[15] In many cases, the symmetry holds without loss of generality, as players with the same degree face the same payoffs as a function of their actions and will often have a unique best response.

The degrees of a player's neighbor are drawn from a degree distribution $\widetilde{P}$. Recall that $\widetilde{P}(d) = \frac{P(d)d}{\langle d \rangle}$ approximates the distribution over a neighbor's degree from the configuration model with respect to a degree sequence represented by P. Under $\widetilde{P}$ there is a well-defined probability that a neighbor takes action 1:

$$p_\sigma = \sum_d \sigma(d)\widetilde{P}(d).$$

Thus the probability that exactly m out of the d_i neighbors of player i choose action 1 is given by the binomial formula $\binom{d_i}{m} p_\sigma^m (1 - p_\sigma)^{d_i - m}$. The expected utility of a player of degree d_i who takes action x_i is then

$$U_{d_i}(x_i, p_\sigma) = \sum_{m=0}^{d_i} u_{d_i}(x_i, m) \binom{d_i}{m} p_\sigma^m (1 - p_\sigma)^{d_i - m}, \qquad (9.4)$$

where $u_{d_i}(x_i, m)$ is the payoff corresponding to an underlying graphical game. One can then think of this as a sort of graphical game, in which players choose their strategies knowing how many links they will have but not knowing which network will be realized. However, the above formulation does not require a specification of the precise set of players or even how many players there will be. Players just need to know their own degrees and have beliefs about their neighbors' behavior; they do not need to have a fully specified model of the world.

The formulation above presumes independence of neighbors' degrees. However, the results extend to allow for correlation among neighbors' degrees, which is important since, as we have seen, many networks exhibit such correlations. Here I limit the discussion to the independent case, since it makes the exposition more transparent. See Section 4.5.7 and Galeotti et al. [274] for details on the appropriate definitions for extensions of these results.[16]

A specification of a utility function u_d for each d and a distribution of neighbors' degrees $\widetilde{P}$ is referred to as a *network game*. It is now easy to define a (Nash)

---

15. The equilibrium definition below allows any player to deviate in any way, and so this symmetry in behavior is an equilibrium phenomenon and is payoff-maximizing for the players; it is not a constraint on behavior.

16. Effectively the generalization allows each degree to have a different anticipated distribution over vectors of neighbors' degrees. What is required in the case of strategic complements is that higher-degree players have a distribution over neighbors' degrees that leads them to expect (weakly) higher degrees among their neighbors than a lower-degree player would. This situation is reversed for substitutes. Comparing joint distributions over different-sized vectors is based on concepts discussed in Section 4.5.7.

equilibrium of a network game.[17] An equilibrium in a network game is a strategy σ such that for each d,

- if $\sigma(d) > 0$, then $U_d(1, p_\sigma) \geq U_d(0, p_\sigma)$, and
- if $\sigma(d) < 1$, then $U_d(1, p_\sigma) \leq U_d(0, p_\sigma)$.

9.4.1 Degree and Behavior

Players' strategies can be ordered as a function of their degrees. The idea is that players with higher degrees have more neighbors, and hence an expectation of having more neighbors choosing 1. In games of strategic complements, if having more total activity among one's neighbors leads one to prefer the higher action, then higher-degree players prefer a higher action compared to the preferences of a lower-degree player. This process reverses itself for substitutes.

The conclusion that players with higher degree have more of an incentive to take higher actions is not guaranteed simply by having strategic complementarities, since that condition examines a given player's incentives as his or her neighbors' behavior is changed. It does not make comparisons of how incentives vary with degree. To make comparisons across degrees, let us focus on the case in which payoffs depend on absolute numbers of neighbors taking action 1, so that

$$u_d(x_i, m) = u_{d+1}(x_i, m) \tag{9.5}$$

for each $m \leq d$ and x_i. Equation (9.5) is not necessary for the results that follow, which hold for much more general payoff settings, including those in which players are influenced by the percentage of neighbors taking a given action rather than by the absolute number of neighbors doing so. Exercise 9.9 addresses this payoff scheme. However, using (9.5) simplifies the exposition and conveys the basic ideas. What is needed for the following results is a payoff structure such that if a higher-degree individual is faced with the same typical behavior by any given neighbor, then he or she prefers to choose action 1 over 0 when a lower-degree player would prefer to choose action 1 over 0. Under such conditions, it is straightforward to deduce the existence of an equilibrium where higher types take higher actions, and similarly for substitutes and lower actions, which leads to the following proposition. Proposition 9.5 is quite useful in deducing how behavior varies with network structure.

Proposition 9.5 (Galeotti et al. [274]) *Consider a network game in which payoff functions satisfy (9.5). If it is a game of strategic complements, then there exists an equilibrium that is nondecreasing in degree;[18] and if it is a game of strategic substitutes, then there exists an equilibrium that is nonincreasing in degree. If the game is one of strict strategic complements, then all equilibria are*

---

17. Such an equilibrium, where players' strategies depend on a "type" (here, their degree) and players are not sure of the other players' types when they choose their action, is also known as a *Bayesian equilibrium*.

18. *Nondecreasing* refers to $\sigma(d)$ being a nondecreasing function of d.

nondecreasing in degree; analogously if it is of strict strategic substitutes then all equilibria are nonincreasing in degree.

The proof of Proposition 9.5 follows the logic of a variety of game-theoretic analyses in the presence of strategic complementarities (e.g., see Topkis [630], Vives [642], and Milgrom and Roberts [469]), here adapted to the network setting. The idea is that if we begin with some σ that is nondecreasing in degree, then under (9.5) there is a best response for the players that is nondecreasing in degree. An equilibrium is a fixed point of the best response correspondence. The set of nondecreasing strategies is convex and compact (with appropriate definitions), and so a fixed point exists by any of a variety of theorems on fixed points; the same holds for substitutes. Let me sketch a more direct and intuititive proof. Let σ^t be such that $\sigma^t(d) = 1$ if $d \geq t$ and $\sigma^t(d) = 0$ for $d < t$ (allowing for $t = \infty$ to allow all players to play 0). Consider the case of strategic complements. Note that if action 1 is a best response to some strategy σ for a player of degree d, then it is a best response for all higher-degree players; and similarly if action 0 is a best response to some σ for a player of degree d, then it is a best response for all lower-degree players. Begin with σ^1. If this is an equilibrium, then stop. Otherwise, a degree-1 player must prefer to play action 0 in response to σ^1. So consider σ^2. Now action 0 must still be a best response for the degree-1 players, as there is less aggregate action by other players. So if this configuration is not an equilibrium, degree-2 players must prefer to play 0 to 1. Iterating on this logic, either the process eventually stops at some σ^t, where the degree-t players do not wish to change from action 1 to 0, or else this process continues for all degrees, in which case all players playing 0 (σ^∞) is an equilibrium.

Note that this logic shows that for the case of complements, there is actually an equilibrium of pure strategies. The case of strategic substitutes is slightly more complicated, as lowering the action of a given type of player might actually reverse the incentives for that type. For instance, consider a situation in which all players have degree d in a best-shot public goods game. The equilibrium will then involve mixing, since if players were to choose 1, then any given player would prefer to play 0, and vice versa.[19] Nevertheless, there is still nice structure to incentives across degrees, so that if 1 is a best response to some strategy σ for a player of degree d, then action 1 is a best response for all lower-degree players; similarly if 0 is a best response to some σ for a player of degree d then it is a best response for all higher-degree players. One can then follow a similar algorithm as above, but starting at σ^∞. As a first step examine whether degree-1 players would prefer to change to action 1. If they do, then raise their action but this time raise the action continuously from $\sigma(1) = 0$ to $\sigma(1) = 1$. Given the continuity of preferences, the difference in utility for such players between action 0 and 1 will change continuously. Either at some point degree-1 players are indifferent between the two actions in response to this mixture and action 0 by other degree players, or else action 1 is their strict best response to a situation in which they play action 1 and all others play action 0. Continue in this manner.

19. Recall that strategies are specified as a function of degree, and so in a regular setting all players must take the same action.

The claim that all equilibria are nondecreasing when the strategic complements or substitutes are strict follows from noting that players with higher degrees expect to have more neighbors choosing action 1 (in terms of first-order stochastic dominance) and is the subject of Exercise 9.11.

The fact that players' incentives vary monotonically with their degree does not necessarily guarantee that their payoffs vary monotonically with their degree. That relation depends on how the actions of others affect a given player. Note that just because the incentives of a player to choose action 1 increase when more neighbors choose action 1 does not mean that the players are better off making this choice. For example, consider a game that involves athletes' choices of whether to engage in doping (e.g., taking illegal drugs or undergoing blood transfusions) to give them an advantage in competition. A player's neighbors are other athletes against whom the player competes. The strategy in the game is either to dope or not to dope. Doping improves an athlete's performance, but also has ethical costs, health costs, and potential costs of detection and punishment. Regardless of exactly how these different factors weigh on a given player's payoffs, as more neighbors are doping, a player is faced with tougher competition and has greater incentives to dope himself or herself just to maintain competitiveness. This game will generally be one of (strict) strategic complements. Nevertheless, all players would be better off if nobody doped compared to everyone doping. So conditions on incentives, such as strategic complementarities, do not necessarily imply orderings of overall payoffs without knowing more about the structure of the game. If the game is such that increased choices of action 1 by neighbors lead to higher payoffs, then indeed, higher-degree players will get higher payoffs (see Exercise 9.10); but the situation could also be reversed, as in the doping example, so that increased choices of action 1 by neighbors decrease payoffs.

While Proposition 9.5 is relatively straightforward to prove and quite intuitive, it concludes that the more-connected members of a society take higher action in situations with complementarities and lower action in those with substitutes. This observation is consistent, for instance, with the data of Coleman, Katz, and Menzel [165] discussed in Section 3.2.10. In the case of strict strategic complements, it also means that the equilibrium can be characterized in terms of a threshold degree, such that all players with degree above the threshold take action 1, and players with degree below it take action 0; the reverse holds for substitutes. It could be that players right at the threshold degree randomize. Figure 9.9 illustrates this threshold for a particular degree distribution. The figure shows a possible frequency distribution of neighbors' degrees, $\widetilde{P}$, as a function of degree. The threshold degree is such that players with higher degree take action 1 and those with lower degree take action 0. Thus in Figure 9.9, the probability of a neighbor taking action 1 is just the sum of the distribution of neighbors' degrees under $\widetilde{P}$ to the right of the threshold degree (adjusting for any mixing by the players exactly at the threshold).

9.4.2 Changes in Networks and Changes in Behavior

This analysis also leads to predictions of how behavior changes as we modify the distribution of neighbors' degrees. Suppose that we compare the equilibrium

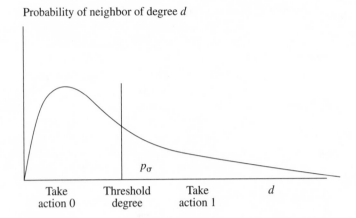

FIGURE 9.9 Behavior as a function of degree with complementarities.

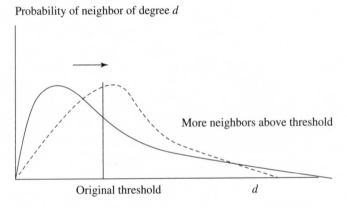

FIGURE 9.10 A shift in the degree distribution leads to more action.

behavior for the distribution pictured in Figure 9.9 with that for a different distribution, such as the dashed distribution in Figure 9.10. The new (dashed) degree distribution places more weight above the original threshold degree. If the equilibrium strategy does not change, then the shift would lead to a higher probability that any given neighbor plays action 1 and results in a higher expected number of neighbors taking action 1. Thus in the case of complements, players of any given degree now have a weakly higher incentive to play action 1 versus 0 than they did before. The threshold should then move down. As it decreases even more players have an incentive to play action 1, and so we move to a new equilibrium where even more neighbors play action 1.

This intuition is formalized in the following proposition. It is stated for the case of strategic complements, but also holds for the case of strategic substitutes, with an appropriate reversal of the direction of the shifts of thresholds and probabilities of action.

Proposition 9.6 (Galeotti et al. [274]) *Consider a network game of strict strategic complements that satisfies (9.5) and has a distribution of neighbors' degrees given by $\widetilde{P}$ and an equilibrium with threshold t. If the distribution of neighbors' degrees is changed to $\widetilde{P}'$ such that $\sum_{d \le t} \widetilde{P}'(d) \le \sum_{d \le t-1} \widetilde{P}(d)$, then there is an equilibrium threshold under $\widetilde{P}'$ that is at least as low as t, and the probability that any given neighbor chooses action 1 (weakly) increases. If instead $\sum_{d \ge t} \widetilde{P}'(d) \ge \sum_{d \le t+1} \widetilde{P}(d)$, then there is an equilibrium threshold under $\widetilde{P}'$ that is at least as high as t, and the probability that any given neighbor chooses action 1 (weakly) decreases.*

Note that Proposition 9.6 effectively allows us to compare any two degree distributions. The only complication is if the two distributions are very close and differ only at the threshold.[20]

Proof of Proposition 9.6. Let σ denote the equilibrium under $\widetilde{P}$, and consider the case $\sum_{d \le t} \widetilde{P}'(d) \le \sum_{d \le t-1} \widetilde{P}(d)$, as the other case is analogous. If σ is played under $\widetilde{P}'$, then there is a new probability $p'_\sigma \ge p_\sigma$ that any given neighbor will choose action 1. It is then easily verified that for any given degree d, this new distribution leads to a first-order stochastic dominance shift in the distribution of m, the number of neighbors who choose action 1. Given strict strategic complements, $u_d(1, m) - u_d(0, m)$ is an increasing function of m, and so given the first-order stochastic dominance shift, $U_d(1, p_\sigma) - U_d(0, p_\sigma)$, is at least as large as it was before for any d under the new distribution of neighbors' degrees. Thus $\sigma(d)$ is still a best response to σ for all $d > t$. If action 1 is a best response for degree-t players, then set their strategy to action 1. So, following the notation of the discussion after Proposition 9.5, we have strategy σ^t. Note also that if 1 is a best response to some strategy for a player of degree d, then action 1 is a best response for all higher-degree players; similarly if 0 is a best response for a player of degree d, then it is a best response for all lower-degree players. Then consider the best response of players of degree $t - 1$ to σ^t under $\widetilde{P}'$. If it is action 0, then σ^t is an equilibrium and the conclusions of the proposition hold. Otherwise, move to strategy σ^{t-1}, and then consider the best responses of players of degree $t - 2$. Continue in this manner until either stopping at some $\sigma^{t'}$ with $t' < t$, or hitting σ^1, in which case all players choosing action 1 is an equilibrium. In either case, the conclusions of the proposition hold. ∎

Proposition 9.6 concludes that the probability of a neighbor choosing action 1 increases when the distribution of neighbors' degrees is shifted to place more weight above the threshold, but it does not claim that the probability that the overall fraction of players choosing action 1 increases at the new equilibrium. There is an important distinction between fractions of neighbors and fractions of players, which stems from the distinction between neighbors' degrees and players' degrees. Neighbors are more likely to be higher-degree players. The conclusion that the overall fraction of players choosing action 1 increases is valid if it is also

---

20. If the starting equilibrium involves no mixing by the threshold-degree players, then the conclusion also holds under the weaker conditions that $\sum_{d \le t-1} \widetilde{P}' \le \sum_{d \le t-1} \widetilde{P}$, or $\sum_{d \ge t} \widetilde{P}' \le \sum_{d \ge t} \widetilde{P}$, which then covers all possible comparisons.

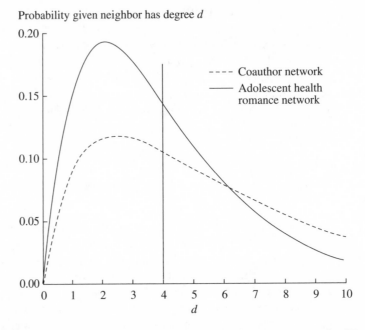

FIGURE 9.11 Distribution of neighbors' degrees for the romance network of Bearman, Moody, and Stovel [51] is first-order stochastically dominated by the distribution of neighbors' degrees for the coauthorship network of Goyal, van der Leij, and Moraga-González [304].

true that the weight that P' places below t is less than the weight that P places below $t - 1$, where P' and P are the degree distributions corresponding to $\widetilde{P}'$ and $\widetilde{P}$, respectively. In many instances this condition holds, but one can find counter examples (e.g., see Galeotti et al. [274]).

To see the potential usefulness of Proposition 9.6 consider Figure 9.11. This figure shows two degree distributions from empirical studies. The first degree distribution is that of neighbors' degrees for the romance network of Bearman, Moody and Stovel [51] (from Figure 1.2; see Section 1.2.2). The second is a distribution of neighbors' degrees for a coauthorship network studied by Goyal, van der Leij, and Moraga-González [304]. This second distribution first-order stochastically dominates the first distribution and thus has greater weight above any threshold degree. While these distributions are from different applications, they show that examining empirically generated degree distributions allows for comparisons of the type treated in Proposition 9.6.

9.5 ▪ Richer Action Spaces

The analysis to this point has focused on situations with two actions. While two-action scenarios capture many applications and offer broad insights, there are settings in which the intensity of activity plays a substantial role. I present two such models. The first is a public goods model in which there are interesting implications for specializing in activities. The second is a model with complementarities that exhibits an interesting relationship between levels of activity and network centrality. For this section I return to the graphical games formulation.

9.5.1 A Local Public Goods Model

The following model analyzed by Bramoullé and Kranton [103] is a variation on a local public goods graphical game, like the one in Example 9.5, but the action space for each player is $X_i = [0, \infty)$. Players benefit from their own action plus the actions of their neighbors with payoffs described by

$$u_i(x_i, x_{N_i(g)}) = f(x_i + \sum_{j \in N_i(g)} x_j) - cx_i,$$

where f is a continuously differentiable, strictly concave function, and $c > 0$ is a cost parameter.

The interesting case occurs when $f'(0) > c > f'(x)$ for some large enough x, as otherwise optimal action levels are 0 or ∞. In this case, in every equilibrium each player's neighborhood has some production of the public good (at least with positive probability). Letting x^* be such that $f'(x^*) = c$, it is easy to see that any pure strategy equilibrium must have at least x^* produced in each player's neighborhood (so that $x_i + \sum_{j \in N_i(g)} x_j \geq x^*$ for each i); otherwise a player could increase his or her payoff by increasing his or her action. In fact, a strategy profile $(x_1, \ldots, x_n)$ is an equilibrium if and only if the following holds for each i:

- If $x_i > 0$, then $x_i + \sum_{j \in N_i(g)} x_j = x^*$; and
- If $x_i = 0$, then $\sum_{j \in N_i(g)} x_j \geq x^*$.

So a pure strategy equilibrium is any profile of actions such that every player's neighborhood produces at least x^*, a player only chooses a positive activity level if his or her neighbors produce less than x^* in aggregate, and in that case the player produces just enough to bring his or her aggregate neighborhood activity level to x^*. Figure 9.12 pictures some pure strategy equilibria for three-person networks in which $x^* = 1$.

There is a class of equilibria, which Bramoullé and Kranton [103] refer to as *specialized equilibria*, where players either choose an action level of x^* or 0. Thus there are players who specialize in providing the information or public good (e.g., the opinion leaders of Katz and Lazarsfeld discussed in Section 8.1), and others who free-ride on their neighbors. Even though the action spaces are richer, the specialized equilibria have the same structure as those of the pure strategy equilibria in the best-shot public goods graphical games. That is, the specialized equilibria are precisely those where the players who specialize in providing the local public good at the level x^* form a maximal independent set and the remaining players choose an action of 0.

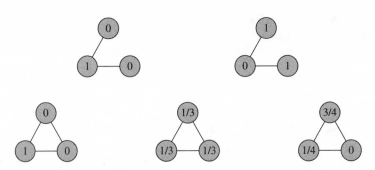

FIGURE 9.12 Examples of equilibrium local public good provision levels of $x^* = 1$. The numbers in each node indicate the node's action in Figures 9.12 and 9.13.

There is a sense in which the specialized equilibria are more robust than other equilibria. As Bramoullé and Kranton [103] point out, only specialized equilibria (and in fact only a subset of them) satisfy the following concept of stability:[21]

- Start with a pure strategy equilibrium profile $(x_1, \ldots, x_n)$.
- Perturb it slightly by adding some small perturbation ε_i to each x_i, with a requirement that $x_i + \varepsilon_i \geq 0$. Denote this perturbation by $x^1 = (x_1 + \varepsilon_1, \ldots, x_n + \varepsilon_n)$.
- Consider the best responses to x^1. That is, for each i, find a level of action that maximizes u_i, presuming that x^1_{-i} will be played by the other players. Let this profile of best-responses be denoted $x^2 = (x^2_1, \ldots, x^2_n)$, where x^2_i is the best response to x^1_{-i}.
- Iterate on the best responses, at each step examining the best responses x^k of the players to the previous step's strategies x^{k-1}.

If there is some $\overline{\varepsilon} > 0$ for which this process always converges back to $(x_1, \ldots, x_n)$ starting from any admissible perturbations such that $|\varepsilon_i| < \overline{\varepsilon}$ for all i, then the original equilibrium is said to be stable.

In this setting, the best responses to some x_{-i} take a simple form: if $\sum_{j \in N_i(g)} x_j \geq x^*$, then the best response is an action of 0; otherwise a best response is the action that raises the neighborhood production to x^*: $x^* - \sum_{j \in N_i(g)} x_j$. It follows fairly directly that the only stable equilibria, if they exist, are specialized equilibria such that each nonspecialist player has at least two specialists in his or her neighborhood, and each specialist has no neighbors choosing x. With at least two specialists in every nonspecialist's neighborhood, even a slight perturbation leads to a best-response dynamic in which the nonspecialists return to an action of 0 and the specialists return to an action of x^*. So such equilibria are stable. If there are fewer than two specialists in some nonspecialist's neighborhood, then the equilibria are unstable. The proof of this takes a bit more argument, but

---

21. This is a classic notion of stability that has been used in a variety of settings. See Chapter 1 in Fudenberg and Tirole [263] for more discussion and references.

to see the basic idea, consider a dyad for which there is no equilibrium with two specialists. It is easy to see that there are no stable equilibria: consider any pure strategy equilibrium, which must be such that $x_1 + x_2 = x^*$. At least one of the two strategies is larger than 0, so suppose that $x_2 > 0$. Then perturb the strategies to $x_1 + \varepsilon$, $x_2 - \varepsilon$. This configuration is also an equilibrium for any $\varepsilon \leq x^2$, and so the best responses do not converge back to the original point and no equilibrium is stable. In Figure 9.12, only the upper right-hand equilibrium is a stable one, and there are no stable equilibria for the complete network.

It is also worth noting that all equilibria in this public goods game are inefficient, in the sense that they stop short of maximizing total utility. Each player is maximizing his or her own payoff, and yet his or her action also benefits other agents. For example, in a dyad the total production is x^*, which maximizes $f(x) - cx$, while overall societal utility is $2f(x) - cx$, which will generally have a higher maximizer if f is smooth and strictly concave. This inefficiency is endemic to public goods provision, and players generally underprovide public goods, because they do not fully account for the benefits that their actions bestow on others. In this setting, there are also some differences across equilibria in terms of the total utility they generate. For example, if we consider the two different equilibria in the two top networks in Figure 9.12, they result in different aggregate payoffs. The one with one specialist on the left results in a payoff of $3f(1) - c$, while the one with two specialists on the right results in a payoff of $f(2) + 2f(1) - 2c$. Here we can rank these two equilibria, as the one on the left generates more total utility. We see this ranking by noting that the difference between the one on the left and that on the right is $c - f(2) + f(1)$. Since $x^* = 1$ and f is strictly concave, it follows that $c > f(2) - f(1)$ (otherwise, a player would prefer to increase the production to 2 even by himself or herself), and so this difference is positive. For more general networks, the comparisons across equilibria depend on the specific configurations and payoffs, but we can conclude that equilibria are generally inefficient, so that the stable equilibria are not always the most efficient equilibria.

This analysis of local public goods, although stylized, provides us with some basic insights into the emergence of individuals who provide local public goods, such as information, and who might act as opinion leaders, while other individuals free-ride, benefiting from this activity while not providing any benefit themselves. Significantly, the only stable equilibria are actually the asymmetric ones, even in very symmetric networks, and so this heterogeneity among individuals emerges because of the network interactions, even when there is no other a priori difference among individuals. While there can exist multiple equilibria, this analysis does not always offer useful predictions as to who will become providers or opinion leaders. Introducing heterogeneity into costs and benefits across individuals can help cut down the multiplicity of equilibria.

Before leaving this model, I comment on an aspect of the predictions of specialization and equilibria in graphical games more generally. Consider a simple example with four players and $x^* = 1$, and the individuals are connected in a circle network, as in Figure 9.13. In this case, there are only three pure strategy equilibria. According to the stability notion in Section 9.5.1, the two specialist equilibria are the only stable ones. However, if there is any cost to maintaining a link, neither of those networks would be pairwise stable in the sense of Section 6.1, as a player would not maintain a link to a neighbor who provides no public good. In contrast,

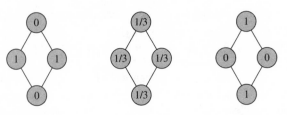

FIGURE 9.13   Examples of equilibrium public good provision choices.

there are situations in which the middle network in Figure 9.13 is pairwise stable (in particular, when the cost of a link is less than $f(1) - f(2/3)$ and greater than $f(4/3) - f(1)$).

The implication is that the graphical games analyses that we have been conducting are affected by considering the network to be endogenous. Indeed, people form relationships with others who provide local public goods, and they are expected to reciprocate in some fashion.[22] This is not to say that the insights behind specialization drawn from the analysis above are flawed, but that they need to be explored in a larger context. For example, if players were involved in two separate local public goods problems at the same time, with some players specializing in becoming informed about political campaigns and others about good local restaurants, then it could be possible to support specialized equilibria in conjunction with each other and an endogenous network. Some of the interplay between network formation and behavior on networks is examined in Section 11.4.3, but is still a largely underexplored subject.

9.5.2   Quadratic Payoffs and Strategic Complementarities

The public goods model described in the previous section is one in which activities are strategic substitutes. Let us now examine a different model, in which actions are again continuously adjustable, but there are strategic complementarities between players' actions. The following is a variation of the model of Ballester, Calvó-Armengol, and Zenou [34], which admits strategic complementarities.

Each player chooses an intensity with which he or she undertakes an activity. Let $x_i \in \mathbb{R}_+$ indicate that intensity, so that higher x_i corresponds to greater action. A player i's payoff is described by

$$u_i(x_i, x_{-i}) = a_i x_i - \frac{b_i}{2} x_i^2 + \sum_{j \neq i} w_{ij} x_i x_j,$$

where $a_i \geq 0$ and $b_i > 0$ are scalars, and w_{ij} is the weight that player i places on j's action. If $w_{ij} > 0$, then i and j's activities are strategic complements, so that more activity by j leads to increased incentives for activity by i; while if $w_{ij} < 0$,

---

22. One can also consider explicit transfers as a means for sustaining specialist equilibria. For more on transfers and stability, see Section 11.6.

then i and j's activities are strategic substitutes and increased activity by j lowers i's activity. The expression $\frac{b_i}{2}x_i^2$ leads to diminishing returns from the activity for player i, so that player i sees some trade-off to taking the action, ensuring a well-defined optimal strategy.

The payoff-maximizing action for player i is found by setting the derivative of the payoff $u_i(x_i, x_{-i})$ with respect to the action level x_i equal to 0, which leads to a solution of

$$x_i = \frac{a_i}{b_i} + \sum_{j \neq i} \frac{w_{ij}}{b_i} x_j. \tag{9.6}$$

The interdependence between the players' actions is quite evident.

Let $g_{ij} = w_{ij}/b_i$ (and set $g_{ii} = 0$). We can think of g as a weighted and directed network.[23] This variable captures the relative dependence of i's choice of action on j's choice. The vector of actions that satisfy (9.6) is described by[24]

$$x = \alpha + gx, \tag{9.7}$$

where x is the $n \times 1$ vector of x_i and α is the $n \times 1$ vector of a_i/b_i. If $a_i = 0$ for each i, then (9.7) becomes $x = gx$, so that x is a unit right-hand eigenvector of g. Otherwise,

$$x = (\mathbb{I} - g)^{-1}\alpha, \tag{9.8}$$

where $\mathbb{I}$ is the identity matrix, provided $\mathbb{I} - g$ is invertible and the solution is nonnegative. These two conditions hold if the b_is are large enough so that the entries of g are small.[25] From (9.7), by substituting for x repeatedly on the right-hand side, we also see that

$$x = \sum_{k=0}^{\infty} g^k \alpha. \tag{9.9}$$

Equations (9.8) and (9.9) have nice interpretations. They are variations on the centrality indices discussed in Section 2.2.4. The intuition is similar. Being linked to players who are more active (have higher levels of x_i) leads a player to want to increase his or her level of activity, presuming g nonnegative. Correspondingly, the more active a player's neighbors' neighbors are, the more active the player's neighbors are, and so forth. Thus, the activity levels in the system depend on activity levels. The fact that the payoffs are quadratic in the Ballester, Calvó-Armengol, and Zenou [34] model leads to a precise relationship to centrality measures, but even more generally, we would expect similar effects to be present.

---

23. It could even allow for some negative weights, depending on the values of w_{ij}.
24. Finding solutions to this problem is related to what is known as the *linear complementarity problem*. See Ballester and Calvó-Armengol [33] for a discussion of the relation and more general formulations of such games.
25. A sufficient condition is that the sum of the entries of each row of g be less than 1 and the sum of entries in each column of g be less than 1, in the case in which they are nonnegative.

To develop this concept a bit further, Ballester, Calvó-Armengol, and Zenou [34] also examine the case in which $a_i = a$ and $b_i = b$ for all i, so that the only heterogeneity in the society comes through the weights w_{ij} in the network of interactions. In that case, the equilibrium levels of actions in (9.8) can be written as

$$x = (II - \frac{1}{b}w)^{-1}\frac{a}{b}\,\mathbb{1},\qquad(9.10)$$

where $\mathbb{1}$ is a vector of 1s. Equation (9.10) looks very much like the equations for Katz prestige-2 (2.9) and Bonacich centrality (2.10). In fact, we can write[26]

$$x = \frac{a}{b}\left(\mathbb{1} + P^{K2}\left(w, \frac{1}{b}\right)\right),\qquad(9.11)$$

where $P^{K2}(w, \frac{1}{b})$ is the Katz prestige-2 from (2.9) (which is the same as the Bonacich centrality $Ce^B(w, \frac{1}{b}, \frac{1}{b})$ from (2.10)). To ensure that x is well defined, the term $1/b$ has to be small enough so that the Katz prestige-2 measure is well defined and nonnegative. There are various sufficient conditions, but ensuring that the rows (or columns) of w/b each sum to less than 1 (presuming they are all nonnegative) is enough to ensure convergence.

There are some clear comparative statics. If we decrease b or increase a, then the solution x increases, and the action of every player increases. There is a direct effect of making higher levels of x_i more attractive, fixing the level of the other player's actions, which then feeds back to increase other neighbors' actions, which further increases incentives to increase player i's action, and so forth. Presuming that w is nonnegative, increasing an entry of w, say w_{ij}, increases the equilibrium actions of all players who have directed paths to i in w. This observation takes a bit of proof and can be shown via different methods. A direct technique is to start at an equilibrium x, increase w, and then consider each player's best response to x at the new w. Player i's best response will be higher, as he or she has an increased benefit from neighbors' actions. Iterating on the best responses, any player ℓ such that $w_{\ell i} > 0$ will increase his or her action in response to i's higher actions, and those having links to ℓ will increase their actions, and so forth. Actions will only move upward, and so convergence is monotone upward to a new equilibrium, provided that an equilibrium is still well defined.[27]

---

26. To see (9.11) note that (9.9) implies that

$$x = \left(1 + (\frac{1}{b})w + (\frac{1}{b})^2w^2\ldots\right)\frac{a}{b}\,\mathbb{1},$$

whereas the corresponding Katz prestige-2 from (2.7) is

$$P^{K2}(w, \frac{1}{b}) = \left(\frac{1}{b}w + (\frac{1}{b})^2w^2\ldots\right)\mathbb{1}.$$

27. Another way to see the increase in the equilibrium action levels is to examine (9.9), noting that the entries of g^k increase in some row j for some large enough k if and only if there is a directed path from j to i in g; note that no entries decrease.

This model provides a tractable formulation that shows how actions relate to network position in a very intuitive manner. Its tractability also allows the equilibrium to be studied in more detail, given its closed form. For example, one interpretation of the above model that Ballester, Calvó-Armengol, and Zenou [34] pursue is that players are choosing levels of criminal activity. A player sees direct benefits $(a_i x_i)$ and costs $(-b_i x_i^2)$ to crime, and there are also interactive effects if there are complementarities with one's neighbors $(\sum_{j \neq i} w_{ij} x_i x_j)$, so that more criminal activity by player's neighbors leads to greater benefits from crime to that player. This enhancement can be due to coordination, if cooperation results in more effective criminal activity, or it might reflect activities such as learning from neighbors. In the context of criminal activity, a natural question is which player should be removed to have the maximal impact on actions? For instance, if some police authority wants to lower criminal activity and can remove a single player, which player should it target? If the a_i and b_i terms coincide, then the interactive effects boil down to the centrality measure and the structure of w. The problem is then to compare players' activity levels when all players are present to equilibrium activity levels when a player is removed. By (9.11), this problem is equivalent to asking how removing one player affects the Katz prestige-2 measures. Ballester, Calvó-Armengol, and Zenou [34] show that the largest reduction in total activity comes from removing the player with the highest value of a variation on the Katz measure, which adjusts for the extent to which a player's prestige comes from paths back to himself or herself.

9.6 ▪ Dynamic Behavior and Contagion

The analyses of behavior up to this point are static in that they examine equilibrium behavior. In many situations, we are interested in the extent to which a new behavior diffuses through a society. For instance, if a new movie opens or a new product becomes available, how many people will take advantage of it? If there are complementarities in the product, so that a person is more likely to want to purchase it if another does, then the system may well have multiple equilibria, but simply examining them does not give us a full picture of which behavior is likely to emerge.

A powerful way of answering such questions is by examining the best-response behavior of a society over time. We have already seen some uses of iterating on best responses in checking for stability and identifying equilibria, but such iteration has been a prominent dynamic for more general analyses. That is, start by having some small portion of the population adopt a new action, say action 1. Then in a situation with complementarities, we can see how their neighbors respond. How many of them buy the product in response? This response then leads to further waves of adoption or diffusion. This process has been examined in variations on network settings by Ellison [221] and Blume [81], among others. An analysis that ties directly to the graphical games setting is one by Morris [487], which helps illustrate some interesting ideas.

Morris [487] considers a semi-anonymous graphical game with strategic complements. He examines a case in which each player responds to the fraction of

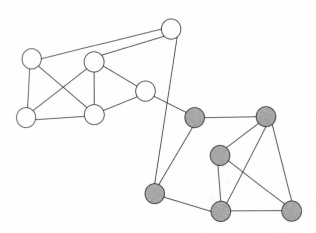

FIGURE 9.14 The sets of light and shaded nodes are each 2/3-cohesive.

neighbors playing action 1 versus 0. There is a threshold fraction q such that action 1 is a best response for a given player if and only if at least that fraction q of the player's neighbors choose 1. This fraction is the same for all players, independent of their degrees.[28] In the nontrivial case for which q lies strictly between 0 and 1, this is effectively a coordination game, and there are at least two equilibria, one where all players choose action 0 and the other where all players choose action 1.

What else can we deduce about equilibrium structure? For example, when is it possible that both actions coexist in a society, so that there is an equilibrium where some nonempty strict subset of the society plays action 1 and the rest plays action 0? Let S be the subset of the society that plays action 1. Each player in S must have at least a fraction q of his or her neighbors in the set S. It must also be that every player outside S has a fraction of no more than q of his neighbors in S, or equivalently, has a fraction of at least $1 - q$ of his neighbors outside S.

To capture these conditions, given $1 \geq r \geq 0$, Morris [487] defines the set of nodes S to be *r-cohesive* with respect to a network g if each node in S has at least a fraction r of its neighbors in S. That is (recalling (2.15)), S is r-cohesive relative to g if

$$\min_{i \in S} \frac{|N_i(g) \cap S|}{d_i(g)} \geq r,$$

where $0/0$ is set to 1. Figure 9.14 illustrates this definition with disjoint sets of nodes that are each 2/3-cohesive.

If a set is such that each player has at least some fraction r of his or her neighbors within the set, then it is easy to see that each player must have at least a fraction r' of his or her neighbors within that set when $r' < r$. So define the *cohesiveness*

---

28. This is a special case of complements games in which the threshold in terms of numbers of neighbors of degree d is simply qd.

of a given set S relative to a network (N, g) to be the maximum r such that S is r-cohesive. Then we have the following proposition, for which the proof is direct.

Proposition 9.7 (Morris [487]) *Consider a network (N, g) and a coordination game such that action 1 is a best response for any player if and only if a fraction of at least q of his or her neighbors play action 1. Both actions are played by different subsets of the society in some pure strategy equilibrium if and only if there exists some nonempty and strict subset of players S that is q-cohesive and such that its complement $N \setminus S$ is $(1 - q)$-cohesive.*

An obvious sufficient condition for both actions to be played in equilibria is to have at least two separate components, as then different actions can be played on each component. The cohesiveness of a component is 1, and thus it is also q- and $(1 - q)$-cohesive for any q.

Beyond components, cohesiveness provides enough of a separation in a network for different equilibria to exist adjacent to one another. For example, in Figure 9.14, the gray and white sets of nodes are connected to each other, but both are sufficiently inward-looking so that they can each sustain different equilibria in any game with q between $1/3$ and $2/3$.

Morris [487] also asks the following question.[29] Consider a given network (N, g) and start with all nodes playing action 0. "Infect" some number m of the nodes by switching them to play action 1 (and they can never switch back). Let players (other than the initially infected) best respond to the current action of their neighbors, switching players to action 1 if their payoffs are at least as good with action 1 as with action 0 against the actions of the other players. Repeat this process starting from the new actions, and stop when no new players change to action 1. If there is some set of m nodes whose initial infection leads to all players taking action 1 under the best response dynamic, then we say that there is *contagion from m nodes*. Let us say that a set S is *uniformly no more than r-cohesive* if there is no nonempty subset of S that is more than r-cohesive. We then have the following proposition.

Proposition 9.8 *Consider a network (N, g) and a coordination game such that action 1 is a best response for any player if and only if a fraction of at least q of his or her neighbors play action 1. Contagion from m nodes occurs if and only if there exists a set of m nodes such that its complement is uniformly no more than $(1 - q)$-cohesive.*

Proof of Proposition 9.8. Consider a set S of m nodes. If its complement has some subset A that is more than $1 - q$-cohesive, then that set A of nodes will all play 0 under the process above, at every step. Thus it is necessary for the complement to be uniformly no more than $(1 - q)$-cohesive to have contagion to all nodes. Next let us show that this condition is sufficient. Since the complement is uniformly no more than $(1 - q)$-cohesive, then it is no more than $(1 - q)$-cohesive. Thus there must be at least one player in the complement who has at least a fraction of q of his or her neighbors in S. So, at the first step, at least one player changes

---

29. Morris [487] works with infinite networks. I have adapted his formulation and results to a finite setting to be comparable to graphical games.

strategies. Subsequently, at each step, the set of players who have not yet changed strategies is no more than $(1-q)$-cohesive, and so some player must have at least q neighbors who are playing 1 and will change. Thus as long as some players have not yet changed, the process will have new players changing, and so every player must eventually change. ▪

A set is uniformly no more than $(1-q)$-cohesive if every subset of at least one node has more than q of its neighbors outside that subset. Thus such a network is quite dispersed in terms of its connections and does not have any highly segregated groups.

Corollary 9.1 *Consider a network (N, g) and a coordination game such that action 1 is a best response for any player if and only if a fraction of at least q of his or her neighbors play action 1. If a set S of nodes is uniformly no more than r-cohesive, then there will be contagion starting from the complement of that set, provided $q \leq 1 - r$.*

While cohesion is an easy concept to state and provides for compact and intuitive characterizations of contagion, it is not always an easy condition to verify. Part of this difficulty is simply because there are 2^n different subsets of players in any given network of n players, and so checking the cohesion of each subset becomes impractical even with relatively few players. Thus verifying whether a given network is r-cohesive needs to take advantage of some structural characteristics of the network. To obtain some feel for the cohesion of different network structures, let us examine the cohesion of a few types of networks, beginning with the simplest case of the complete network. A subset of players S of a complete network is $\frac{|S|-1}{n-1}$-cohesive, since each player in S has $|S| - 1$ neighbors in S and the remaining $n - 1$ neighbors outside S. As the size of the set S grows, so does its cohesion. As the number of players becomes large and the S becomes large relative to N, the cohesion approaches 1, which makes contagion impossible except for q approaching 0. Clearly, this type of network is extraordinary in at least two ways: the degree of each player is large, and every subset of agents forms a clique so that the network is highly clustered. Let us examine the opposite extreme of a tree network in which all agents have degree of no more than some d. To keep things simple, consider a tree in which all agents have degree d or 1. Here it is easy to hit the upper bound of a strict subset being $\frac{d-1}{d}$-cohesive. To see this, simply pick a subtree, as in Figure 9.15, so that there is only one link from one player to the rest of the network. That player has $\frac{d-1}{d}$ of his or her neighbors in the subtree, and the other players in the subtree have all their neighbors in the subtree. As we know that many random networks have some subsets of nodes that are nearly tree-like in their structure when n is large (e.g., see Exercise 1.2), there are many networks in which the cohesion of some subsets is quite high, and so contagion requires a low threshold.[30]

---

30. Morris [487] examines only connected networks on an infinite set of nodes, and for those he shows that an upper bound on the contagion threshold is 1/2 (by showing that every co-finite set—a set with a finite complement—contains an infinite subset that is 1/2 cohesive). These networks are not good approximations for (even very) large finite networks, as we see from the high cohesiveness of various finite networks.

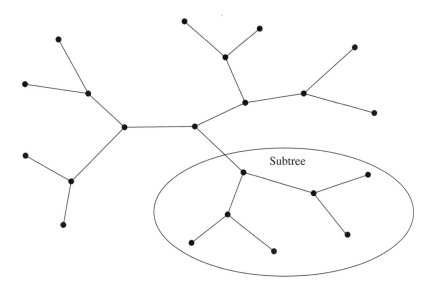

FIGURE 9.15   A subtree that is 2/3-cohesive.

When bridge links are present in a network (so that deleting that link would lead to a new component structure), then it is clear that the two sets of nodes that are bridged each have relatively high cohesion of $\frac{d_i-1}{d_i}$, where i is the bridging node. This cohesion makes contagion difficult and the support of different actions in equilibrium relatively easier.

Contagion is demanding in that it requires that all nodes be reached. Thus a network that has even a few players who are very cohesive among themselves will fail to be susceptible to contagion using the above definition, even though action 1 might spread to almost the entire network. The ideas behind cohesion still provide partial characterizations, as it must be that there is some large set that has low cohesion in nodes to have substantial contagion. A precise characterization of partial contagion is the subject of Exercise 9.17. To better grasp the multiplicities of equilibria and the partial diffusion of actions, let us return to the network games setting.

9.7 ▪ Multiple Equilibria and Diffusion in Network Games*

The various models we have explored often have multiple equilibria, and some of the analyses consider aspects of stability and contagion such as movement from one equilibrium to another. Indeed, the multiplicity of equilibria is important in many applications, and there are many case studies (e.g., see Rogers [564]) in which for some cases a product of behavior diffuses broadly while in other cases it does not. The analysis of contagion in Section 9.6 gives us some feel for the notion but is extreme in requiring that an action be adopted by the entire population. As we have seen in the network games setting and many case studies, there is often some heterogeneity in a population, with some people adhering to one behavior and others adopting a different behavior. Introducing some heterogeneity among players, beyond their degrees, can actually help in producing a more tractable analysis of the structure and stability of multiple equilibria and the diffusion of behavior.

9.7.1 Best-Response Dynamics and Equilibria

An analysis of such diffusion is performed by Jackson and Yariv [364] (see also [362]) in the following context. The setting is similar to the network game setting with a couple of modifications.

Players all begin by taking action 0, which can be thought of as a status quo, for instance, not having bought a product, not having learned a language, or not having become educated. Players are described by their degrees, which indicate the number of future interactions they might undertake. Players' preferences are as in network games, described by (9.4), with one variation. The players also can have some idiosyncratic cost of taking action 1, which is described by c_i. This cost captures the idea that some players might have a personal bias toward buying a given product, or a proclivity or aversion to learning a language, or becoming educated, or the like. If the c_is are all 0, the model reduces to the network games that we considered before. But in the more general model, the payoff to a player is

$$U_{d_i}(0, p)$$

if the player stays with action 0, where p is the probability that any given neighbor chooses action 1, U is the network game payoff as described by (9.4), and

$$U_{d_i}(1, p) - c_i$$

if the player switches to adopt action 1.

Without loss of generality, normalize the payoff to adhering to action 0 to be 0, so that $U_{d_i}(1, p) - c_i$ captures the change in payoffs from switching to action 1 for a given player of degree d_i with idiosyncratic cost c_i and faced with a probability of neighbors' adoptions of p. Thus player i prefers to switch to action 1 when

$$U_{d_i}(1, p) \geq c_i.$$

Let us focus on the case of strategic complements, so that the player's payoff from switching to action 1, $U_d(1, p)$, increases with the probability p of neighbors taking action 1. The case of strategic substitutes is examined in Exercise 9.18.

Let F describe the distribution function of costs, so that $F(c)$ is the probability that any given player's cost c_i is less than or equal to c. Then the probability that a player of degree d prefers action 1 is the probability that his or her cost is less than the benefit from adopting action 1 of $U_d(1, p)$ and so that probability is

$$F(U_d(1, p)).$$

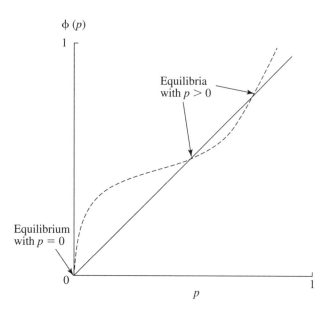

FIGURE 9.16 Resulting fraction of neighbors who choose action 1 ($\phi(p)$) as a best response to a fraction of neighbors who choose action 1 (p).

Now consider the following dynamic.[31] Start with some beginning probability that a player's neighbors will choose action 1, say, p^0. Players then best respond to p^0, which results in a new fraction of players who wish to adopt action 1, p^1, and so forth. Under strategic complements, this process is monotone, so that players never wish to switch back to action 0 as the adoption increases over time. In particular, given a probability that a neighbor chooses action 1, p^t, the new probability that a neighbors chooses action 1 at $t+1$ is

$$p^{t+1} = \phi(p^t) \equiv \sum_d \tilde{P}(d) F(U_d(1, p^t)). \tag{9.12}$$

An equilibrium corresponds to a probability p of neighbors' choosing action 1 such that $p = \phi(p)$. Figure 9.16 pictures a hypothetical function $\phi(p)$, indicating the best-response levels of action 1 as a function of the starting level of action 1.[32] Figure 9.16 shows three different equilibria. There are also situations, as in Figure 9.17, in which a unique equilibrium occurs. In that figure, $\phi(0) > 0$, so that there are some players who choose action 1 regardless of whether any other players do.

---

31. This "dynamic" has various interpretations. It can explicitly be a dynamic, or it might also simply be a tool to define stability of equilibria and study the properties of various equilibria.

32. This sort of analysis of the multiplicity of, dynamics leading to, and stability of equilibria draws from quite standard techniques for analyzing the equilibria of a system. For example, see Fisher [250] for a survey of the analysis of the equilibria of economic systems and Granovetter [306] for an application of such techniques to a social setting.

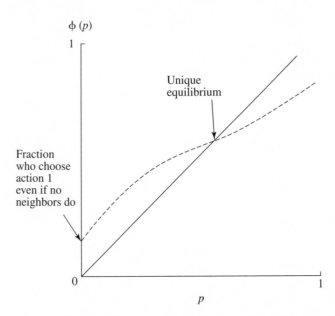

FIGURE 9.17   A unique equilibrium.

9.7.2 Stability

As discussed above, there are various notions of the stability of equilibria. Using dynamics described by best responses and the function ϕ, we can define stability as in Section 9.5.1. That is, start at some equilibrium $p = \phi(p)$, and then perturb p to $p + \varepsilon$ or $p - \varepsilon$ for some small ε, with the constraint that the perturbed probability lie in $[0,1]$. If iterating on ϕ from this point always converges back to p for small enough ε, then the equilibrium is stable; if it does not for arbitrarily small ε, then it is unstable. A stable equilibrium is pictured in Figure 9.18.

Generally, if ϕ cuts the 45-degree line from above, then the equilibrium is stable; if it cuts the 45-degree line from below, then it is unstable.[33] Figure 9.19 shows a multiplicity of equilibria, some stable and others not. The figure shows some interesting aspects of equilibria. The equilibrium at 0 is unstable, and the next higher one is stable. There is a tipping phenomenon such that if p is pushed above 0, then the best-response dynamics leads behavior upward to the higher stable equilibrium. Thus if some initial adoption occurs, then behavior diffuses up to the higher stable equilibrium.

---

33. It is possible to have ϕ be tangent to the 45-degree line, in which case the equilibrium is unstable.

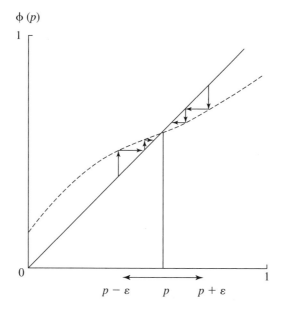

FIGURE 9.18 A stable equilibrium.

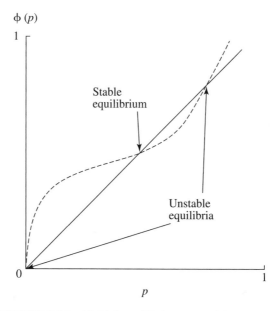

FIGURE 9.19 Multiple equilibria: some stable, some not.

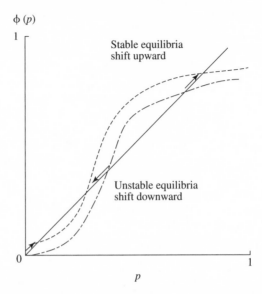

FIGURE 9.20   Change in equilibria due to a shift in best responses.

9.7.3 Equilibrium Behavior and Changes in the Environment

A change in the environment, such as an alteration to the cost of adopting action 1, to the relative attractiveness of the two actions, or to the network structure, leads to a change in the best-response function ϕ.

If the shift is systematic, so that $\phi(p)$ shifts up (or down) for every p, then we can deduce how equilibria change. For instance, in Figure 9.20 the best response to any p is higher under the dashed curve. The figure shows that as ϕ shifts upward, the unstable equilibria move down and the stable equilibria move up. This reconfiguration makes it easier to reach tipping points (the unstable equilibrium) and leads dynamics to reach higher equilibria, so that the diffusion of behavior is more prevalent in a well-defined way. Thus any changes in the setting that result in systematic shifts in ϕ lead to concrete conclusions about how equilibrium behavior will respond, even in the presence of multiple equilibria.

To be careful, these conclusions about changes in equilibria due to an upward shift in ϕ hold for sufficiently small shifts in ϕ and presume that ϕ is a continuous function, which is true if F and U are continuous. Given the continuity of ϕ, any stable equilibrium is locally unique; however, unstable equilibria may not be unique and may even be such that ϕ is tangent to the 45-degree line at some point. If ϕ is continuous, then if we shift $\phi(p)$ up at every point, then for every stable equilibrium there is a new equilibrium that is higher than the old one. In addition, for an unstable equilibrium p, it is possible that there is no longer any equilibrium at or below p.[34] In what follows, the conclusions based on shifts in the best-response function ϕ should be interpreted with these consequences in mind.

---

34. When $p = 1$ is initially an equilibrium, then $\phi(1)$ cannot increase, but any shift up of ϕ at other points would still leave 1 as an equilibrium. Note also the new equilibrium above a given

From (9.12) the best response to a given p is described by

$$\phi(p) = \sum_d \widetilde{P}(d) F(U_d(1, p)). \qquad (9.13)$$

As Jackson and Yariv [364] point out, (9.13) makes clear that some types of changes systematically shift ϕ up. Let us examine such changes.

Lowering the Cost of Changing Actions Changing F, so that costs of adopting action 1 are lower (e.g., increasing $F(c)$ for each c), leads to a shift up in ϕ and so to lower tipping points and higher stable equilibrium choices of action 1.[35] Lowering the costs of taking action 1 corresponds to increasing the probability that the cost of choosing action 1 is below its benefit $U_d(1, p)$, so this change corresponds to increasing $F(U_d(1, p))$ for any given d and p. The result is an increase on the right-hand side of (9.13), so that indeed ϕ shifts up at every point.

Changes in Network Structure Beyond lowering costs, other alterations can lead to the same sorts of systematic changes observed above. For example, consider a first-order stochastic dominance shift from $\widetilde{P}$ to some new distribution of neighbors' degrees $\widetilde{P}'$. If the payoff to choosing action 1, $U_d(1, p)$, is increasing in degree d for any positive proportion $p > 0$ of neighbors taking action 1 and the distribution of costs F is increasing (which is true when the distribution corresponds to a continuous density function on the relevant range), then the result is again an increase on the right-hand side of (9.13) at every positive p and an upward shift of ϕ at every positive p. This shift then leads to an increase in stable equilibria and a lowering (or disappearance) of the unstable equilibria or tipping points, and similarly to the case of lowering of costs, we should expect higher overall diffusion in the sense that there are lower thresholds for diffusion and higher equilibria.

Note that we can compare this situation to Proposition 9.6, which did not distinguish between stable and unstable equilibria but did deal with shifts. Here, by accounting for all equilibria and seeing how ϕ adjusts, we have a more complete picture of how equilibria change with modifications to the environment. Proposition 9.6 only concluded that for any equilibrium under the old distribution, there is at least one that has moved up under the new distribution, so that the highest equilibrium is (weakly) higher than it was before.

As Jackson and Yariv [362] point out, we can also examine changes in terms of mean-preserving spreads of the degree distribution, recalling how such spreads affect the expectation of a convex function (see Section 4.5.5). The impact of such a change depends on the convexity of $F(U_d(1, p))$ as a function of d. If F is convex and increasing in d, then the change in equilibria can be well ordered, as $\phi(p)$

---

stable equilibrium might no longer be stable, as the new ϕ may be tangent to the 45-degree line at the higher equilibrium.

35. This statement presumes that ϕ is continuous after the change.

increases for every p. That is, a mean-preserving spread in the degree distribution, such as a change from a Poisson degree distribution to a scale-free degree distribution, leads to more diffusion of action 1 (in the sense of shifting stable equilibria to be higher and unstable equilibria to be lower). This increased diffusion might apply in the case of strategic complements, provided there are sufficient complementarities. In contrast, if the compound function $F(U_d(1, p))$ is concave and decreasing in d, then the shift is reversed. Thus if $F(U_d(1, p))$ is increasing and convex in d, then power, Poisson, and regular degree distributions with identical means generate corresponding best-response functions ϕ^{power}, ϕ^{Poisson}, and ϕ^{regular} such that $\phi^{\text{power}}(p) \geq \phi^{\text{Poisson}}(p) \geq \phi^{\text{regular}}(p)$ for all p.

The various graphical game and network game models examined provide a basis for understanding how equilibria behave as a function of payoffs, player degree or position in the network, and network characteristics. The multiplicity of equilibria make systematic conclusions a challenge, and we also see some sensitivity of the conclusions to the fine details of the setting. Nonetheless, such concepts as strategic complementarities provide powerful tools that allow us to draw some fairly specific conclusions about equilibrium properties and how they vary with network structure.

9.8 ▪ Computing Equilibria*

Beyond the analysis of graphical games and network games, it is also important to know how to compute an equilibrium and to appreciate the difficulty of finding one. This knowledge is not only useful for researchers or scientists exploring the behavior of a society but also important in determining whether the society will reach an equilibrium. There are a variety of ways that we might posit that players adjust their behavior, including deductive reasoning, communicating with others, updating strategies over time in response to what actions have been played in the past, or responding to evolutionary or other selective pressures over time. If computing an equilibrium using the full description of the game is so complicated that it is not expected to be done in finite time, it is hard to expect a system to lead to equilibrium (as otherwise we could mimic that system to compute equilibria). Also, knowing something about the multiplicity of equilibria is also important, for at least two reasons. Having many equilibria lowers predictive power, as more profiles of behavior are consistent with the model. And having many equilibria can make it difficult for players to coordinate or reach equilibrium, even if they can communicate with one another.

Obviously, these issues have not escaped game theorists' attention, and a good deal of attention has been devoted to understanding the multiplicities of equilibria; computing equilibria; modeling how societies might learn or evolve to play equilibria; refining equilibrium predictions; and studying focalness, social norms, and other methods of coordination. I will not try to distill such a breadth of material here and instead I refer the reader to standard game-theory texts, such as Binmore [70], Fudenberg and Tirole [263], Myerson [497], and Osborne and Rubinstein [514], to learn more about these issues. It is important, however, to see how these issues manifest themselves in graphical games. So let us begin by computing equilibria.

Computing equilibria for the threshold games described in Example 9.2 is quite easy and takes advantage of the strategic complementarities of the game, as we have already seen in special cases. Here is a method that takes at most $(n+1)n/2$ steps for any such threshold game.

Set all players' actions to 1. Now consider player 1. If player 1 would improve by changing to action 0, then do so and otherwise leave the profile of actions as it is. Do the same for player 2, given the new profile of actions. Continue iterating in this manner. After passing through all players, repeat the procedure. Stop adjusting actions when all players have been considered and no actions have changed.

This algorithm takes advantage of the fact that players' actions only change from 1 to 0, and given the strategic structure of the game, such changes can only lead other players to change from 1 to 0, but never to reverse their decisions. Actually, this algorithm finds the maximum equilibrium in the sense that there is no other equilibrium where any player ever takes a higher action (see Exercise 9.22). Moreover, the technique also finds the maximum equilibrium in a wider class of games in which strategies are ordered and there are such complementarities between strategies.[36]

Next, consider the best-shot public goods game in the case of an undirected network.[37] As noted, the pure strategy equilibria occur when the players who take action 1 form a maximal independent set. Maximal independent sets are easy to find. For instance, some of the maximal independent sets on a tree can be found as follows. The following set A is a maximal independent set:

$$A = \bigcup_{m : m \text{ is even}} D_i^m(g),$$

where $D_i^m(g)$ are the nodes at distance m from some i in g (with $D_i^0(g) = \{i\}$ being one of the sets where m is even).

Even when the network is not a tree, there still exist obvious (and fast) methods of finding maximal independent sets. Here is an algorithm that finds an equilibrium for a connected network (N, g). By applying it to each component, this method can be used for any network. At step k, the algorithm constructs sets A_k and B_k; the eventual maximal independent set will be the final A_k. In terms of finding an equilibrium for the best-shot game, the final A_k is the list of players who take action 1, and the final B_k is the set of players who take action 0:

Step 1: Pick some node i and let A_1 be i and B_1 be i's neighbors ($A_1 = \{i\}$ and $B_1 = D_i^1(g)$).

Step 2: Pick some node j at distance 2 from i ($j \in D_i^2(g)$), and let $A_2 = \{i, j\}$ and $B_2 = B_1 \cup D_j^1(g)$.

36. For more background on games with strategic complementarities, see Topkis [630], Vives [642], and Milgrom and Roberts [469].

37. For a directed network in the best-shot public goods game the analysis is a bit different, as outlined in Exercise 9.6.

Step k: Iterate by selecting one of the players j' who has a minimal distance
            to i out of those players not yet assigned to a set A_{k-1} or B_{k-1}. Let
            $A_k = A_{k-1} \cup \{j'\}$ and $B_k = B_{k-1} \cup D_{j'}^1$.[38]

End: Stop when $A_k \cup B_k = N$.

Although varying the starting player i and the order in which new players are chosen in each step k results in finding several different equilibria, there can be many more that are not found by this algorithm (see Exercise 9.23).

For the general class of graphical games, computing equilibria becomes more challenging. For example, consider a game in which players have thresholds for choosing action 1 but are also concerned about congestion, so that they do not take action 1 if too many neighbors choose action 1. In particular, a player has a lower threshold and an upper threshold, so that the player prefers action 1 if and only if the number of neighbors playing 1 lies between the two thresholds. Moreover, allow the lower and upper thresholds to differ across players. In such a setting, which might not admit any pure strategy equilibria, it could be hard to find even one equilibrium pure or mixed. Each time one player's strategy is adjusted, we may have to adjust the strategies of previously considered players in response, as these can feed back on one another. In the best-shot and threshold examples, changing a player's strategy in one direction had clear implications for how others respond, but more generally, as in the example with multiple thresholds, the feedback and interaction can be complex. The details of defining what is meant by a "hard to find" equilibrium are beyond the scope of this text, but here I sketch the basic ideas.

For any algorithm that computes equilibria, there are some inputs that describe the game. In the case of a graphical game, the inputs are the number of players, the network that connects them, and each player's payoff function. The number of players is simply n, and the information about the network can be coded in an $n \times n$ matrix, so that there are n^2 bits of information (although this number can be lowered in some classes of games). If each player has a degree of at most d, then each player's payoff matrix has 2^{d+1} entries, indicating the payoff as a function of each vector of choices of 0 or 1 for each neighbor and the player. Thus the full description of the game involves on the order of n^2 plus n times a constant (related to the maximum d) bits of information. Given this information, we construct an algorithm for finding an equilibrium. This involves prescribing a series of steps that use the information about the game to do some calculations and eventually spit out a list of strategies for each player. How many steps will it take to terminate with the determination of an equilibrium? The method of counting steps that is generally followed is to look at the upper bound, or worst possible performance. So the performance measure is to find a game and payoff structure that would lead to the most steps before the given algorithm finds an equilibrium and to keep track of this number as a function of n. Algorithms are considered to be relatively quick if the upper bound on the number of steps needed is at most some polynomial function of n. Slow algorithms require more steps than polynomial in n for some games, for example, when worst-case scenarios use a number of

---

38. Note that these sets are well defined, since no neighbors of j' can be in A_{k-1}, as otherwise j' would have been in B_{k-1}.

steps that grows exponentially with n. It is difficult to show that a problem is such that all algorithms sometimes require more than a polynomial number of steps. There is a deep and long-standing open question on which hinges the answer to complexity for a number of problems, including equilibrium computation in a class of graphical games.[39]

There are several side issues of interest here. For instance, is it reasonable to measure the performance of an algorithm for computing equilibria based on worst cases? Is an exponential number of steps really that much more than a high-order polynomial for some n? How large does n have to be before there is a serious distinction between polynomial and exponential time for the problem in question? Is this the right accounting for complexity, given that we are not considering the complexity of the calculation at each step? Are there large classes of graphical games for which the task is much easier, and are the graphical games for which computing an equilibrium is difficult very interesting? What happens when we look for approximate equilibria (so that players nearly maximize their payoff) instead of exact equilibria? Are there other definitions of equilibria for which computation is easy? These are all difficult and open questions that have received attention. There are also other difficulties that we face with graphical games. For instance, if players do not have some maximal degree, but the maximal degree grows with n, then describing payoffs could take up to 2^n bits of information. Part of the reason that the threshold and best-shot games were easy to handle is that the payoffs were quite simple to describe.[40]

Here I summarize what is known about finding equilibria of graphical games. When the network is a tree, there are algorithms that involve a polynomial in n number of steps and find an equilibrium of any graphical game on a tree, as shown by Littman, Kearns, and Singh [437]. Once we venture beyond trees, however, the strong conjecture is that no such algorithm exists with a number of steps that is always polynomial in n (see Daskalakis, Goldberg, and Papadimitriou [185]).[41] Although this conjecture is somewhat pessimistic with regards to being able to make predictions of behavior in the broad class of graphical games, there is often much more structure to the games of interest compared to the worst-case scenarios. As we have seen, strategic complementarities make finding equilibria easy and fast.[42]

---

39. For more background on algorithms and complexity, see Papadimitriou [523] and Roughgarden [570].

40. For more about the complexity of describing payoffs and representing such games see Daskalakis and Papadimitriou [183].

41. The strong conjecture is based on the fact that the problem of computing Nash equilibria in a graphical game has been shown to be equivalent to a problem (lying in a class called *PPAD-complete*) that is conjectured to have no polynomial time algorithm for finding a solution. That conjecture is among a class of long-standing open problems regarding the complexity of algorithms, which have received a great deal of attention.

42. Note also that while computing Nash equilibria in general graphical (and other large) games can be hard, there are polynomial-time algorithms for finding correlated equilibria (which are a generalization of Nash equilibria that admit correlation in the players' strategies) in certain graphical games that have nice representations. See Kakade et al. [373] and Papadimitriou and Roughgarden [525].

9.9 ▪ Appendix: A Primer on Noncooperative Game Theory

This appendix discusses what is known as *noncooperative game theory,* in which agents act in self-interested ways to maximize their own payoffs and equilibrium notions are applied to predict outcomes. Cooperative game theory examines coalitions and how payoffs might be allocated within coalitions. It is examined in Section 12.1.

The basic elements of performing a noncooperative game-theoretic analysis are (1) framing the situation in terms of the actions available to players and their payoffs as a function of actions, and (2) using various equilibrium notions to make either descriptive or prescriptive predictions. In framing the analysis, a number of questions become important. First, who are the players? They may be people, firms, organizations, governments, ethnic groups, and so on. Second, what actions are available to them? All actions that the players might take that could affect any player's payoffs should be listed. Third, what is the timing of the interactions? Are actions taken simultaneously or sequentially? Are interactions repeated? The order of play is also important. Moving after another player may give player i an advantage of knowing what the other player has done; it may also put player i at a disadvantage in terms of lost time or the ability to take some action. What information do different players have when they take actions? Fourth, what are the payoffs to the various players as a result of the interaction? Ascertaining payoffs involves estimating the costs and benefits of each potential set of choices by all players. In many situations it may be easier to estimate payoffs for some players (such as yourself) than others, and it may be unclear whether other players are also thinking strategically. This consideration suggests that careful attention be paid to a sensitivity analysis.

Once we have framed the situation, we can look from different players' perspectives to analyze which actions are optimal for them. There are various criteria we can use.

9.9.1 Games in Normal Form

Let us begin with a standard representation of a game, which is known as a *normal form* game, or a game in *strategic form:*

The set of players is $N = \{1, \ldots, n\}$.

Player i has a set of actions, X_i, available. These are generally referred to as *pure strategies.* This set might be finite or infinite.

Let $X = X_1 \times \cdots \times X_n$ be the set of all profiles of pure strategies or actions, with a generic element denoted by $x = (x_1, \ldots, x_n)$.

Player i's payoff as a function of the vector of actions taken is described by a function $u_i : X \to \mathbb{R}$, where $u_i(x)$ is i's payoff if the x is the profile of actions chosen in the society.

Normal form games are often represented by a table. Perhaps the most famous such game is the *prisoners' dilemma,* which is represented in Table 9.3. In this game

TABLE 9.3
A prisoners' dilemma game

		Player 2	
		C	D
Player 1	C	−1, −1	−3, 0
	D	0, −3	−2, −2

Note: C, cooperate; D, defect.

TABLE 9.4
A rescaling of the prisoners' dilemma
game

		Player 2	
		C	D
Player 1	C	4, 4	0, 6
	D	6, 0	2, 2

Note: C, cooperate; D, defect.

there are two players who each have two pure strategies, where $X_i = \{C, D\}$, and C stands for "cooperate" and D stands for "defect." The first entry indicates the payoff to the row player (or player 1) as a function of the pair of actions, while the second entry is the payoff to the column player (or player 2).

The usual story behind the payoffs in the prisoners' dilemma is as follows. The two players have committed a crime and are now in separate rooms in a police station. The prosecutor has come to each of them and told them each: "If you confess and agree to testify against the other player, and the other player does not confess, then I will let you go. If you both confess, then I will send you both to prison for 2 years. If you do not confess and the other player does, then you will be convicted and I will seek the maximum prison sentence of 3 years. If nobody confesses, then I will charge you with a lighter crime for which we have enough evidence to convict you and you will each go to prison for 1 year." So the payoffs in the matrix represent time lost in terms of years in prison. The term *cooperate* refers to cooperating with the other player. The term *defect* refers to confessing and agreeing to testify, and so breaking the (implicit) agreement with the other player.

Note that we could also multiply each payoff by a scalar and add a constant, which is an equivalent representation (as long as all of a given player's payoffs are rescaled in the same way). For instance, in Table 9.4 I have doubled each entry and added 6. This transformation leaves the strategic aspect of the game unchanged.

There are many games that might have different descriptions motivating them but have a similar normal form in terms of the strategic aspects of the game. Another example of the same game as the prisoners' dilemma is what is known as a *Cournot duopoly*. The story is as follows. Two firms produce identical goods. They each have two production levels, high or low. If they produce at high production, they will have a lot of the goods to sell, while at low production they have less to sell. If they cooperate, then they agree to each produce at low production. In this case, the product is rare and fetches a very high price on the market, and they each make a profit of 4. If they each produce at high production (or defect), then they will depress the price, and even though they sell more of the goods, the price drops sufficiently to lower their overall profits to 2 each. If one defects and the other cooperates, then the price is in a middle range. The firm with the higher production sells more goods and earns a higher profit of 6, while the firm with the lower production just covers its costs and earns a profit of 0.

9.9.2 Dominant Strategies

Given a game in normal form, we then can make predictions about which actions will be chosen. Predictions are particularly easy when there are dominant strategies. A dominant strategy for a player is one that produces the highest payoff of any strategy available *for every possible action by the other players.*

That is, a strategy $x_i \in X_i$ is a *dominant* (or weakly dominant) strategy for player i if $u_i(x_i, x_{-i}) \geq u_i(x_i', x_{-i})$ for all x_i' and all $x_{-i} \in X_{-i}$. A strategy is a *strictly dominant strategy* if the above inequality holds strictly for all $x_i' \neq x_i$ and all $x_{-i} \in X_{-i}$.

Dominant strategies are powerful from both an analytical point of view and a player's perspective. An individual does not have to make any predictions about what other players might do, and still has a well-defined best strategy.

In the prisoners' dilemma, it is easy to check that each player has a strictly dominant strategy to defect—that is, to confess to the police and agree to testify. So, if we use dominant strategies to predict play, then the unique prediction is that each player will defect, and both players fare worse than for the alternative strategies in which neither defects. A basic lesson from the prisoners' dilemma is that individual incentives and overall welfare need not coincide. The players both end up going to jail for 2 years, even though they would have gone to jail for only 1 year if neither had defected. The problem is that they cannot trust each other to cooperate: no matter what the other player does, a player is best off defecting.

Note that this analysis presumes that all relevant payoff information is included in the payoff function. If, for instance, a player fears retribution for confessing and testifying, then that should be included in the payoffs and can change the incentives in the game. If the player cares about how many years the other player spends in jail, then that can be written into the payoff function as well.

When dominant strategies exist, they make the game-theoretic analysis relatively easy. However, such strategies do not always exist, and then we can turn to notions of equilibrium.

9.9.3 Nash Equilibrium

A pure strategy Nash equilibrium[43] is a profile of strategies such that each player's strategy is a best response (results in the highest available payoff) against the equilibrium strategies of the other players.

A strategy x_i is a *best reply*, also known as a *best response,* of player i to a profile of strategies $x_{-i} \in X_{-i}$ for the other players if

$$u_i(x_i, x_{-i}) \geq u_i(x_i', x_{-i})$$

for all x_i'. A best response of player i to a profile of strategies of the other players is said to be a *strict best response* if it is the unique best response.

A profile of strategies $x \in X$ is a *pure strategy Nash equilibrium* if x_i is a best reply to x_{-i} for each i. That is, x is a Nash equilibrium if

$$u_i(x_i, x_{-i}) \geq u_i(x_i', x_{-i})$$

for all i and x_i'. This definition might seem somewhat similar to that of dominant strategy, but there is a critical difference. A pure strategy Nash equilibrium only requires that the action taken by each agent be best against the actions taken by the other players, and not necessarily against all possible strategies of the other players.

A Nash equilibrium has the nice property that it is stable: if each player expects x to be the profile of strategies played, then no player has any incentive to change his or her action. In other words, no player regrets having played the strategy that he or she played in a Nash equilibrium.

In some cases, the best response of a player to the strategies of others is unique. A Nash equilibrium such that all players are playing strategies that are unique best responses is called a *strict* Nash equilibrium. A profile of dominant strategies is a Nash equilibrium but not vice versa.

To see a Nash equilibrium in action, consider the following game between two firms that are deciding whether to advertise. Total available profits are 28, to be split between the two firms. Advertising costs a firm 8. Firm 1 currently has a larger market share than firm 2, so it is seeing 16 in profits while firm 2 is seeing 12 in profits. If they both advertise, then they will split the market evenly 14 each but must bear the cost of advertising, so they will see profits of 6 each. If one advertises while the other does not, then the advertiser captures three-quarters of the market (but also pays for advertising) and the nonadvertiser gets one-quarter of the market. (There are obvious simplifications here: just considering two levels of advertising and assuming that advertising only affects the split and not the total profitability.) The net payoffs are given in the Table 9.5.

To find the equilibrium, we have to look for a pair of actions such that neither firm wants to change its action given what the other firm has chosen. The search

---

43. On occasion referred to as *Cournot–Nash equilibrium,* with reference to Cournot [179], who first developed such an equilibrium concept in the analysis of oligopoly (a set of firms in competition with one another).

TABLE 9.5
An advertising game

		Firm 2	
		Not	Adv
Firm 1	Not	16, 12	7, 13
	Adv	13, 7	6, 6

Note: Adv, advertise; Not, do not advertise.

TABLE 9.6
A coordination game

		Player 2	
		A	B
Player 1	A	5, 5	0, 3
	B	3, 0	4, 4

is made easier in this case, since firm 1 has a strictly dominant strategy of not advertising. Firm 2 does not have a dominant strategy; which strategy is optimal for it depends on what firm 1 does. But given the prediction that firm 1 will not advertise, firm 2 is best off advertising. This forms a Nash equilibrium, since neither firm wishes to change strategies. You can easily check that no other pairs of strategies form an equilibrium.

While each of the previous games provides a unique prediction, there are games in which there are multiple equilibria. Here are three examples.

Example 9.8 (A Stag Hunt Game) *The first is an example of a coordination game, as depicted in Table 9.6. This game might be thought of as selecting between two technologies, or coordinating on a meeting location. Players earn higher payoffs when they choose the same action than when they choose different actions. There are two (pure strategy) Nash equilibria: (A, A) and (B, B).*

This game is also a variation on Rousseau's "stag hunt" game.[44] The story is that two hunters are out, and they can either hunt for a stag (strategy A) or look for hares (strategy B). Succeeding in getting a stag takes the effort of both hunters, and the hunters are separated in the forest and cannot be sure of each other's behavior. If both hunters are convinced that the other will hunt for stag, then hunting stag is a strict or unique best reply for each player. However, if one turns out to be mistaken and the other hunter hunts for hare, then one will go hungry. Both hunting for hare

---

44. To be completely consistent with Rousseau's story, (B, B) should result in payoffs of (3, 3), as the payoff to hunting for hare is independent of the actions of the other player in Rousseau's story.

TABLE 9.7
A battle of the sexes game

| | | Player 2 | |
		A	B
Player 1	A	3, 1	0, 0
	B	0, 0	1, 3

is also an equilibrium and hunting for hare is a strict best reply if the other player is hunting for hare. This example hints at the subtleties of making predictions in games with multiple equilibria. On the one hand, (A, A) (hunting stag by both) is a more attractive equilibrium and results in high payoffs for both players. Indeed, if the players can communicate and be sure that the other player will follow through with an action, then playing (A, A) is a stable and reasonable prediction. However, (B, B) (hunting hare by both) has properties that make it a useful prediction as well. It does not offer as high a payoff, but it has less risk associated with it. Here playing B guarantees a minimum payoff of 3, while the minimum payoff to A is 0. There is an extensive literature on this subject, and more generally on how to make predictions when there are multiple equilibria (see, e.g., Binmore [70], Fudenberg and Tirole [263], Myerson [497], and Osborne and Rubinstein [514]).

Example 9.9 (A "Battle of the Sexes" Game) *The next example is another form of coordination game, but with some asymmetries in it. It is generally referred to as a "battle of the sexes" game, as depicted in Table 9.7. The players have an incentive to choose the same action, but they each have a different favorite action. There are again two (pure strategy) Nash equilibria: (A, A) and (B, B). Here, however, player 1 would prefer that they play equilibrium (A, A) and player 2 would prefer (B, B). The battle of the sexes title refers to a couple trying to coordinate on where to meet for a night out. They prefer to be together, but also have different preferred outings.*

Example 9.10 (Hawk-Dove and Chicken Games) *There are also what are known as anticoordination games, with the prototypical version being what is known as the hawk-dove game or the chicken game, with payoffs as in Table 9.8. Here there are two pure strategy equilibria, (Hawk, Dove) and (Dove, Hawk). Players are in a potential conflict and can be either aggressive like a hawk or timid like a dove. If they both act like hawks, then the outcome is destructive and costly for both players with payoffs of 0 for both. If they each act like doves, then the outcome is peaceful and each gets a payoff of 2. However, if the other player acts like a dove, then a player would prefer to act like a hawk and take advantage of the other player, receiving a payoff of 3. If the other player is playing a hawk strategy, then it is best to play a dove strategy and at least survive rather than to be hawkish and end in mutual destruction.*

TABLE 9.8
A hawk-dove game

| | | Player 2 | |
		Hawk	Dove
Player 1	Hawk	0, 0	3, 1
	Dove	1, 3	2, 2

TABLE 9.9
A penalty-kick game

| | | Goalie | |
		L	R
Kicker	L	−1, 1	1, −1
	R	1, −1	−1, 1

Note: L, left; R, right.

9.9.4 Randomization and Mixed Strategies

In each of the above games, there was at least one pure strategy Nash equilibrium. There are also simple games for which no pure strategy equilibrium exists. To see this, consider the following simple variation on a penalty kick in a soccer match. There are two players: the player kicking the ball and the goalie. Suppose, to simplify the exposition, that we restrict the actions to just two for each player (there are still no pure strategy equilibria in the larger game, but this limitation makes the exposition easier). The kicking player can kick to the left side or to the right side of the goal. The goalie can move to the left side or to the right side of the goal and has to choose before seeing the kick, as otherwise there is too little time to react. To keep things simple, assume that if the player kicks to one side, then she scores for sure if the goalie goes to the other side, while the goalie is certain to save it if the goalie goes to the same side. The basic payoff structure is depicted in Table 9.9. This is also the game known as "matching pennies." The goalie would like to choose a strategy that matches that of the kicker, and the kicker wants to choose a strategy that mismatches the goalie's strategy.[45]

It is easy to check that no pair of pure strategies forms an equilibrium. What is the solution here? It is just what you see in practice: the kicker randomly picks left versus right, in this particular case with equal probability, and the goalie does the same. To formalize this observation we need to define randomized strategies, or

---

45. For an interesting empirical test of whether goalies and kickers on professional soccer teams randomize properly, see Chiappori, Levitt, and Groseclose [150]; and see Walker and Wooders [646] for an analysis of randomization in the location of tennis serves in professional tennis matches.

what are called *mixed strategies*. For ease of exposition suppose that X_i is finite; the definition extends to infinite strategy spaces with proper definitions of probability measures over pure actions.

A mixed strategy for a player i is a distribution μ_i on X_i, where $\mu_i(x_i)$ is the probability that x_i is chosen. A profile of mixed strategies $(\mu_1, \ldots, \mu_n)$ forms a mixed-strategy Nash equilibrium if

$$\sum_x \left(\prod_j \mu_j(x_j) \right) u_i(x) \geq \sum_{x_{-i}} \left(\prod_{j \neq i} \mu_j(x_j) \right) u_i(x_i', x_{-i})$$

for all i and x_i'.

So a profile of mixed strategies is an equilibrium if no player has some strategy that would offer a better payoff than his or her mixed strategy in reply to the mixed strategies of the other players. Note that this reasoning implies that a player must be indifferent to each strategy that he or she chooses with a positive probability under his or her mixed strategy. Also, players' randomizations are independent.[46] A special case of a mixed strategy is a pure strategy, where probability 1 is placed on some action.

It is easy to check that each mixing with probability 1/2 on L and R is the unique mixed strategy of the matching pennies game above. If a player, say the goalie, places weight of more than 1/2 on L, for instance, then the kicker would have a best response of choosing R with probability 1, but then that could not be an equilibrium as the goalie would want to change his or her action, and so forth.

There is a deep and long-standing debate about how to interpret mixed strategies, and the extent to which people really randomize. Note that in the goalie and kicker game, what is important is that each player not know what the other player will do. For instance, it could be that the kicker decided before the game that if there was a penalty kick then she would kick to the left. What is important is that the kicker not be known to always kick to the left.[47]

We can begin to see how the equilibrium changes as we change the payoff structure. For example, suppose that the kicker is more skilled at kicking to the right side than to the left. In particular, keep the payoffs as before, but now suppose that the kicker has an even chance of scoring when kicking right when the goalie goes right. This leads to the payoffs in Table 9.10. What does the equilibrium look like? To calculate the equilibrium, it is enough to find a strategy for the goalie that

---

46. An alternative definition of correlated equilibrium allows players to use correlated strategies but requires some correlating device that only reveals to each player his or her prescribed strategy and that these are best responses given the conditional distribution over other players' strategies.

47. The contest between pitchers and batters in baseball is quite similar. Pitchers make choices about the location, velocity, and type of pitch (e.g., whether various types of spin are put on the ball). If a batter knows what pitch to expect in a given circumstance, that can be a significant advantage. Teams scout one another's players and note any tendencies or biases that they might have and then try to respond accordingly.

TABLE 9.10
A biased penalty-kick game

		Goalie	
		L	R
Kicker	L	$-1, 1$	$1, -1$
	R	$1, -1$	$0, 0$

Note: L, left; R, right.

makes the kicker indifferent, and a strategy for the kicker that makes the goalie indifferent.[48]

Let μ_1 be the kicker's strategy and μ_2 be the goalie's strategy. It must be that the kicker is indifferent. The kicker's payoff from L is $-\mu_2(L) + \mu_2(R)$ and the payoff from R is $\mu_2(L)$, so that

$$-\mu_2(L) + \mu_2(R) = \mu_2(L),$$

or $\mu_2(L) = 1/3$ and $\mu_2(R) = 2/3$. For the goalie to be indifferent, it must be that

$$\mu_1(L) - \mu_1(R) = -\mu_1(L) + \mu_1(R),$$

and so the kicker must choose $\mu_1(L) = 1/2 = \mu_2(R)$.

Note that as the kicker gets more skilled at kicking to the right, it is the *goalie's* strategy that adjusts to moving to the right more often! The kicker still mixes evenly. It is a common misconception to presume that it should be the kicker who should adjust to using his or her better strategy with more frequency.[49]

While not all games have pure strategy Nash equilibrium, every game with a finite set of actions has at least one mixed strategy Nash equilibrium (with a special case of a mixed strategy equilibrium being a pure strategy equilibrium), as shown in an important paper by Nash [498].

---

48. This reasoning is a bit subtle, as we are not directly choosing actions that maximize the goalie's payoff and maximize the kicker's payoff, but instead are looking for a mixture by one player that makes the other indifferent. This feature of mixed strategies takes a while to grasp, but experienced players seem to understand it well (e.g., see Chiappori, Levitt, and Groseclose [150] and Walker and Wooders [646]).

49. Interestingly, there is evidence that professional soccer players are better at playing games that have mixed strategy equilibria than are people with less experience in such games, which is consistent with this observation (see Palacios-Huerta and Volij [521]).

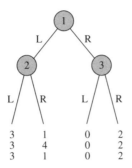

FIGURE 9.21   A game tree with three players and two actions each.

9.9.5 Sequentiality, Extensive Form Games, and Backward Induction

Let us now turn to the question of timing. In the above discussion it was implicit that each player was selecting a strategy with beliefs about the other players' strategies but without knowing exactly what they were.

If we wish to be more explicit about timing, then we can consider what are known as games in *extensive form,* which include a complete description of who moves in what order and what they have observed when they move.[50] There are advantages to working with extensive form games, as they allow for more explicit treatments of timing and for equilibrium concepts that require credibility of strategies in response to the strategies of others.

Definitions for a general class of extensive form games are notationally intensive. In this book, we mainly look at a special class of extensive form games—finite games of perfect information—which allows for a treatment that avoids much of the notation. These are games in which players move sequentially in some pre-specified order (sometimes contingent on which actions have been chosen), each player moves at most a finite number of times, and each player is completely aware of all moves that have been made previously. These games are particularly well behaved and can be represented by simple trees, where a node is associated with the move of a specified player and an edge corresponds to different actions the player might take, as in Figure 9.21. I will not provide formal definitions, but simply refer directly to games representable by such finite game trees.

Each node has a player's label attached to it. There is an identified *root node* that corresponds to the first player to move (player 1 in Figure 9.21) and then subsequent nodes that correspond to subsequent players who make choices. In Figure 9.21, player 1 has a choice of moving either left or right. The branches in the tree correspond to the different actions available to the player at a given node.

---

50. One can collapse certain types of extensive form games into normal form by simply defining an action to be a complete specification of how an agent would act in all possible contingencies. Agents then choose these actions simultaneously at the beginning of the game. But the normal form becomes more complicated than the two-by-two games in Sections 9.9.3 and 9.9.4.

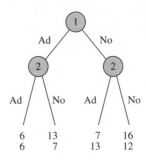

FIGURE 9.22   Advertising choices of two competitors.

In this game, if player 1 moves left, then player 2 moves next; while if player 1 moves right, then player 3 moves next. It is also possible to have trees in which player 1 chooses twice in a row, or no matter what choice a given player makes it is a certain player who follows, and so forth. The payoffs are given at the end nodes and are listed for the respective players. The top payoff is for player 1, the second for player 2, and the bottom for player 3. So the payoffs depend on the set of actions taken, which then determines a path through the tree. An equilibrium provides a prediction about how each player will move in each contingency and thus makes a prediction about which path will be taken; we refer to that prediction as the *equilibrium path*.

We can apply the concept of a Nash equilibrium to such games, which here is a specification of what each player would do at each node with the requirement that each player's strategy be a best response to the other players' strategies. Nash equilibrium does not always make sensible predictions when applied to the extensive form. For instance, reconsider the advertising example discussed above (Table 9.5). Suppose that firm 1 makes its decision prior to firm 2, and that firm 2 knows firm 1's choice before it chooses. This scenario is represented in Figure 9.22. To apply the Nash equilibrium concept to this extensive form game, we must specify what each player does at each node. There are two Nash equilibria of this game in pure strategies. The first is where firm 1 advertises, and firm 2 does not (and firm 2's strategy conditional on firm 1 not advertising is to advertise). The other equilibrium corresponds to the one identified in the normal form: firm 1 does not advertise, and firm 2 advertises regardless of what firm 1 does. This is an equilibrium, since neither wants to change its behavior, given the other's strategy. However, it is not really credible in the following sense: it involves firm 2 advertising even after it has seen that firm 1 has advertised, and even though this action is not in firm 2's interest in that contingency.

To capture the idea that each player's strategy has to be credible, we can solve the game backward. That is, we can look at each decision node that has no successor, and start by making predictions at those nodes. Given those decisions, we can roll the game backward and decide how player's will act at next-to-last decision nodes, anticipating the actions at the last decision nodes, and then iterate. This is called *backward induction*. Consider the choice of firm 2, given that firm 1 has decided not to advertise. In this case, firm 2 will choose to advertise, since 13 is larger than 12. Next, consider the choice of firm 2, given that firm 1 has decided to advertise. In this case, firm 2 will choose not to advertise, since 7 is larger than 6. Now we can collapse the tree. Firm 1 will predict that if it does not advertise, then firm 2 will advertise, while if firm 1 advertises then firm 2 will not. Thus when making its choice, firm 1 anticipates a payoff of 7 if it chooses not to advertise and 13 if it chooses to advertise. Its optimal choice is to advertise. The backward induction prediction about the actions that will be taken is for firm 1 to advertise and firm 2 not to.

Note that this prediction differs from that in the simultaneous move game we analyzed before. Firm 1 has gained a first-mover advantage in the sequential version. Not advertising is no longer a dominant strategy for firm 1, since firm 2's decision depends on what firm 1 does. By committing to advertising, firm 1 forces firm 2 to choose not to advertise. Firm 1 is better off being able to commit to advertising in advance.

A solution concept that formalizes the backward induction solution found in this game and applies to more general classes of games is known as *subgame perfect equilibrium* (due to Selten [586]). A subgame in terms of a finite game tree is simply the subtree that one obtains starting from some given node. Subgame perfection requires that the stated strategies constitute a Nash equilibrium in every subgame (including those with only one move left). So it requires that if we start at any node, then the strategy taken at that node must be optimal in response to the remaining specification of strategies. In the game between the two firms, it requires that firm 2 choose an optimal response in the subgame following a choice by firm 1 to advertise, and so it coincides with the backward induction solution for such a game.

I close this appendix by noting that moving first is not always advantageous. Sometimes it allows one to commit to strategies which would otherwise be untenable, which can be advantageous; but in other cases the information that the second mover gains from knowing which strategy the first mover has chosen may be a more important consideration. For example, when playing the matching pennies game sequentially, it is clearly not good for a player to move first.

9.10 ■ Exercises

9.1 *Fashionable Ants*   Consider the model described in Section 9.1.2. Suppose that a player has a probability $\varepsilon > 0$ of flipping a coin to choose an (binary) action, and a probability of $1 - \varepsilon$ of matching the action being taken by the majority of other individuals. Taking n to be even, so that the number of other individuals is always odd, describe p_s for any s. Next, pick a value of n, and for several values of ε plot the steady-state probability of there being s individuals taking action 1 as a function of s (similar to Figure 9.1).

9.2 *Proof of Proposition 9.1*   Prove Proposition 9.1.

9.3 *Steady-State Probabilities of Action*   Consider the following variation on a model of social interaction by Calvó-Armengol and Jackson [131] and discussed in Section 9.1.2. Let $p_s = q$ for some $1 > q > 0$ when $s \geq \tau$ and $p_s = 1 - q$ when $s < \tau$, where $\tau \in \{0, \ldots, n - 1\}$ is a threshold. Thus individuals choose action 1 with probability q if at least τ others have, and with probability $1 - q$ otherwise. Solve for the steady-state probability μ_s as a function of μ_0.

9.4 *Another Pure Strategy Equilibrium for the Game in Figures 9.2–9.4* Find a pure strategy equilibrium of the game in Figures 9.2–9.4 that is not pictured there.

9.5 *The Lattice Structure of Equilibria in Semi-Anonymous Games of Complementaries* Show that if $x = (x_1, \ldots, x_n)$ and $x' = (x'_1, \ldots, x'_n)$ are pure strategy equilibria of a semi-anonymous graphical game with strategic complementarities, then there exists a pure strategy equilibrium $\bar{x}$ such that

$$\bar{x}_i \geq \max\left(x_i, x'_i\right)$$

for all i, as well as a pure strategy equilibrium such that

$$\bar{x}_i \leq \min\left(x_i, x'_i\right)$$

for all i. This property means that the set of equilibria form a lattice. (In fact, the lattice is complete, so that for any set of pure strategy equilibria we can find a pure strategy equilibrium which is greater than or equal to each set member and another pure strategy equilibrium that is less than or equal to each member.)

9.6 *Possible Nonexistence of Pure Strategy Equilibria in Best-Shot Graphical Games on a Directed Network* Provide an example of a directed network with three players for which the only equilibria to a best-shot game played on that network are mixed strategies. Identify a mixed strategy equilibrium.

9.7 *Existence of Pure Strategy Equilibria in Semi-Anonymous Graphical Games of Strategic Complements with Infinite Action Spaces*[*] Consider a graphical game on a network (N, g) in which player i has a compact action space $X_i \subset [0, M]$. Let $u_i(x_i, x_{N_i(g)})$ be continuous for each i. A graphical game exhibits *strategic complements* if

$$u_i(x'_i, x'_{N_i(g)}) - u_i(x_i, x'_{N_i(g)}) \geq u_i(x'_i, x_{N_i(g)}) - u_i(x_i, x_{N_i(g)}) \quad (9.14)$$

for every i, $x'_i > x_i$ and $x'_{N_i(g)} \geq x_{N_i(g)}$.[51]

Show that there exists a pure strategy equilibrium in such a game. Show that the set of pure strategy equilibria form a complete lattice (see Exercise 9.5). Show that there are examples for which each of these conclusions fails if we set $X_i = \mathbb{R}_+$.

9.8 *Graphical Games of Complements*[*] Consider a graphical game as in Exercise 9.7 but that is also semi-anonymous, so that all agents have the same action space and payoffs depend only on the vector $x'_{N_i(g)}$ up to a relabeling of the agents.[52]

---

51. The inequality $x'_{N_i(g)} \geq x_{N_i(g)}$ indicates that each coordinate of $x'_{N_i(g)}$ is at least as large as the corresponding coordinate of $x_{N_i(g)}$.

52. That is, if there exists a bijection π from $N_i(g)$ to $N_i(g)$ such that jth coordinate of $x'_{N_i(g)}$ is equal to the $\pi(j)$th coordinate of $x_{N_i(g)}$ for each j, then $u_i(\cdot, x_{N_i(g)}) = u_i(\cdot, x'_{N_i(g)})$.

Suppose also that for any z in $[0, M]^{d_i}$

$$u_{d_i}(x_i, z) = u_{d_i+1}(x_i, (z, 0)). \tag{9.15}$$

Thus if we add a link from one player to a second player who is choosing action 0, then the payoff is as if the second player were not there.

(a) Show that if $g \subset g'$, then for every pure strategy equilibrium x under g there exists an equilibrium $x' \geq x$ under g'.

(b) Consider a case in which the inequality in (9.14) is strict when $x'_{N_i(g)} \neq x_{N_i(g)}$ and X_i is connected for each i. Show that if an equilibrium x relative to g is such that $x_i < M$ for each i, then there exists an equilibrium x' under $g + ij$ in which all players in the component of i and j play strictly higher actions.

(c) Show that the conclusions of (a) and (b) can fail if (9.15) is violated.

9.9 *Payoffs Increase with Degree** Galeotti et al. [274] state that a network game exhibits *degree complementarity* if

$$U_d(1, \sigma) - U_d(0, \sigma) \geq U_{d'}(1, \sigma) - U_{d'}(0, \sigma) \tag{9.16}$$

when $d > d'$. Equation (9.16) states that facing the same behavior by other players, a player with a higher degree has at least as big an incentive to take action 1 compared to a player with a lower degree.[53]

(a) Show that if (9.5) holds and the network game is one of strategic complements, then degree complementarity holds. Show that degree complementarity also holds in the case of strategic complements when a player cares about the fraction of neighbors taking action 1, so that $u_d(1, m) = \frac{m}{d} - c$ and $u_d(0, m) = b\frac{m}{d} - a$, with $b \leq 1$.

(b) Show that a network game that satisfies the condition in (a) has an equilibrium that is nondecreasing in degree.

9.10 *Payoffs Increase with Degree* Consider the setting of Proposition 9.5. Suppose that there are *positive externalities*, so that for each d and x_i, $u_d(x_i, m)$ is nondecreasing in m.[54] Show that for every equilibrium in a game with either strategic complements or substitutes, the payoff to a player with degree d', where $d' > d$, is at least as high as the payoff to a player with degree d.

9.11 *All Equilibria of a Network Game Are Monotone** Consider the setting of Proposition 9.5. Prove the last claim that if the game is one of strict strategic complements, then all equilibria are nondecreasing in degree.

9.12 *A Local Public Goods Graphical Game in Which Players Have Heterogeneous Costs* Bramoullé and Kranton [103] consider a variation on the model of Section

---

53. Analogously, payoffs exhibit *degree substitution* if the inequality above is reversed, and the following statements hold as well.

54. This setting is different from that of network formation, for which externalities were defined relative to network structure. Here the actions considered are those in the graphical game.

9.5.1 in which players can have different costs of providing the public good. That is, payoffs are given by

$$u_i(x_i, x_{N_i(g)}) = f(x_i + \sum_{j \in N_i(g)} x_j) - c_i x_i,$$

where c_i can differ across players. Let the function f be increasing and strictly concave, let x_i^* denote the maximizer of $f(x) - c_i x$, and suppose that x_i^* is well defined and nonzero for every player. Provide an algorithm that finds an equilibrium in a setting with $c_1 < c_2 < \cdots < c_n$. Provide an example in which there is more than one stable equilibrium.

9.13 *Convex Costs in a Local Public Goods Graphical Game* Consider the following variation on the local public goods graphical game of Bramoullé and Kranton [103] from Section 9.5.1. Payoffs are given by

$$u_i(x_i, x_{N_i(g)}) = f(x_i + \sum_{j \in N_i(g)} x_j) - c(x_i),$$

where f is strictly concave and c is strictly convex, and there exists $x^* > 0$ such that $f'(x^*) = c'(x^*)$, which is the action level that an individual chooses if he or she is the only provider.

(a) Find a pure strategy equilibrium on a complete network and show that it is the unique pure strategy equilibrium and that all players choose positive actions.

(b) Consider a circle network with an even number of players, and suppose that $f'(2x^*) < c'(0)$. Describe a specialized equilibrium where only some players choose positive actions.

9.14 *Cohesiveness* Find a partition of the set of nodes in Figure 9.14 into two sets such that one set is 2/3-cohesive and its complement is 3/4-cohesive.

9.15 *Labelings of Nodes and Cohesion* Consider a network (N, g) and a coordination game such that action 1 is a best response for any player if and only if a fraction of at least q of his or her neighbors play action 1.

(a) Show the following result from Morris [487]. Let a labeling of nodes be a bijection (a one-to-one and onto function) ℓ from N to N. Let $\alpha_\ell(i)$ be the fraction of $\ell(i)$'s neighbors who have labels less than $\ell(i)$. Show that there is a contagion from m nodes if and only if there exists a labeling ℓ such that $\alpha_\ell(i) \geq q$ for all $\ell(i) \geq m + 1$.

(b) From (a) show that there exists a set S that is uniformly no more than r-cohesive if and only if there is a labeling ℓ such that $\alpha_\ell(i) \geq 1 - r$ for all $\ell(i) \geq m + 1$, where m is the cardinality of the complement of S.

9.16 *A Sufficient Condition for the Failure of Contagion* Consider a network (N, g) and a coordination game such that action 1 is a best response for any player if and only if a fraction of at least q of his or her neighbors play action 1. Show that a sufficient condition for never having contagion from any group of m nodes is to have at least $m + 1$ separate groups that are each more than $(1 - q)$ cohesive.

9.17 *Contagion to a Subset of Nodes*   Consider a network (N, g) and a coordination game such that action 1 is a best response for any player if and only if a fraction of at least q of his or her neighbors play action 1. Show that $B \cup A$ is the eventual set of nodes playing 1 under the contagion system described in Section 9.6 if and only if the complement of $B \cup A$, denoted by C, is more than q-cohesive and for every nonempty subset D of B, $D \cup C$ has a cohesiveness of no more than q.

9.18 *Diffusion of Behavior in Network Games of Strategic Substitutes*   Consider the network games setting from Section 9.7.1 and suppose that $U_d(1, p)$ is decreasing and continuous in p for each d, and suppose that F is increasing and continuous on the entire range of U_d for each d. Show that there is a unique equilibrium p and that it is a stable equilibrium.

9.19 *Adoption Patterns by Degree: Diffusion of Behavior in Network Games*   Consider the network games setting from Section 9.7.1 in a case such that $U_d(1, p) = pd$ and F is uniform on $[0, 5]$, so that $F(U_d(1, p)) = \min\left[pd, 5\right]/5$.

   (a) Suppose that the network game is regular so that all players have degree d. What is the unique equilibrium p for $d < 5$? What are the two equilibria p when $d > 5$? What are the equilibrium ps when $d = 5$?

   (b) Consider a degree distribution that has equal weights on degrees $\{1, 2, \ldots, 10\}$ (so you need to use the corresponding $\widetilde{P}$ that is biased toward higher degrees with weight $d/55$ on degree d to obtain the distribution of neighbors' degrees). Using a simple spreadsheet or other program of your choosing, start with an initial $p^0 = .1$ and trace the evolution of the proportion of degree-d types that have chosen action 1 at a sequence of dates $t = 1, 2, \ldots$ until you have some sense of convergence. Plot the resulting adoption curves for $d = 1$, $d = 5$, and $d = 10$ versus time.

9.20 *S-Shaped Adoption Curves: Diffusion of Behavior in Network Games*
S-shaped adoption curves have been found in a variety of studies of diffusion. For such curves, adoption starts slowly, then increases its rate of diffusion, and then eventually slows down again.[55] In terms of diffusion in the network games setting from Section 9.7.1, we can track $p^{t+1} - p^t = \phi(p^t) - p^t$ as a proxy for the rate of diffusion as in Jackson and Yariv [362].

   Let $H(d, p) = F(U_d(1, p))$, which lies between 0 and 1 for every d and p, since F is a distribution function. Suppose that $H(d, 0) > 0$ for some d such that $\widetilde{P}(d) > 0$, and that H is twice continuously differentiable and increasing in both variables and strictly concave in p. Show that ϕ will be S-shaped. That is, show that there exists $p^* \in [0, 1]$ such that $\phi(p) - p$ is increasing when $p < p^*$ and then decreasing when $p > p^*$ (when $\phi(p) < 1$).

---

55. See Bass [46] for a discussion of this behavior, Rogers [564] for more detailed references, and Young [669] for alternative learning-based models.

9.21 *The Expected Number of Equilibria in a Generic Graphical Game* *[56]   Consider an arbitrary network (N, g) as the basis for a graphical game. Define the payoffs for players as a function of their actions as follows. For each player i and configuration of strategies $(x_i, x_{N_i(g)}) \in \{0, 1\}^{d_i(g)+1}$, assign the payoff $u_i(x_i, x_{N_i(g)})$ according to an atomless distribution F on $\mathbb{R}$. Do this independently for each player and profile of strategies. Once we have specified every u_i, the graphical game is well defined. It might have one pure strategy Nash equilibria, it might have several, or it might not have any, depending on the values of the u_is. Show that the expected number of pure strategy Nash equilibria is 1.

Hint: What is the probability that $x_i = 1$ is a best reply to some $x_{N_i(g)} \in \{0, 1\}^{d_i(g)}$? Then what is the probability that some profile of actions $(x_1, \ldots, x_n)$ is an equilibrium?

9.22 *Finding Equilibria in Graphical Games of Strategic Complements*   (a) Show that the algorithm for threshold games described in Section 9.8 finds the maximal equilibrium x, in the sense that $x_i \geq x_i'$ for all other equilibrium x' and all i.

(b) Describe an algorithm for finding the minimal equilibrium x such that $x_i \leq x_i'$ for all other equilibrium x' and all i.

(c) Argue that the claims in (a) (and (b)) are true even when considering mixed strategy equilibria so that x_i is at least (atmost) as large (small) as the maximum (minimum) of the support of the strategy of player i in any alternative equilibrium.

(d) Show that your algorithm also works for any graphical game of strategic complements with an action space of $\{0, 1\}$.

9.23 *Finding All Equilibria in Best-Shot Graphical Games*   Provide an example of an equilibrium in a best-shot public goods graphical game that would not be found by the algorithm for best-shot games described in Section 9.8.[57]

9.10.1 Exercises on Games

9.G1 *Product Choices*   Two electronics firms are making product-development decisions. Each firm is choosing between the development of two alternative computer chips. One system has higher efficiency but requires a larger investment and is more costly to produce. Based on estimates of development costs, production costs, and demand, the present-value calculations shown in Table 9.11 represent the value of the alternatives (high-efficiency chips or low-efficiency chips) to the firms. The first entry in each cell in Table 9.11 is the present value to firm 1 and the second entry is the present value to firm 2. The payoffs in the table are not symmetric. Firm 2 has a cost advantage in producing the higher-efficiency chip, while firm 1 has a cost advantage in producing the lower-efficiency chip. Overall profits are largest when the firms choose different chips and do not compete head to head.

(a) Firm 1 has a dominant strategy. What is it?

(b) Given your answer to (a), what should firm 2 expect firm 1's choice to be? What is firm 2's optimal choice, given what it anticipates firm 1 will do?

---

56. This exercise is based on a result by Daskalakis, Dimakis, and Mossel [184].

57. For an algorithm that finds all maximal independent sets, see Johnson, Papadimitriou, and Yannakakis [368].

TABLE 9.11
A production-choice game

		Firm 2	
		High	Low
Firm 1	High	1, 2	4, 5
	Low	2, 7	5, 3

(c) Do firm 1's strategy (answer to (a)) and firm 2's strategy (answer to (b)) form an equilibrium? Explain.

(d) Compared to (c), firm 1 would make larger profits if the choices were reversed. Why don't those strategies form an equilibrium?

(e) Suppose that firm 1 can commit to a product before firm 2. Draw the corresponding game tree and describe the backward induction/subgame perfect equilibrium.

9.G2 *Hotelling's Hotels*   Two hotels are considering a location along a newly constructed highway through the desert. The highway is 500 miles long with an exit every 50 miles (including both ends). The hotels may choose to locate at any exit. These will be the only two hotels available to any traveler using the highway. Each traveler has a most-preferred location along the highway (at some exit) for a hotel, and will choose to go to the hotel closest to that location. Travelers' most-preferred locations are distributed evenly, so that each exit has the same number of travelers who prefer that exit. If both hotels are the same distance from a traveler's most-preferred location, then that traveler flips a coin to determine which hotel to stay at. Each hotel would like to maximize the number of travelers who stay at it.

(a) If hotel 1 locates at the 100-mile exit, where should hotel 2 locate?

(b) Given hotel 2's location in (a) where would hotel 1 prefer to locate?

(c) Which pairs of locations form Nash equilibria?

9.G3 *Backward Induction*   Find the backward induction solution to Figure 9.21 and argue that there is a unique subgame perfect equilibrium. Provide a Nash equilibrium of that game that is not subgame perfect.

9.G4 *The Colonel Blotto Game*   Two armies are fighting a war. There are three battlefields. Each army consists of six units. The armies must each decide how many units to place on each battlefield. They do this without knowing the number of units that the other army has committed to a given battlefield. The army who has the most units on a given battlefield wins that battle, and the army that wins the most battles wins the war. If the armies each have the same number of units on a given battlefield, then there is an equal chance that either army wins that battle. A pure strategy for an army is a list (u_1, u_2, u_3) of the number of units it places on battlefields 1, 2, and 3, respectively, where each u_k is in $\{0, 1, \ldots, 6\}$ and the sum of the u_ks is 6. For example, if army A allocates its units (3,2,1), and army B allocates its units (0,3,3), then army A wins the first battle, army B wins the second and third battles, and army B wins the war.

(a) Argue that there is no pure strategy Nash equilibrium in this game.

(b) Show that mixing uniformly at random over all possible configurations of units is not a mixed strategy Nash equilibrium. (Hint: placing all units on one battlefield is not a good idea).

(c) Show that each army mixing with equal probability between (0,3,3), (3,0,3), and (3,3,0) is not an equilibrium.[58]

9.G5 *Divide and Choose*   Two children must split a pie. They are gluttons and each prefers to eat as much of the pie as possible. The parent tells one child to cut the pie into two pieces and then allows the other child to choose which piece to eat. The first child can divide the pie into any multiple of a tenth (e.g., 1/10 and 9/10, 2/10 and 8/10). Show that there is a unique backward induction solution.

9.G6 *Take It or Leave It Bargaining*   Two players are bargaining over a pie. The first player can suggest a division x_1, x_2 such that $x_1 + x_2 = 1$ and each share is nonnegative. Thus the game is in extensive form with an infinite number of strategies for the first player. The second player can then say either "yes" or "no." If the second player says "yes," then they each get their proposed share. If the second player says "no," then they both get nothing. Player's payoffs are their share of the pie. Argue that there is a *unique* subgame perfect equilibrium to this game.

9.G7 *Information and Equilibrium*   Each of two players receives an envelope containing money. The amount of money has been randomly selected to be between 1 and 1,000 dollars (inclusive), with each dollar amount equally likely. The random amounts in the two envelopes are drawn independently. After looking in their own envelope, the players have a chance to trade envelopes. That is, they are simultaneously asked if they would like to trade. If they both say "yes," then the envelopes are swapped and they each go home with the new envelope. If either player says "no," then they each go home with their original envelope. What is the highest amount for which either player says "yes" in a Nash equilibrium? (Hint: Should a player say "yes" with 1,000 dollars in his or her envelope?)

---

58. Finding equilibria in Colonel Blotto games is notoriously difficult. One exists for this particular version, but finding it will take you some time.

CHAPTER 10

Networked Markets

Classic theories of supply and demand, and the competitive models that underlie them, are built on the trade of precisely defined and known commodities. This allows for markets that are largely anonymous and unmodeled. The idea that modern societies can produce regular goods and deliver them across large and effectively anonymous markets was championed by Adam Smith [605] and has become a cornerstone of modern economic analysis. Indeed, many goods and services are such that the quality and reliability are predictable, delivery is easy and contractable, and transactions can occur between parties who need not know each other outside a single interaction. These conditions hold for many final goods, such as consumer products. However, even in modern societies there are many goods and services that are not so uniform across large numbers of buyers and sellers. They might involve specific features tailored to a particular situation or involve smaller numbers of buyers and sellers. Labor markets are an example. Because of these nonuniformities, many markets operate through decentralized networks or at least partly through networked interactions. Many intermediate goods and services are handled in this way, with parts or inputs supplied from one firm to another, or one individual to another. The importance of social networks in the market is evident to anyone who has ever searched for employment, and the extent and importance of the embeddedness of economic activity in social networks in developed markets is cogently argued by Granovetter [307].[1]

This chapter explores the role of social networks in various markets. A series of questions arise. To what extent are social and economic networks used in the exchange of goods and services? Which markets tend to be networked and why? Does the use of networked markets affect the terms of trade, prices, and efficiency of a market? How do social networks affect wages and employment patterns? How does position in a network affect the trade and welfare of an individual or firm? Which networks are likely to emerge in the context of networked markets? This chapter addresses these questions and also provides the groundwork for further research on networked markets.

---

1. For an early discussion of the social embeddedness of economic activity, see Polanyi [542].

10.1 ▪ Social Embeddedness of Markets and Exchange

Social networks have been the primary fabric of many economic interactions for centuries, if not millennia (e.g., recall the discussion of fifteenth-century Florence in Section 1.2.1), and detailed research has been conducted on the embeddedness of various markets in social networks for more than six decades. To begin this chapter, I provide an overview of some of the empirical work and a discussion of which markets we should expect to be networked and why.

10.1.1 The Use of Job Contacts in Labor Markets

The importance of social contacts in obtaining information about jobs and in the referral process is so prevalent that it has come to embody part of what we understand by the term *networking*. In fact, a definition of *networking* in the Merriam-Webster Dictionary (online, 2007) is "the cultivation of productive relationships for employment or business." Indeed, a substantial portion of jobs are filled through referrals from social contacts of current employees of a firm. This observation is not only important as an example of the role of social networks but also because it has implications for employment, wages, and the efficiency of labor markets. It shows how understanding social networks can help us to gain new and deeper insights into the workings of a market.

To see the extent to which social networks play a role in labor markets, consider the statistics from some studies.[2] An early study by Myers and Shultz [495] of textile workers found that 62 percent of interviewed workers found out about their job through a social contact; only 23 percent found the job by a direct application, and 15 percent through an employment agency, advertisement, or some other means. This sort of pattern is not unique to the textile industry. For instance, Rees and Shultz [558] interviewed workers in a Chicago neighborhood and kept track of what percentage of them found out about their current jobs through friends or relatives. They considered 12 different occupations and found rates of jobs obtained through social contacts ranging from a low of 23.5 percent (for accountants) to a high of 73.8 percent (for material handlers). The array of occupations they considered was broad, including typists (37.3 percent from social contact), janitors (65.5 percent), electricians (57.4 percent), and truck drivers (56.8 percent), to take a few examples. While these are mostly manual labor jobs, similar patterns are exhibited across different types of work. Granovetter's [305], [308] interviews of residents of Amherst, Massachusetts, found similar patterns across various types of occupations. He found that 44 percent of technical workers found their jobs through a social contact, as did 56 percent of professional workers and 65 percent of managerial workers. Corcoran, Datcher, and Duncan [177] examine the Panel Study in Income Dynamics data set (PSID) and compare across race and gender. They found the following percentages for finding jobs by social contact: black males 58.5 percent; black females, 43 percent; white males, 52 percent; and white females, 47.1 percent.

---

2. See Montgomery [479] and Ioannides and Datcher Loury [345] for more references and discussion.

These numbers suggest the extent to which social networks play a role in labor markets. There are also a variety of studies that examine how the role of social networks in labor markets varies across different groups and professions (see Ioannides and Datcher Loury [345] for more of an overview). As an example, Pellizari [535] examines data from European countries and finds a range of the prevalences for social contacts in the labor market, as well as for whether jobs obtained through social means lead to higher or lower wages. While going into the details of those differences is beyond the scope of this text, clearly it is very important to develop models to understand why and when employers use referrals as a means of filling job vacancies, and how social network structure figures in this process. While there are some models that shed light on these issues, this area is still underdeveloped.

10.1.2 Features of Networked Markets

To provide further background, it is helpful to discuss several sample studies of networked markets and some of the features that those markets exhibit. I have chosen several studies that highlight the interaction between social structure and economic outcomes, but there are many studies of networks and markets. (For example, see the edited volume by Rauch and Casella [555].) An influential study is Uzzi's [632] investigation of the importance of social relationships in the apparel industry in New York. Uzzi interviewed the executive officers of 43 "better dress" firms in New York City with annual sales between five hundred thousand and one billion dollars. The firms are basically divided into two groups: manufacturers and contractors. For instance, a contractor might have a design for a particular garment and want a given number of them produced and delivered. A manufacturer would then take the design and produce and deliver the garments. The firms were selected through a stratified random sample. Uzzi's focus was on the type of relationships that firms had with one another. One type of relationship he categorized as "market" or "arm's length" relationships, which included many one-time transactions; the other he categorized as "special" or "close" relationships, which included many relationships with repeated interaction and those that involved idiosyncracies in products or special investments. Based on the interviews, Uzzi identified three main ingredients associated with close ties but not with arm's length ties: trust, fine-grained information transfer, and joint problem solving. This view is derived from interpretations of the interviews and associated anecdotes. Uzzi quotes an example in which trust is associated with dealing with an unforeseen problem, such as when a fabric does not produce the desired garment, and then reaching an agreement on how to deal with this problem rather than leaving the manufacturer at a loss. The fine-grained information transfer refers to passing along useful information about, for instance, a new design, fabric, or technique, to agents in a close relationship. Firms involved in close relationships have more of an incentive to pass along such information, and can also do so in a credible manner. The joint problem solving refers to being able to quickly deal with problems that arise, which often hinges on knowing what the capabilities and situations of both firms in a relationship are.

Using these insights on how close relationships might help firms, Uzzi [632] builds a hypothesis that firms with more embedded relationships have a higher probability of survival. To test this hypothesis, he examines 1991 data from the

International Ladies Garment Workers' Union, which keeps detailed records of inter-firm transactions in the industry. More than 80 percent of the firms are unionized, and these data cover most of the active firms in New York that year. Based on these data, Uzzi examines the contractors who failed versus those who did not. Out of the 479 contractors with complete records in the data set, 125 failed during 1991.[3] Uzzi measures how embedded a contractor i is by examining $\sum_j P_{ij}^2$, where P_{ij} is the percentage of contractor i's output contracted with manufacturer j. If a contractor deals with only one manufacturer, then this measure will be 1, whereas if the contractor spreads its business out among many manufacturers, the measure will tend toward 0. A total of 15 percent of the contractors send 100 percent of their business to one manufacturer, 45 percent send at least 50 percent of their business to one manufacturer, while almost 20 percent send less than 25 percent to any one manufacturer. Thus there is some variation in the data, but also substantial embeddedness according to this measure. Uzzi then regresses the survival variable on this embeddedness measure and on other background variables (geographic location, age of the firm, size of the firm, some network centrality measures, and other neighborhood variables). He finds a positive and significant relationship (at the 95 percent confidence level) between survival and embeddedness. Based on the fitted regression line, a typical firm (setting other controls to average levels) with an embeddedness score of 0 will fail by the end of the year with a probability of .27, while a firm with an embeddedness score of 1 will fail by the end of the year with a probability of .14.

As with any cross-sectional statistical analysis, it is difficult to draw causal conclusions from an observed correlation. Embeddedness may help firms to survive, or perhaps firms that are near failure have a harder time establishing close relationships. Uzzi argues that the former is true. There could also be many other unobserved factors that contribute to the type of arrangements that firms have. For example, perhaps more complicated garments require specialized relationships and are in a less competitive part of the business. Regardless of the precise explanation for the observed correlations or the direction of causation, Uzzi's [632] study provides evidence that many firms in this industry do a substantial portion of their business with just a few partners, and in many cases with just one partner. The study also provides some insight into the potential differences between a network of close relationships versus a more distant market-style relationship.

Another example of a partly networked market is that of the Marseille fish market as analyzed by Kirman [399] and Weisbuch, Kirman, and Herreiner [661]. The fish market has several critical features. First, fish are perishable. Thus the ability of buyers and sellers to smooth inventory fluctuations is quite limited, as "fresh" fish need to be consumed soon after they are caught.[4] Second, the supply of fish is variable, as the catch is random and is affected by various factors including

---

3. Only 8 of 89 manufacturers in the data set failed, and so Uzzi does not report the relationship between survival and embeddedness for manufacturers.

4. This constraint is changing over time, as methods of freezing and transporting fish advance. "Fresh" is in quotes, as in fact some types of fish that are considered fresh are frozen on board the fishing boats. Nevertheless, there is a demand for local fish in the Marseille market (e.g., as ingredients in Marseille's famous bouillabaisse).

weather and water conditions, and fish populations. Third, the buyers of the fish differ in their demand elasticities for the fish; that is, they differ in the importance of fish to them and their willingness to pay as a function of quantity. A famous restaurant that has built a reputation on serving Marseille's finest bouillabaisse has to be able to consistently buy fish of a reliable quantity. Another local restaurant that does not specialize in bouillabaisse can adjust its menu and ingredients as availability changes.

Weisbuch, Kirman, and Herreiner [661] examine the market between 1988 and 1991, considering buyers and sellers in the market for more than 8 months. There are 45 sellers and more than 1,400 buyers for a wide range of fish. The buyers are restaurants and retailers; there are no posted prices, so that each price transaction is decided bilaterally between the buyer and seller. In terms of a network structure, the researchers find that a sizable fraction of buyers are loyal to a single seller, while other buyers purchase from many sellers. For example, for cod, almost half (48 percent) of the buyers purchase more than 95 percent of their cod from just one seller. For whiting and sole, more than half of the buyers buy more than 80 percent of their fish from just one seller. The researchers break these data down by the patterns of buyers and note that buyers purchasing large quantities are significantly more likely to be loyal than are other buyers. For example, dividing buyers into those who bought more than 2 tons per month and those who bought less, the ones who bought more transact on average 85 percent of the time with their most visited seller while the rate drops to 56 percent of the time for those who buy less. So there are some established relationships and some shopping around. Weisbuch, Kirman, and Herreiner [661] examine a simple model of this market, in which the driving forces are predictability of available fish and of demand for fish. The model builds on what might be thought of as a complicated version of musical chairs: buyers must choose to visit just one seller each day, and the seller is either able or not to meet the buyer's demand. Buyers follow an adaptive updating rule, with a higher tendency to visit sellers who have met their demands in the past. This model is a sort of metaphor that captures the idea that as a buyer shops around during the day, time is lost and fewer fish are available. On a day when there is a relatively large catch, this shopping around might result in a better price, but when there is a relatively small catch, it might result in a relatively high price or no fish at all. Thus there are some frictions in this type of market, so that it does not simply clear as a classic supply and demand model all at once, but requires some search by buyers and price-setting by individual and only partly informed sellers. Buyers face coordination issues and also learn about the available supply of different sellers over time. Having a larger demand, among other things, tilts a buyer toward establishing with just one (large) seller, partly because the predictability of the supply.

10.1.3 Which Markets Should Be Networked?

The examples in Section 10.1.2 point to some of the issues that underlie why specific relationships might emerge in the trade of goods and services. The garment industry analysis highlights trust as central to close relationships. The fish market analysis highlights predictability as central to that market's repeat relationships.

There are a variety of situations in which repeat or close relationships are advantageous, basically having to do with difficulties in contracting. The advantages of close relationships are related to a theory developed by Williamson [664] explaining the organization of firms. Williamson examines a variety of different frictions that can make it difficult for two parties to completely contract on the exchange of goods or services. Williamson then argues that if a single organization or firm is on both sides of the transaction, then it can internalize the difficulty and overcome the obstacle to contracting. Granovetter [307] critiques Williamson's argument, pointing out that it is not so clear that integration within a firm provides a better solution than an external relationship. For example, reputations and repeated interactions can help discipline transactions, and this process could take place in a network instead of being internal to a firm. Also, contracting parties being within the same firm does not imply that their incentives are aligned.

Drawing from the above examples and studies of incomplete contracting, let us examine the features of transactions that might favor placing them in the context of a network of transactions in which reputations and/or repeat relationships help circumvent the difficulties inherent in a given transaction.

In many situations, there are unforeseen contingencies that arise between two parties involved in a contractual relationship. An input may not be available, or a new regulation is mandated that requires the redesign of a product and some delay, and so forth. One can try to write a contract that completely covers all possible contingencies, but for complex transactions (e.g., the construction of a building) this might not be possible. If the relationships between the parties are repeated, so that they deal with each other on an ongoing basis, then these problems are not viewed as a one-time expense on one of the two parties' side, but instead can be balanced over time. These considerations can facilitate the bargaining over unforeseen issues, as suggested by some of Uzzi's [632] interviews.

Contractual incompleteness can arise not only because of unforeseen contingencies, but also because of specific investments that might need to be made for particular transactions and require long-term use to realize their full value. Contractual incompleteness can also arise from asymmetries in information. Asymmetries in information manifest themselves in the form of moral hazard problems (one party to a contract cannot fully observe the actions of the other) and adverse selection problems (one party does not fully observe some attributes of the good being traded). Once again long-term relationships can help resolve these issues. For example, for moral hazard problems, having a repeat interaction allows one party to examine the long-run performance of the other. If a customer takes a car to a local mechanic on a regular basis, and the mechanic frequently claims that the car requires extensive and expensive repairs, the customer can look for a new mechanic. The fact that current performance can influence future business helps temper the moral hazard problem. In contrast, if a car needs a repair far from home, and it is a one-time interaction, the incentives for the mechanic to suggest a more expensive repair than is necessary can be substantially larger. Similar reasoning favors long-term relationships in the face of adverse selection. If a firm buys parts whose reliability or longevity cannot be observed except with the passage of time, then repeat transactions can help provide incentives for the supplier to deliver a specified quality of parts.

There are tradeoffs to maintaining such closed relationships, as it limits one's ability to shop around for alternative prices. A given firm might want to work with two or three suppliers of the same parts over time. The transmission of information through a network can also temper asymmetric information problems. Maintaining a reputation for providing high-quality parts or service can provide incentives that overcome some of the difficulties arising from asymmetries in information, if word of mouth can spread information about outcomes to other potential future business partners. Thus we might see more complex network relationships for various reasons.

Beyond these asymmetric information considerations, we also saw the issues of predictability and more basic uncertainty as potential explanations for the relationships in the Marseille fish market.[5] As an incentive for a continued relationship, a regular customer might be given access to better produce or a higher chance at obtaining a desired quantity of produce. In situations in which the uncertainties of the crop or production totals are not fully insurable, risk aversion can then favor repeat interactions.

There are other aspects of networks that can be valuable in the trade of goods and services beyond those which arise between the parties directly involved in the transaction. Networks also serve as an integral part of many markets by bringing different buyers and sellers together. Labor markets serve as an excellent example. A firm wanting to hire a new employee might ask its existing employees for referrals. This could happen for a variety of reasons.[6] The firm may simply want to hire people similar to the employees it already has. Given the homophily in many social networks, a firm can take advantage of its existing workforce to find other people with similar characteristics. For example, if a fast-food restaurant wants to hire someone willing to work part-time on weeknights and for minimum wage with low benefits, it might ask its employees if they know of anyone else who would be available in such circumstances. This procedure can save the time and the cost of advertising and then sifting through applications. In addition, using current employees might reach potential hires that might not respond to advertising. Or perhaps current employees are good at communicating with potential employees regarding whether a potential job is a good match. In addition to the benefits of locating potential hires who fit well with the firm, current employees might be credible sources of information about the quality of a potential hire. A recommendation coming from a current employee or other acquaintance could carry more weight than that of a stranger.

This discussion has sketched the potential benefits and reasoning behind networked relationships. Let us now examine some models of such interactions to

---

5. See also the study by Podolny [539] of investment banking, which shows an increased concentration of transaction relationships when market uncertainty increases.

6. See Fernandez, Castilla, and Moore [246] for discussion of factors favoring the use of referrals and evidence from a study of hiring practices in a phone bank that there can be economic benefits to firms who hire through referrals. In addition to the benefits from better matching, they also examine the extent to which hiring friends of current employees affects turnover in the firm, which might be related to the social environment of the firm.

gain a fuller understanding of the implications of network structure for economic transactions and welfare, and also for incentives to form such networks.

10.2 ▪ Networks in Labor Markets

The pervasiveness of networks in labor markets makes them a leading example of networked markets and a source of a variety of insights. So let us begin by analyzing different aspects of networked labor markets.

10.2.1 Strong and Weak Ties

As discussed in Section 3.2.7, the role of social networks in finding jobs was central to Granovetter's [305], [308] influential research that distinguished between strong and weak social ties. Granovetter measured the strength of a tie by the number of times that individuals had interacted in a past year (strong = at least twice a week, medium = less than twice a week but more than once a year, and weak = once a year or less). Of the 54 interviewees who had found their most recent job through a social contact, Granovetter found that 16.7 percent had found their jobs through a strong tie, 55.7 percent through a medium tie, and 27.6 percent through a weak one.[7]

Building on a distinction between strong and weak ties, Boorman [92] modeled individuals' decisions of how to allocate their time between maintaining these two different forms of ties in one of the first economic models of social networks. Boorman's model is based on the following structure. An individual has to divide his or her time between maintaining strong and weak ties. Strong ties take more time, and so the individual is faced with a tradeoff between having more but weaker ties or fewer but stronger ones. Boorman represents this tradeoff by requiring that an individual have T units of time spent maintaining relationships. If W is the number of weak ties that an individual has and S is the number of strong ties, then they must satisfy:

$$W + \lambda S = T,$$

where $\lambda > 1$ is a factor indicating how much more time must be spent to maintain a strong tie.

Boorman's ties also lead to different benefits. Strong ties have priority in obtaining job information from social contacts, which operates as follows. Time elapses in discrete periods. In any period, with probability μ an individual has need of a job. This probability is the same across individuals and independent of history and the state of the system in the previous period. Effectively, the system restarts with each period. If an individual needs a job, then he or she can find one in two ways. First, the individual can hear about a job directly, which again happens at

---

7. There is also a series of studies that have examined how strong and weak ties affect labor outcomes and some debate about the relative effectiveness of weak ties. For example, see Lin, Ensel, and Vaughn [435], and Bridges and Villemez [108] and the literature that followed.

some exogenous rate δ, independent of the state of the system. In that case, the individual takes the job. Second, the individual might not hear directly, but instead might have a friend who is employed who happens to randomly hear about a job. In that case, the employed friend looks around at his or her strong and weak ties. If some of the strong ties are unemployed, then the employed friend passes the job information to one of them uniformly at random. If all of the strong ties are employed, then the employed friend passes the information to one of the weak ties uniformly at random. Thus strong ties take priority in obtaining information about a job. Boorman examines networks that are trees with infinite numbers of nodes, so that one does not have to worry about the issue of two neighbors being neighbors of each other. He also considers situations in which all individuals choose the same allocation of ties, so that the network is regular in a strong sense. Let $q_s \le q_w$ be the probability that an agent does *not* hear about a job through a given strong tie or weak tie, respectively, when in need. The chance of getting a job when in need can then be written as

$$\delta + (1 - \delta) \left(1 - q_s^S q_w^W\right).$$

A given agent thus trades off the higher probability that strong ties lead to job information against being able to maintain fewer of them. One can derive the expressions for q_s and q_w as a function of the parameters of the model. For instance, q_s depends on the probability that a given strong friend will be employed and on the number of other strong ties that might be competing for job information at a given time. With expressions for q_s and q_w in hand, we can then look for an equilibrium of the system where individuals are optimally choosing S and W given the anticipated q_s and q_w, and the anticipated q_s and q_w correspond to the ones generated by the choices that individuals have made concerning S and W. This model is hard to solve for directly and can involve multiple equilibria, but one can at least work out simulations for some parameter values, as Boorman does.

There are several intuitive results that Boorman reports from simulations. First, as λ increases, the relative cost of strong versus weak ties increases, and so the equilibrium involves fewer strong and more weak ties. Second, as μ decreases, so that an agent is less likely to need a job, the relative value of weak ties increases. The agent only receives job information through a weak tie when all of the weak acquaintance's strong neighbors are employed, which is more likely when μ is low.

The Boorman model makes important strides in terms of considering strong and weak ties to be choice variables and deriving tradeoffs among different forms of ties. However, the model lacks a number of features that might be of interest. As a model of strong and weak ties, it is missing one of the critical ingredients that was the basis for Granovetter's theory: the idea that weak ties are more likely to bridge parts of a network not accessed more directly, whereas strong ties are more likely to link to nodes that are already at a short path distance. Such aspects of network architecture are missing from Boorman's model. Also, the model is missing the correlation in employment and time series implications that might be useful for explaining how labor markets work. The fact that the state of the network is history independent simplifies the model (with the need for jobs being independent of history and the previous state of the network), but then we miss interesting dynamics and patterns of employment as a function of social structure.

10.2.2 A Networked Model of Employment

Calvó-Armengol and Jackson [130], [132] examine a model that is similar to Boorman's [92] in having job information arrive directly and through neighbors, but it brings network structure to the forefront to see how network structure can affect employment and wage dynamics and distribution across a society.[8] Before discussing the model in detail, I start with a brief overview of it. In the simplest version of the model, workers are connected by an undirected network. An employed worker can randomly lose his or her job. If that happens, then the worker looks for work. Information about new job openings arrives randomly to the workers in the network. A worker can hear about a job directly. If the worker is unemployed, then he or she takes the job. Similarly to the Boorman model, if the worker is employed, then the worker randomly picks an unemployed neighbor to receive the information; but in this case treating all unemployed neighbors with equal weight. The system operates over time, so that the starting state at the beginning of one period is the state of the system at the end of the previous period; thus dynamic patterns can be studied. The network becomes important in several ways. Having more neighbors gives a worker potential access to more job information and so a higher average employment rate and higher average wages (in the version of the model with wages). The network also leads to correlation in neighbors' employment status. An unemployed worker who has an employed neighbor is more likely to hear about a job opening than a similar worker with an unemployed neighbor, leading to a higher probability that a worker becomes employed in any given period as a function of how many of his or her neighbors are employed, which leads to a positive correlation in employment.

Beyond this basic structure, there are a number of features that can be analyzed. The model described above leads to a Markov chain, so that one can keep track of what tomorrow's employment pattern is likely to look like, given today's employment pattern. One can then deduce long-run steady-state distributions of employment and how these vary with network structure. In addition to things like the correlation structure, one can also examine the time series of employment. For instance, a prevalent observation in the labor economics literature is what is known as *duration dependence:* workers who have been unemployed for longer times are less likely to find work in a given period than workers who are just recently unemployed. There are various partial explanations for this phenomenon, but the network model exhibits this quite naturally. Conditional on a worker being unemployed for a longer time, it is likely that many of the worker's neighbors are also unemployed, and hence have not been passing information along. Thus it is less likely that such a worker will hear about a job in the next period, compared to a worker whose neighborhood has a higher employment level. One can also enrich

---

8. Such information arrival processes are also described in Diamond [199], without networks. Calvó-Armengol [127] developed a networked variation of Diamond's [199] and Boorman's [92] models of information passing to study incentives for network formation. Calvó-Armengol and Jackson [130], [132] developed a richer model by bringing in wages and variations on passing job information and used it to study employment and wage patterns and dynamics.

the model to allow for different types of jobs, different wage levels, and decisions (e.g., whether to pursue education).

Model Description   I present the simplest version of the model. For variations with heterogeneous jobs and multiple wage levels, see Calvó-Armengol and Jackson [132]. In the version presented here, all jobs are identical, and there is just one wage level.

There are n workers or agents who are connected by an undirected network, represented by the $n \times n$ symmetric matrix g, which has entries in $\{0, 1\}$. Time evolves in discrete periods indexed by $t \in \{1, 2, \ldots\}$. The n-dimensional vector s_t describes the employment status of the agents at time t. If agent i is employed at the end of period t, then $s_{it} = 1$; i is unemployed, then $s_{it} = 0$.

Period t begins with some agents employed and others unemployed, as described by the state s_{t-1}. The first thing that happens in a period is that information about new job openings arrives. Each agent directly hears about a job opening with a probability $a \in [0, 1]$. This job arrival process is independent across agents. If an agent i is unemployed ($s_{i,t-1} = 0$) and hears about a job, then he or she takes that job and becomes employed. If an agent i is employed ($s_{i,t-1} = 1$) and hears about a job, then he or she picks an unemployed neighbor ($j \in N_i(g)$ such that $s_{j,t-1} = 0$) and passes the job information to that neighbor. If agent i has several unemployed neighbors, then the agent picks one uniformly at random. If agent i's neighbors are all employed, then the job information is lost.

The probability of the joint event that agent i learns about a job and this job ends up in agent j's hands is described by $p_{ij}(s_{t-1})$, where

$$
p_{ij}(s_{t-1}) = \begin{cases}
a & \text{if } s_{i,t-1} = 0 \text{ and } i = j; \\
\dfrac{a}{\sum_{k:s_{k,t-1}=0} g_{ik}} & \text{if } s_{i,t-1} = 1, s_{j,t-1} = 0, \text{ and } g_{ij} = 1; \text{ and} \\
0 & \text{otherwise.}
\end{cases}
$$

At the end of a period some employed agents lose their jobs. This loss happens randomly according to an exogenous breakup probability, $b \in [0, 1]$, independently across agents.

Some Simple Examples   It is useful to start with some simple examples to see how the model evolves.

Example 10.1 (An Isolated Agent)   *Consider an isolated agent as a benchmark. Let μ be the long-run steady-state probability that the agent is employed.[9] This probability must satisfy*

$$
\mu = (1 - b)\left(\mu + a(1 - \mu)\right). \tag{10.1}
$$

---

9. This probability may be thought of in two ways. First, regardless of what the agent's initial state is, it is the limit of the probability that the agent will be employed in a distant period in the future. Second, if one starts by randomly setting the agent's initial state with this probability, then it is the probability that the agent will be employed tomorrow and at any date in the future.

Equation (10.1) keeps track of the two different ways the agent could be employed. First, at the end of last period the agent might be employed, which has a probability μ, and then the agent did not lose his or her job at the end of this period, which happens with probability $1 - b$. Second, the agent might have been unemployed at the end of last period and then heard about a job in the beginning of this period, which has a probability $a(1 - \mu)$, and subsequently the agent did not lose his or her job at the end of this period, which happens with probability $1 - b$. Solving (10.1) for μ leads to

$$\mu = \frac{(1 - b)a}{b + (1 - b)a} = \frac{1}{1 + \frac{b}{(1-b)a}}. \tag{10.2}$$

As one would expect, the steady-state employment probability increases with the probability of hearing about a job, a, and decreases with the probability of losing a job, b. It is no more than the probability of retaining a job $(1 - b)$, but approaches $(1 - b)$ as a approaches 1. Moreover, it is not the absolute values of a and b that matter, but their relative values. In particular, it is how $b/(1 - b)$ compares to a that is critical.

Example 10.2 (A Dyad) *Consider a dyad. Here $n = 2$ and $g_{12} = g_{21} = 1$. Given the symmetry of this setting, the steady-state distribution can be tracked simply through the probability that no agents are employed, μ_0, one agent is employed, μ_1, and both agents are employed, μ_2.*

We can then track the transitions as follows. The situation with two employed workers can result in three ways. It could be that they were both employed at the end of last period and neither lost a job, which happens with probability $\mu_2(1 - b)^2$; or perhaps only one was employed at the end of the last period and at least one heard about a job and neither lost a job, which happens with probability $\mu_1(1 - (1 - a)^2)(1 - b)^2$; or perhaps neither started out employed and both heard about jobs and then both kept those jobs, which happens with probability $\mu_0 a^2(1 - b)^2$. Similar reasoning applies to the other states, and we can characterize the steady states as the solutions to

$$\begin{pmatrix} \mu_0 \\ \mu_1 \\ \mu_2 \end{pmatrix} = \begin{pmatrix} (1 - a + ab)^2 & (1 - a)^2 b(1 - b) + b^2 & b^2 \\ 2a(1 - b)(1 - a + ab) & (1 - b)\left((1 - a)^2(1 - 2b) + 2b\right) & 2b(1 - b) \\ a^2(1 - b)^2 & \left(1 - (1 - a)^2\right)(1 - b)^2 & (1 - b)^2 \end{pmatrix} \begin{pmatrix} \mu_0 \\ \mu_1 \\ \mu_2 \end{pmatrix}. \tag{10.3}$$

We solve (10.3) (noting that the vector μ is a unit eigenvector of the transition matrix) to find

$$\begin{pmatrix} \mu_0 \\ \mu_1 \\ \mu_2 \end{pmatrix} = \begin{pmatrix} b^2 \left(1 + (1-b)(1-a)^2\right) / X \\ 2ab(1-b) \left(1 + (1-b)(1-a)\right) / X \\ a^2(1-b)^2 \left((1-a)(3-a)(1-b) + 1\right) / X \end{pmatrix}, \qquad (10.4)$$

where

$$X = b^2 \left(1 + (1-b)(1-a)^2\right) + 2ab(1-b) \left(1 + (1-b)(1-a)\right)$$

$$+ a^2(1-b)^2 \left((1-a)(3-a)(1-b) + 1\right).$$

If we let $a = 1$, so that workers are sure to hear about jobs in any period, then the probability of having nobody employed tends to b^2, the probability of having one employed tends to $2b(1-b)$, and the probability of having both employed tends to $(1-b)^2$, as we should expect. As b tends to 0, μ_2 tends to 1; and as b tends to 1, μ_0 tends to 1.

As these expressions are cumbersome, we can also examine the model when the time between periods becomes small. Then, a and b both go to 0, and it is only the relative rates that matter. In particular, the chance that more than one change occurs in a period (i.e., having more than one piece of information and/or loss of a job) tends to 0 relative to the probability that one change occurs in a period. For instance, let us replace a and b with a/T and b/T. As T becomes large, second-order and higher terms become negligible relative to single changes, and then (10.3) is approximated by

$$\begin{pmatrix} \mu_0 \\ \mu_1 \\ \mu_2 \end{pmatrix} = \begin{pmatrix} 1 - \frac{2a}{T} & \frac{b}{T} & 0 \\ \frac{2a}{T} & 1 - \frac{2a}{T} - \frac{b}{T} & \frac{2b}{T} \\ 0 & \frac{2a}{T} & 1 - \frac{2b}{T} \end{pmatrix} \begin{pmatrix} \mu_0 \\ \mu_1 \\ \mu_2 \end{pmatrix}. \qquad (10.5)$$

The solution to (10.5) is[10]

$$\begin{pmatrix} \mu_0 \\ \mu_1 \\ \mu_2 \end{pmatrix} = \begin{pmatrix} \frac{b^2}{b^2 + 2ab + 2a^2} \\ \frac{2ab}{b^2 + 2ab + 2a^2} \\ \frac{2a^2}{b^2 + 2ab + 2a^2} \end{pmatrix}. \qquad (10.6)$$

From (10.6) we can deduce a few things. First, the probability that a given agent is employed in any period is

$$\mu_{\text{dyad}} = \mu_2 + \frac{\mu_1}{2} = \frac{2a^2 + ab}{b^2 + 2ab + 2a^2}.$$

If we compare this probability to the same limit for (10.2), which is $\mu_{\text{isolate}} = a/(a + b)$, we see that it is larger. Indeed, simplifying the above expression leads to

$$\mu_{\text{dyad}} = \frac{a}{a + b - \frac{ba}{2a+b}} > \mu_{\text{isolate}} = \frac{a}{a + b}. \qquad (10.7)$$

---

10. We can also obtain (10.6) by examining the limits in (10.4) directly.

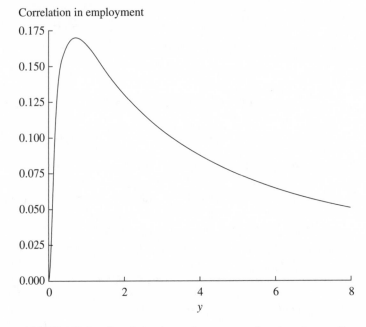

FIGURE 10.1   Correlation in employment as a function of $y = a/b$.

It is clear that (10.7) should hold, since having a neighbor increases the opportunities for hearing about employment.

We can also examine the correlation of employment across the two agents:

$$\text{Corr}_{\text{dyad}} = \frac{E[s_{1t}s_{2t}] - E[s_{1t}]E[s_{2t}]}{E[s_{it}^2] - E[s_{it}]^2} = \frac{\mu_2 - \mu_{\text{dyad}}^2}{\mu_{\text{dyad}} - \mu_{\text{dyad}}^2}.$$

Substituting from (10.6) and (10.7), the correlation is

$$\text{Corr}_{\text{dyad}} = \frac{ab}{3ab + 2a^2 + b^2} = \frac{1}{3 + 2\frac{a}{b} + \frac{b}{a}}. \tag{10.8}$$

Equation (10.8) reaches a maximum when a/b is $1/\sqrt{2}$ and is always positive. It is graphed in Figure 10.1 as a function of $y = a/b$. The correlation of two neighbors should clearly be positive, as one's probability of finding a job goes up when his or her neighbor is employed. It is less clear whether larger groups of agents should have positively correlated employment outcomes.

Complete Networks   Let us next examine settings in which individuals live in cliques of n individuals who are all connected to one another. There is a full symmetry, so that we can keep track of the state simply in terms of how many agents are employed. Let μ_k be the steady-state probability that exactly k agents are employed in a given period.

Let us again examine situations in which the time periods become small, so the arrival rates are a/T and b/T for some large T. Thus the probability of having two or more events occur in a period (e.g., two people hearing about a job) becomes infinitely less likely than having just one event occur, as the former is on the order of $1/T^2$ and the latter is on the order of $1/T$. To calculate what happens at the limit as T grows, we need only track transitions between neighboring states. This permits us to solve in closed form for the steady-state probability of any state.

Proposition 10.1 *Consider a complete network of n agents, with arrival rate a/T and breakup rate b/T. As T grows, the steady-state probability of having k agents employed converges to*

$$\mu_k = \frac{\frac{n!}{k!}\left(\frac{b}{na}\right)^{n-k}}{\sum_{j=0}^n \frac{n!}{j!}\left(\frac{b}{na}\right)^{n-j}} = \frac{1}{\sum_{j=0}^n \frac{k!}{j!}\left(\frac{b}{na}\right)^{k-j}}. \qquad (10.9)$$

And for $k' > k$,

$$\frac{\mu_{k'}}{\mu_k} = \left(\frac{na}{b}\right)^{k'-k}\frac{k!}{k'!}. \qquad (10.10)$$

The proof is straightforward but helps to illustrate how to derive the properties of such a Markov chain.

Proof of Proposition 10.1. In steady state, for $1 < k < n$, the system can have exactly k employed agents from three different states in the previous period. It could be that $k - 1$ agents were employed in the previous period, and then the probability of transitioning to having k employed agents is the probability that some agent heard about a job (as then information certainly reaches an unemployed agent in a completely connected clique), which is approximately na/T for large T. Or perhaps $k + 1$ agents were employed in the previous period, and then an agent lost a job, which happens with probability approximately $(k + 1)b/T$. It could also be that in the previous period k agents were employed and nobody heard about a job or lost one, which happens with probability $1 - na/T - kb/T$. Thus for large T we approximate the steady-state probability of being in state k by

$$\mu_k = \mu_{k-1}n\frac{a}{T} + \mu_k(1 - n\frac{a}{T} - k\frac{b}{T}) + \mu_{k+1}(k + 1)\frac{b}{T}. \qquad (10.11)$$

The zero employment state satisfies

$$\mu_0 = \mu_0(1 - n\frac{a}{T}) + \mu_1\frac{b}{T},$$

or

$$\mu_0 = \frac{b}{na}\mu_1. \qquad (10.12)$$

Substituting (10.12) into (10.11) with $k = 1$, we can solve to find

$$\mu_1 = \frac{2b}{na}\mu_2. \tag{10.13}$$

Iterating on (10.13), we find that for any $k < n$

$$\mu_k = \frac{(k+1)b}{na}\mu_{k+1}. \tag{10.14}$$

Equation (10.14) implies that

$$\mu_k = \frac{n!}{k!}\left(\frac{b}{na}\right)^{n-k}\mu_n. \tag{10.15}$$

Equation (10.10) then follows.

Noting that $\sum_{j=0}^{n}\mu_j = 1$, (10.15) implies that

$$\mu_n = \frac{1}{\sum_{j=0}^{n}\frac{n!}{j!}\left(\frac{b}{na}\right)^{n-j}},$$

and more generally that

$$\mu_k = \frac{\frac{n!}{k!}\left(\frac{b}{na}\right)^{n-k}}{\sum_{j=0}^{n}\frac{n!}{j!}\left(\frac{b}{na}\right)^{n-j}} = \frac{1}{\sum_{j=0}^{n}\frac{k!}{j!}\left(\frac{b}{na}\right)^{k-j}}.$$

Thus we have the claimed expression (10.9) for the probabilities of the states. ▪

Using (10.9) and (10.10) for the steady-state probabilities allows computation of the average employment rates as well as correlations (Table 10.1). Although the correlation decreases between any two agents as the network size increases, that does not mean that the network effect is decreasing. In fact, quite the contrary is true. The society swings more to situations in which many agents are employed or many agents are unemployed at the same time. Although any two of them in a large society will have low correlations, there is still a large overall effect. One way of measuring the impact of the network effect is to examine the variance of the total employment under the steady-state distribution of the network and then normalize it to see how it compares to the variance of a society with the same average employment but where each agent's employment follows a binomial distribution, independent of the employment of other agents, as pictured in Figure 10.2.[11] The figure shows how the normalized variation in steady-state total employment increases with network size.

Other Networks and Correlation in Employment   For more complex network structures, solving for the steady-state employment rates and correlations in employment becomes difficult analytically but is still possible numerically

---

11. I thank Toni Calvó-Armengol for suggesting this illustration.

TABLE 10.1

Average employment and correlation in employment in complete networks

Size n	Ratio of job arrival to breakup: a/b		
	1/2	1	2
1	.333 (—)	.500 (—)	.667 (—)
2	.400 (.167)	.600 (.167)	.769 (.133)
4	.452 (.135)	.689 (.139)	.851 (.099)
8	.485 (.099)	.764 (.111)	.910 (.067)
16	.498 (.06)1	.825 (.087)	.948 (.043)
32	.500 (.032)	.871 (.066)	.972 (.025)
64	.500 (.016)	.907 (.049)	.985 (.014)
128	.500 (.008)	.933 (.036)	.992 (.007)
256	.500 (.004)	.952 (.026)	.996 (.004)
512	.500 (.002)	.966 (.019)	.998 (.002)

Note: Entries are averaged with correlation coefficients in parentheses.

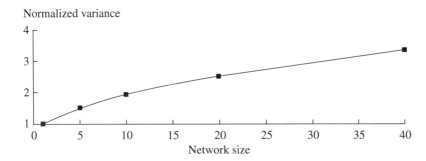

FIGURE 10.2   Variance of steady-state total employment divided by the variance of total employment if each agent were independently employed with the same average probability, as a function of network size, with $a = .5 = b$.

through simulations. To get a feel for some comparative statics, Calvó-Armengol and Jackson [130] present a few examples. All of the following examples use an arrival rate of $a = .100$ and a breakup rate of $b = .015$. Based on a time period of a week, then, an agent loses a job about once every 67 weeks, and hears about a job directly on average once every 10 weeks. Figure 10.3 presents unemployment rates and the correlation in employment for four different four-person networks.

As the network becomes more connected, the unemployment rate falls, because information about jobs has a lower probability of being lost. The correlation is higher for agents who are directly connected than for those who are indirectly

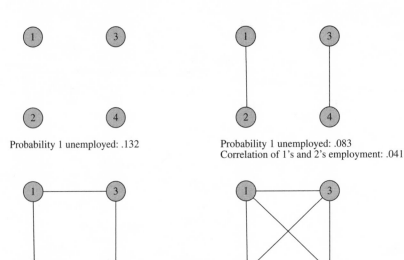

Probability 1 unemployed: .132

Probability 1 unemployed: .083
Correlation of 1's and 2's employment: .041

Probability 1 unemployed: .063
Correlation of 1's and 2's employment: .025
Correlation of 1's and 4's employment: .019

Probability 1 unemployed: .050
Correlation of 1's and 2's employment: .025
Correlation of 1's and 4's employment: .025

FIGURE 10.3 Unemployment rates and correlation in employment, with $a = .100$ and $b = .015$, in examples from the Calvó-Armengol and Jackson [130] model.

connected. The correlation between the employment of any two agents falls as the number of ties that they have increases, as there are more sources of information affecting their employment.

Figure 10.4 shows that asymmetries in network position can lead to differences in steady-state employment rates even when the degree of agents is identical. In the figure, the agents whose link forms a bridge (agents 1 and 6) have higher employment rates. The effect here comes about because those two agents are more diversified in their social connections than the other agents: none of their respective neighbors are linked to one another. In contrast, each of the other agents has some clustering in his or her neighborhood, and has higher correlation with neighbors' employment. The correlation in the employment of neighbors makes it more likely that an agent will hear about either no jobs or multiple jobs at once, while an agent would rather instead have a higher probability of hearing about (at least) one job.

The correlation observed in the networks in Figures 10.3 and 10.4 is not unique to these examples, but holds more generally. In fact, Calvó-Armengol and Jackson [130] show that as a and b each converge to 0, but a/b converges to some positive limit, the correlation in the employment of any two path-connected agents is positive. The interpretation of this limit is that the time between periods is shrinking.[12]

---

12. In the short run, two indirectly connected agents might be competitors for a mutual neighbor's job information and so might have negatively correlated employment conditional on some states (see Exercise 10.7). Examining this limit allows one to look more directly at longer-term effects.

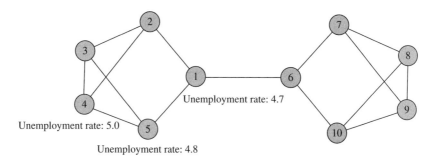

FIGURE 10.4 Unemployment rates as a function of position in a network, with $a = .100$ and $b = .015$, in an example from the Calvó-Armengol and Jackson [130] model.

10.2.3 Duration Dependence

An important aspect of networked interaction models of this type is that they generate specific correlation patterns over time and can help us to understand time-series patterns of behavior observed in various data that have not been well understood in the absence of the social setting.

A good example is what is called *duration dependence* in the labor economics literature. If we examine an unemployed worker and ask what the probability is that the worker will be employed in the next month, that probability goes down conditional on the worker having a longer history of unemployment, holding all else equal. This duration dependence has been found in a variety of studies, including Schweitzer and Smith [583], Heckman and Borjas [324], Flinn and Heckman [251], and Lynch [445]. As an illustration, Lynch [445] computes average probabilities of a typical worker finding employment on the order of .30 after one week of unemployment, .08 after eight weeks of unemployment, and .02 after a year of unemployment, after correcting for other observable characteristics, such as skill level and local employment rates.

A standard explanation for duration dependence is that there must be some features of workers that we cannot observe in the data but that are observable to firms, and the workers who are unemployed for long times have unattractive features from the firms' perspectives or other characteristics that make them less likely to become employed. However, the magnitude of the residual effects on employment (e.g., a 15-fold difference in the probability of becoming employed after one week of unemployment compared to one year), even after including a wide variety of characteristics, has been a puzzle. As pointed out by Calvó-Armengol and Jackson [130], a networked model of employment generates duration dependence as a general proposition. Let us illustrate this effect with some examples.

While the networks in Figure 10.5 are small, they show that understanding a worker's social context can account for some of the observed duration dependence. The idea behind a networked labor market exhibiting duration dependence is as follows. The longer a worker has been unemployed, the greater the probability of the worker's neighbors being unemployed. This correlation reflects (1) that the worker has not been able to pass to the neighbors any job information, and (2) that

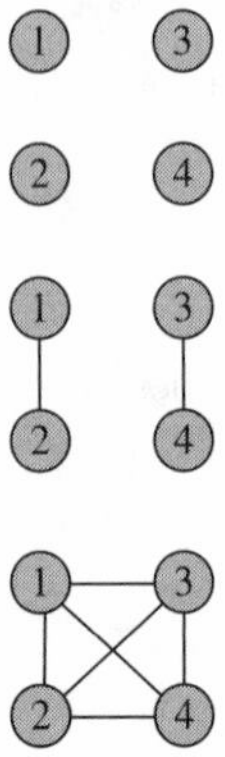

Probability employed next period
if unemployed for at least:

	1 period	2 periods	10 periods
	.099	.099	.099
	.176	.175	.170
	.305	.300	.278

FIGURE 10.5 Probability of becoming employed as a function of network and previous periods unemployed, with $a = .100$ and $b = .015$, in examples of the Calvó-Armengol and Jackson [130] model.

the worker has not heard about a job from the neighbors and so it is less likely that the neighbors are employed. As more of a worker's neighbors become unemployed, it becomes less likely that the worker will hear about a job in the coming period. For instance, if a worker has been unemployed for only a week, it is still quite possible that many of the worker's neighbors are employed, and so the worker will hear about a job shortly. However, if he or she has been unemployed for a year, then many of the worker's neighbors may also be unemployed, which makes it unlikely that he or she will hear about a job from a neighbor. Thus an unobserved feature of a worker (the status of his or her neighbors) could affect the employment probability of the worker. This explanation of duration dependence complements the usual unobserved characteristics explanation.

This feature is not unique to networks in a labor market, but also appears in other settings in which behaviors among individuals are complementary. Observing the behavior of one individual tells us something about the likely behavior of the neighbors and vice versa. When we look at the time series of individual behavior in isolation, it will take on a greater dependence on history than one would expect without a network model in the background.

10.2.4 Education and Drop-Out Decisions

Networked labor markets also provide interesting incentives for investment in education and human capital, which ties back to the discussion of strategic complementarities and graphical games in Chapter 9.

Suppose that agents start out in a network g. Each agent has an idiosyncratic cost c_i of investing in education. If the agent invests in education, then he becomes

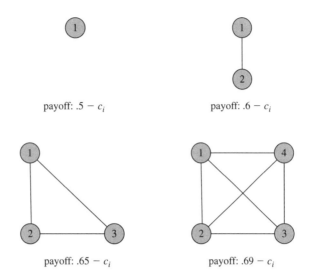

payoff: $.5 - c_i$

payoff: $.6 - c_i$

payoff: $.65 - c_i$

payoff: $.69 - c_i$

FIGURE 10.6 Expected payoffs for agents who become educated as a function of the job network, with $a = b$.

eligible for jobs; otherwise he simply has a payoff of 0.[13] Let $x_i \in \{0, 1\}$ be 1 if agent i invests in education and 0 if not. If an agent is educated, then the labor market described in the previous sections applies, and the agent can hear about jobs directly and from an employed neighbor j, who has also chosen to be educated.

Thus, we start with network g and end up with network $g(x)$, which is the subnetwork of g restricted to the nodes i such that $x_i = 1$. Based on $g(x)$, each agent will have a long-run employment rate. We can then examine Nash equilibria of the game, where the payoff to investing in education is the long-run expected employment rate minus the cost c_i. If this difference is greater than 0, then the agent invests.

Note that this is a graphical game of strategic complements. To explore the structure of the game, consider a dyad with $a = b$ and look at the limiting process as in Proposition 10.1. Figure 10.6 illustrates the complementarities well. The employment probability, and hence the payoff, of an agent goes up as he or she has more neighbors, since they pass job information. Thus there are multiple equilibria and contagion effects in actions. To see the possibility of multiple equilibria explicitly, consider four agents and the payoffs pictured in Figure 10.6. If each agent has a cost c_i of .6, then one equilibrium occurs when no agent obtains an education, and another occurs when all agents obtain one. There is a complementarity in their incentives. The better equilibrium for all involved is for all of them to get an education, but it is possible to end up at the other equilibrium.

---

13. This model is clearly a simplification. It could be that if the agent does not invest, then she obtains information about unskilled jobs from neighbors who did not invest; while if she does invest, then she obtains information about skilled jobs from neighbors who also invested. The important thing is that neighbors' decisions to invest affect her decision.

We can also see contagion effects. Suppose that the four agents have costs of education that are $c_1 = .51$, $c_2 = .61$, $c_3 = .66$, and $c_4 = .70$. Then there is a unique equilibrium such that no agents become educated. Agent 4 would not find it worthwhile no matter what the other agents do, which leaves at most three agents to become educated. Under that circumstance, agent 3 does not find it worthwhile to become educated, and so forth. So agent 4's decision has wide consequences. If we lower agent 4's cost just slightly (e.g., $c_4 = .68$), then a new equilibrium appears in which all four agents become educated.[14]

This model also provides some intuition for poverty traps. The idea is that initial conditions can be very important, especially if there is even the slightest sequentiality in agents' decisions. For instance, if historically no agents have become educated, and then we ask whether some agent wants to become educated, he or she has to be willing to do so without any neighbors having invested. Such complementarities can lead whole groups to stay under-invested relative to what the best equilibrium for the group would be. Complementarity can also lead to dramatic differences in behavior among different groups of agents embedded in highly segregated networks; similar to what we saw in Chapter 9, so that some subgroup invests highly while another does not. Bringing homophily into the picture shows how these ideas can help explain dramatic differences in the pursuit of higher education across different ethnicities. If most of one's friends are of the same ethnicity and almost none go on to higher education, then that tends to be a best response; while if almost all attain a higher education, then that tends to be a best response. These feedback effects can exacerbate those that arise in part from other socioeconomic factors.

10.2.5 A Labor Market in a Homophilous Network

The model of networked labor markets of Calvó-Armengol and Jackson [130] discussed above only analyzes half of the market. That is, firms play no role in the analysis, as jobs simply appear at an exogenous rate.

A model of labor markets in which firms play a richer role (and the structure of workers' networks plays a lesser role) was developed by Montgomery [479]. This model provides a reason for firms to use referrals as a means of hiring, as workers' social ties are homophilous in a sense associated with their productivity. It also provides different insights into wage dispersion than those found in the Calvó-Armengol and Jackson [130], [132] model. Montgomery's [479] model is described as follows:

- The economy lasts two periods.
- There are N workers in each period and they live for only one period.
- Half of the workers produce no output and half produce one unit of output.
- Firms cannot observe the workers' types until output is delivered.
- Firms employ at most one worker in any period.
- A firm's profit is the output minus any wage paid.

---

14. It still remains an equilibrium for none of them to invest. One can also create examples in which slight changes lead that equilibrium to disappear.

- Wages are paid upon hiring the worker and cannot be contingent on the output.
- There is free entry into the market so that firms can enter the market in either period.

The potential workers are connected in a network which is formed as follows.

- Each worker from the first period knows one second-period worker with probability τ and does not know any second-period worker with probability $1 - \tau$.
- If a first-period worker knows a second-period worker, then the latter is a worker of the same "type" (productivity) with a probability $a > 1/2$.
- Each first-period worker with a link has that tie assigned to a second-period worker by first choosing whether the link is to a worker of the same or different type (with probability a and $1 - a$, respectively) and then choosing uniformly at random from workers of the selected type. Thus second-period workers can have multiple ties.

The timing of decisions is as follows:

- Firms hire first-period workers at an (equilibrium) wage w_1.
- A firm that hired a first-period worker observes that worker's (and only that worker's) first-period output. If that firm desires, it may then make a "referral" wage offer to its worker's social tie (if the worker has a social tie).
- Second-period workers who receive offers from the firms of their first-period friends may accept one of those offers or decide to go on the second-period market.
- The second-period market is such that any second-period workers who have not accepted a job through the referral process are hired at an (equilibrium) wage of w_2.

The equilibrium notion that Montgomery employs is a variation on a competitive equilibrium, requiring that no firm wants to enter or exit, that firms optimize given their information, and workers take the best offer made to them. It also involves aspects of a Nash equilibrium, since in offering a referral wage, a firm is entering an auction against other potential employers who might also be making a referral offer to the same worker. Montgomery proves the following result.

Proposition 10.2 (Montgomery [479]) *A firm makes a referral offer if and only if it has a productive worker in the first period, and it then randomly picks a wage to offer from an interval with lower bound of w_2 and an upper bound below 1.*[15]

---

15. In fact, one can reason that the upper bound on bids will be no more than a, which is the expected value of a second-period worker conditional on having a tie to a first-period worker. Conditional on the worker accepting the wage offer, it is less likely that the worker has ties to other high-output first-period workers, and so the conditional expectation of the worker's value is generally less than a.

The idea behind the proposition is as follows. Let us work backward from the second period. In the second-period market for the workers who have not accepted an offer, the wage is equal to the expected value of those workers, given their distribution of types in equilibrium. Firms all have the same information about those workers' values, will not overpay, and cannot underpay given that new firms can enter the market. The claim in Proposition 10.2 allows us to conclude that the expected value of such a worker is less than $1/2$, as these workers have not received and accepted an offer, and so are conditionally more likely to either be connected to a low-output first-period worker or to lack connections. Given that a second-period worker can obtain a wage of w_2 by waiting for the open market, any nondegenerate referral offer has to be at least w_2. The equilibrium involves a mixed strategy, because offering a wage through a referral is like bidding in an auction with an unknown number of other bidders. The second-period worker could have other ties to first-period workers and thus could be receiving other wage offers. The worker's expected value is generally more than w_2, given that the worker is tied to a high-output first-period worker and that the wage w_2 is less than $1/2$. If referrals were all hired at a given fixed wage below the expected value of the worker, then by slightly raising the wage, one firm could hire the second-period worker for certain. If the wage were at or above the expected value of the worker, then a firm could lower the offer and still win if there are no other bidders.

The important aspects of the equilibrium are:

- Having more social ties leads to higher expected wages for second-period workers. Each additional tie has some chance of being a high-output worker and thus resulting in an extra wage offer.

- A low-output second-period worker with social ties has a higher expected wage than a high-output second-period worker who has no social ties. A high-output worker without social ties has to go on the second-period market, while any worker (regardless of actual productivity) with social ties has some probability of being connected to a high-output first-period worker and getting a wage offer above the second-period open-market wage.

- There is a dispersion of wages in the second period. This follows from the randomization in referral wages and the fact that workers differ in the number of social connections they have.

- Firms earn positive profits in the second period from using referrals. Firms have a higher chance of finding high-output workers through referrals, and as a lower bound, there is at least some chance that they can hire the worker with a wage just higher than w_2 when the worker has no other social ties.[16]

---

16. The full argument here is a bit tricky, as the conditional expectation of a worker's value depends on what wage is offered and accepted. Hiring the worker with a lower wage provides some indication that the worker received few other offers. But in equilibrium, the expected profit is the same at all wages that are offered, and the workers who do receive offers are biased toward being more productive.

While highly stylized, this model provides insights into several aspects of networks in labor markets: why referrals can be attractive for firms, why they can lead to dispersion in wages,[17] and how workers who are more connected can fare better.

10.2.6 Evidence and Effects of Networked Labor Markets

Topa [629], Conley and Topa [167], and Bayer, Ross, and Topa [48] fit models of social interactions and employment that have similar features to those described in Section 10.2.2, in which networked workers should exhibit correlated employment and wages. Topa [629] and Conley and Topa [167] examine census data in Chicago, focusing on data from the 1980 and 1990 censuses. As a proxy for network neighborhood relationships, Topa uses geographic neighborhood relationships. He examines the correlation of unemployment across census tracts and finds statistically significant correlation patterns between adjacent census tracts. He also finds significantly positive correlation between tracts that are not immediately adjacent but are still both adjacent to a common tract.

As it is possible that social connections are not simply related to geographic proximity, Conley and Topa [167] also examine other distance measures. In addition to census tract distance measures, they examine travel-time distance, ethnic distance, occupational distance, and education measures, as well as other socioeconomic covariates. However, these measures may be related to other characteristics that drive employment patterns, and so the similar outcomes in employment that seem to indicate the role of social relationships in influencing employment might simply result from a correlation of social measures to other characteristics. To deal with this issue, Conley and Topa perform different exercises. First they examine how the raw unemployment rate correlations depend on combinations of these proximity measures. When combining various measures, being close on the ethnic dimension seems to explain the majority of covariation in unemployment rates. In view of this correlation, they next examine the residual employment rates when adjusting for a number of observable tract level characteristics. These residuals are obtained by subtracting out the variation in employment that can be explained directly by tract-level characteristics. They then examine how these residuals correlate with the distance measured in tracts or some other characteristic (e.g., ethnicity, occupation). The correlation patterns across these measures largely disappear as a result, indicating that it is not that people have similar outcomes because they are nearby and thus socially related. Instead nearby tracts may be quite similar in their characteristics and hence their workers tend to be employed or unemployed at the same time for other reasons.

The Conley and Topa [167] study could cast significant doubt on social proximity being important in employment outcomes. However, failing to find a relationship when looking at census-tract data does not imply that social relationships are irrelevant to employment. Social relationships are not so obviously related to census tracts, especially when aggregated. Here a study by Bayer, Ross, and Topa [48]

---

17. See Arrow and Borzekowski [20] for another model of wage dispersion based on the number of ties and a calibration to wage data.

looks more closely at social relationships and employment. They examine census data from Boston where they can pinpoint residence down to the block level. This degree of localization allows them to examine whether people living on the same block have more highly correlated employment outcomes than do people living on nearby blocks with similar characteristics. They find that living on the same block compared to a nearby block with similar demographics significantly increases the probability that two individuals work together, and this effect is magnified when the individuals are of similar ages and backgrounds. Then, examining pairs of individuals who have a strong predicted referral effect, they find a substantial effect on employment and wages. They also examine other questions. For instance, they find that there is assortative matching along education, income, and age, so that similarity along these dimensions with those in one's city block improves labor market outcomes significantly. The study is also careful to examine "reverse causation" explanations: people end up on the same block because they are similarly employed.

There are other ways of studying social network effects on labor outcomes, such as examining immigrant populations and the social networks that they move into, or other exogenous factors that affect social networks. The difficulty is in finding extensive social network data together with rich measures of employment outcomes. Some examples of clever proxies for social networks appear in Munshi [493], Laschever [423], and Beaman [50]. Laschever [423] examines the formation of military units via the U.S. draft in World War I. He examines units that were formed at random and then studies subsequent employment outcomes using the 1930 census. If the friendships formed within a given military unit did not matter, then there would not be any correlation among employment outcomes in the later period (after correcting for other factors). He finds statistically significant effects, so that a 10 percent decrease in the employment rate of a veteran's unit decreases that veteran's employment rate by more than 3 percentage points. Munshi [493] examines Mexican immigrants to the United States, using rainfall in Mexico to estimate the number of immigrants during various time periods[18] and then shows that having a larger number of immigrants arrive more than 3 years prior to one's own arrival leads to a significant increase in the probability of one's employment. Beaman [50] finds a more direct measure of the size of waves of specific immigrant groups, with a rich variation in sizes, countries of origin, and locations, as she examines the assigned relocation of political refugees into the United States. The network size proxy is the number of refugees from the same country who did not have prior family members in the United States and who are relocated to the same city. She finds several effects. First, the larger the number of political refugees relocated to a particular location at the same time or within a year of each other, the lower their average employment rate; which is consistent with the new arrivals competing for jobs and job information. More pointedly, in terms of evidence of the effect of social networks, the larger the number of refugees who were relocated

18. This might, at first, seem to be a strange method; but one wants a factor that influences immigration but is not correlated with employment possibilities in the United States. Adverse agricultural conditions can lead to emigration from Mexico and yet are unlikely to be correlated with job opportunities in the United States.

to a given area at least 2 years prior, the higher the employment rate and wages. For instance, a standard deviation increase in the number of refugees arriving 2 years prior leads to an increase in the probability of becoming employed by 4.6 percent, along with a 50-cent increase in the average hourly wage, while a similar increase in the number of refugees arriving in the prior year leads to a decrease of 4.9 percent in the probability of becoming employed and a 70-cent decrease in average hourly wages.

10.3 ▪ Models of Networked Markets

The above studies were tailored to labor markets. Beyond labor settings, it is important to have a more general understanding of how the structure of the network interactions affects the terms of trade.

10.3.1 Exchange Theory

An area of research known as *exchange theory* (see Cook and Whitmeyer [173] for an overview) is concerned with how the structure of relationships among agents affects exchanges among them. Such exchanges could be economic transactions of goods and services, the trading of favors, communication of information, or a variety of social interactions that convey direct benefits and costs to the involved agents. The term *exchange theory* has its origins in work by Homans [336], [337] on "social behavior as exchange," which initiated a theory of socially embedded behavior based on the idea of psychological reinforcement applied to dyadic exchanges. This view was complemented by work by Thibaut and Kelley [627], and more direct connections to economic ideas of exchange were brought into the picture by Blau's [73] influential work. The ideas have been extensively developed and applied to a range of economic interactions that involve explicit relationships, such as decentralized markets, the formation of corporate boards, and international relations. Networks have played an increasing role in exchange theory, especially since the work of Emerson [224], [225]. Emerson considered explicitly networked interactions to explain the power and dependencies that underlie exchange. Critical to Emerson's theory is the idea that the exchange that occurs between two agents depends on their outside options and influences and thus on their other relationships. Thus one cannot examine an exchange between two individuals without studying the influences of the network on their behavior.

To develop a feel for this theory, it is useful to discuss the work of Cook and Emerson [172], who laid out hypotheses about how power derives from social network structure. They also examined the role of equity considerations in exchanges and conducted some of the first experiments on this subject. These ideas provide a nice background for the predictions and observations that we will see in Section 10.3.2 (and Chapter 12) when we consider models of economic transactions in social networks.

Cook and Emerson [172] operationalize the idea of equity through a condition that two agents involved in an exchange should equilibrate their respective profits or net gains from a given transaction. They examine this condition in a context in

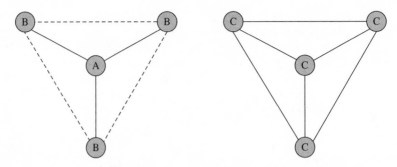

FIGURE 10.7   Exchange networks in the Cook and Emerson [172] experiments. A solid line indicates a potential transaction worth 24 units, and a dashed line indicates a potential transaction worth 8 units.

which agent 1 holds good X and agent 2 holds good Y, agent 1 has a higher per unit value for Y than for X, and agent 2 has a higher value for X than for Y. The idea is that if x units of good X are given by agent 1 to 2 in exchange for y units of good Y, and v_{iz} represents the marginal value to agent i for good z, then equity requires that

$$v_{1y}y - v_{1x}x = v_{2x}x - v_{2y}y.$$

Cook and Emerson [172, p. 723] define the power of 1 over 2 as "the potential of agent 1 to obtain favorable Y minus X outcomes at agent 2's expense."[19] They do not provide a formal recipe for how to evaluate power as a function of the network, but they do distinguish between how relationships affect each other. Consider agent 1 who has links to both agent 2 and agent 3 in a social network. The researchers state that these relationships have a positive connection if a transaction across one link is contingent on a transaction across the other, and have negative connection if a transaction on one link precludes a transaction on the other. To analyze how power depends on the network of relationships and potential exchanges, Cook and Emerson examine the networks in Figure 10.7.

In this figure Cook and Emerson state that A has power over each B, but the Bs have equal power relative to one another, and the Cs have equal power relative to one another. In terms of a specific measure of the power that A has over a B, Cook and Emerson reason based on the comparison levels that the agents have. In particular, they argue that A's comparison level, in terms of the expected value of the transaction, should be 20 units. The idea is that A can trade with any of the Bs. The Bs that do not trade with A are balanced and so should end up splitting their 8 units equally for a value of 4 units, which is their comparison level. Thus if the B who transacts with A gets more than 4 units, then another B would have an incentive to offer to split 24 with A in a more favorable way.[20] Cook and Emerson conceive of the comparison levels as some indication of power used, and then the excess of power of an A (with a comparison level of 20) over that of a B (with comparison level of 4) is 16 units. So power is a measure of the difference between the resources A receives from a transaction with B and what B receives.

Cook and Emerson [172] ran a series of experiments on these networks, working with human subjects playing the roles of network agents for cash earnings in proportion to the units of transaction. The subjects interacted through computers and were not aware of the identities of other agents. The bargaining protocol was such that players could make direct offers of units (up to 8 between two Bs and up to 24 between any other pair) to agents in their network, and if that agent agreed, then they would complete that transaction, with the offering player keeping the total less the offer. Subjects played the game (with fixed position) 40 times. For the first 20, players were only aware of the values of their own potential transactions, and not of the transactions of the players other than those they were involved with. For the next 20 games, players observed one another's cumulative earnings. Cook and Emerson hypothesize that having knowledge of others' payoffs leads to more equitable behavior and less exercise of power. The Cook and Emerson [172] data provide insight into several different things: first, whether there was an exercise of power so that As earned more than Bs; second, how this power exercise compared with the behavior of the evenly balanced Cs; and, third, how the knowledge of others' payoffs affected behavior. Furthermore, Cook and Emerson had a pool of 56 male and 56 female subjects, and so could compare behavior across genders. A brief summary of the results is that the As did exercise power, beginning by earning between 2 and 4 units more than Bs and tending up to 10 to 12 units more than Bs just before the cumulative earnings were shown, but exercising less than the full 16 units of power that they had.[21] Among the Cs, there were imbalances between the even partners, with the average imbalances beginning at 7 or 8 units (out of 24) and tending downward to 4 to 5 units by the end of the 40 periods. When looking across genders, Cook and Emerson found significant differences between how males and females act in position A in the periods after the cumulative histories of payoffs are revealed. Males exhibit a short-term lowering of their exercise of power, but eventually return back to about 12 units, while the females (significantly) lower their power usage to around 4 units by the end of the experiment. These experiments provide evidence that network position matters in bargaining and in helping to operationalize notions of power. The data show that network-based bargaining power ends up being exercised even when agents are not fully aware of the values of the possible transactions that can occur, but also that some equity concerns can mitigate the exercise of power. The differences between female and male behavior provide some interesting puzzles.

---

20. This allocation (20 to A and 4 to each B) turns out to be the unique core allocation in this problem, as defined using a standard cooperative game-theory concept. See Section 12.4.2 and Exercise 12.3. It differs from the Shapley value or Myerson value allocations for this problem.
21. These numbers actually look more consistent with the Myerson value predictions than the core predictions for this network. See Exercise 12.3.

10.3.2 Bilateral Trading Models

The Cook and Emerson [172] exchange studies provide insight into how bargaining power might be exercised and suggest that agents are sensitive to the network of potential transactions in which they are embedded. To examine this aspect in more detail, let us consider model networks of buyers and sellers. The following analyses use game-theoretic models of the bargaining on networks to make predictions about which networks will form as agents try to maximize the value of their transactions.

A Networked Trading Model Based on Alternating Offers Bargaining

A natural starting point is a simple model of networks with bilateral bargaining that is due to Corominas-Bosch [178]. Each seller has a single object to sell that has no value to the seller. Buyers have a valuation of 1 for an object and do not care from whom they purchase it. If a buyer and seller exchange at a price p, then the buyer receives a payoff of $1 - p$ and the seller a payoff of p. A link in the network represents the opportunity for a buyer and a seller to bargain and potentially exchange a good.

Corominas-Bosch models the bargaining process explicitly by an alternating move game between the various buyers and sellers. That game leads to a particular solution. A link is necessary between a buyer and seller for a transaction to occur, but if an individual has several links, then there are several possible trading patterns. Thus the network structure essentially determines the bargaining power of various buyers and sellers.

The game that Corominas-Bosch examines to predict the prices and transactions is described as follows. In the first period sellers simultaneously call out prices. A buyer can only select from the prices that she has heard called out by the sellers to whom she is linked. Buyers simultaneously respond by either choosing to accept a single price offer received or rejecting all price offers received. If several sellers have called out the same price and/or several buyers have accepted the same price, and there is any discretion under the given network connections as to which trades should occur, then a careful protocol is used to determine which trades occur (the protocol is designed to maximize the number of eventual transactions). At the end of the period, trades are made, and buyers and sellers who have traded are cleared from the market. In the next period the situation reverses and buyers call out prices. These are then either accepted or rejected by the sellers connected to them in the same way as described above. Each period the roles of proposer and responder switch, and this process repeats itself until only buyers and sellers who are not linked to one another remain. Buyers and sellers are impatient and discount according to a common discount factor $0 < \delta < 1$. So a transaction at price p in period t is worth $\delta^t p$ to a seller and $\delta^t(1 - p)$ to a buyer.

In an equilibrium with very patient agents (so that δ is close to 1), there are effectively three possible outcomes for any given agent: either he or she gets most of the available gains from trade, or roughly half of the gains from trade, or a small portion of the available gains from trade. Which of these three cases ensues depends on that agent's position in the network. Some easy special cases are as follows. Consider a seller linked to two buyers, who are only linked to that seller. Competition between the buyers to accept the price leads to an equilibrium price of close to 1 if agents are sufficiently patient. So the payoff to the seller in such a network will be close to 1, while the payoff to the buyers will be close 0. These payoffs are reversed for a single buyer linked to two sellers.

More generally, which side of the market outnumbers the other is a bit tricky to determine, as it depends on the overall link structure, which can be much more complicated than that described above. Corominas-Bosch [178] describes a clever algorithm that has roots in Hall's theorem (recall Theorem 2.1) and subdivides any network into three types of subnetworks: those in which a set of sellers are collectively linked to a larger set of buyers (sellers obtain payoffs close to 1, and buyers receive payoffs near 0); those in which the collective set of sellers is linked to a same-sized collective set of buyers (each receives a payoff of about 1/2); and those in which sellers outnumber buyers (sellers receive payoffs near 0, and buyers obtain payoffs close to 1). The limiting payoffs, as the discount factor approaches 1, are found using the following algorithm.

Step 1a: Identify groups of two or more sellers who are all linked only to the same buyer. Regardless of that buyer's other connections, eliminate that set of sellers and buyer; that buyer obtains a payoff of 1 and the sellers all receive payoffs of 0.

Step 1b: On the remaining network, repeat step 1a but with the role of buyers and sellers reversed.

Step k: Proceed inductively in k, each time identifying subsets of at least k sellers who are collectively linked to some set of fewer-than-k buyers, or some collection of at least k buyers who are collectively linked to some set of fewer-than-k sellers.

End: When all such subgraphs are removed, the buyers and sellers in the remaining network are such that every subset of sellers is linked to at least as many buyers and vice versa, and the buyers and sellers in that subnetwork earn payoffs of 1/2.

The limiting payoffs found by this algorithm are illustrated in Figure 10.8. To see how the algorithm works on the last network, consider Figure 10.9. First, step 1a is concluded with no sets identified (top left panel in Figure 10.9). Next, in step 1b, the set of two buyers linked to just one seller is eliminated (top right panel). The remaining two buyers and two sellers are each linked to the same number on the other side (bottom left panel), and so the algorithm concludes with the remaining subset of four agents each receiving a payoff of 1/2 (bottom right panel).

The intuition behind why the algorithm identifies the unique equilibrium outcome is as follows. If there are two or more sellers who are linked to just one buyer, they will compete in bargaining and the buyer can obtain a price of 0 (in the limit with high patience). As a result any other sellers linked to that buyer cannot expect the buyer to bid for their goods. At a later step, when three sellers are linked to just two buyers, then it must be that each of the buyers is linked to at least two of the sellers (as otherwise two of the sellers would have been linked to just one buyer and removed at an earlier step). In this case, the game is a sort of "musical chairs" among the sellers. At most two can sell an object, and so sellers who are quoting the highest price in some round have an incentive to cut their price slightly to avoid being left with an object; if it is the buyers who are quoting prices, then no seller

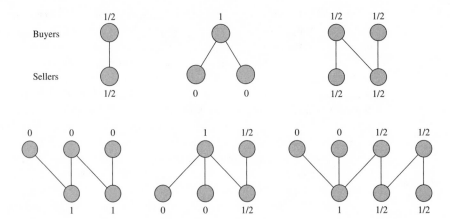

FIGURE 10.8   Limit payoffs in the Corominas-Bosch [178] model for selected networks.

wants to be the only one not accepting, as he or she would be left without selling an object. This competition again leads to an unraveling of the price, and so the buyers obtain all of the gains from trade.

For quite involved networks, the logic of this derivation relies heavily on induction. However, the process itself naturally involves induction, just as the algorithm does. That is, with some patience, several sellers linked to just one buyer tend to compete away their surplus. As such transactions occur, the more complicated games of musical chairs begin to play out. A series of experiments by Charness, Corominas-Bosch, and Frechette [146] examine the extent to which human subjects play as predicted in such games. They examine a game of bargaining on a fixed network that proceeds just as in the Corominas-Bosch model, except that there are only a finite (but uncertain) number of rounds. While the payoffs in those experiments rarely reach the extremes that are predicted under the limit of full patience and potentially infinitely many periods of bargaining, the payoffs do share patterns similar to those predictions. Figure 10.10 pictures the predicted payoffs for one of the networks that was tested. This figure reports the percentage of potential pies to be split that each role received (so the total sums to 3, as there were three possible transactions in all) and also provides the benchmark prediction for the same network in the case with infinite patience and repetition.

Critical aspects of limiting payoffs under the bargaining are:

1. if a buyer obtains a payoff of 1, then some seller linked to that buyer must receive a payoff of 0, and similarly if the roles are reversed;

2. a buyer and seller who are only linked to each other receive payoffs of $1/2$; and

3. the subnetwork restricted to the set of agents who receive payoffs of $1/2$ is such that any subgroup of k buyers in the subnetwork is linked with at least k distinct sellers in the subnetwork and vice versa for any k.

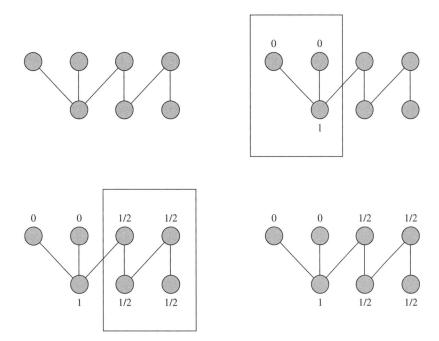

FIGURE 10.9 An illustration of the algorithm in the Corominas-Bosch [178] model.

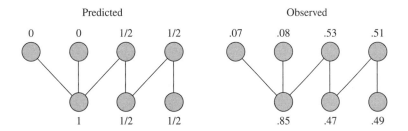

FIGURE 10.10 Predicted and observed payoffs in the Charness, Corominas-Bosch, and Frechette [146] experiments.

These observations about the payoffs, coupled with a cost per link, lead to the following sharp predictions concerning network formation using pairwise stability (as defined in Section 6.1).

Proposition 10.3 *Consider a version of the Corominas-Bosch model in which agents receive the limiting payoffs as described above. If the cost of a link for each* *agent involved lies strictly between 0 and 1/2, then the pairwise stable networks coincide with the set of efficient networks.*[22]

The proof of Proposition 10.3 is straightforward, so I sketch it here, and the details are left for Exercise 10.1. An individual receiving a payoff of 0 cannot have any links, as by severing a link he or she could save the link cost and not lose any benefit. Thus all individuals who have links must obtain payoffs of 1/2. One can then show that if there are extra links in such a network (relative to the efficient network, which consists of a maximal number of disjoint linked pairs), some could be severed without changing the bargaining payoffs, thus saving link costs. This argument builds on the fact that for payoffs to be 1/2, buyers and sellers must be evenly balanced. It cannot be that there is a buyer and seller who each have no links, as by linking they would both be better off. So the network must consists of pairs, and the maximum number of potential pairs must form.

The conclusion of an equitable split of the value in each transaction contrasts with the conclusion from a supply and demand model. If there are more buyers than sellers, then in a competitive model, the sellers collect all of the surplus. What accounts for the difference between these models? In the model of Proposition 10.3, there is a cost to connecting, and connecting occurs before the bargaining. Buyers and sellers only connect if they expect to obtain a positive payoff. The model is extreme, so that slight imbalances in the network (mismatches of buyers to sellers) lead to extreme outcomes and 0 payoffs to some agents. Thus the equilibrium entry into the market results in a balanced set of buyers and sellers. The limited set of outcomes, so that transactions are either even splits or completely favor one side of the exchange, is critical to the result. As we shall now see, if we enrich the set of possible transaction outcomes, more complex networks emerge.

A Networked Trading Model Based on Auctions   While the model outline in Proposition 10.3 provides a benchmark, the conclusion that agents will form an efficient network relies on agents having complete information and the fact that buyers are identical, as are sellers. If heterogeneity and uncertainty in valuations exist, there can be benefits from having a more connected network so that the set of potential transactions is larger. This generalization leads to a more complicated analysis and can lead to inefficient networks being stable. Let us consider a simple example, based on a model of Kranton and Minehart [414]. It is similar to the Corominas-Bosch [178] model described above except that the valuations of the buyers for an object are random and the determination of prices is made through an auction.[23]

The potential sources of inefficiencies can be seen from a situation with one seller and two buyers; so I focus on that setting and refer the reader to Kranton and Minehart [414] for a fuller analysis. The buyers each have a valuation for the good that is uniformly and independently distributed on [0, 1]. The good is sold by

---

22. This result contrasts with Corominas-Bosch's [178] analysis, which considers a formation process in which no cost is saved by severing a link. That lack of cost can lead to players having links even when they know that they receive a 0 payoff from the bargaining and trade.

23. In many applications one would also see random and heterogeneous valuations among the sellers, but the main ideas can be seen without introducing such complications.

second-price auction.[24] This auction is one in which the highest bidder obtains the object and pays the highest bid made among the bidders who are not getting the object (with ties for the highest bid broken uniformly at random). If only one buyer links to the seller, then he or she gets the object for a price of 0. If both buyers are linked to the seller, then it is a dominant strategy for each buyer to bid his or her value, and the corresponding revenue to the seller is the minimum value of the two buyers.

In this model the potential value of a transaction is random, depending on the realized valuations of the buyers. The auction on the network with just one link has an expected transaction value of 1/2, and an equilibrium price of 0, so that the full expected gains from trade go to the buyer. In the auction with two links, the ex ante expected payoff to each buyer (before he or she sees his or her value for the object) is 1/6. Each buyer has a 1/2 chance of having the high value, an the expected valuation of the highest bidder out of 2 draws from a uniform distribution on [0,1] is 2/3, and the expected price is the expected second-highest valuation, which is 1/3. So the ex ante total gains from trade in the two-link network is 2/3, with 1/3 going to the seller and 1/6 to each buyer.

Now consider a situation with a cost of links c to each individual. Efficient network structures are an empty network if $c \geq 1/4$, a one-link network if $1/4 \geq c \geq 1/12$, and a two-link network if $1/12 \geq c$.

Let us consider whether the pairwise stable networks are efficient. Given the seller's payoffs as a function of the network, the seller will only link to both buyers or to neither. Thus if link costs lie between 1/2 and 1/4, then the efficient network is not pairwise stable. If the cost is more than 1/6 and less than 1/4, then the only pairwise stable network is empty, as buyers do not expect a high enough payoff to maintain a link in a two-link network and the seller will not maintain a link in a one-link network. If the cost is between 1/12 and 1/6, then the two-link network is pairwise stable, but it is overconnected relative to the efficient network. Thus an efficient network is only pairwise stable when costs are less than 1/12 or greater than 1/4.[25]

To see why there is inefficiency in this setting, note that the increase in expected price to the seller from adding a link comes from two sources. One is the expected increase in willingness to pay on the part of the winning bidder, since the sale is to the highest valuation out of a set of independent draws from the same distribution, and one more draw is added when a link is added. This increase is of social value, as it means that the good is going to someone who values it more. The other source of price increase to the seller from adding a link comes from the increased competition

---

24. Any of a variety of auctions will have the same property in this example, as there is an equivalence between any two different auctions and corresponding equilibria, provided that the equilibrium lead to the same allocation of the good as a function of valuations and are such that buyers with 0 values pay 0 if they "win." Each buyer has the same expected payment in the two auctions, and the seller expects the same revenue.

25. This conclusion contrasts with that in Kranton and Minehart [414]. However, they analyze a case in which link costs are 0 for sellers and positive for buyers. If sellers bear no costs of links, then the efficient networks are pairwise stable. Kranton and Minehart do discuss the fact that costly investment by the seller can lead to inefficiency.

among the bidders in the auction. This source of price increase is not of social value since it only increases the proportion of value that is transferred to the seller.

While the pairwise stable networks in this example are not efficient (or even constrained efficient), they are Pareto efficient. This is not true with more sellers as shown in Exercise 10.3, which exhibits pairwise stable networks that are Pareto inefficient.

10.3.3 Price Dispersion on Networks

Beyond the strategic formation of networks in exchange settings, there are also studies that examine how the terms of trade depend on network structure when networks are (exogenously) generated by random graph models. For instance, Kakade et al. [374] (as well as Kakade, Kearns, and Ortiz [372]) examine a model of exchange on random graph–generated networks.

Buyers have cash endowments and a constant marginal value for a consumption good. Sellers have unit endowments of the consumption good (which they do not value) and desire cash. Buyers buy from the least expensive seller(s) with whom they are connected until they have exhausted their cash budgets. Prices are seller-specific and determined to clear markets. An equilibrium is a set of prices and transactions such that the market clears. In this setting, market clearing implies that each buyer who is connected to at least one seller exhausts his or her budget, and each seller who is connected to at least one buyer sells all of his or her endowment. The simplest version of this model is such that all buyers have the same marginal valuation and endowments, say each normalized to 1. In that case, there is an equilibrium of the following form:

- A buyer purchasing from multiple sellers sees the same price from each of those sellers.
- The price of a given seller can be found by computing, for each buyer, the fraction of the buyer's total purchases that come from that seller and then summing across buyers.
- The price of a given seller is no higher than the seller's degree and no lower than one over the minimum degree of the buyers connected to the seller.

The outcomes for a few networks are pictured in Figure 10.11. The pattern of trades in the figure is similar to that in the model of Corominas-Bosch and Figure 10.8. There is a richer variation in prices in the competitive model, but the basics of which buyers and sellers do well and which do poorly are qualitatively similar. The richer variation in prices reflects that buyers and sellers are assumed not to exercise any monopoly power they might have; instead they adjust prices to clear markets. This added motivation also results in differences if one examines the networks that are pairwise stable, as shown in Exercise 10.5. In this competitive model, there are still gains from trade to be earned by both sides of the market, even when there are large imbalances among buyers and sellers. As a result, too many agents connect to one another relative to the efficient network, which involves pairing agents.[26]

---

26. This simple form of efficient network reflects the linear utility functions. With concavities, more interesting architectures emerge as being efficient.

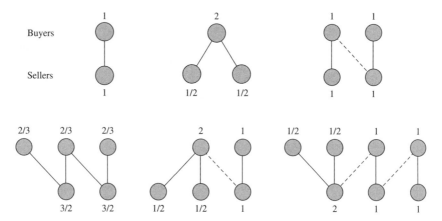

FIGURE 10.11 Equilibrium prices and trades for some network configurations in the competitive model of Kakade et al. [374]. The number under each seller is the price that the seller charges and the total cash that the seller ends up with. The number above each buyer is the amount of the consumption good that the buyer purchases. The solid links indicate trades, whereas there are no trades along the dashed links.

Figure 10.11 demonstrates that the configuration of prices that emerges as a function of the network is very network specific, and so deriving general conclusions for complex networks is difficult. There are simple observations, such as the fact that adding (costless) connections weakly benefits the agents involved with that connection, but there is no general relationship between degree and welfare, as terms of trade depend on the full network configuration. Even though the model is difficult to solve analytically, Kakade et al. [374] show that if links are formed uniformly at random, the probability of forming a link is high enough, and the number of agents grows, then there is no limiting price dispersion. In the case of a network formed by preferential attachment, there is a greater asymmetry in the degrees of nodes, and such networks can maintain some price dispersion.[27]

10.3.4 Collaboration Networks among Firms

Most of the discussion in this chapter has been either about labor markets and information transmission or about exchange networks, where trades occur between linked agents. There are also other ways in which networks play a role in markets. For example, firms collaborate in research and development, merge, produce joint products and ventures, and contract on specific supply relationships.[28]

---

27. For more on how network shape affects trading behavior, see Judd and Kearns [369] for a set of experiments of networked trade.
28. See Bloch [75] for an overview of the related literature.

To get a feel for how various relationships among firms might affect their actions in the market, let us examine an example from Goyal and Joshi [298]. They examine a setting in which a link between two firms lowers their respective costs of production. Since firms eventually compete in the market, the costs of production affect the overall market outcome and profits.

In particular, each firm produces identical goods. If firm i produces q_i units of the good and the network structure is g, then firm i's cost is

$$q_i \left(a - bd_i(g) \right), \tag{10.16}$$

where $a > (n-1)b > 0$ and n is the number of firms, so that costs are always positive. So a firm's marginal cost of production is decreasing in the number of collaborative links it forms with other firms, which in this model is its degree $d_i(g)$.

The profits to a firm then depend on how much each firm produces and what the resulting price in the market is. To model this market, consider the textbook case of Cournot competition, which works as follows. The market price is described by an (inverse) demand function, such as

$$p = \alpha - \sum_j q_j,$$

where $\alpha > 0$ is scalar. Thus the price decreases as firms produce more. So firm i's profits are

$$q_i p - q_i \left(a - bd_i(g) \right) = q_i \left(\alpha - \sum_j q_j - a + bd_i(g) \right).$$

If q_i maximizes this equation, then the derivative with respect to q_i must be 0, or

$$-q_i + \alpha - \sum_j q_j - a + bd_i(g) = 0. \tag{10.17}$$

Solving (10.17) simultaneously across i leads to[29]

$$q_i = \frac{\alpha - a + nbd_i(g) - b\sum_{j \neq i} d_j(g)}{n+1}.$$

A firm's profits are $(p - c_i)q_i$ (where $c_i = a - bd_i(g)$ is the marginal cost of firm i), and so, noting from (10.17) that $q_i = p - c_i$, it follows that each firm's Cournot equilibrium profits are q_i^2, where q_i is given above.

This quantity equation makes it easy to deduce pairwise stable networks. The profits of a firm are increasing in the equilibrium q_i, and the network enters q_i in proportion to $nd_i(g) - \sum_{j \neq i} d_j(g)$. Thus firm i gains $n-1$ (noting that d_j increases for some j) with each link that it adds. If the link costs are the same across links, then the set of pairwise stable networks falls into one of two extremes: it is either the complete network or else the empty network, depending on the cost of a link. If the link costs are heterogeneous across firms, then the pairwise

---

29. A sufficient condition for all quantities to be positive is that α is large, or that $\alpha - a - (n-1)(n-2)b > 0$.

stable network is a complete network among the subset of firms whose costs are lower than $n - 1$ (presuming no firm has costs exactly at $n - 1$). More interesting configurations require some nonlinearities in link costs.

In terms of efficiency, there are a variety of different benchmarks to consider. In particular, efficiency depends on whether firm profits alone are considered or the welfare of the consumers buying the firms' products is taken into account; it also matters whether competition is restricted to Cournot equilibrium or can take some other form (see Exercise 10.6). From the standpoint of industry profit, in most cases the highest industry profits actually involve a star network, where the center firm enjoys a low cost and also sees higher costs, and thus prices, from its competitors. A star is the efficient network structure if link costs are small and only firms' profits are considered. If we also include the consumers' welfare, and link costs are small enough, then it is best to see a low price and high production, which emerges when the complete network forms.[30]

10.4 ▪ Concluding Remarks

The models, empirical analyses, and experiments on networked markets described in this chapter offer insights that are fairly general. In the context of labor markets, we saw how links lead to correlated outcomes across neighbors both, at a given time and across time. Such linked outcomes lead to complementarities in incentives to make investments in things like education. These patterns are not unique to labor markets but also hold for some other economic transactions and behaviors when there are complementarities between the states of neighbors. In the context of exchange, we saw that the relation between network structure and the terms of trade can be complex, but that an important determinant of favorable terms of trade is having connections to agents whose other trading options are somewhat limited. So it is not simply direct connections that are important for terms of trade, but instead being connected to others who are not well connected. This contrasts with other sorts of applications, such as information networks, in which having well-connected neighbors is desirable. Moreover, high connection to low-degree neighbors leads to good outcomes for an agent not only in the exchange setting, but also in settings of collaboration among firms or of competition among agents.

While we have seen that networked labor markets are pervasive and general insights can be gained from observing and modeling networked markets, there is much left to be learned in this extensive area of application where networks play such a central and critical role.[31]

---

30. This conclusion, however, depends on how firms compete. If they compete via prices, then two low-cost firms suffice to push prices down (again, see Exercise 10.6). In such a case, the efficient network is what Goyal and Joshi [298] call "interlocking stars," such that two firms, i and j, are each linked to every other firm, and firms other than i and j are only linked to i and j.
31. I am not implying that this chapter provides an exhaustive survey of the literature, as there are areas that I did not cover, including international trade (e.g., Furusawa and Konishi [264]), financial contagion and the role of networks in financial markets (e.g., Allen and Babus [13]), collusion and market-sharing agreements among firms (e.g., Belleflamme and Bloch [52]), and models of competition among buyers and sellers with added heterogeneity (e.g., Wang and Watts [647]).

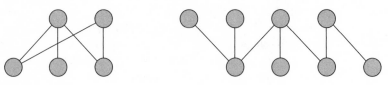

FIGURE 10.12   Find the payoffs for the pictured network using the model of Corominas-Bosch [178].

10.5 ▪ Exercises

10.1 *Proof of Proposition 10.3*   Provide a full proof of Proposition 10.3.

10.2 *Payoffs in the Corominas-Bosch [178] Model*   Find the predicted payoffs in the networks in Figure 10.12 using the Corominas-Bosch [178] model.

10.3 *Pareto Inefficient Pairwise Stable Networks with Trade by Auction*   The following example from Jackson [347] shows that it is possible for (nonempty) pairwise stable networks in the Kranton-Minehart model to be Pareto inefficient. For this we need more than one seller, and the auction works as follows. Prices rise simultaneously across all sellers. Buyers drop out when the price exceeds their valuations. As buyers drop out, there emerge sets of sellers such that the set of buyers still linked to those sellers is no larger than the set of sellers. Those sellers transact with the buyers still linked to them. (The exact matching of who trades with whom given the link pattern is done carefully to maximize the number of transactions.) Those sellers and buyers are cleared from the market, the prices continue to rise among remaining sellers, and the process repeats itself.

Consider a population with two sellers and four buyers. Let individuals 1 and 2 be the sellers and 3–6 be the buyers. Let the cost of a link to a seller be $c_s = 5/60$ and the cost of one to a buyer be $c_b = 1/60$. Expected payoffs to buyers and sellers in some of the relevant network configurations are:

$g^a = \{13\}$: $u_1(g^a) = -\frac{5}{60}$ and $u_3(g^a) = \frac{29}{60}$.

$g^b = \{13, 14\}$: $u_1(g^b) = \frac{10}{60}$ and $u_3 = u_4(g^b) = \frac{9}{60}$.

$g^c = \{13, 14, 15\}$: $u_1(g^c) = \frac{15}{60}$ and $u_3 = u_4 = u_5(g^c) = \frac{4}{60}$.

$g^d = \{13, 14, 15, 16\}$: $u_1(g^d) = \frac{16}{60}$ and $u_3 = u_4 = u_5(g^d) = \frac{2}{60}$.

$g^e = \{13, 14, 25, 26\}$: $u_1 = u_2(g^e) = \frac{10}{60}$ and $u_3 = u_4 = u_5 = u_6(g^e) = \frac{9}{60}$.

$g^f = \{13, 14, 15, 25, 26\}$: $u_1(g^f) = \frac{13}{60}$, $u_2(g^f) = \frac{8}{60}$, and $u_3 = u_4(g^f) = \frac{6}{60}$, while $u_5(g^f) = \frac{10}{60}$ and $u_6(g^f) = \frac{11}{60}$.

$g^g = \{13, 14, 15, 24, 25, 26\}$: $u_1 = u_2(g^g) = \frac{9}{60}$ and $u_3 = u_4 = u_5 = u_6(g^g) = \frac{8}{60}$.

Show that out of these networks and the empty network, the pairwise stable networks are the empty network, g^d, and g^g. Show that none of the pairwise stable networks is efficient. Show that g^g is not Pareto efficient.

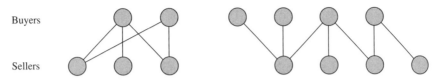

Buyers

Sellers

FIGURE 10.13   Find the equilibrium prices and trades for these network configurations using the competitive model of Kakade et al. [374].

10.4   *Competitive Trades on a Network*   Find the equilibrium prices and allocations for the networks in Figure 10.13 under the competitive model of Kakade et al. [374] discussed in Section 10.3.3.

10.5   *Pairwise Stable Networks with Competitive Trades*   Consider the competitive trading model of Kakade et al. [374] discussed in Section 10.3.3, when there are N_B buyers and N_S sellers. Each seller has a single unit for sale, and buyers have valuations of 1 and endowments of 1. Show that if N_B is an integer multiple of N_S, then there is a pairwise stable network in which each buyer links to one seller and each seller links to N_B/N_S sellers, and the reverse is true if N_S is an integer multiple of N_B. Assume that there is a cost $c > 0$ per link, which is less than the inverse of the integer multiple and is paid out of the final goods (so in cash for sellers and in the consumption good for buyers).

   Is such a network structure Pareto efficient when $N_B > N_S$? Show that an efficient network is pairwise stable when $N_B > N_S$.

10.6   *Collaboration Networks among Firms with Bertrand Competition*   Consider a collaboration network among firms as in Section 10.3.4 in which production costs are given by (10.16). Let firms compete for the sale of their products by pure Bertrand competition: each firm simultaneously quotes a price, and the firms charging the lowest price evenly split the market. Consider a situation in which the total amount purchased is Q independently of the price.[32] If at least two firms have the lowest cost level (highest degree), then there is a Nash equilibrium of the Bertrand game where all firms quote prices equal to their per unit costs and do not make any profits. If a single firm has the lowest cost, then there is an equilibrium where that firm charges the second-lowest per unit cost and sells amount Q.[33]

   If there is a positive cost to a link, show that the empty network is the only pairwise stable network.

10.7   *Negative Correlation in Short-Run Labor Networks*   Consider a triad (three-agent, completely connected network) in the model of Section 10.2.2. Suppose that at the end of one period, agents 1 and 2 are unemployed and agent 3 is employed.

---

32. The shape of the demand curve is not important for this exercise.

33. The precise equilibrium with asymmetric costs in Bertrand competition involves mixed strategies in which the firm with the lowest cost charges the second-lowest cost, and then the firm(s) with the second-lowest cost employs a mixed strategy that has some atomless weight on prices just above the second lowest cost. See Blume [81] for details.

Show that the next-period employment states of agents 1 and 2 are negatively correlated.

10.8 *Association** ⁣   Consider a network in which each agent has an employment state $s_i \in \{0, 1\}$. Suppose that the state of i in period t is a function of the previous vector of all agents' states and an undirected network g. In particular, the probability that $s_{it} = 1$ depends on the states $s_{j,t-1}$ for $j \in N_i(g) \cup \{i\}$, it lies strictly between 0 and 1 and is increasing in $s_{j,t-1}$ for each $j \in N_i(g) \cup \{i\}$ holding the other agents' states constant, and it is independent of the states $s_{j,t-1}$ for each $j \notin N_i(g) \cup \{i\}$.[34] Show that the steady-state distribution of the vector $(s_1, \ldots, s_n)$ exhibits strong association (see Section 4.5.7) relative to the components of the network.

---

34. The same conclusion holds under weaker conditions, but involves substantial complications in the proof. See Calvó-Armengol and Jackson [130] for details.

PART IV

METHODS, TOOLS, AND EMPIRICAL ANALYSES

CHAPTER 11

Game-Theoretic Modeling of Network Formation

Chapter 6 showed the importance of understanding strategic network formation, highlighting the tension between social welfare and individual incentives to maintain relationships. As discussed in the introduction to that chapter, there are many settings in which links are formed in a cognizant manner, especially for applications in which the individual nodes are firms, organizations, or countries, which have explicit objectives when pursuing their relationships. And even friendships and other more purely social relationships exhibit costs and benefits that influence which ones emerge and endure. These considerations lead to a rich set of questions regarding the modeling of network formation:

- There are issues of how to define equilibrium and stability: Can players adjust many relationships at a time? Can players coordinate their choices?

- There are issues of sophistication: Are players farsighted or myopic? Do players take into account how the links that they form influence others? Do they make errors?

- There are issues of dynamics: Can players revise links over time? Are there evolutionary pressures on their choices? Do random forces of opportunity determine which relationships can be formed?

- There are issues of bargaining and transfers: Can players compensate others for the relationships that they do (or do not) maintain, either through negotiated payment or through favors? Are there compensations bargained over at the time of network formation?

- There are issues of the formation of directed networks: How should we model network formation if links can be formed unilaterally? How does network formation depend on whether one or both players involved in a directed link benefit from its presence?

- There are issues associated with the strength of links: How do we model link strength? What happens if players can allocate effort or resources to maintaining different links? How will the outcome depend on the context?

Given the diversity of these questions, there is no simple message emerging from this chapter. Instead, this chapter examines foundational questions concerning network formation. There is, however, a pervasive question that ties the analyses of a range of network-formation models together: Under what circumstances do incentives lead to the formation of efficient networks?

11.1 ▪ Defining Stability and Equilibrium

In Chapters 1 and 6 we used the concept of pairwise stability as a method for modeling network formation. Why should we use that concept rather than an explicit game? This was discussed briefly in Section 6.1, but let us consider that question while examining possible noncooperative games of network formation and alternative notions of equilibrium and stability.

11.1.1 An Extensive Form Game of Network Formation

Aumann and Myerson [23] provided an early model of network formation. More specifically, they were interested in the formation of a communication graph that served as a basis for a cooperative game (as discussed in Section 12.2). The formation game they examined extends to serve as a model of network formation and is described in our setting as follows.

Players move sequentially and propose links, which are then accepted or rejected. The extensive form game is based on an ordering over all possible links, denoted by $(i_1 j_1, \ldots, i_K j_K)$. When the link $i_k j_k$ appears in the ordering, the pair of players $i_k j_k$ decide on whether to form that link, knowing the decisions of all pairs coming before them and forecasting the play that will follow them. Player i_k moves first and says "yes" or "no," and then player j_k says "yes" or "no," and the link forms if both say "yes." A decision to form a link is binding and cannot be undone. However, if a pair $i_k j_k$ decide not to form a link, but some other pair coming after them forms a link, then $i_k j_k$ are later allowed to reconsider their decision. This feature allows player 1 to make a threat to 2 of the form "I will not form a link with 3 if you do not. But if you do form a link with 3, then I will also do so." The threat is captured as follows. The game initially moves through all links. If at least one link forms, then the game starts again with the same ordering, moving this time only through the links that have not yet been formed. The game continues through the remaining unformed links in order, until either all links are formed or there is a round such that all of the links that have not yet formed have been considered, and no new links have formed.

This approach has the advantage of always having a pure strategy subgame perfect equilibrium.[1] Its main drawbacck is that the game can be very difficult to solve, even in simple settings with only a few players. Moreover, the ordering of links can have a substantial impact on which networks emerge, and it is not clear what a natural ordering is.

---

1. It is a finite game of perfect information, and so has at least one solution found by backward induction.

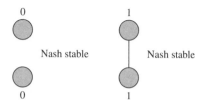

FIGURE 11.1 Two networks that are Nash equilibria of the link-announcement game.

11.1.2 A Simultaneous Link-Announcement Game

Given the intractability of the sequential ordering and its inherent asymmetries, Myerson [496] suggests another game in the context of the formation of communication graphs, which also extends to the formation of networks. It is probably the most natural simultaneous-move game of network formation. I refer to it as the *link-announcement game*. Each player simultaneously announces the set of players with whom he or she wishes to be linked. The links that are formed are those such that both of the players involved in the link named each other.

More formally, the strategy space of player i is $S_i = 2^{N \setminus \{i\}}$.[2] If $s \in S_1 \times \cdots \times S_n$ is the profile of strategies played, then link ij forms if and only if both $j \in s_i$ and $i \in s_j$. The network that forms is

$$g(s) = \{ij | i \in s_j \text{ and } j \in s_i\}.$$

In modeling the networks that emerge from the link-announcement game, we can use any of a variety of game-theoretic solutions, such as Nash equilibrium.

The payoffs in the link-announcement game are described by a profile of utility functions, $u = (u_1, \ldots, u_n)$, which indicate the payoffs of each player as a function of the network. A network $g \in G(N)$ is *Nash stable* if it results from a pure strategy Nash equilibrium of the link-announcement game, in which player i's payoff as a function of the profile of strategies is $u_i(g(s))$.

This game is much easier to describe than the Aumann and Myerson [23] extensive form, and it avoids inducing a priori asymmetries between the players or links. Arguably, any network that is stable over time (in the sense that no players want to delete any links) would be an equilibrium of this game. The main drawback of the game that it has too many Nash equilibria, including some that are easily seen to be unreasonable. In particular, $s_i = \emptyset$ for all i is always a Nash equilibrium, regardless of the payoffs. Each player refuses to link with any other player, because he or she correctly forecasts that the other players will do the same. This drawback is seen most starkly in the dyadic case, as pictured in Figure 11.1. Here both networks are equilibria, although clearly the network with the link is the only reasonable one.

Thus, while the link-formation game may at first seem to be a natural way to model network formation, it is not reasonable when using Nash equilibrium alone

---

2. 2^A is a notation for the set of all subsets of A, also known as the power set of A.

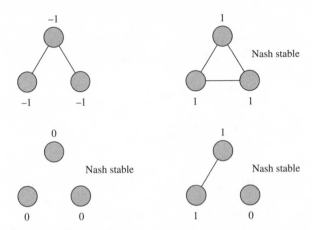

FIGURE 11.2   All networks except two-link networks are Nash stable, and all strategies in the link-announcement game are undominated.

as a solution concept. Basically, Nash equilibrium allows players to refuse to form links and thus effectively to "delete" links, but it does not capture the fact that it may be mutually advantageous for two players to form a new relationship. We need to move beyond Nash equilibrium to capture this.

In the example pictured in Figure 11.1, it is a dominant strategy for each player to propose to link with the other player. Thus one way around the shortcoming of this equilibrium in modeling network formation might be to use a refinement of this equilibrium in which players do not play weakly dominated strategies. However, a slight enrichment of the example in Figure 11.1 shows that such a refinement does not work. Consider a triad such that the empty network leads to a payoff of 0 for all players, a single link leads to a payoff of 1 for each of the linked players (and 0 for the other), a two-link network leads to a payoff of -1 for all players, and the complete network leads to a payoff of 1 for all players (pictured in Figure 11.2). In this example, all strategies in the link-announcement game are undominated. As a result, the empty network is an outcome of a Nash equilibrium that only uses undominated strategies, where every player announces the empty set of players.[3]

To address the fact that the consent of both players is needed to form a link in an undirected network, one has to explicitly consider coordinated actions on the part of pairs of players. This forces one to move beyond Nash equilibrium and standard refinements of it, and somehow coalitional considerations (at least for pairs of players) have to be considered. That is the reasoning behind pairwise stability.

---

3. This is also a trembling hand perfect equilibrium (for a definition, see Fudenberg and Tirole [263]). It is not a strict Nash equilibrium. However, requiring strictness in this game leads to very general existence problems, as outlined in Exercise 11.4.

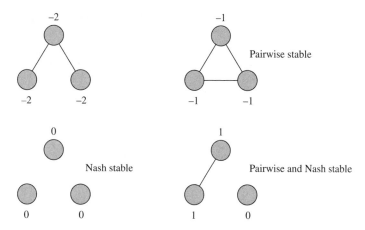

FIGURE 11.3 An overconnected pairwise stable network. Payoffs on permutations of these networks are the permuted payoffs.

11.1.3 Pairwise Nash Stability

Although pairwise stability overcomes the difficulties inherent in examining Nash equilibria of the link-announcement game, it restricts attention to changes of one link at a time. But then overconnected networks can be pairwise stable, even when some player might benefit from deleting multiple links at once, as pictured in Figure 11.3. In the figure, the reasonable network is the one that is both Nash stable and pairwise stable. These considerations lead to a concept of pairwise Nash stable networks: A network is *pairwise Nash stable* if it is both Nash stable and pairwise stable.[4]

As people who have worked with game-theoretic solution concepts are aware, given any equilibrium or stability concept one can find some setting where it makes a questionable prediction or is deficient in some way, because as applications vary, so do the sorts of deviations that are feasible or salient. Can players communicate and coordinate their actions? Can they make multiple changes at once? Can more than two players coordinate at a time? Do players move simultaneously, or are we really just looking for the stable points of some process with sequential timing? Can players revise their actions? Although finding a single solution concept that works well in all settings is a futile activity,[5] we should not give up on studying variations of solutions, since different ones can be more appropriate and/or useful

----

4. This refinement was first discussed by Jackson and Wolinsky [361] and has been used in various studies (e.g., Goyal and Joshi [298] and Belleflamme and Bloch [52]). For more detailed studies of the relationships among Nash stability, pairwise Nash stability, and pairwise stability, as well as refinements of Nash equilibria that justify pairwise Nash stability for a wide class of settings, see Calvó-Armengol and Ilkilic [129], Gilles and Sarangi [281], Bloch and Jackson [78], and Ilkilic [341].

5. This view is not universally held among game theorists.

in different settings. We should not give up modeling simply because a universal solution concept is not available.

Several variations on stability concepts are discussed in the exercises of this chapter, but here I discuss a few of the basic considerations and some solutions that have been used to capture them.

11.1.4 Strong Stability

In some settings, players have open lines of communication and more than two players can coordinate their link formation decisions. Alternatives to pairwise stability and pairwise Nash stability that consider larger coalitions of players were first considered by Dutta and Mutuswami [213].[6] The following is a slight variation on Dutta and Mutuswami's definition, from Jackson and van den Nouweland [356]. It always selects from among the pairwise Nash stable networks.

A network $g' \in G$ is obtainable from $g \in G$ through deviations by $S \subset N$ if

1. $ij \in g'$ and $ij \notin g$ implies $\{i, j\} \subset S$, and
2. $ij \in g$ and $ij \notin g'$ implies $\{i, j\} \cap S \neq \emptyset$.

The above definition identifies changes in a network that can be made by a coalition S without the consent of any players outside S. Part (1) requires that any new links only involve players in S, in line with the consent of both players being needed to add a link. Part (2) requires that at least one player of any deleted link be in S, in line with the idea that either player in a link can unilaterally sever the relationship.

A network g is *strongly stable* with respect to a profile of utility functions $u = (u_1, \ldots, u_n)$ if for any $S \subset N$, g' that is obtainable from g through deviations by S, and $i \in S$ such that $u_i(g') > u_i(g)$, there exists $j \in S$ such that $u_j(g') < u_j(g)$.

The relationship between this definition and the definition of Dutta and Mutuswami [213] is examined in Exercise 11.6; it relates to whether a blocking coalition must merely have some members be strictly better off and others be weakly better off, as above, or must have all members of a blocking coalition be strictly better off. The definition given here is consistent with pairwise stability, as the strongly stable networks are always a subset of pairwise stable networks (and in fact a subset of the pairwise Nash stable networks).

The implications of strong stability are shown in Figure 11.4. In that example, a one-link network is pairwise Nash stable, as is the complete network. However, only the complete network is strongly stable.

Strongly stable networks are necessarily Pareto efficient, as one of the groups that can potentially deviate to form a better network is the society as a whole. Thus if some network is Pareto dominated by another network, so that the second network is weakly better for all players and strictly better for some, then that will provide a viable deviation and so the dominated network will not be strongly stable. In addition to this efficiency property, strongly stable networks are immune to all

---

6. There was also early discussion of core-based allocations in the exchange network literature (e.g., by Bienenstock and Bonacich [63]).

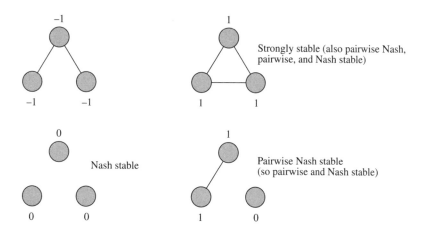

FIGURE 11.4   Strong stability. An example with multiple pairwise stable, Nash stable, and pairwise Nash stable networks, but a unique strongly stable network. Permutations of these networks lead to permuted payoffs.

sorts of coordinated deviations by players, and so they are very robust. However, they only make sense as a predictive tool when such coordination is feasible and thus might be limited to situations in which players have substantial knowledge about the opportunities for network formation and the payoffs, and can also readily communicate with each other. Also, while strongly stable networks are robust and Pareto efficient, there are many contexts where they fail to exist. The issue of existence of various sorts of stable networks is an important one to which I now turn, before returning to discuss other ways of modeling network formation.

11.2 ▪ The Existence of Stable Networks

While the link-announcement game always has an equilibrium, this is due to the fact that there is always a trivial equilibrium where no links form because no player expects any other player to be willing to form a link. Once we move to refinements for which the empty network is not always stable, such as pairwise stability, pairwise Nash stability, or strong stability, existence is not always guaranteed. Let us explore when stable networks exist.

11.2.1 Improving Paths, Dynamics, and Cycles

In studying the existence of various forms of stable networks, it is useful to consider some simple dynamics. The idea is to examine the sequences of networks that might emerge as players add or delete links to improve their payoffs. The resting points of these processes will be stable points, and so understanding these sequences helps in understanding when stable networks exist and what might happen when stable networks do not exist.

Let us say that two networks are *adjacent* if they differ by only one link. That is, g and g' are adjacent if either $g' = g + ij$ for some $ij \notin g$ or $g' = g - ij$ for some $ij \in g$.

A network g' *defeats* an adjacent network g if either

- $g' = g - ij$ and $u_i(g') > u_i(g)$, or
- $g' = g + ij$ and $u_i(g') \geq u_i(g)$ and $u_j(g') \geq u_j(g)$, with at least one inequality holding strictly.

A network is pairwise stable if and only if it is not defeated by an (adjacent) network.

The following definition from Jackson and Watts [358] captures this notion of sequences of networks in which each network defeats the previous one.

An *improving path* is a sequence of distinct networks $\{g_1, g_2, \ldots, g_K\}$, such that each network g_k with $k < K$ is adjacent to and defeated by the subsequent network g_{k+1}.

This usage of *path* refers to a sequence of networks and should not be confused with a path inside a network. The idea here is to examine the sequences of networks that can emerge as players add and delete links in a way that makes them better off. Clearly, the resting points of such a process are the pairwise stable networks. That is, a network is pairwise stable if and only if it has no improving paths emanating from it. Improving paths are illustrated in Figure 11.5.

The notion of improving paths is a myopic one, in that the agents involved in adding or deleting links are doing so without forecasting how their actions might affect the evolution of the process. This is a natural variation on best-response dynamics and has some experimental justification (e.g., see Pantz and Ziegelmeyer [522]), but nevertheless it exhibits some forms of bounded rationality. I return to discuss farsighted network formation in Section 11.5.

If no pairwise stable network exists, then there must exist at least one *improving cycle*—that is, a sequence of adjacent networks $\{g_1, g_2, \ldots, g_K\}$ such that each defeats the previous one and such that $g_1 = g_K$. The possibility of cycles and nonexistence of a pairwise stable network is illustrated in the following example from Jackson and Watts [358].

Example 11.1 (Nonexistence of a Pairwise Stable Network)   *There are $n \geq 4$ players who obtain payoffs from trading with one another. The players have random endowments, and the benefits from trading depend on the realization of these random endowments. The more players who are linked, the greater the gains from trade, but with diminishing marginal returns.*

In particular, there is a cost of a link of $c = 5$ to each player involved in the link. The utility of being alone is 0. Not accounting for the cost of links, the benefit to each player in a dyad is 12, the benefit for being connected (directly or indirectly) to two other players is 16, and that of being connected to three other individuals is 18.[7]

7. In terms of the economic background behind these payoffs, they can be derived as follows. There are two consumption goods, and players each have a Cobb-Douglas utility function for the two goods of $u(x, y) = 96xy$, where x is the consumption of the first good and y is the consumption of the second good. A player's endowment is either (1,0) or (0,1), each with

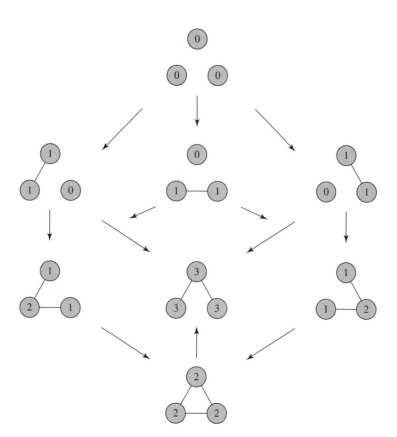

FIGURE 11.5 Improving paths. Payoffs are listed in the nodes, and the arrows point toward a network that defeats the one from which the arrow emanates. Following the arrows provides improving paths. There is a unique pairwise stable network.

The resulting payoffs for several of the key network configurations are pictured in Figure 11.6.

Any network with more than three links in this example is defeated by a network with fewer links, as some players will save the link cost by severing the link, and yet the full trading benefits are already realized with just three links. The critical aspect of the example is that two separate dyads gain by forming a link between

---

probability 1/2, and the realizations are independent across players. Players in each component trade to a Walrasian equilibrium within their component, regardless of the precise set of links in the component. For example, the networks {12, 23} and {12, 23, 13} lead to the same expected trades, but different costs of links. In a dyad there is a 1/2 probability that one player has an endowment of (1,0) and the other has an endowment of (0,1). They then trade to the Walrasian allocation of $(\frac{1}{2}, \frac{1}{2})$ each, and so their utility is 24 each. There is also a 1/2 probability that the players have the same endowment, so then there are no gains from trade and they each receive a utility of 0. Taking expectations over these two situations leads to an expected utility of 12. Similar calculations for more players lead to the claimed payoffs.

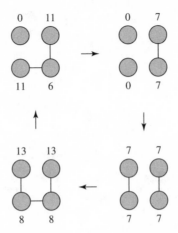

FIGURE 11.6   Nonexistence of a pairwise stable network. Payoffs pictured for one-, two-, and three-link networks and an improving cycle that includes the four networks. Networks with more than three links are defeated by networks with fewer links.

them, expanding the network from two to four players. However, in any network of four players that has just three links, one of the players who has more than one link will save 5 units in cost by severing a link and only lose 2 units in trading benefits.

Clearly, the nonexistence of pairwise stable networks also implies the non-existence of pairwise Nash stable networks. Moreover, pairwise Nash stable networks can fail to exist even when the sets of pairwise stable networks and Nash stable networks are both nonempty (see Exercise 11.9).[8]

While Example 11.1 shows that pairwise stable networks may not exist in some settings, there are settings in which they always exist. We have already seen several, including distance-based generalizations of the connections model, the coauthor model of Section 6.4, and a variety of market settings from Chapter 10.

A set of sufficient conditions for the existence of pairwise stable and pairwise Nash stable networks comes from ruling out improving cycles. Having no improving cycles also means that any dynamics that follow improving paths will find stable networks. The absence of improving cycles is related to the existence of what is known as an (ordinal) potential function.

An *ordinal potential function* for a society N with payoff functions $u = (u_1, \ldots, u_n)$ is a function $f : G(N) \to \mathbb{R}$ such that g' defeats g if and only if $f(g') > f(g)$ and g' and g are adjacent.[9]

For a society N, the payoff functions $u = (u_1, \ldots, u_n)$ *exhibit no indifference* if for any two adjacent networks, one defeats the other.

Proposition 11.1 (Jackson and Watts [357]) *If a society (N, u) has an ordinal potential function, then there are no improving cycles. Conversely, if a society is such that payoffs exhibit no indifference, then there are no improving cycles only if there exists an ordinal potential function.*

This proposition echoes results on ordinal potentials in noncooperative games, and a corollary is that if society admits an ordinal potential function then there exists a pairwise stable network. This follows since any network that maximizes f must be undefeated.

Proof of Proposition 11.1. The proof that there are no improving cycles if a society has an ordinal potential function is straightforward and left to the reader. So let us show that the converse holds when payoffs exhibit no indifference.

Suppose that there are no improving cycles and payoffs exhibit no indifference. Define $f(g)$ to be the number of networks g' such that there exists an improving path from g' to g.[10] We need to show that $f(g)$ is an ordinal potential function. Consider two adjacent networks g and g'. If g defeats g', then every network that has an improving path leading to g' also has an improving path leading to g. Moreover, g' has an improving path leading to g, but the reverse is not true, as otherwise there would be an improving cycle. Therefore $f(g) > f(g')$. Conversely, if $f(g) > f(g')$ and g' defeated g, then we reach a contradiction by a similar argument. Thus $f(g) > f(g')$ implies that g' does not defeat g, which by the no-indifference condition implies that g defeats g'. Therefore, f is an ordinal potential function. ∎

The no-indifference condition is needed for the proposition's conclusions and is the subject of Exercise 11.11.

We can easily extend this analysis to the case of pairwise Nash stability with some modifications to the definitions above. Let us say that the networks $g \neq g'$ are *weakly adjacent* if g' is either obtained from g by the addition of a single link or obtained by the deletion of some set of links such that there is some agent involved in all of the deleted links. If we then redefine *defeats, improving path, improving cycle,* and *ordinal potential function* accordingly, Proposition 11.1 still holds, and the existence of an ordinal potential function with those definitions implies the existence of pairwise Nash stable networks.

The existence of an ordinal potential function is a demanding condition, which emphasizes that the absence of improving cycles is also a demanding condition. Nevertheless, there exist situations in which these conditions are satisfied. In fact, there are always transfers that lead to payoffs for which there exists an ordinal potential function (see Section 12.3.7). In some cases, simply examining the sum of

---

is adapted to network-formation settings and pairwise stability. For a more detailed look at the relationship between various sorts of potential functions and the existence of equilibria in network-formation games, see Gilles and Sarangi [285].

10. The idea behind this construction for finding ordinal potential functions is due to Milchtaich [467].

all payoffs leads to an ordinal potential function. This method works in the case of the Corominas-Bosch's [178] model of buyer-seller networks from Section 10.3.2 (see Exercise 11.12). It also works for the symmetric connections model when $\delta > c > n(\delta - \delta^{n-1})$ or when c is very large or small. More generally, in cases where potential functions exist, the analysis can be greatly simplified.

11.2.2 The Existence of Strongly Stable Networks

Let us now examine the existence of strongly stable networks. I start by showing that strong stability demands that certain patterns in payoffs be present.

A profile of utility or payoff functions $u = (u_1, \ldots, u_n)$ is *anonymous* if for every permutation π on N (a one-to-one function mapping the set of agents N to N), it follows that $u_{\pi(i)}(g^\pi) = u_i(g)$, where $g^\pi = \{\{\pi(i), \pi(j)\} | ij \in g\}$ is the network obtained from g by permuting the positions of agents according to π.

Anonymity requires that payoffs depend only on players' positions within the network and not on their labels. Payoff relevance is captured through the network structure and not through other innate characteristics of the players.

Recall the definition of component decomposability from Section 6.6.2: A profile of utility functions or payoff functions $u = (u_1, \ldots, u_n)$ is *component decomposable* if $u_i(g) = u_i(g')$ when $C_i(g) = C_i(g')$. Component decomposability requires payoffs to players to depend only on the structure of their components and not on the structure of other components. This dependence allows for externalities within components but precludes externalities across components. It holds in some settings, such as those in which payoffs only depend on communication patterns within components, but not when separate components interact with each other.

Proposition 11.2 (Jackson and van den Nouweland [356]) *Consider a society with anonymous and component-decomposable payoffs. If there exists a strongly stable network $g \in G(N)$ that is not connected, then all players must get an equal payoff.*

This proposition is a variation on what is known as an *equal treatment condition,* which is implied in a wide variety of settings when requiring stability with respect to deviations by groups of agents. The proof is straightforward and only sketched here. The idea is that if all players do not obtain the same payoff, then there is a player in one component who receives less than the payoff of a player in another component. By replacing the higher payoff player j with the lower payoff player i, the payoffs to the other players in the new component of player i do not change, but i's payoff goes up (as implied by the anonymity and decomposability of payoffs, which ensure that i's new payoff is j's old payoff), which is an improving deviation.

Proposition 11.2 shows that the existence of strongly stable networks imposes stringent requirements. For instance, in the symmetric connections model, it implies that (for generic choices of parameters) the only networks that could possibly be strongly stable are networks with strong symmetry properties: those such that the cardinality of each extended neighborhood of every player is identical.

It is clear that if payoffs are equal across all players at every network, then strongly stable networks exist and coincide with the efficient networks. In that case, players' payoffs are perfectly aligned with society's total payoff. However, such alignment demands that transfers be made across components in many contexts. Sufficient conditions for the existence of strongly stable networks when there are no transfers across components require some definitions from the next chapter and are explored in Exercise 12.8.

11.3 ▪ Directed Networks

The modeling of network formation with directed networks differs from that with undirected networks, as links can be formed unilaterally, and the Nash equilibrium of a formation game becomes an appropriate modeling tool.

Clearly, whether a network is directed or undirected is not just a modeling choice but instead depends on the application. Although many social and economic relationships involve the consent of both parties, there are applications for which links can be formed unilaterally. For example, one article can cite another without the consent of the first, and a web page can link to another without its consent. In those applications, one needs to adjust the network-formation model to account for the unilateral nature of the formation process.[11]

In the case of directed networks, we can still write the payoffs as a function of the network that is formed, where now (fixing a society N) g is a directed network and $u_i(g)$ represents the utility to player i if g is the directed network that is formed. If players can form a directed link unilaterally, one way to model network formation is for each player to list the set of directed links that he or she wishes to form (and the player can only list links from him- or herself to another player) and then form the network from the union of the listed links. This method was suggested by Bala and Goyal [31]. More formally, we model the formation as follows.

A network g' is obtainable from a network g by player i if $g'_{kj} \neq g_{kj}$ implies that $k = i$.

Thus a network g' is obtainable from a network g by player i if the only changes in the network involve links that are directed from i to other players.

A directed network g is *directed Nash stable* if $u_i(g) \geq u_i(g')$ for each i and all networks g' that are obtainable from network g by player i.

When it is clear that directed networks are meant, I omit the "directed" from "directed Nash stability" and simply refer to a network as being Nash stable. Thus a directed network g is Nash stable if and only if it is the outcome of a Nash equilibrium of a game in which the players simultaneously announce lists of directed links from themselves to other players and the network that forms is the union of those lists.

---

11. A temptation is to mention things like phone calls or other sorts of broadcasting as falling into the directed case. However, those fall into a different category altogether. A phone call involves an asymmetry in the process since one person initiates the action, but it also requires that both people be willing to hold a conversation, which is a costly activity. Access to many people and organizations is guarded.

11.3.1 Two-Way Flow

There are several aspects to consider in a directed network in terms of how benefits accrue. For instance, in the case of a citation network, different values come from being cited as opposed to citing, which is similar to the case of web pages. Having links to other web sites can enhance the value of a web page and make it more attractive to visit, and being linked to by other sites makes a site easier to find and thus more likely to be visited. Thus, it might be that one side initiates a link, and yet both sides benefit from the link being present. As a first approximation of this asymmetry, we can keep track of who forms the link, as that might involve specific costs (e.g., space on a web page or time), but then allow the benefits of a link to be bilateral. This is what Bala and Goyal [31] term *two-way flow*.

11.3.2 Distance-Based Utility

To get some feel for the formation of directed networks, let us start by considering a variation on the distance-based utility model from Section 6.3, adjusted to allow for two-way flow and directed networks. Given a directed network g, let $\widehat{g}$ denote the undirected network obtained by allowing an (undirected) link to be present where there is a directed link present in g. That is, let $\widehat{g}_{ij} = \max(g_{ij}, g_{ji})$.

Recall that in the distance-based model, players obtain benefits from connections and indirect connections to other agents, where the value that they obtain from indirect connections is a decreasing function of the distance to the other player.

Let $b : \{1, \ldots, n-1\} \to \mathbb{R}$ denote the net benefit that a player receives from indirect connections as a function of the distance between the agents. The *distance-based utility model* is one in which an agent's utility can be written as

$$u_i(g) = \sum_{j \neq i : j \in N^{n-1}(\widehat{g})} b(\ell_{ij}(\widehat{g})) - d_i(g)c,$$

where $\ell_{ij}(\widehat{g})$ is the shortest path length between i and j in the undirected network obtained from g and $d_i(g)$ is i's *out-degree*. Let $b(k) > b(k+1) > 0$ for any k and $c \geq 0$.

This definition embodies the idea that a player obtains greater benefits for being closer to other players. A special case of the distance-based utility model, analyzed by Bala and Goyal [31], is a directed adaptation of the symmetric connections model, where $b(k) = \delta^k$.

Proposition 6.1, characterizing efficient networks in the undirected distance-based utility model, generalizes directly and shows that efficient networks in a directed version of the distance-based utility model share features with the symmetric connections model. The only difference is an adjustment that reflects the fact that only one player bears the cost of a link instead of two.

Let us say that a directed network g is a *directed star* if the associated undirected network $\widehat{g}$ is a star, and if $g_{ij} = 1$, then $g_{ji} = 0$, so that links between two players only go in one direction.

Proposition 11.3 *The efficient networks in the directed version of the distance-based utility model*

1. *consist of one directed link between each pair of players if* $c < 2(b(1) - b(2))$,

2. *are directed stars encompassing all nodes if* $2(b(1) - b(2)) < c < 2b(1) + (n - 2)b(2)$, *and*

3. *consist of the empty network if* $2b(1) + (n - 2)b(2) < c$.

Proposition 11.4 *Consider the directed version of the distance-based utility model.*

1. *If* $c < b(1) - b(2)$, *then the directed Nash stable networks are those that have one directed link between each pair of players.*

2. *If* $b(1) - b(2) < c < b(1)$, *then any directed star encompassing all nodes is directed Nash stable, and for some parameters there are other directed Nash stable networks.*

3. *If* $b(1) < c < b(1) + \frac{(n-2)}{2}b(2)$, *then peripherally sponsored stars*[12] *are Nash stable and so are other networks (e.g., the empty network).*

4. *If* $b(1) + \frac{(n-2)}{2}b(2) < c$, *then only the empty network is directed Nash stable.*

The proof is straightforward and the subject of Exercise 11.14.

Here we see very similar results to those of the nondirected case. With very high or very low costs to links, the efficient and stable networks coincide, while otherwise they may not. Again, efficient networks take the form of variations on stars or the complete network. The most interesting difference arises in the case of peripherally sponsored stars. Instability of stars in the undirected case can arise because the hub of the star has to bear some costs and sees only direct, not indirect, benefits from connections. With directed links, it is possible for the outside players only to direct the links, so that the hub does not have to bear any costs. Nevertheless, there are still inefficiencies, most notably since only one player bears the cost of a link while many players can benefit from its existence. Indeed, even when one can impose transfers, for a variety of settings there still exist conflicts between stability and efficiency. That is, variations on the results that we saw in Section 6.3 hold in the directed case, as explored by Dutta and Jackson [210].

11.3.3 One-Way Flow

While the two-way flow directed setting has much in common with the undirected setting, a one-way flow directed setting introduces some twists. If we look at one extreme of the distance-based model, then a simple and intuitive characterization of both efficient and stable networks emerges. In particular, consider a one-way flow directed version of the symmetric connections model with $\delta = 1$. In this benchmark, an arbitrarily distant connection provides the same benefit as a direct connection. The model was analyzed by Bala and Goyal [31]. In particular,

---

12. This is a directed star with no link formed by the center.

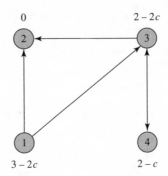

0   2 − 2c

3 − 2c   2 − c

FIGURE 11.7   Payoffs in a one-way flow model with no decay.

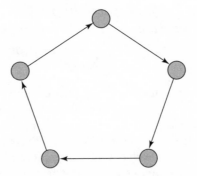

FIGURE 11.8   A wheel on five nodes.

let $R_i(g)$ denote the number of other players who can be reached from i by a directed path in g. Then i's payoff is

$$u_i(g) = R_i(g) - cd_i(g), \tag{11.1}$$

where $d_i(g)$ is i's out-degree. While this model is clearly extreme, since a player at a great distance is just as beneficial as a direct neighbor, it still provides some insight into the one-way flow setting. The payoffs are illustrated in Figure 11.7. In this setting, the characterizations of efficient networks and strict Nash stable networks are simple. First, we need a couple of definitions.

A network is an n-player *wheel* if it consists of n directed links and has a single directed cycle that involves n players. A wheel is illustrated in Figure 11.8.

A directed Nash stable network is *strictly Nash stable* if any change in the directed links from some player leads to a strictly lower payoff for that player.

Proposition 11.5 (Bala and Goyal [31])   *The unique efficient network structure in a one-way flow model in which there is no decay and payoffs are as in (11.1) is an n-player wheel if $c < n − 1$ and an empty network if $c > n − 1$. Moreover, if $c < 1$, then n-player wheels are the (only) strictly Nash stable networks; if $1 < c < n − 1$, then n-player wheels and empty networks are the (only) strictly*

Nash stable networks; and if $c > n - 1$, then the empty network is the unique strictly Nash stable network.

Proof of Proposition 11.5. First, let us show that a k-player wheel is the unique total-payoff-maximizing network among those that are nonempty, involve at least k links, and for which exactly k players have a link in or out. The cost is at least kc for any such network, which is the cost of a wheel. To have at least as high a payoff as a wheel, it must be that $R_i(g) = k - 1$. Thus there is a directed cycle containing all k players. If the network is not a wheel, and it contains a directed cycle with k players, it has more than k links, and so leads to a lower payoff than the k-player wheel. Thus the only possibilities besides wheels for efficient networks are the empty network and networks that involve k players but have fewer than k links. In the latter case, it must have $k - 1$ links to include k players in a component. So there must be a player i who has at least one link in but no links out, and another player j who has at least one link out but no links in. Given that it is efficient for j to link to some k (and since j has no links in, only j benefits from that link), adding a link from i to j would increase payoffs by even more than the link from j to k does on the margin (since $R_i(g + ij) - R_i(g) \geq 1 + R_j(g) - R_j(g - jk)$, which follows since i reaches j and $R_i(g) = 0$), which is a contradiction.

Thus different wheels (and combinations of wheels) and the empty network are the only possible efficient networks. The remainder of the claim is straightforward, noting that if the value of a wheel with less than n players is positive, then the value of a wheel involving all players generates a higher per capita payoff.

Next, let us characterize the strictly Nash stable networks. If $c > n - 1$ it follows directly that the only (strict) Nash network is the empty network, since a link can lead to a marginal payoff of at most $n - 1 - c$. If $1 < c < n - 1$, the empty network is still a strict Nash equilibrium, as each link that a player adds changes that player's payoff by $1 - c < 0$. If $c < 1$, it is clear that the empty network is not Nash stable. The proof is then completed by showing that when $c < n - 1$, any nonempty strictly Nash stable network must be a wheel involving all players, as it is clear that such a network is strictly Nash stable.

So, let $c < n - 1$ and consider a nonempty strictly Nash stable network. First, note that all players have an out-degree of at least 1. Suppose not. There is at least one player j who strictly benefits from a link ij, since the network is nonempty. By duplicating that link, a player with no outlinks would also strictly benefit, which contradicts equilibrium. Next, note that each player must have at most one link coming in. Suppose to the contrary that players i and j both have links to player k. By deleting the link to k and adding a link to j (or keeping the link to j if i already has one), i's payoff can only increase, as there is still a path to k (and hence to all other players reached through k), and i has not increased the number of links. Thus i benefits weakly from such a change, and so the network cannot have been a strictly Nash stable network, which is a contradiction. Hence we have a network such that every player has at least one link out and at most one link in, and hence every player has one link in and one link out. This network must be a wheel including all players. ▪

The strict aspect of equilibrium is a useful device here, as it narrows down the set of stable networks dramatically. Moreover, in some experiments on this sort of model, Callander and Plott [122] find evidence that strictness is a useful predictor of behavior. Part of the reason for the predictive power is the absence of any decay (since $\delta = 1$), which leads to many instances of indifference over which links form. More generally, when there is decay, many indifferences are naturally eliminated, and a refinement to strict equilibrium is not useful.

11.4 ▪ Stochastic Strategic Models of Network Formation

Recall that in Section 6.3.2 we considered the following process for growing a network. At each point in time a link is randomly chosen, with equal weight on all links. If the link is not in the network, then the two players involved in the link have the choice to add it to the network, and they add it if it makes each of them weakly better off in terms of payoffs and makes at least one of them strictly better off. If the link is already in the network, then either of the players involved in the link can choose to delete it, and it is deleted if that would increase the payoff for either player. If this process comes to rest, then it results in a pairwise stable network. If there do not exist any pairwise stable networks, then it will keep cycling.

While such a process models the dynamics of network formation, it can get stuck at networks that are pairwise stable, but such that we would expect players to deviate from. To see this issue more starkly, consider a three-player society in which the payoffs to different networks are as pictured in Figure 11.9. For the payoffs pictured in the figure, the empty network and the complete network are both pairwise stable. However, a process simply following improving paths, as indicated below, can end at the empty network. There are two different ways in which this process might avoid this dead end. One variation would be to allow for trembles or some exogenous events that cause links to be added or deleted with some (small) probability ε. Once one link forms, then there is a good chance that another will be formed, and then the process would reach the complete network. Another variation is to consider farsighted players. Players might realize that if they add one link then other links would subsequently form. Thus, even though a single link would lead to negative payoffs, they might add it, anticipating that it will lead to other links being formed. These are both quite natural variations on the process, but from very different perspectives.[13] One simply introduces some randomness into the process, while the other relies on rational and forward-looking players. They also rely on different assumptions about the knowledge and behavior of the players. In a farsighted process, players understand the incentives of other players, and they forecast the subsequent evolution of the network, while the perturbed myopic process does not require knowledge on the part of the players other than whether a given link is beneficial on the margin. These are different types of arguments and thus might be more or less appropriate, depending on the setting. Let us consider each in turn.

---

13. There are other perspectives, including that of strong stability. The arguments here are most pertinent to settings in which such coordination between larger coalitions of players is not possible.

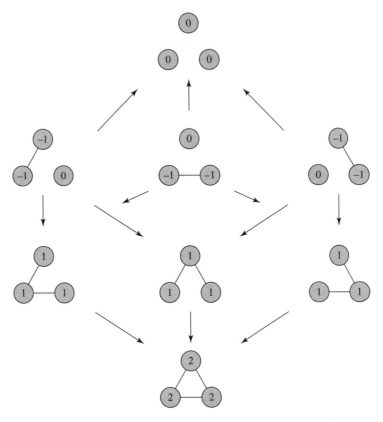

FIGURE 11.9 Example with two pairwise (Nash) stable networks, in which improving paths can get stuck at the empty network.

11.4.1 Random Improving Paths and Stochastic Stability*

Exogenous randomness in the network-formation process, so that links occasionally are added or deleted even though the benefits do not outweigh the costs, leads to a network-formation process that can yield sharp predictions about which networks are likely to emerge. Such a variation on the improving path process was introduced by Jackson and Watts [358].

The process can be described starting at any network $g \in G(N)$. At each time $t \in \{1, 2, \ldots\}$ a link ij is randomly identified, with each link having an equal probability of being identified and with the randomness being independent across time.[14] For an improving path, if the link is not in the network and the players in question would like to add the link (both weakly and at least one strictly), then the

---

14. The results described here extend to more general processes in which several links are identified at once or the links that are identified depend on the current network or the history of links that have been considered. However, this process is a useful one for the purposes of illustration.

link is added, while if the link is already in the network, then it is deleted if either of the two players strictly prefers to delete it. There is an added randomness to the process. With a probability of $1 - \varepsilon$ the intent of the players (to add a link, delete a link, or leave the network as it is) is carried out, and with probability $\varepsilon > 0$ the reverse occurs.

Thus at each time some link is examined and with some probability the link is added, deleted, or ignored, depending on what the players would like to do under the concept of an improving path; and with some probability there is a perturbation and exactly the opposite occurs. Effectively, this process introduces small probabilities that some exogenous events happen, which might be errors on the part of players or other interventions that break up beneficial relationships or introduce nonbeneficial relationships. There are many possible perturbations that could account for such randomness.

Given these random perturbations, the process now continues (with probability 1) to have the network change indefinitely. Moreover, it continues to visit each network over time. In fact, this process is now a finite-state, aperiodic, irreducible Markov chain (recalling definitions from Section 4.5.8).[15] Thus it has a steady-state distribution.

To see how this process unfolds, let us reconsider the example pictured in Figure 11.9. If the process is at the empty network at some time, then it will change to another network only if there is an error. Thus there is a ε chance that the process leads away from the empty network.

If the process is at a one-link network, then regardless of which link is identified, the players want to change the network, either adding a new link or deleting the link. Thus there is a $1 - \varepsilon$ chance that the process changes the network and only a ε chance that it does not change. If the network does change, it is twice as likely that a two-link network will result as the empty network.

For a two-link network, the players choose to change the network only if the missing link is the one identified. In this case, there is a $1/3$ chance that the probability of leaving the network will be $1 - \varepsilon$ and a $2/3$ chance that the probability of leaving the network will be ε. In particular, there is a $(1 - \varepsilon)/3$ chance of changing to the complete network, a $(2 - \varepsilon)/3$ chance of staying at the same network, and a $2\varepsilon/3$ chance of changing to a one-link network.

For the complete network, no player wishes to make any changes. So the process changes networks only if an error occurs, and so changes to a two-link network with probability ε and stays put otherwise. Viewing this process as a Markov chain, let the state of this system simply be defined as the number of links that the network has, and consider the probability of transitioning from one state to another. The transition probabilities are described in the following matrix, where the ijth entry is the probability that the network will change from a network with i links to one with j links:

$$\Pi(\varepsilon) = \begin{pmatrix} 1 - \varepsilon & \varepsilon & 0 & 0 \\ \frac{1-\varepsilon}{3} & \varepsilon & \frac{2(1-\varepsilon)}{3} & 0 \\ 0 & \frac{2\varepsilon}{3} & \frac{2-\varepsilon}{3} & \frac{1-\varepsilon}{3} \\ 0 & 0 & \varepsilon & 1 - \varepsilon \end{pmatrix}.$$

---

15. Players' behavior, and that of the system, depends only on the current network and not on its history.

From this matrix, it is easy to deduce the steady-state distribution of the process. It is a 1×4 vector μ such that $\mu \Pi = \mu$. This is the left-hand unit eigenvector, which in this case is:

$$\mu(\varepsilon) = \left(\frac{\varepsilon(1-\varepsilon)}{1+2\varepsilon}, \frac{3\varepsilon^2}{1+2\varepsilon}, \frac{3\varepsilon(1-\varepsilon)}{1+2\varepsilon}, \frac{(1-\varepsilon)^2}{1+2\varepsilon} \right).$$

Let us examine the properties of this process. As ε tends to 0, $\mu(\varepsilon)$ tends to $(0, 0, 0, 1)$, and the time that the process spends in the complete network tends to 1, while the time that the process spends in any other network tends to 0.

To understand this process, note that for small ε, once the process reaches the empty network, it stays there until an error occurs, and so it can stay there for a long time as ε becomes small. In contrast, if this process is at either a one- or two-link network, it leaves that state with very high probability. When it is at the complete network, it will stay there with very high probability. To see why the process spends almost all of its time in the complete network rather than the empty network when ε is small, note the following. If it is at the complete network, even if an error occurs, the process moves to a two-link network. All two-link networks lead back to the complete network with high probability. An error is required to transition from a two-link to a one-link network. Thus, moving from the complete network to the empty network requires at least two errors to occur, with a probability on the order of ε^2. In contrast, moving from the empty network to the complete network only takes one error. In particular, once a one-link network is reached, then there is a nontrivial probability of transitioning to a two-link network and then to the complete network. Thus the probability of transitioning from the empty network to the complete network is on the order of ε. As ε becomes small, the process is much more likely to transition from the empty network to the complete network than the other way around. Although the process can still stay at the empty network for many periods after reaching it, asymptotically it will spend much more time at the complete network.

Note that it is necessary to look at the limit of $\mu(\varepsilon)$ to find μ. If we examine the limiting improving path process directly without any errors, then that process is described by the transition matrix

$$\Pi = \begin{pmatrix} 1 & 0 & 0 & 0 \\ \frac{1}{3} & 0 & \frac{2}{3} & 0 \\ 0 & 0 & \frac{2}{3} & \frac{1}{3} \\ 0 & 0 & 0 & 1 \end{pmatrix}.$$

While $\mu = (0, 0, 0, 1)$ is a steady state of that system, so is $(a, 0, 0, 1-a)$ for any $a \in [0, 1]$. That is, the process without any mutations does not discriminate between the empty and complete networks. One needs the error process to discover how stable various networks are to perturbations.

11.4.2 Stochastically Stable Networks*

With more than a few players, working directly with the perturbed improving path process $\Pi(\varepsilon)$ can be cumbersome, and so it is important to discover the set of stochastically stable networks more directly and generally. Here, we make use of a powerful theorem by Freidlin and Wentzell [257] that characterizes the steady-state distribution of Markov processes, which was adapted by Kandori, Mailath, and Rob [377] and Young [668] to understand how perturbed Markov chains behave as the probability ε of an error tends to 0. Jackson and Watts [358] show how such a general stochastic stability analysis can adapted to a network setting to derive insights about the evolution and dynamics of network formation. Here I outline some of the central tools and techniques. Consider an improving path process with added errors as described in Section 11.4.1. Let $\mu(g, \varepsilon)$ be the steady-state probability that the process is at network g when the process has error rate ε.

A network $g \in G(N)$ is *stochastically stable* if its steady-state probability is bounded below as the error rate ε tends to 0; that is, g is stochastically stable if $\mu(g, \varepsilon)_{\varepsilon \to 0} \to a > 0$.

When pairwise stable networks exist, any stochastically stable network must be pairwise stable. This is clear, since the process leaves any non-pairwise stable network with a probability that is bounded below as ε goes to 0, while a pairwise stable network is left only if an error occurs.[16] When pairwise stable networks do not exist, the stochastically stable networks only include networks that lie on improving cycles of networks that are randomly visited over time. The process could pick certain cycles and not others, as many more errors may be required to leave one cycle than to leave another.

Thus stochastic stability selects from among the pairwise stable networks, when they exist, and can provide a more refined prediction based on a sort of robustness argument. We already saw such a selection in the example above with two pairwise stable networks and only one stochastically stable one. Stochastic stability identifies the most robust or easy-to-reach networks in a particular sense. The disadvantage of this approach is that the limit points of the dynamics can be difficult to identify in some applications. Nevertheless, there are many settings in which it provides a meaningful refinement, as discussed below.

The characterization of stochastically stable networks follows from results characterizing the limiting properties of perturbed Markov chains. Consider a Markov chain on a finite state space S with transition matrix Π. For networks, the state space is the set of networks, and the transition matrix is determined by randomly selecting a link and then following an improving path (adding or deleting the link).

A set of *mutations* of Π is a set of transition matrices $\Pi(\varepsilon)$, one for each ε in a range $a > \varepsilon > 0$ for some $a > 0$, such that

1. $\Pi(\varepsilon)$ is aperiodic and irreducible for each ε,

2. $\Pi(\varepsilon)$ converges to Π as $\varepsilon \to 0$, and

3. $\Pi(\varepsilon)_{ss'} > 0$ implies that there exists $r \geq 0$ such that $0 < \lim_{\varepsilon \to 0} \dfrac{\Pi(\varepsilon)_{ss'}}{\varepsilon^r} < \infty.$

Part 1 ensures that the mutations add noise in such a way that any state can eventually be reached from any other state in an aperiodic way. Part 2 ensures that for small ε the mutated matrix is close to the original matrix. The number r in

---

16. It is possible for a (unique) stochastically stable network to be pairwise stable but not pairwise Nash stable. See Exercise 11.15.

(3) is the *resistance* of the transition from state s to s', and can be thought of as quantifying the level of error or mutation needed to get from state s to s'.

In the application to networks, the perturbations are found by including an ε error in the addition or deletion of the randomly identified link. Since then any network can lead to any adjacent network, as well as back to itself, the process satisfies (1), and (2) is also clearly satisfied. To verify that (3) is satisfied, first note that $\Pi(\varepsilon)_{gg'} > 0$ implies that g and g' are adjacent. There is a resistance of 0 in the case in which g' defeats g (i.e., g' lies on an improving path with only one link), and so $\Pi_{gg'} > 0$ and setting $r = 0$ satisfies (3). Otherwise, the transition only occurs if there is an error, and so $\Pi(\varepsilon)_{gg'}$ is simply ε divided by the number of links which could be identified by the process. Thus $\Pi(\varepsilon)_{gg'}$ tends to 0 at the rate of ε, and so $r = 1$ satisfies (3).

I now give a theorem from Young [668] characterizing the states with positive probability in the limit of the steady-state distributions of the mutations of the process.

Given any state s, an s-tree is a directed graph with a vertex for each state and a unique directed path leading from each state $s' \neq s$ to s. The *resistance* of s is the minimum across all s-trees of the summed resistance over directed edges in that tree.

Theorem 11.1 (Young [668]) *Let Π be the transition matrix associated with a Markov chain on a finite state space with an associated set of mutations $\{\Pi(\varepsilon)\}$ and with corresponding unique stationary distributions $\{\mu(\varepsilon)\}$. Then the steady-state distributions $\mu(\varepsilon)$ converge to a stationary distribution μ of Π. Moreover, a state s has positive probability under μ (and is thus stochastically stable) if and only if s has minimum resistance.*

From Theorem 11.1 it is easy to see that if a state s has positive probability under μ and there is an adjacent state s' that can be reached from s with no resistance, then the state s' also has minimum resistance and thus also has positive probability under μ. To see this, simply start with an s-tree with minimum resistance, and construct an s'-tree with at least as low a resistance as follows. Cut the directed link out from s' and form a directed link from s to s'. The new link has resistance 0, and so this tree has at least as low a resistance, and since s has minimum resistance, then s' must also.

By this reasoning, we can consider whole sets of states that can reach one another. The stochastically stable networks will either be pairwise stable networks or lie on cycles, where they can reach (and be reached by) other stochastically stable networks through an improving path. In terms of the results on stochastic stability of Markov chains, this result is stated as follows.

The *recurrent communication classes* of Π, denoted by $S_1, \ldots, S_J$, are disjoint subsets of states (not necessarily including all states) such that

- from each state there exists at least one path of zero resistance leading to some state in one of the recurrent communication classes,
- any state in a recurrent communication class can reach any other state in the same recurrent communication class by a path of zero resistance, and

▪ for any recurrent communication class S_j and states $s \in S_j$ and $s' \notin S_j$ such that $\Pi(\varepsilon)_{ss'} > 0$ for some ε, the resistance of the transition from s to s' is positive.

For networks, the recurrent communication classes are either singletons consisting of a pairwise stable network, or a closed improving cycle, where closure refers to the third item above and means that there is no improving path leading out from a network in the cycle to a network that is not part of the cycle.

For two recurrent communication classes S_i and S_j, since $\Pi(\varepsilon)$ is irreducible for each ε, it follows that there is a sequence of states $s_1, \ldots, s_k$ with $s_1 \in S_i$ and $s_k \in S_j$ such that the resistance of transition from each consecutive state to the next in the sequence (e.g., from s_h to s_{h+1}) is defined by (3) in the definition of mutations and is finite. Let this be denoted by $r(s_h, s_{h+1})$. The resistance of transition from recurrent communication class S_i to recurrent communication class S_j is the minimum over all such sequences of $\sum_h r(s_h, s_{h+1})$ and is denoted by $r(S_i, S_j)$.

Given a recurrent communication class S_i, an S_i-tree is a directed graph with a vertex for each communication class and a unique directed path leading from each recurrent communication class $S_j \neq S_i$ to S_i. The *stochastic potential* of a recurrent communication class S_i is then defined by finding an S_i-tree that minimizes the summed resistance over directed edges and setting the stochastic potential equal to that summed resistance.

With these definitions in hand, we can relate resistance to stochastic stability.

Theorem 11.2 (Young [668]) *Let Π be the transition matrix associated with a Markov chain on a finite state space with an associated set of mutations $\{\Pi(\varepsilon)\}$ and with corresponding unique stationary distributions $\{\mu(\varepsilon)\}$. Then the steady-state distributions $\mu(\varepsilon)$ converge to a stationary distribution μ of Π, and a state s has positive probability under μ (and thus is stochastically stable) if and only if s is in a recurrent communication class of Π that achieves the minimal stochastic potential. This condition is equivalent to s having minimum resistance.*

Theorem 11.2 is similar to the previous theorem, except that it establishes that stochastically stable states can be identified by working with the stochastic potential of the recurrent communication classes rather than keeping track of the resistance state-by-state. Thus to identify stochastically stable networks, one need only focus on pairwise stable networks and closed improving cycles of networks, which can substantially simplify the analysis. To get a feel for this and see an example of how stochastic stability can refine the set of pairwise (Nash) stable networks, consider the coauthor model from Section 6.4. Recall that the payoff to player i in network g is

$$u_i(g) = \sum_{j : ij \in g} \left(\frac{1}{d_i(g)} + \frac{1}{d_j(g)} + \frac{1}{d_i(g)d_j(g)} \right)$$

for $d_i(g) > 0$, and $u_i(g) = 1$ if $d_i(g) = 0$.

When $n = 7$ there are 22 different pairwise (Nash) stable networks: the complete network and each network with five completely connected players and a separate dyad. As Jackson and Watts [358] point out, only the complete network is stochastically stable, which can be seen as follows. Each pairwise stable network other than the complete network has a resistance of 1 to the complete network. Indeed, it is easily checked that deleting the link in the dyad leads to a network that lies on an improving path to the complete network. Thus when g is the complete network, we can construct a g-tree that has a stochastic potential of 21 by pointing every other network directly to the complete network. In contrast, it takes several errors to move from the complete network to one that lies on an improving path to some other pairwise stable network. If just one link in the complete network is severed, then the only improving path leads back to the complete network. Thus the resistance from the complete network to any other pairwise (Nash) stable network is more than 1. Constructing a g-tree for any other network leads to a stochastic potential of more than 21.[17]

Interestingly, the complete network is Pareto dominated by any of the other pairwise (Nash) stable networks, yet it is the unique stochastically stable network, which offers a further illustration of the tension between efficiency and stability. In some cases, there are weakenings and generalizations of stochastic stability that can be easier to work with, as shown by Tercieux and Vannetelbosch [623].

11.4.3 Stochastic Stability Coupled with Behavior

It is possible to extend the apparatus of stochastic stability to include other considerations. For instance, Jackson and Watts [359] examine a graphical game when strategies co-evolve with the network. That is, players choose between two actions in a coordination game in which their payoffs depend on the play of their neighbors. At the same time as choosing their actions, they can also decide on adding or deleting links. If there are costs to links, they prefer to be linked to other players with whom they coordinate their play. Their payoffs are thus affected both by whom they link to and by what strategies they play. Analyzing these features together provides for interplay between the network structure and play of the game. This type of analysis can lead to stochastically stable outcomes that differ from the behavior predicted when simply fixing the network structure (players who play more efficient actions can be more attractive). It can also lead to different predictions in terms of network structure than having fixed play. It can also result in differences in the speed of convergence of play, as the evolving network structure can more rapidly diffuse certain types of play than in a static network. The analysis becomes sensitive to the details of the setting but shows the importance of analyzing the co-evolution of behavior and network structure.[18]

11.5 • Farsighted Network Formation

A radically different perspective on network formation from that of random errors and myopic behavior is one in which players are forward-looking and make no

---

17. There are no improving cycles, so the only recurrent communication classes are the ones that each consist of a different pairwise stable network.

18. That analysis has been extended to directed networks by Goyal and Vega-Redondo [302] and Hojman and Szeidl [332]; to anticoordination games by Bramoullé et al. [104]; and to settings with geography by Droste, Gilles, and Johnson [206]. See also Skyrms and Pemantle [596] for a reinforcement-based evolutionary analysis of games played on networks.

errors. This perspective applies when players have a good idea of the setting and the incentives that drive various players to form or sever links. The topic has been explored from various vantage points, some of which involve explicit variations on network-formation games (see Dutta, Ghosal, and Ray [214]) as well as other approaches that capture farsightedness directly through variations on improving paths (e.g., Page, Wooders, and Kamat [519]). Let us examine the latter approach in more detail.

In the definition of improving path, changes from one network to the next are improving for the players involved, but without anticipating the subsequent changes that will occur along the path. In contrast, the idea of a farsighted improving path captures the notion that the players anticipate further changes along the path and compare the ultimate network to the current one.

Let us say that a network g' is *improving* for S relative to g if it is weakly preferred by all players in S to g, with strict preference holding for at least one player in S.

Consider a sequence of networks $g_1, \ldots, g_K$, and a corresponding sequence $S_1, \ldots, S_{K-1}$, such that g_{k+1} is obtainable from g_k by deviations by S_k. Such a sequence is a *farsighted improving path* if, for each k, the ending network g_K is improving for S_k relative to g_k.

If consecutive networks in the sequence are required to be adjacent, then this sequence is a farsighted analogue of an improving path, while more generally it allows for large coalitional deviations.

Now we can say that a network g is *farsightedly pairwise stable* if there is no farsighted improving path from g to some other network g' such that each pair of consecutive networks along the sequence are adjacent.

Similarly, a network g is *farsightedly strongly stable* if there is no farsighted improving path from g to some other network g'.

These are both demanding requirements. They refine the set of pairwise stable and strongly stable networks, respectively. That is, any network that is farsightedly pairwise stable is necessarily pairwise stable, and a network that is farsightedly strongly stable is necessarily strongly stable. These definitions require that networks be immune to both immediate deviations and farsighted sequences of deviations of arbitrary length.

In Figure 11.4 the complete network is farsightedly strongly stable and thus farsightedly pairwise stable. But more generally, such definitions are difficult to satisfy. The definitions are too strong in that they do not require that a farsighted improving path end at a network that is stable. That is, if players are farsighted, then they would not follow some farsighted improving path unless they anticipated that the endpoint is justified as the stopping point of the process. Such a definition of stability, however, is circular, as it requires the endpoint to be farsightedly stable. This sort of existence problem is nicely handled through a concept developed by Chwe [158] and adapted to networks by Page, Wooders, and Kamat [519].

The idea is based on a self-consistent set-based definition. In particular, a set of networks is said to be consistent if all deviations away from the network are expected to lead (in a farsighted manner) back to some network in the set that is not improving for the original deviating coalition.

More formally, a set $A \subset G(N)$ is *consistent* if for each $g \in A$, and g' obtainable from g through deviations by some $S \subset N$, either $g' \in A$ and g' is not improving for S, or there exists a farsighted improving path from g' to some $g'' \in A$ such that g'' is not improving for S.[19]

It is easily checked that a union of consistent sets is consistent. Thus Chwe suggests examining the largest consistent set, which he shows is always nonempty. The idea of a consistent set A is that any network in the set is justified as being stable as follows. Consider $g \in A$, and g' obtainable from g through deviations by some $S \subset N$. If g' is improving for S, then there must be some expectation that discourages S from deviating to g'. In particular, there must exist a farsighted improving path moving away from g' that S anticipates will be followed and will lead to some $g'' \in A$ such that g'' is not improving for S. The presence of g'' in A implies that it is also a justifiable resting point, and so S can expect the process to stay there. This anticipation then deters the original deviation. If every deviation away from g can be deterred in this way, then it is a viable resting point, and the set is consistent in that its various elements are used in justifying each other as resting points.

To see how the largest consistent set can result in different predictions from other solutions, let us re-examine a variation on Figure 6.1, which involved a bargaining network. Allowing for coalitional deviations, we find that no network is strongly stable. The difficulty is that the efficient network is defeated by a network with a link added. This in turn is defeated by the addition of another link, but then all players would be better off moving back to the efficient network.

The largest consistent set relative to the payoffs in Figure 11.10 includes the efficient network and two other networks. Without going through the full set of calculations needed to verify that this set is the largest consistent one, let us examine a few of the key deviations to check that the set is consistent. The reasoning is a bit subtle, but is as follows. Suppose that we start at the efficient network. To check that it is part of a consistent set, we need to check that if some group deviates to change the network, they could then anticipate some farsighted improving path that leads away from the deviation and to another network in the consistent set that would not be improving. A possible problematic deviation is players 1 and 2 threatening to add the top link, which would lead them to a higher payoff of 3.25 each. To check the condition for this to be a consistent set, we need to find a farsighted improving path away from this network leading to another network in the consistent set that is not improving for 1 and 2 relative to their original payoffs of 3. Indeed, players 3 and 4 could add a link, which would be improving for players 3 and 4 and lead to another network in the consistent set that offers a lower payoff to players 1 and 2 (2.5 each). Thus the possible anticipated continuation reasoning deters 1 and 2 from adding a link in the first place and is one way to justify the original efficient network being in the set. Next consider the network with the three links in which players 1 and 2 receive a payoff of 3.25. Suppose that players 3 and 4 deviate to add a link, resulting in a network in which all players receive 2.5. For the network with three links to be in the consistent set, we need to check that there is a farsighted improving path that leads to a consistent

---

19. The definitions here differ slightly from Chwe's [158] definitions, as here the definition of improving only requires weak improvement for some players and strict for at least one, while Chwe's definition requires improvement for all players.

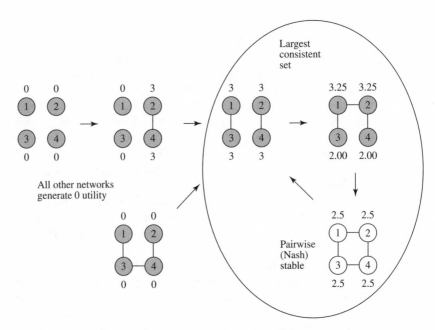

FIGURE 11.10 No network is strongly stable and pairwise (Nash) stable networks are inefficient, while the largest consistent set includes an efficient network.

network with a payoff for players 3 and 4 that deters them from making this initial change. There is a farsighted improving path that goes to the three-link network at the bottom left of the figure (which results in payoffs of 0 for all players) and then continues to the original three-link network at the top right (in which players 3 and 4 have a payoff of 2 each). Again we have found a subsequent farsighted improving path that can deter the original deviation. Finally, we need to check that the network at the bottom right is in the consistent set. An obvious deviation is to the efficient network. From that network there is a farsighted improving path that leads to the three-link network at the top right, which is in the consistent set and leads to a lower payoff for players 3 and 4, thus deterring their original deviation.

This example shows the reasoning behind the consistent set, and why it is a set-based notion. Various networks in the set are sustained because of anticipated deviations leading to other consistent networks. The consistency refers to the fact that various networks in the set are used to sustain others, and the reasoning has to be fully consistent so that each is sustainable via the set itself. Considering the largest such set implies that nothing outside the set could be sustained through such reasoning.

Proposition 11.6 (Chwe [158]) *Consider any N and profile of preferences. There exists a unique largest consistent set (so that every consistent set is a subset of it and it is consistent), and this set is nonempty.*

The ideas behind the proof are relatively straightforward. First, it is easily checked that the union of consistent sets is consistent, as there are fewer potential deviations that lead outside the set and more deterents. Thus, given the finite setting, there exists a largest consistent set simply by taking the union of all consistent sets. Showing that the largest consistent set is nonempty is the harder part of the proof. It relies on showing that a sequence of networks with a farsighted improving path from any network to any subsequent network in the sequence must be finite in length. This follows in our setting since an infinite path must repeat some networks, which cannot be improving relative to themselves.

The basic ideas behind the largest consistent set are not particular to the specific notions of improvement or deviation that we have been working with. Indeed, Chwe's original definitions do not mention networks at all. It is a quite versatile idea that can be adapted to other settings, including networks with multiple relations between pairs or groups of players, directions, and other considerations. Such extensions are explored by Page, Wooders, and Kamat [519] and Page and Wooders [518].[20]

11.6 ▪ Transfers and Network Formation

As discussed in Chapter 6, the tension between efficiency and stability in network formation stems from externalities. Players do not take into account the indirect impact that their link-formation decisions have on others. At least in some contexts, this disregard can be rectified if transfers are allowed so that players can pay one another for forming or severing links, or if players can bargain over how the value of relationships is allocated when forming links. Such transfers are applicable in many settings in which links represent economic or business relationships, and transactions are occurring as part of the relationship. If the relationship is more advantageous for one party, then that can be contracted upon at the time of the relationship formation. Such agreements not only arise in purely economic relationships but also in social ones. For instance, we even see transactions in some marriages in the form of dowries. Even with no wealth or goods exchanged, there can be an allocation of tasks or favors that is either implicitly or explicitly agreed to in order to maintain a relationship.

11.6.1 Forming Network Relationships and Bargaining

Because transfers can affect the network, it can be important to model the use of transfers as part of the network-formation process. Such a process was first investigated by Currarini and Morelli [181] by examining a sequential network-formation game that is described as follows.[21] The game is defined for cases in which the utilities are component decomposable (recalling the definition from Section 6.6.2), so that the payoffs to a given component depend only on that component's configuration and not on the remainder of the network. Players are

---

20. See Dutta, Ghosal, and Ray [214] and Herings, Mauleon, and Vannetelbosch [328] for other approaches to defining farsighted stability.
21. See Mutuswami and Winter [494] for elaboration on a similar model.

ordered exogenously (labeled in the order of their moves), so that player 1 moves first, then player 2, and so forth. At his or her turn, player i announces the set of players with whom he or she is willing to be linked ($S_i \subset N \setminus \{i\}$) and a payoff demand $v_i \in \mathbb{R}$—which is interpreted as the net payoff that the player desires to get. The outcome of the game is then as follows. First one examines the network that could potentially form by including the links such that both players involved in the link announced each other. That is, the potential network that might form is $g(S)$, where $ij \in g(S)$ if and only if $j \in S_i$ and $i \in S_j$. This is not the final network, as one has to see whether the payoff demands that the agents made can all be satisfied. The network that is eventually formed is determined by checking which components of $g(S)$ are actually feasible in terms of the demands submitted. That is, if g' is a component of g, then g' is actually formed if $\sum_{i \in N(g')} v_i \leq \sum_i u_i(g)$, and otherwise none of the links in g' are formed.

The presumption is that if a component forms, then each player in that component receives a payoff of v_i that was demanded. Exactly how transfers are made or what needs to be done to convert the initial utilities into this final allocation of payoffs is not specified and might involve some complicated transfers that travel some distance in the network. Currarini and Morelli then show the following result for a class of settings in which payoffs satisfy a condition that they call *size monotonicity*.

Payoffs $u = (u_1, \ldots, u_n)$ satisfy size monotonicity if $\sum_i u_i(g) > \sum_i u_i(g - kj)$ for every g and every bridge $kj \in g$ (such that $g - kj$ has more components than g).

Proposition 11.7 (Currarini and Morelli [181]) *If payoffs satisfy size monotonicity, then every (subgame perfect) equilibrium of the Currarini and Morelli bargaining and network-formation game leads to an efficient network.*

The intuition behind the result is as follows. It is helpful to consider a simple dyad to see the idea. Suppose that if the dyad is formed, player 1 obtains a utility of 2 while player 2 will receive a utility of -1. This relationship is beneficial for player 1 but costly for player 2. There is a total utility of 1 to be had, and so it is efficient for the link to form; however, without any reallocation of value, player 2 would not be willing to form the relationship. Given that player 1 moves first, the equilibrium here is easy to see: player 1 states $S_1 = \{2\}$ and sets $v_1 = 1$; and then 2 responds with $S_2 = \{1\}$ and $v_2 = 0.$[22] This is the only subgame perfect equilibrium outcome. Effectively, player 1 pays player 2 a unit of utility to form the link. If the player roles were reversed and player 2 moved first, then player 2 would demand $v_2 = 1$ and leave player 1 with no value. The important aspect is that the players are now able to ensure that each gets enough value to form the efficient network. When there are more than two players the analysis becomes more complicated, as the first movers have to forecast how much utility they can extract from the network, how much will be left, and what options the later players have in terms of forming links; but the basic idea is that they will correctly forecast exactly how much they can

---

22. In terms of the full specification of how player 2 behaves, he or she declares $S_2 = \{1\}$ and $v_2 = 1 - v_1$ when $v_1 \leq 1$ and otherwise says $S_2 = \emptyset$.

extract in equilibrium by maximizing with respect to the foreseeable equilibrium strategies of the subsequent players.[23]

This game has features that significantly facilitate obtaining results but are also somewhat artificial. In particular, the fact that players move exactly once in a fixed, forecastable order allows for backward induction and for determining efficiency. Suggesting links that lead to an efficient network maximizes the payoff that each player is able to extract at his or her turn, given the previous demands of the other players and given how the remaining players will be forced to react. It is clear that the ordering of moves can provide some players with big advantages or disadvantages in terms of the network and payoffs that eventually emerge, and that efficiency hinges on the fixed sequential ordering.

11.6.2 A Network-Formation Game with Transfers

As an alternative to the game in Section 11.6.1, Bloch and Jackson [79] suggest a simultaneous-move game in which players can directly offer transfers to one another to form links. The motivation is not that a simultaneous-move game describes how networks are formed. To the contrary, the idea is that if one models a richer setting in which players can go back and forth and bargain, in the end they must reach a point where none of them wants to change their proposed transfers or links. Thus the resting point of a more open and dynamic process has to reach an equilibrium point where no player would gain from changing his or her action given the actions of the others. This is modeled as follows.

Let each player i announce a vector $t_i = (t_{i1}, \ldots, t_{in}) \in \mathbb{R}^n$ such that $t_{ii} = 0$; t_{ij} is the amount that i is willing to transfer to j to form a link, or, if this is a negative number, the amount that i requests in order to form the link. A link ij is formed if and only if $t_{ij} + t_{ji} \geq 0$. In equilibrium this relation holds with equality. Let $g(t) = \{ij | t_{ij} + t_{ji} \geq 0\}$ denote the network that forms. Player i's payoff is then

$$\pi_i(t) = u_i(g(t)) - \sum_{j:ij \in g(t)} t_{ij}.$$

Nash equilibrium can be used to solve this game; however, the same difficulties emerge as are faced with Nash equilibrium in the basic network-formation game. For instance, if a player expects all other players to demand enormous amounts to form a link (i.e., to state large negative t_{ij}s), then it is in his or her interest to do the same. Thus the empty network is always an equilibrium, regardless of how attractive other networks are. Then the problem arises that nobody forms any links since they all correctly forecast that no one else will form a link. To deal with this issue, Bloch and Jackson [79] adapt pairwise Nash equilibrium to this setting.

A *pairwise transfer equilibrium* is a profile of vectors of proposed transfers $(t_1, \ldots, t_n)$ such that t is a Nash equilibrium; that is,

$$\pi_i(t) \geq \pi_i(\hat{t}_i, t_{-i})$$

---

23. To see how this works in some richer examples, see Jackson [347]; the proof is provided in Currarini and Morelli [181].

for all i and $\hat{t}_i \in \mathbb{R}^n$ (such that $\hat{t}_{ii} = 0$); and, for any $ij \notin g(t)$

$$u_i(g(t) + ij) + u_j(g(t) + ij) \leq u_i(g(t)) + u_j(g(t)).$$

Note that the latter stipulation is equivalent to requiring that there does not exist any $\hat{t}_{ij}$ and $\hat{t}_{ji}$ such that $\hat{t}_{ij} + \hat{t}_{ji} \geq 0$ and such that both i and j would be weakly better off with these new announcements and the addition of the link between them with at least one of them strictly better off. Thus the definition parallels that for pairwise Nash equilibrium but allows for transfers between players.[24]

Proposition 11.8 (Bloch and Jackson [79]) *In the distance-based utility model, for each efficient network, there exists a pairwise transfer equilibrium that results in that network and in balanced transfers (so that the transfers sum to 0).*

Bloch and Jackson do not prove Proposition 11.8 directly, but instead as part of a more general characterization that provides the necessary and sufficient conditions for a network to be supportable as an equilibrium of this transfer game. The direct proof in the case of the distance-based utility model is quite intuitive, as it deals mainly with the star, and works by showing that the peripheral players can offer sufficient transfers to the center player to sustain the star as an equilibrium when it is efficient. That proof is left as Exercise 11.16.

The pairwise transfer equilibria in this transfer game do not necessarily include efficient networks. There are various reasons for this omission. One is that there can be indirect externalities, so that the efficient network involves relationships that affect players who are not directly involved. A player might want to pay a neighbor to undertake more or fewer other relationships but cannot, as the types of transfers described only affect the given link in question. For example, a player might benefit from having friends with more contacts or from having friends who are less distracted. Also, it could be that players would like to subsidize links that are far away in the network, so that they need to make transfers to players who are not their neighbors. Furthermore, players might be hurt by (distant) links that others wish to form. Then they would like to pay other players not to form such relationships. This is the case in some research and development settings, where a firm would be willing to pay other firms not to collaborate with one another. Bloch and Jackson [79] also consider two other variations of such transfer games, one in which players can offer to subsidize links they are not involved with and can make transfers contingent on the network. They characterize which types of externalities can be overcome by which sorts of transfers.

11.7 ▪ Weighted Network Formation

Most of the literature on strategic network formation has examined discrete linking decisions. That is, relationships are modeled as either being present or not, or either weak or strong, but without much richer choices. In many contexts decisions

---

24. For a notion of equilibrium incorporating transfers for the case of directed network formation, see Johari, Mannor, and Tsitsiklis [366].

are much richer: for instance, we decide how much time to spend with different friends and how much effort to devote to various collaborations. Allowing for richer choices leads to new insights into how relationships form and what incentives players have to maintain efficient relationships, as illustrated in the following model from Rogers [563].

There is a finite society of $n \geq 3$ players who form a weighted and directed network. Each player has a budget of time that he or she can spend with other players. Let the budget for player i be denoted $B_i > 0$. A feasible strategy for player i is a vector $(g_{i1}, \ldots, g_{in})$ such that $\sum_{j \neq i} g_{ij} = B_i$, $g_{ij} \geq 0$ for all $j \neq i$, and $g_{ii} = 0$. The quantity g_{ij} is the amount of time that i spends with j. This network is directed, and these values need not be reciprocal.

Each player i has some natural intrinsic base utility $v_i > 0$ that would be his or her payoff in the absence of any network interactions. In addition to that natural utility, the player benefits from other players' payoffs by an amount that depends on the time that is spent with other players multiplied by their payoffs. So spending time with a "happier" player leads to greater utility, all else held equal. However, there is a diminishing return to the time spent with any other player. In particular, in Rogers's [563] model, the payoff to player i is

$$u_i = v_i + \sum_{j \neq i} f(g_{ij}) u_j, \qquad (11.2)$$

where f is a nonnegative and continuously differentiable function such that $f(0) = 0$ and $\lim_{x \to 0} f'(x) = \infty$.

In (11.2) payoffs are self-referential, as the payoff to a given player depends on the payoffs to others, which in turn depend on the given player's payoff. Such payoffs relate to the eigenvector-based centrality measures discussed in Chapter 2, which had a similar self-referential formulation. So, the model can be thought of as a network-formation problem using endogenous centrality measures. The last condition on the derivative of f ensures that $g_{ij} > 0$ for all i and $j \neq i$, which makes the model easier to work with.

Letting $f(g)$ denote the $n \times n$ matrix with ijth entry $f(g_{ij})$, we can write

$$u(g) = v + f(g)u(g),$$

where $u(g)$ and v are $n \times 1$ column vectors. This equation has the solution

$$u(g) = (\mathbb{I} - f(g))^{-1} v = A(g)v,$$

when $A(g) = (\mathbb{I} - f(g))^{-1}$ is well defined. In this setting a natural notion of equilibrium is simply a Nash equilibrium, where each player i is choosing $g_i = (g_{i1}, \ldots, g_{in})$ to maximize $u_i(g)$. The following is a strengthening of the results by Rogers [563].

Proposition 11.9 (Rogers [563]) *Suppose that $\max_i f(B_i)$ is small enough so that $A(g) = (\mathbb{I} - f(g))^{-1}$ is well defined, continuous, and nonnegative for all feasible g (and is described by $A(g) = \sum_p f(g)^p$).*

- *All Nash equilibrium are interior ($B_i > g_{ij} > 0$ for all i and $j \neq i$).*
- *Any best response for player i to a feasible and interior g_{-i} (and thus a Nash equilibrium strategy) is such that for each j and h,*

$$f'(g_{ij})u_j(g) = f'(g_{ih})u_h(g). \tag{11.3}$$

- *If for each i and feasible and interior g_{-i} there is a unique g_i that satisfies (11.3), then g is a Nash equilibrium if and only if it is feasible, interior, and satisfies (11.3) for each i, j, and h.*
- *If for each i and feasible and interior g_{-i} there is a unique g_i that satisfies (11.3), then the network strategy g_i that maximizes $u_i(g_i, g_{-i})$, given a feasible and interior g_{-i}, also maximizes $u_k(g_i, g_{-i})$ given g_{-i} for each k, and so g_i maximizes the total sum of utilities $\sum_j u_j(g_i, g_{-i})$ given g_{-i}.*

Proof of Proposition 11.9. Given that the limit, as x goes to 0, of the derivative of $f(x)$ is infinite, and that $u_j(g) \geq v_j > 0$ for each j, by (11.2) it follows that every Nash equilibrium (and every maximizer of the total sum of utilities) is such that $g_{ij} > 0$ for all i and $j \neq i$.

Consider any player i. Given that A is well defined and given interior strategies of other players, it follows that regardless of i's strategy, all entries of A are strictly positive.[25]

To maximize u_i it must be that

$$\frac{\partial u_i}{\partial g_{ij}} = \frac{\partial u_i}{\partial g_{ih}}$$

for each j and h other than i. For any k we can write

$$u_k = \sum_\ell A_{k\ell}(g)v_\ell.$$

Therefore

$$\frac{\partial u_k}{\partial g_{ij}} = \sum_\ell \frac{\partial A_{kl}(g)}{\partial g_{ij}} v_\ell.$$

To develop an expression for $\partial A_{kl}(g)/\partial g_{ij}$, we follow Rogers [563], who shows that differentiating $AA^{-1} = I$ leads to

$$\frac{\partial A(g)}{\partial g_{ij}} A(g)^{-1} = -A(g)\frac{\partial A(g)^{-1}}{\partial g_{ij}},$$

---

25. We can write $A(g) = \sum_p f(g)^p$. Given that $g_{kj} > 0$ for all $k \neq i$ and $j \neq k$, it follows that for any i and j that the ijth entry of $f(g)^p$ is positive for large enough p. To see this, recall that $f(g)^p$ will be positive if there is a directed walk of length p from i to j. Given that there are at least two other players, there is a path of some length from any other player to every player. Regardless of whom i connects to, i also reaches all other players.

and so

$$\frac{\partial A(g)}{\partial g_{\ell k}} = -A(g)\frac{\partial f(g)}{\partial g_{ij}}A(g).$$

Therefore

$$\frac{\partial A_{kl}(g)}{\partial g_{ij}} = f'(g_{ij})A_{ki}(g)A_{j\ell}(g). \tag{11.4}$$

Substituting from (11.4), it follows that

$$\frac{\partial u_k}{\partial g_{ij}} = \sum_{\ell} f'(g_{ij})A_{ki}(g)A_{j\ell}(g)v_\ell = f'(g_{ij})A_{ki}(g)\sum_{\ell} A_{j\ell}(g)v_\ell$$

$$= f'(g_{ij})A_{ki}(g)u_j(g). \tag{11.5}$$

Setting $k = i$ implies that

$$\frac{\partial u_i}{\partial g_{ij}} = f'(g_{ij})A_{ii}(g)u_j(g).$$

This equation and the relations $A_{ii} > 0$ and $u_j(g) \geq v_j > 0$ imply that for every $j \neq i \neq h$,

$$f'(g_{ij})u_j(g) = f'(g_{ih})u_h(g).$$

Given that $A_{ki}(g) > 0$, (11.5) implies that the same condition characterizes the maximization of $u_k(g)$. The claims in the proposition follow directly. ∎

An important implication of Proposition 11.9 is that (when best responses are unique) any network that maximizes the total sum of utilities is a Nash equilibrium network. The choice to maximize a given player's utility is the same choice as a society would make to maximize overall welfare.

What is special about this setting that leads to the congruence of efficiency and stability, in contrast to the more general conflict between stability and efficiency that we have seen? There are several important factors, and understanding them helps to understand this conflict more generally. First, the problem faced by any given player in the Rogers model is to allocate a given budget of time or effort on different relationships. Thus the problem is solely one of allocating the budget across different relationships, rather than deciding on how many relationships to have in total. That is, generally we can think of a network-formation problem faced by a player as having two main components: the total quantity of effort or relationships to maintain, and how to distribute that effort among the different relationships. The first component is missing here and is generally the problematic one. Usually the inefficiency of network formation stems from the failure of a given player to properly account (with respect to social welfare) for the additional costs or benefits to others of his or her relationships beyond his or her own private benefit. So the player either under- or overinvests in the total number of relationships relative to what is socially valuable. The decision faced by players in the Rogers's model is solely allocative: Given the fixed budget, how should a player distribute it? The player wants to allocate effort in a way that maximizes his or her payoff, which is the same distribution that maximizes indirect payoffs, since in this model indirect payoffs come through a given player's utility. That is, a player obtains utility from his or her neighbors' utility, and so whatever makes them happy also makes the player happy.

This linkage points to the second distinctive aspect of the Rogers model. A player obtains indirect utility precisely through increases in neighbors' utility. Consider a network with player 1 connected to both 2 and 3, who are each connected to player 4. In the Rogers model, the benefits that player 1 obtains from the indirect connection to player 4 come from both the utility that player 2 receives from being connected to 4 (independently of whether 3 is connected) and the value that player 3 receives from being connected to player 4 (independently of whether 2 is connected). In many contexts, the marginal benefit to player 1 of having a second indirect path to player 4 might be lower than the marginal benefit of having the first path. Finally, the Rogers model also displays symmetry, in that all players obtain the same direct or indirect utility from any given connection, and they do not have any heterogeneity in their preferences for connections.

While this model has special features, and they are responsible for a congruence between equilibrium networks and efficient networks, the model still provides a useful benchmark in terms of understanding the tradeoffs that players face in deciding how to allocate their time or effort among different relationships. There are many variations on the model, including some by Rogers [563], a model by Brueckner [112] in which effort translates into the probability that a link forms, and a model by Bloch and Dutta [76] in which players do not face a budget constraint.

11.8 ▪ Agent-Based Modeling

When modeling network formation (or behavior on networks, or some combination of the two), one difficulty is that complex networks and/or patterns of behavior can emerge from simple specifications, especially when even minimal heterogeneities (e.g., in geography, age, costs, or preferences) are introduced. Although many insights can be derived analytically, there are some properties that cannot be directly derived. In many cases, it is more expedient to examine the behavior of large computer-simulated societies.

Such analyses have become more pervasive in the literature and are often referred to as *agent-based modeling*. These techniques can be very useful. As mentioned above, they can be used to analyze systems in which equilibrium or dynamics cannot be determined analytically. They are useful as tools to illustrate systems or for exploratory analyses that help in formulating hypotheses and conjectures. Such techniques are also useful in empirical analyses for generating distributions of behaviors that emerge under a model, which can then be compared to or fitted to observed data.

As with any form of analysis, there are important considerations of how sensitive or robust the conclusions are to the specification. In agent-based modeling there are also issues of how many simulations to run, how long to run them, how large a society to consider, and so forth. As there are already a number of good sources on this subject, I do not discuss it here.[26]

11.9 ▪ Exercises

11.1 *Nash Stability and Pairwise Stability* Provide an example of a society of individuals and utility functions such that the set of Nash stable networks is a strict subset of the set of pairwise stable networks. Provide an example where the reverse is true.

11.2 *Nash Stable Networks* Show that a nondirected network g is Nash stable if and only if no player wishes to delete a set of his or her links.

11.3 *Pairwise Nash Stability* Provide an example of a society of individuals and utility functions such that the set of pairwise Nash stable networks is a strict subset of the set of pairwise stable networks and also a strict subset of the set of Nash stable networks.

11.4 *Strict Nash Equilibria in the Link-Announcement Game and Nonexistence* Consider a potential dyad. Suppose that the payoff to having the link is negative for each player, while the payoff to not having the link is 0 for each. Show that there is no strict Nash equilibrium of the link-announcement game. Recall that a strict Nash equilibrium is a pure strategy Nash equilibrium in which the actions played are the unique best responses to each other.

11.5 *Strongly Stable Networks and the Connections Model* Consider the symmetric connections model (see Section 1.2.4) with $\delta < c$ and $n \geq 4$. Identify a network that is pairwise Nash stable but not strongly stable for some choice of parameters. Find an example of a strongly stable network that is not efficient for a case in which $n \geq 5$.

11.6 *Deviations and Strongly Stable Networks* The following definition follows one in Dutta and Mutuswami [213]. A network g is *strongly stable** with respect to u if for any $S \subset N$ and g' that is obtainable from g through deviations by S there exists $j \in S$ such that $u_j(g') \leq u_j(g)$.

Find an example of a network that is strongly stable* but not pairwise stable and hence not strongly stable.

11.7 *Existence of Pairwise Stable Networks* Consider any N and any profile of utility functions, one for each player. Show that either there exists a pairwise stable network or a closed cycle (as defined in Section 11.2.1).

11.8 *Improving Paths for Pairwise Nash Stability* Develop a definition of *improving path** that allows pairs of agents adding one link, or a single agent deleting multiple

---

26. Some starting references include: Axelrod [24], [25]; Bonabeau [89]; Brately, Fox, and Schrage [106]; Epstein and Axtell [226]; Gilbert and Troitzsch [279]; Grimm and Railsback [312]; Tesfatsion [625]; and Tesfatsion and Judd [626].

links, and relate it to the existence of pairwise Nash stable networks. Provide an example in which all improving paths* are part of cycles, even though there exists a pairwise stable network.

11.9 *Existence of Pairwise Stable and Pairwise Nash Stable Networks*   Find an example in which payoffs are anonymous[27] and a pairwise stable network exists, but there does not exist a pairwise Nash stable network. Does there exist such an example with $n = 3$ in which isolated players receive a payoff of 0?

11.10 *Simultaneous Stability*   Consider the following variation on strong stability. A network g is *simultaneously stable* if for any $S \subset N$ such that $|S| \leq 2$, g' that is obtainable from g through deviations by S, and $i \in S$ such that $u_i(g') > u_i(g)$, there exists $j \in S$ such that $u_j(g') < u_j(g)$. Thus a network is simultaneously stable if no single player strictly prefers to delete some set of his or her links, and no two players would each weakly benefit (with at least one benefiting strictly) by deleting some of their links and/or adding a link between them. This requirement is stronger than pairwise Nash stability but weaker than strong stability, because only coalitions of two players are considered.

Consider a setting in which players have "types" in some finite set Θ, and let player i's type be denoted by θ_i. Let $s_i(g)$ be the number of players of i's type to whom i is linked (so $\theta_i = \theta_j$ and $ij \in g$), and $o_i(g)$ be the number of players of types other than i's type to whom i is linked. Suppose that the payoffs are:

- $u_i(g) = o_i(g) + 2s_i(g)$, if $o_i(g) + s_i(g) \leq d_i$, and
- $u_i(g) = 0$, if $o_i(g) + s_i(g) > d_i$,

where $d_i \geq 1$ is the number of links that i can maintain. So players benefit more from links to their own types.

Show that the pairwise Nash networks are those such that (1) no player exceeds his or her capacity d_i and (2) there is at most one player with fewer links than his or her capacity. Characterize the set of simultaneously stable networks. Show that if there are at least two players who have the same type, then the set of simultaneous stable networks is a strict subset of the set of pairwise Nash networks.

11.11 *Ordinal Potential Functions and the Absence of Indifference*   Provide an example in which the payoffs exhibit indifference and there are no improving cycles, but there does not exist an ordinal potential function (three players will suffice).

11.12 *An Ordinal Potential Function for a Buyer-Seller Network*   Consider the Corominas-Bosch model from Section 10.3.2. Show that the sum of all payoffs is an ordinal potential function.

11.13 *Existence of Strongly Stable Networks and Top Convexity*   Payoffs are *top-convex* if

---

27. A profile of utility functions is anonymous if for every π that is a permutation on N (a one-to-one function mapping the set of agents N to N), it follows that $u_{\pi(i)}(g^\pi) = u_i(g)$, where $g^\pi = \{\{\pi(i), \pi(j)\} | ij \in g\}$ is the network obtained from g by permuting the positions of agents according to π.

$$\max_{g \in G(N)} \frac{\sum_{i \in N} u_i(g)}{|N|} \geq \max_{g \in G(S)} \frac{\sum_{i \in S} u_i(g)}{|S|}$$

for all $S \subset N$.[28]

(a) Suppose that payoffs are component decomposable and are such that any two players in the same component get the same payoff. Show that the set of strongly stable networks is nonempty if and only if payoffs are top-convex.

(b) Show that if we reallocate utility so that players within a component get an equal split of the total utility within a component, then payoffs in the symmetric connections model are top-convex.

11.14 *Proof of Proposition 11.4: Directed Nash Stable Networks in a Distance-Based Model* Prove Proposition 11.4.

11.15 *A Stochastically Stable Network That Is Not Pairwise Nash Stable* Provide an example of a stochastically stable network that is pairwise stable but not pairwise Nash stable. Describe a variation on the random process that would instead select from pairwise Nash stable networks.

11.16 *Proof of Proposition 11.8* Prove Proposition 11.8.

11.17 *Complementarities in Link Efforts with Convexities* Consider the following variation of a model by Bloch and Dutta [76]. Each agent has a unit of effort to allocate on different relationships. In particular, agent i chooses a vector of efforts, where $x_{ij} \in [0, 1]$ is the effort that i invests on a relationship with agent j, and $\sum_{j \neq i} x_{ij} = 1$. The strength of the overall relation between i and j is then $s_{ij} = \phi(x_{ij}) + \phi(x_{ji})$, where ϕ is an increasing function.

The payoff to i as a function of s (the matrix of s_{ij}s) is $u_i(s) = \sum_{k \neq i} v_{ik}(s)$, where $v_{ik}(s)$ is determined as follows. Let P_{ik} be the set of the potential paths between i and k that could occur in any network. For a path $p = i_1 i_2, i_2 i_3, \ldots, i_{m-1} i_m$ let $v(p, s) = s_{i_1 i_2} \times s_{i_2 i_3} \times \cdots \times s_{i_{m-1} i_m}$. Then $v_{ik} = \max_{p \in P_{ik}} v(p, s)$.

Show that if ϕ is a strictly convex function then any Nash equilibrium choice of x_{ij}s is such that, for each i, $x_{ij} = 1$ for some j and $x_{ik} = 0$ for all $k \neq j$. What are the equilibrium configurations for a three-agent system when ϕ is strictly convex?

11.18 *Schelling's Tipping and Segregation Models: An Agent-Based Computation Exercise* Consider the following simple model of segregation from Schelling [578], [579]. A finite set of agents live on a line segment.[29] Agents are either red or blue. An agent prefers to be in a neighborhood where at least half of his or her neighbors are of the same color. Agents have no further preferences. An agent considers his or her neighborhood to consist of the k nearest agents to his left plus the k nearest agents to his right. (For agents near the end of the segment, simply wrap the segment around to form a circle when defining closest neighbors.) At any point in time each agent can be labeled as either "contented" or "discontented,"

---

28. This definition and result are due to Jackson and van den Nouweland [356].
29. There are two-dimensional versions of the model, which are known as *Schelling's checkerboard model*.

depending on whether at least 50 percent of their neighbors are of their color. Schelling describes the following algorithm:

1. Start with n agents randomly positioned on a line segment (each agent at a distinct point), their colors randomly assigned with independent and equal probability of red or blue.

2. Identify the discontent agents and label this set $D(0)$.

3. Starting with the leftmost discontent agent, move that agent to a point where the agent is content, and do so in a manner that minimizes the number of agents leapfrogged. Break ties by moving to the left.

4. If the second agent who was discontented is now content (due to the change induced by the first discontented agent's move), then leave that second initially discontented agent in place. Otherwise, move that agent to a point where he or she will be content (again leapfrogging as few agents as possible and moving to the left in the case of a tie).

5. Repeat step 4 until all agents in $D(0)$ have been considered.

6. Repeat steps 1–5, identifying a new set of discontented agents $D(1)$, and moving discontented agents as before.

7. Iterate on steps 1–6 until all agents are contented.

Write a program to run this algorithm when $k = 4$ and $n = 100$. Run the program 100 times. For each run, record how many changes in color there are as one moves along the segment from left to right at the starting configuration and at the ending configuration. This record gives an idea of how many different segmented groups of agents of the same color there are on the line. Also record the average fraction of neighbors of the same color at both the starting and ending configurations.

CHAPTER 12

Allocation Rules, Networks, and Cooperative Games

Throughout this book we have seen different sorts of predictions of the power, centrality, or influence of different agents, the terms of trade they might obtain in an exchange, and the utility they might acquire in a game played on a network. These predictions were derived using a variety of tools, ranging from specific measures of power and centrality to models of the spread of information, bargaining on networks, or strategic interaction on a network. Beyond these tools that are directed at specific applications, there is a more general perspective that builds on properties of how the aggregate payoff behaves as a function of the network and then deduces allocations of payoffs from those properties. It is an offshoot of cooperative game theory, which examines how productive value is split among the members of a society based on the relative contributions of different coalitions of players. This approach has both normative (how the total payoff should be split) and positive (how value is split) sides to it.

In this chapter I examine how the tools of cooperative game theory have been extended and adapted to network settings. I begin with a brief background discussion of cooperative game theory and then turn to the extension of various methods and concepts to networks. As we shall see, the concepts provide a useful basis for making predictions about the outcomes of multilateral bargaining on networks, as well as more generally for analyzing the power or influence of various players in a network. There are different ways to extend and adapt cooperative tools to network settings, and the "right" extension depends on the context.

Let me say a few words about the rationale behind modeling allocation rules for networks, which is related to that underlying cooperative game theory. In many applications, especially economic ones, there are resources, services, or other beneficial forms of interaction that can be used to transfer payoffs throughout a network. For example, if links represent friendships, then there are favors that can be exchanged between friends, while if links represent business alliances, then there are monetary payments that can be made. In general, the pattern of these transfers results in an ultimate payoff to each player in the network. What these transfers and resulting payoffs are depends on the structure of the network, players' positions in the network, players' productive values as a function of the network, and various aspects of the bargaining process. Allocation rules provide a prediction of how the ultimate payoffs to players in a network will depend on the details of the setting. In making such payoff predictions there are various approaches that can be taken. We could write down a noncooperative game that completely models the bargaining process through which ultimate payoffs are determined. This application is generally intractable because of the large number of players and the rather open-ended bargaining protocol in many settings. Alternatively, we can reason at a more abstract level, examining properties that we expect the bargaining process or resulting allocations to satisfy and then reasoning about the implications of those properties, which is the approach taken here. This approach has two very different perspectives. One is a positive perspective in which the properties are meant to encapsulate the essential elements of the actual processes that govern the allocation of value on a network. The other is a normative perspective in which the properties are meant to represent elements that might be appealing from a moral or ethical viewpoint, in terms of how we think the allocation of value should take place.

12.1 ▪ Cooperative Game Theory

Cooperative game theory starts with a society of players, just as a network setting does, but instead of thinking about how the players might be structured in terms of social networks, the foundation is built on simpler structures. In particular, the primitives are how the players might be grouped into subsets or coalitions.

There are several branches of the theory, and an important distinction is made between transferable utility (TU) and nontransferable utility (NTU) games. In TU games the payoffs can be freely reallocated among players (so that payoffs are transferable), while in NTU games the payoffs consist of given configurations. While both of these branches have natural cousins in the network setting, I focus on the more developed branch of TU.[1]

---

1. The terminology of cooperative and noncooperative games is perhaps no longer as useful as it was when first defined. The idea of a cooperative game began as an offshoot of a noncooperative game. The value that could be generated by a group of players is what they could guarantee themselves by coordinating their strategies in the noncooperative game. (See Luce and Raiffa [444] for an overview of these ideas.) That is, the value of a given coalition of players was derived by looking at the maximum value (in terms of the sum of their payoffs) that they could obtain across their choices of strategies when the remaining players react by collectively choosing their strategies to minimize the coalition's payoff. The idea of a value of a group of agents extends far beyond the original cooperation as defined in a noncooperative game, and so the terminology is no longer so pertinent. Also, in many applications there is a well-defined value generated by a group of agents without requiring them to cooperate, especially when the theory is applied normatively. Nevertheless, the terminology lingers because of its history.

Cooperative game theory provides a prediction (or prescription, depending on the interpretation) of how the total value generated in a society will (or should) be split among its members. It takes into account the relative values that every possible subset of players could generate, and is then based on certain properties of how the allocation of values reacts (or should react) to these values. The resulting allocation can be interpreted as a prediction of the outcome if the players bargain over how to allocate the total value generated in the society, taking into account the values of various coalitions of players. Again, as multilateral bargaining is difficult to model noncooperatively, cooperative techniques can be quite useful in this regard.

12.1.1 Transferable Utility (TU) Cooperative Games

The society of players is denoted by $N = \{1, \ldots, n\}$. The productive values of different coalitions are captured through a *characteristic function* $w : 2^N \to \mathbb{R}$, where the value of a coalition $S \subset N$ is denoted by $w(S)$. Let us normalize a characteristic function so that $w(\emptyset) = 0$. Together (N, w) are referred to as a *TU game*. The set of all such games on a society N is denoted $W(N)$. The term *transferable utility* refers to the idea that the value of a coalition can be transferred its members.

It helps to keep a few examples in mind.

Example 12.1 (Divide the Dollar) *Consider a legislature that operates by majority rule and has a budget to split among its members. Normalize the value of the budget to 1. A proposal of how to split the budget can be passed if it receives the votes of a majority of the legislature's members. Thus, any coalition of at least $n/2$ members, and only such coalitions, can generate a value of 1. This process is represented by a TU game such that $w(S) = 1$ if $|S| \geq n/2$ and $w(S) = 0$ otherwise.*

Example 12.2 (Simple Games) *The divide-the-dollar game is an example of a more general class of games such that there are "winning" coalitions that can generate a value of 1. In the divide-the-dollar setting, those coalitions contain a majority of the players. More generally, one can consider other possible rules for which coalitions can generate value.*

A *simple game* is a TU game such that

- $w(S) \in \{0, 1\}$,
- $w(S) = 1$ and $S \subset S'$ implies that $w(S') = 1$,
- $w(N) = 1$, and
- $w(S) = 1$ implies that $w(N \setminus S) = 0$.

Thus a simple game specifies which coalitions generate a value of 1, is such that larger coalitions are at least as powerful as smaller coalitions, and is such that there cannot exist disjoint coalitions that each simultaneously generate a value of 1.

Example 12.3 (A Quota Game with Veto Players) *Another interesting class of TU games is a subset of the simple games in which a specific player is needed to generate value. That is, a coalition is worthless unless it contains some specific player. Let there be a quota $q \geq n/2$ and a set of veto players $C \subset N$ such that $w(S) = 1$ if and only if $|S| \geq q$ and $C \subset S$.*

Thus a coalition generates a value of 1 if and only if it meets the size quota, and all of the veto players are included. An example of such a setting is the United Nations Security Council, which has 15 members and 5 players with the power of veto (the 5 permanent members: China, France, Russia, the United States, and the United Kingdom). A resolution passes only if it receives yes votes from at least two-thirds of its members and none of the 5 veto-wielding players votes no.[2]

12.1.2 Allocating the Value

The values of different coalitions form the foundation for the analysis of a cooperative game, as they determine the value to be split among the players and also the extent to which different groups of players are responsible for the generation of the value. The heart of the analysis is then how the value of the society is (or should be) allocated among its members. This distribution is captured by an imputation.

An *imputation* is a function $\phi : W(N) \to \mathbb{R}^n$ such that $\sum_i \phi_i(w) = w(N)$.

An imputation thus indicates how much of the value generated by the full society is allocated to each player. Generally it is presumed that the grand coalition of the full society generates the maximum possible value.[3]

An imputation can model different aspects of a game. It might capture the result of a bargaining process, or it might be a normative analysis of how the value should be allocated. It might also be a measure of the relative power of different members of the society.

A prominent imputation rule is the Shapley value, introduced by Shapley [591]. It has a number of interesting properties and can be interpreted as being based on the relative marginal contributions of players toward productive value.

12.1.3 The Shapley Value

The Shapley value is an imputation defined by

$$\phi_i^{SV}(w) = \sum_{S \subset N \setminus \{i\}} (w(S \cup \{i\}) - w(S)) \left(\frac{\#S!(n - \#S - 1)!}{n!} \right).$$

A standard interpretation of the Shapley value is as follows. Uniformly at random choose an ordering of the players and let it be $\{i_1, i_2, \ldots, i_n\}$. Consider building the

---

2. The actual rules of the Security Council are a bit more complicated than a simple game, as countries are allowed to abstain, and they sometimes do. In addition, different voting rules are used depending on the issue, but the basic structure is built on a quota game with veto players.
3. There is a specification that examines more general settings in which the value generated depends on how a society is partitioned and is such that it might be efficient to have partitions other than the one with all players grouped together. Such games are called *games in partition function form*, and can also be seen as special cases of the network setting discussed below.

society by adding one player at a time in this order. A player obtains the marginal contribution that he or she makes to the society when added to the players who preceded him or her. So a player i whose place in the order follows a coalition S receives value $w(S \cup \{i\}) - w(S)$. There are $\#S!(n - \#S - 1)!$ such orderings, and averaging over all such orderings leads to the Shapley value.

In the divide-the-dollar game the Shapley value allocates $1/n$ to each player because of the full symmetry of the game. This allocation holds for most any imputation. If instead we examine a simple majority game with a single veto player, then we see asymmetries and gain a sense of how the Shapley value operates.

Example 12.4 (A Majority Game with One Veto Player) *Consider a quota game with three players and a single veto player, player 1. In particular suppose that $w(S) = 1$ if $|S| \geq 2$ and $1 \in S$, and otherwise $w(S) = 0$. The Shapley value of this game can be calculated as follows. There are six possible orderings of the players: (1,2,3), (1,3,2), and so forth. Player 1 contributes a marginal value of 1 in four of these six possible orderings (those in which 1 is not the first player included). The other two players each contribute a marginal value of 1 in just one of the six possible orderings (that in which player 1 comes first and he or she comes second). Thus the Shapley value of this game is*

$$\phi^{SV}(w) = \left(\frac{2}{3}, \frac{1}{6}, \frac{1}{6}\right).$$

Player 1 has a larger value than the other two players because of the asymmetric roles in this game. Players 2 and 3 receive some value because they do contribute productive value in that player 1 cannot generate any value without at least one of them.

12.1.4 The Core

Instead of having an imputation, such as the Shapley value, which makes a unique prediction for each cooperative game, we might instead make a set of predictions based on some principles. The most prominent such predictions are based on the "core." The idea is that the allocation of the value of the whole society must be such that no coalition could secede and improve each member's payoff by allocating the value it alone generates for its members.

The *core* of a TU cooperative game is the set of all allocations $x \in \mathbb{R}^n$ such that

- $\sum_{i \in N} x_i = w(N)$, and
- $\sum_{i \in S} x_i \geq w(S)$ for all $S \subset N$.

When the core is nonempty, it makes powerful predictions, as an allocation in the core cannot be blocked by any coalition.

The core is sometimes empty, for instance, in the divide-the-dollar game. Consider any allocation x such that $\sum_{i \in N} x_i = w(N)$. There must exist i such that $x_i > 0$. Thus it must be that $\sum_{j \neq i} x_j < 1$, while $w(\{j : j \neq i\}) = 1$. This allocation then cannot satisfy the second requirement in the definition of the core, as the coalition of players other than i are not receiving the value that they would obtain by excluding player i. So the core of this game is empty.

This example shows the inherent instability of majority rule and also previews the difference between predictions based on an imputation rule such as the Shapley value (which in the divide-the-dollar game gives an equal allocation to each player) and the core. To see further differences, let us re-examine a game with a veto player.

Example 12.5 (The Core in a Majority Game with One Veto Player)

Reconsider the quota game from Example 12.4 with three players and a single veto player, player 1, and recall that the allocation under the Shapley value was $\phi^{SV}(w) = \left(\frac{2}{3}, \frac{1}{6}, \frac{1}{6}\right)$. *The core in this game consists of a single allocation: (1,0,0). To see this, note that a core allocation must satisfy* $x_1 + x_2 \geq 1$, $x_1 + x_3 \geq 1$, $\sum_i x_i = 1$, *and* $x_i \geq 0$ *(given that* $w(\{i\}) = 0$*) for each i.*

This example shows that the core and Shapley value are capturing different properties. The Shapley value is not necessarily in the core. The core is built on ensuring that no coalition could block the allocation with a better allocation for its members, while the Shapley value is derived from calculations of relative contributions.

The Shapley value always lies in the core in certain games, including a subclass of games called *convex games*. That class of games is such that the core is nonempty.

A TU game w is *convex* if

$$w(S \cup \{i\}) - w(S) \leq w(S' \cup \{i\}) - w(S')$$

when $S \subset S'$ and $i \notin S'$.

The convexity refers to the fact that the marginal contribution of a player (weakly) increases as the size of the coalition he or she joins is increased. If the value of a coalition depends only on the number of members it has, then the value must be a convex function of the number of players. In such games, the grand coalition generates enough value to allocate in such a way that each coalition receives at least its value, and so the core is nonempty; in fact, the core contains the Shapley value.

Proposition 12.1 *If a TU game is convex, then the Shapley value of the game is in the core of the game.*

The proof is relatively straightforward and is the subject of Exercise 12.1.

With a brief introduction to cooperative game theory in hand,[4] let us now begin to bring network structures into play.

12.2 ▪ Communication Games

Myerson [496] introduced an interesting subclass of cooperative games called *communication games*.[5] Given a network (N, g), recall that $g|_S$ is the subnetwork

---

4. For more background on cooperative game theory, see Myerson [497] and Osborne and Rubinstein [514].

5. Myerson referred to the network as a "cooperation structure," and such games are also referred to as *games with cooperation structures*.

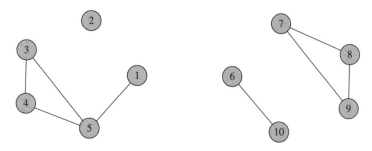

FIGURE 12.1 A communication network.

of g restricted to the nodes in $S \subset N$ and $\Pi(S, g|_S)$ is the partition of S generated by the components of g restricted to S (as defined in Section 2.1.5).

Myerson's definition begins with a convex TU cooperative game $(N, w) \in W(N)$, and augments this by a network $g \in G(N)$ that describes who can communicate with whom. The *communication game* (N, w, g) induces a cooperative game $(N, \widehat{w}_g)$ such that

$$\widehat{w}_g(S) = \sum_{C \in \Pi(S, g|_S)} w(C).$$

The idea is that coalitions can only function to the extent that they can communicate. Consider the network pictured in Figure 12.1. The coalition $\{1, 4, 5\}$ can function because the players are path-connected to one another, each lying in the same component, and thus $\widehat{w}_g(\{1, 4, 5\}) = w(\{1, 4, 5\})$. In contrast, the coalition $\{1, 3, 4\}$ can only partially function, and

$$\widehat{w}_g(\{1, 3, 4\}) = w(\{1\}) + w(\{3, 4\}) = w(\{3, 4\}).$$

In this case, even though 1, 3, and 4 are path-connected in g, player 1 cannot communicate with 3 or 4 without 5 being present. The value of coalition $\{1, 2, 6, 7\}$ is 0, since they cannot communicate at all in g.

12.2.1 The Myerson Value

The Shapley value has a natural extension to communication games.[6] Myerson [496] defined an allocation rule for a communication game (N, w, g):

$$\psi^{MV}(w, g) = \phi^{SV}(\widehat{w}_g).$$

Although one can view a communication game as a specific form of a cooperative game (basically, the induced $(N, \widehat{w}_g)$), the converse is also a reasonable viewpoint. In the case where g is the complete network, a communication game reduces to a

---

6. For more of an overview of the literature on communication games and other allocation rules, see Slikker and van den Nouweland [601].

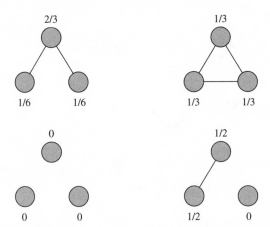

FIGURE 12.2   The Myerson value in a communication game: a three-player divide-the-dollar game.

cooperative game in that $\widehat{w}_g = w$. Moreover, as the network structure varies, one can see how the allocation of value changes. That is, as the network structure is varied, $\widehat{w}_g$ varies and so does the allocation, even though the underlying cooperative game remains fixed.

To see this, consider the divide-the-dollar game of Example 12.1 with three players. In that case the Shapley value allocates 1/3 to each player, which is also the allocation under the Myerson value if g is the complete network. However, if instead the network has only one link, then the two agents involved in the link would each receive a value of 1/2. The most interesting case is a two-link game. This game now looks like a cooperative one in which the middle player is a veto player, since without that player a coalition cannot function. Thus the middle player obtains 2/3 of the value, while the end players each receive 1/6, as pictured in Figure 12.2.

While the communication games introduced by Myerson [496] bring networks into the context of cooperative game theory, they stop short of allowing one to fully analyze the allocation of the values in a network. The difficulty arises because the actual value that a society generates is still based on a characteristic function, and it is mainly the allocation of value that is affected by the network structure, rather than the overall productive possibilities. To see the issue, consider a society of $N = \{1, 2, 3\}$ and any underlying cooperative game. A network of two links, say $g = \{12, 23\}$, and the complete network, $g' = \{12, 23, 13\}$, both lead to the same overall productive value in a communication game since they allow all three agents to communicate. While the Myerson value allocates value differently to the players in the game, it still requires that both networks have the same value to allocate. In most productive situations, including the structure of a firm or any sort of organization, links involve some costs and, more generally, network structure affects productivity. Dealing with this issue requires a richer setting in which the productive value is not based on a cooperative game but is instead directly dependent on the network in place, as discussed next.

12.3 ▪ Networks and Allocation Rules

Jackson and Wolinsky [361] proposed a richer model than that of communication games, where the value that a society generates depends explicitly and directly on the network structure. This model has as special cases both cooperative games and communication games, but uses networks as the primitive. It is described as follows. Throughout, let a society N be given.

12.3.1 Value Functions

The productive value of a society is determined directly by the network structure and is captured by a value function.

A *value function* is a function $v : G(N) \rightarrow \mathbb{R}$.

Let us normalize the value of the empty network to be 0, so that $v(\emptyset) = 0$. The set of all possible value functions for a society N is denoted by $\mathcal{V}(N)$. Note that any profile of utility functions $u = (u_1, \ldots, u_n)$ generates a value function defined by $v(g) = \sum_i u_i(g)$. Thus the utility-based models of network formation that we discussed in Chapter 6, such as the connections model, distance-based utility models, and the coauthor model, give rise to distinct value functions.

A prominent class of value functions is the set of component additive ones.

A value function v is *component additive* if $\sum_{h \in C(g)} v(h) = v(g)$.

Component additivity is a condition that rules out externalities across components but still allows externalities within components. That is, the value of a given component does not depend on how other components are structured. It is quite natural in some contexts, for instance, social interactions, but not in situations in which different components interact with one another. If the value function is derived as the sum of a component decomposable profile of utility functions, then it will clearly be component additive.

Another prominent subclass of value functions is the set of anonymous ones. Given a permutation of players π (a bijection from N to N) and any $g \in G(N)$, let $g^\pi = \{\{\pi(i), \pi(j)\} | ij \in g\}$. Thus g^π is a network that shares the same architecture as g but with the players relabeled according to π.

A value function is *anonymous* if $v(g^\pi) = v(g)$ for any permutation of the set of players π.

Anonymity implies that the value of a network depends only on the structure of the network and not on the labels of the players who occupy various positions. It requires that the critical productive determinant be social structure and abstracts away personal productive differences among individuals.

12.3.2 Allocation Rules

The analog of an imputation in the network setting is an allocation rule. It is a richer object because it not only depends on the value function but also on the network structure.

An *allocation rule* is a function $Y : G(N) \times \mathcal{V}(N) \rightarrow \mathbb{R}^n$ such that $\sum_i Y_i(g, v) = v(g)$ for all v and g.

The definition of an allocation rule has a balance condition, $\sum_i Y_i(g, v) = v(g)$, built into it. An allocation rule analyzes how the total productive value or utility of a society is allocated. In many of the previous chapters, a profile of utility functions was taken as given. However, bargaining might occur over the terms of trade or even over favors within a friendship or an allocation of chores. It could also be that there are taxes or subsidies imposed. An allocation rule keeps track of how the total value is allocated after such a process.

An allocation rule depends on both g and v, which is important, as it can then take account of player i's role in productive value beyond the specific network in place. For instance, consider a network $g = \{12, 23\}$ when the value generated is 1 ($v(g) = 1$). Player 2's allocation might depend heavily on the values of other networks. For instance, if $v(\{12, 23, 13\}) = 0 = v(\{13\})$, then 2 is essential to the network and may receive a large allocation. If, on the other hand, $v(g') = 1$ for all networks, then 2's role is not special. This information can be relevant, especially in bargaining situations, which is why the allocation rule is allowed to depend on it.

12.3.3 Some Properties of Allocation Rules

There are properties of allocation rules that are useful for studying extensions of the Shapley and Myerson values to the network setting.

An allocation rule Y is *component balanced* if $\sum_{i \in S} Y_i(g, v) = v(g|_S)$ for each component additive v, $g \in G$, and $S \in \Pi(g)$.

Component balance requires that the value of a component of a network be allocated to the members of that component when the value of the component is independent of how other components are organized. This condition tends to arise naturally. It also is a condition that an intervening government would like to respect if it wishes to avoid secession by the components of the network.

Given a permutation $\pi : N \to N$, let v^π be defined by $v^\pi(g) = v(g^{\pi^{-1}})$ for each $g \in G$ (recalling the definition of g^π from Section 12.3.1). This is just the value function obtained when agents' names are relabeled through π.

An allocation rule Y is *anonymous* if for any $v \in \mathcal{V}$, $g \in G$, and permutation of the set of players π, $Y_{\pi(i)}(g^\pi, v^\pi) = Y_i(g, v)$.

Anonymity of an allocation rule requires that if players are relabeled, then the allocation must change with the labels.

12.3.4 Egalitarian Allocation Rules

Egalitarian allocation rules spread value equitably among the members of a society. This distribution can be done in different ways. One can simply spread value completely equally among all players, or one might instead distribute the value of a component back to the members of that component.

The *egalitarian allocation rule* Y^e is defined by $Y_i^e(g, v) = v(g)/n$.

The egalitarian allocation rule has useful properties. Any efficient network is pairwise (Nash) stable, and in fact strongly stable, if payoffs are given by the egalitarian rule, since the network maximizes the payoffs of all players. Moreover, there are no improving cycles, as we can see by setting $f(g) = v(g)/n$ and applying Proposition 11.1.

Despite these virtues, the egalitarian rule fails to satisfy component balance in component additive settings in which not all components generate the same value. That is, there are many situations in which the rule makes transfers across components even though there are no externalities across the components. While such transfers might be attractive from a normative perspective, it is not natural from a positive perspective (e.g., as the prediction of the outcome of a bargaining process), especially when the value function is component additive. Moreover, egalitarian allocation can fail a basic stability property by encouraging components to secede, as they might end up being taxed as a whole. A natural variation on the egalitarian rule only equalizes allocations within components.

The *component-wise egalitarian allocation rule,* denoted by Y^{ce}, is defined as follows. For a component additive v and network g, Y^{ce} is such that for any $h \in C(g)$ and each $i \in N(h)$,

$$Y_i^{ce}(g, v) = \frac{v(h)}{\#N(h)}.$$

For a value function v that is not component additive, $Y^{ce}(g, v) = Y^e(g, v)$ for all g.

The component-wise egalitarian allocation rule differs from the egalitarian rule only for situations in which a value can be unambiguously attributed to each component separately from the rest of the network; that is, when component additive value holds. Otherwise, there is no obvious value to attribute to a component, and then this allocation rule coincides with the egalitarian rule.

While the component-wise egalitarian rule is not quite as nice as the egalitarian rule in terms of all efficient networks being pairwise Nash stable and strongly stable, it still has some useful stability properties. For instance, when payoffs are given by the component-wise egalitarian allocation rule, there always exists a pairwise (Nash) stable network, which can be found by a simple algorithm outlined by Jackson [347] for the case of a component additive v (as otherwise it is the same as the egalitarian rule):

- Find a component h that maximizes the payoff $Y_i^{ce}(h, v)$ over i and h, and if there is more than one such component, then choose the one that has the most agents.

- Follow the same policy on the remaining population $N \setminus N(h)$, and iterate.

The collection of resulting components forms the network.[7] While the algorithm identifies a pairwise Nash stable network, it does not always find a strongly stable network. It does, however, always find networks that are nearly strongly stable (see Exercise 12.7). In many contexts the component-wise egalitarian allocation rule is such that efficient networks are pairwise stable (see Exercise 12.9). Beyond egalitarian type allocation rules, there are allocation rules based on a radically different approach, namely, Shapley value–style marginal contribution calculations.

---

7. This statement follows an argument similar to one used by Banerjee, Konishi, and Sönmez [40] to establish existence of core-stable coalition structures in a class of coalition formation games called *hedonic games.*

12.3.5 The Myerson Value in Network Settings

The Shapley and Myerson values have a natural extension in the context of networks, as shown by Jackson and Wolinsky [361]. The allocation rule is expressed as:

$$Y_i^{MV}(g, v) = \sum_{S \subset N \setminus \{i\}} (v(g|_{S \cup i}) - v(g|_S)) \left(\frac{\#S!(n - \#S - 1)!}{n!} \right).$$

The Myerson value in this full network setting again allocates value using Shapley value–style calculations, now based on how the value changes as the players comprising the network are changed. The Myerson value has some nice properties that distinguish it from other rules, as I now discuss.

12.3.6 Equal Bargaining Power, Fairness, and the Myerson Value

An allocation rule satisfies *equal bargaining power* if for any component additive v and $g \in G(N)$,

$$Y_i(g, v) - Y_i(g - ij, v) = Y_j(g, v) - Y_j(g - ij, v)$$

for any link ij.

Equal bargaining power is a variation on a condition called "fairness" by Myerson [496]. Fairness can be thought of as a property from a normative perspective, while equal bargaining interprets the property from the positive viewpoint. Note that equal bargaining power does *not* require that players split the marginal value of a link. It just requires that they equally benefit or suffer from its addition. It is possible (and is generally the case) that

$$Y_i(g, v) - Y_i(g - ij, v) + Y_j(g, v) - Y_j(g - ij, v) \neq v(g) - v(g - ij),$$

so that the marginal value of a link is not solely allocated to the two involved players.

At first sight, equal bargaining power seems like a natural condition. Why should two players involved in a relationship not gain or suffer equally from the addition of that relationship to a network? (This question is answered shortly.) Equal bargaining power is a strong condition that is not satisfied by many rules, including egalitarian ones. In fact, equal bargaining power in conjunction with component balance uniquely determines the allocation rule. The following proposition from Jackson and Wolinsky [361] is a direct extension of Myerson's [496] result from the communication game setting to that of networks.

Proposition 12.2 (Myerson [496], Jackson and Wolinsky [361])
Y satisfies component balance and equal bargaining power if and only if $Y(g, v) = Y^{MV}(g, v)$ for all $g \in G$ and any component additive v.

The proof follows the logic of Myerson's proof but is adapted to a network setting. One can show that there is a unique rule that satisfies equal bargaining power and component balance when v is component additive and that the Myerson value satisfies these conditions.

Without duplicating the proof here, let me illustrate the ideas behind why these two properties uniquely determine the allocation. Start by considering a one-link component. In that case, if the link were deleted, both players would receive a 0 payoff (by component balance and the "normalization" that isolated nodes generate value 0).[8] Equal bargaining power then implies that the players receive the same allocation, and component balance requires that the two players split the entire value of a single link, and so

$$Y_i(\{ij\}, v) = Y_j(\{ij\}, v) = \frac{v(\{ij\})}{2}.$$

Now consider a two-link component $h = \{ij, jk\}$. Component balance requires that

$$Y_i(h, v) + Y_j(h, v) + Y_k(h, v) = v(h). \tag{12.1}$$

Equal bargaining power requires that

$$Y_i(h, v) - Y_i(\{jk\}, v) = Y_j(h, v) - Y_j(\{jk\}, v),$$

and by component balance $Y_i(\{jk\}, v) = 0$, and so:

$$Y_i(h, v) = Y_j(h, v) - \frac{v(\{jk\})}{2}. \tag{12.2}$$

Similarly,

$$Y_k(h, v) = Y_j(h, v) - \frac{v(\{ij\})}{2}. \tag{12.3}$$

Then from (12.1), (12.2), and (12.3) it follows that

$$3Y_j(h, v) - \frac{v(\{ij\})}{2} - \frac{v(\{jk\})}{2} = v(h),$$

or

$$Y_j(h, v) = \frac{v(h)}{3} + \frac{v(\{ij\})}{6} + \frac{v(\{jk\})}{6}.$$

Then (12.2) and (12.3) imply that

$$Y_i(h, v) = \frac{v(h)}{3} + \frac{v(\{ij\})}{6} - \frac{v(\{jk\})}{3},$$

and

$$Y_k(h, v) = \frac{v(h)}{3} - \frac{v(\{ij\})}{3} + \frac{v(\{jk\})}{6}.$$

---

8. This condition is implied by $v(\emptyset) = 0$ and component balance. We now see that this condition is more than a normalization, as it requires that all players generate the same value when isolated.

These three expressions provide the Myerson value for a component of the form $h = \{ij, jk\}$.

We see that the allocations for one- and two-link networks are unique under component balance and equal bargaining power. We can derive the two-link network from two different one-link networks, and so there are two different conditions defining the allocation. Together with component balance, we then have three conditions to determine three allocations. As we examine larger and larger components, an iterative logic can be used to determine the allocation uniquely at each step. The proof of Proposition 12.2 shows that there is at most one rule satisfying these conditions and so builds on this logic, but it does not explicitly derive the allocation as in this exposition.

There are also weighted versions of the Shapley and Myerson values, in which the bargaining power is not equal but instead involves asymmetries among the players, and some players receive systematically larger shares than others. Dutta and Mutuswami [213] extend the characterization to allow for weighted bargaining power and show that a version of a weighted Shapley (Myerson) value results.

12.3.7 Pairwise Stable Networks under the Myerson Value

A nice feature of the Myerson value is that pairwise stable networks always exist under it. In fact, the Myerson value has an ordinal potential function, which then allows us to apply Proposition 11.1 to conclude that there are no improving cycles and that pairwise stable networks exist.

Proposition 12.3 (Jackson [347]) *There exists a pairwise stable network relative to the Myerson value allocation rule Y^{MV} for every value function v. Moreover, following improving paths relative to the Myerson value and any value function and starting from any network eventually leads to a pairwise stable network, and there are no improving cycles under the Myerson value.*

Proof of Proposition 12.3. This is a corollary to Proposition 11.1. Let

$$f(g) = \sum_{S \subset N} v(g|_S) \left(\frac{(\#S - 1)!(n - \#S)!}{n!} \right).$$

Then $Y_i^{MV}(g, v) - Y_i^{MV}(g - ij, v) = f(g) - f(g - ij)$, and so f is an ordinal potential function as required in Proposition 11.1. ∎

Although the Myerson value leads to useful stability properties in terms of the existence of pairwise stable networks and the absence of cycles, it does not guarantee that the stable networks are even Pareto efficient. In fact, it can lead to systematic overconnection, as illustrated by the following example and detailed in Exercise 12.6.

Example 12.6 (Overconnection under the Myerson Value) *Figure 12.3 shows the incentives for players to overconnect under the Myerson value. In this example a dyad generates a value of 6, and any other nonempty component generates a value of 10. It is easy to check that the unique pairwise stable network is the complete network (which is verified by the improving paths pictured in the*

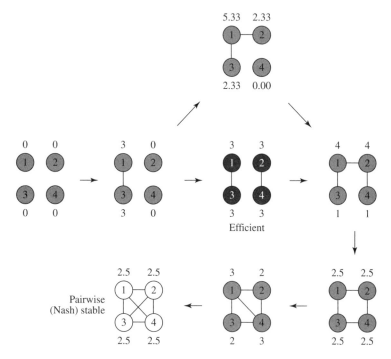

FIGURE 12.3 Overconnection and Pareto inefficiency of pairwise Nash stable networks under the Myerson value. Dyads have value 6 and other nonempty components have value 10.

figure and a few that are not pictured), but this network is Pareto dominated by the efficient network. The reason for the inefficiency is that by having more links, there are more orderings under which a given player is important in contributing to the network. Thus a player wishes to have more connections, as they lead to increased bargaining power (as reflected in the Shapley value calculations).

12.4 • Allocation Rules When Networks Are Formed

If the network can be formed at the players' discretion,[9] then one can argue that the Myerson value is not correct from either a normative standpoint or a positive one, especially if the allocation rule is partly determined by the bargaining of players during the formation process. In particular, values of all networks, and not just subnetworks, should play some role in determining the allocation, as they are all viable alternatives. The Myerson value takes into account subnetworks of a given network when calculating its value, but not other networks. For a

_____

9. Actually, if the network is fixed and cannot be altered, then it is not clear why one would pay attention to marginal contributions to the network, or why just subnetworks would be important. So the criticisms here apply more broadly.

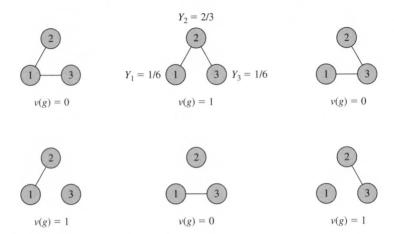

FIGURE 12.4 The Myerson value on v.

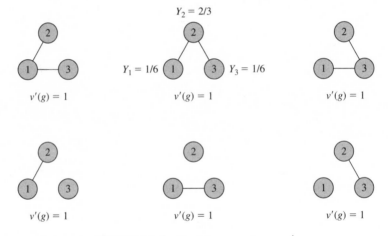

FIGURE 12.5 The Myerson value on v'.

better understanding of these shortcomings, let us examine some examples from Jackson [349].

Example 12.7 (A Criticism of the Myerson Value) *Consider a three-person society with the two different value functions pictured in Figures 12.4 and 12.5. One value function v is $v(\{12\}) = v(\{23\}) = v(\{12, 23\}) = 1$, while $v(g) = 0$ for all other networks (Figure 12.4). The other value function v' is such that all nonempty networks generate the same value of 1 as shown in Figure 12.5.*

The Myerson value assigns the same allocation to the agents in the network $g = \{1, 2, 3\}$ regardless of which of the two value functions is in place. That is,

$$Y^{MV}(\{12, 23\}, v) = Y^{MV}(\{12, 23\}, v') = \left(\frac{1}{6}, \frac{2}{3}, \frac{1}{6}\right).$$

Player 2 is rewarded for being the central player in the network. Although the network is asymmetric, under the value function v', player 2 is not special in any way.

Let us consider the two main perspectives. First, it could be that the network can be adjusted and the allocation needs to take into account the ability of agents to rearrange, or it might be that the allocation is the result of a bargaining process. From this perspective, the fact that player 2 is essential to generating value under v and is not special under v' leads to very different viewpoints that should be reflected in the bargaining, but the Myerson value does not account for this. Second, the network might be fixed, and changing the network can only be relevant from a normative perspective. If that were the case, then why should the allocation rule take into account subnetworks but not other networks? Basically, the criticism here is that the Myerson value takes into account value changes with respect to some, but not all, networks and thus does not fully account for the roles of different players in generating value. This issue manifests itself here but not in the cooperative game setting, because the cooperative game setting generally views the grand coalition as forming, and so all other possible coalitional configurations are subsets. In a network setting, the efficient (or stable) networks will generally not be fully connected, and thus the alternative networks include more than subnetworks.

A related issue can be seen when we examine the conditions that characterize the Myerson value. The next example from Jackson [349] shows some shortcomings of the equal bargaining power condition.[10]

Example 12.8 (A Criticism of Equal Bargaining Power)   *Let $v(\{12\}) = v(\{23\}) = 1$ and $v(g) = 0$ for all other networks. Thus single-link networks that include player 2 result in a value of 1, and other networks result in a value of 0. Any allocation rule, including the Myerson value, that satisfies equal bargaining power and allocates 0 to players on the empty network results in $Y_1(\{12\}, v) = Y_2(\{12\}, v)$, as in Figure 12.6.*

While there might be situations in which the value is split evenly even though player 2 is essential for generating value but neither of the other players is, requiring that player 2 receive the same allocation as the other player in a link is quite strong and should not be expected from a bargaining process. Indeed, this would be inconsistent with the outcomes seen in the exchange experiments of Cook and Emerson [172] and of Charness, Corominas-Bosch, and Frechette [146] as discussed in Section 10.3. Even from a normative point of view, it is not entirely obvious that one should require that the allocation be symmetric in the dyad.

---

10. For a discussion of the shortcomings of the component balance condition and more examples, see Jackson [349].

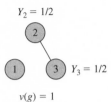

FIGURE 12.6   A critique of equal bargaining power. Player 2 is required to generate any value, while other players are not, but the allocation is necessarily the same for all players under equal bargaining power.

12.4.1 Defining Allocation Rules from Network-Formation Possibilities

To account for the outside options that players and groups of players have when constructing a network, Jackson [349] suggests the following alternative method of deriving allocation rules. From any value function over networks, define an associated cooperative game by[11]

$$w_v(S) = \max_{g \in G(S)} v(g).$$

The idea is that $w_v(S)$ captures the value of a coalition S by measuring the maximal possible value that the coalition members could generate by forming a network exclusively among themselves. This value is a measurement of the coalition's threat value, or alternatively what its members could generate for society without the help of any other agents.[12]

With this measure in hand, we can then allocate value using Shapley value calculations, now based on this auxiliary cooperative game that keeps track of the productive value of different groups of players. So if g is efficient, then set

$$Y(g, v) = \phi^{SV}(w_v), \tag{12.4}$$

which is equivalently written as

$$Y_i(g, v) = \sum_{S \subset N \setminus \{i\}} (w_v(S \cup \{i\}) - w_v(S)) \left(\frac{\#S!(n - \#S - 1)!}{n!} \right).$$

Although this quantity might appear to be similar to the Myerson value, given its Shapley value–like calculations, it is a very different allocation rule. We see the differences immediately by noting that it gives different allocations for the

---

11. There is a slight abuse of notation here, since v is defined on networks on N while g is a network on S, where S is a subset of N. This is easily translated to be a network on N for which players outside S are disconnected.

12. I say "a measurement" as, if there are externalities across components of a network, it could be that the value generated by S depends on how other players are organized. This threat point is unambiguous if v is component additive but is less clear otherwise.

two value functions given in Example 12.7. In that example it provides the same allocations under v as the Myerson value but leads to a completely egalitarian allocation under v', while the Myerson value does not show any difference when the value function changes. The rule also leads to a higher allocation for player 2 in Example 12.8, in contrast to the Myerson value. This allocation rule violates both equal bargaining power and component balance and is characterized by conditions that are violated by the Myerson value, as shown in Jackson [349].[13]

This definition of an allocation rule determines the allocation only for an efficient network. There are many ways to define the allocation on other networks. Jackson [349] suggests one possibility of simply adjusting allocations to be proportional to the allocation that would be obtained on an efficient network. That is, if some inefficient network generates $2/3$ of the value of an efficient network, then each player would receive $2/3$ of the allocation that he or she would obtain under an efficient network. There are other ways of adjusting allocations as well.[14]

12.4.2 The Core in Network Settings

Once we view the network as a flexible or changeable entity and visualize the associated cooperative game, then we can make more use of the cooperative game-theoretic toolbox. For example, there is a natural definition of the core for networks.

A network-allocation pair $g \in G(N)$ and $y \in \mathbb{R}^n$ is in the *core* relative to (N, v) if

- $\sum_i y_i \leq v(g)$, and
- $\sum_{i \in S} y_i \geq w_v(S) = \max_{g \in G(S)} v(g)$ for all $S \subset N$.

The core includes the specification of both a network and an allocation of its value. The requirement is that no coalition could deviate, form a network on its own, and generate a higher value than what they are being allocated in the initial network.[15] So, analogously to its role in cooperative game theory, the core concept captures allocations that are stable to deviations from various groups; it can lead to allocations that differ from those derived from Shapley value calculations.

An allocation rule Y is *core consistent* if for any v such that the core is nonempty, there exists at least one g such that $(g, Y(g, v))$ is in the core.

While the Myerson value is not always in the core and thus is not core consistent, there are allocation rules that are core consistent. The nucleolus (of Schmeidler [581]) is an imputation defined on cooperative games that is core consistent relative

---

13. Jackson called this rule the *player-based flexible network allocation rule*. He also examined link-based variations on such a rule. For such variations, one allocates a value to links based on their importance in providing value, and then players obtain value from their links. Such an idea has roots in communication games as studied by Meessen [463] and Borm, Owen, and Tijs [97].

14. See Navarro [500] for further discussion and other allocation rules.

15. Again, this definition makes the most sense for component additive value functions, as otherwise the threat value generated by S in forming its network could depend on how the other players are organized. This issue has resulted in various core definitions in cooperative settings and appears here as well.

to cooperative games. There is a natural analog of the nucleolus for networks, termed the *networkolus* by Jackson [349], that is core consistent.

Let $B(g, v) = \{y \in \mathbb{R}^n | \sum_i y_i = v(g)\}$ be the balanced allocations for g under v. Let $e_S(y) = \sum_{i \in S} y_i - w_v(S)$ be the excess allocated to coalition S at an allocation y relative to their threat value under v, and let $e(y)$ denote the vector with entries indexed by a list of the nonempty Ss, $S \subset N$. Given an efficient g, let $Y^N(g, v) = y$ be the unique allocation such that $e(y)$ leximin dominates $e(y')$ for all $y' \in B(g, v)$.[16]

The networkolus examines how much various coalitions receive relative to their threat values. If the core is nonempty, so that there is some allocation that gives each coalition at least its threat value, then the networkolus equilibrates (to the extent possible) the excess value given to each coalition. More generally, even when the core is empty, it provides an allocation, and although in that case some of the excesses are negative, the networkolus still minimizes the amount by which any coalition falls below its value.

12.5 ■ Concluding Remarks

As we have seen in this chapter, tools from cooperative game theory can be adapted to provide insight into how the value of a network might be allocated among the players in a society and how this allocation depends on the network in place and the values generated by alternative networks. There are different perspectives that one might take, making this either a question of how value should be allocated or of how an allocation results from some process. Although we can adapt concepts from cooperative game theory to networks, there are issues that arise in the network setting that lead to new questions regarding how value should be allocated. These questions about allocation are still a largely unexplored topic.

12.6 ■ Exercises

12.1 *Convex TU Games*   Prove Proposition 12.1.

12.2 *A Convex Three-Player TU Game*   Consider a three-player TU game in which $w(\{1, 2, 3\}) = 2$ and $w(\{2, 3\}) = 1$, while $w(S) = 0$ for all other S. Find the core allocations and Shapley value of the game.

12.3 *The Core in an Exchange Network*   Consider the society described in the left-hand network in Figure 10.7, where singletons are worthless, a coalition of any two B players is worth 8, a coalition of A with a B is worth 24, a coalition of three players is worth the same as the maximal value across its subsets of size 2,

---

16. A vector e *leximin dominates* a vector e' if there is some scalar x such that for any $x' < x$, e and e' have the same number of entries with value x', while e has fewer entries with value x. The calculation of the networkolus can be a difficult task without some underlying structure on v, but it can be shown that it is well defined through a straightforward extension of results on the nucleolus.

and the grand coalition is worth 32. Show that the unique core allocation is 4 for each B player and 20 for the A player. Find the Shapley value for this game. Show that it differs from the core allocation.

12.4 *Additivity of the Myerson Value*   Let a value function v be such that $v(g) = v_1(g) + v_2(g)$ for two other value functions v_1 and v_2 and every $g \in G(N)$. Show that

$$Y^{MV}(v, g) = Y^{MV}(v_1, g) + Y^{MV}(v_2, g).$$

12.5 *The Myerson Value in the Symmetric Connections Model*
   (a) Consider a star network comprising all players in the symmetric connections model. Find the Myerson value allocation.
   (b) Consider a three-player society, and show that there exists a range of δ and c such that a star is efficient but only the complete network is pairwise stable under the Myerson value allocation rule.

12.6 *Overconnection under the Myerson Value*   Consider a value function v such that $v(g) = b(g) - c \sum_i d_i(g)$, where b represents benefits and $c > 0$ is a cost of maintaining a link. The benefit $b(g)$ is *monotone* if
   • $b(g') \leq b(g)$ if $g' \subset g$, and
   • $b(\{ij\}) > 0$ for any ij.

The following is a special case of a result from Jackson [347].

Proposition 12.4   *Let $n \geq 4$, and consider an anonymous and monotone benefit function b for which there is some efficient network g^* relative to b that is symmetric and is not the complete network. There exists $\bar{c} > 0$ such that for any $c < \bar{c}$, any pairwise stable network relative to Y^{MV} and the value function $v(g) = b(g) - c \sum_i d_i(g)$ is Pareto dominated by some subnetwork.*

   Prove Proposition 12.4.
   Hint: First show that if $ij \notin g$, then $Y_i(g + ij, b) - Y_i(g, b) \geq b(\{ij\}) > 0$, and conclude that the complete network is the unique pairwise stable network under b. Then apply the additivity of the Myerson value from Exercise 12.4 and work with small costs.

12.7 *Possible Nonexistence of Strongly Stable Networks under the Component-Wise Egalitarian Rule*   Show that there exists a component additive value function for which there is no strongly stable network under the component-wise egalitarian rule. Show that if we weaken strong stability to say that a network is stable if there is no deviation by a coalition of agents that is *strictly* improving for *all* members of the deviating coalition, then there exists a strongly stable network under the component-wise egalitarian rule for any v.

12.8 *The Existence of Strongly Stable Networks under the Component-Wise Egalitarian Allocation Rule**   The following condition and result are from Jackson and van den Nouweland [356].

A value function is *top-convex* if

$$\max_{g \in G(N)} \frac{v(g)}{|N|} \geq \max_{g \in G(S)} \frac{v(g)}{|S|}$$

for all $S \subset N$.

(a) Show that if the value function is component additive, then under this condition, the per capita value of each component of an efficient network is equal and is at least as high as the per capita value of any component of any network.

(b) Suppose that payoffs are governed by the component-wise egalitarian allocation rule, and consider an anonymous and component additive value function. Show that the set of strongly stable networks is nonempty if and only if the value function is top convex. Moreover, show that in this case, the strongly stable networks are the efficient ones.

12.9 *The Component-Wise Egalitarian Allocation Rule and Bridges*   A pair of a network and a component additive value function, (g, v), is *bridge-monotonic* if

$$v(C_i(g))/\#C_i(g) \geq \max \left[v(g^1)/\#N(g^1), v(g^2)/\#N(g^2) \right]$$

for every bridge ij in g such that $v(g) \geq v(g - ij)$, where $\#C_i(g)$ is the number of players in i's component of g, and g^1 and g^2 are the components of g bridged by ij.

Prove the following proposition.

Proposition 12.5 (Jackson and Wolinsky [361])   *If g is efficient relative to a component additive v, then g is pairwise stable for Y^{ce} relative to v if and only if (g, v) is bridge-monotonic.*

12.10 *The Shapley Value in the Connections Model*   Consider the symmetric connections model with parameters such that a star is the unique efficient network structure. Let v be defined by $v(g) = \sum_i u_i(g)$, where u_i is from the symmetric connections model. Compare the Myerson value allocation for a star to the Shapley value of w_v, as defined in (12.4), for a star.

12.11 *Anonymity and the Shapley Value in Network Settings*   Let v be an anonymous value function. Show that for any efficient network, the Shapley value of w_v, as defined in (12.4), results in the same allocation as the egalitarian allocation rule from Section 12.3.4.

12.12 *The Monotonic Cover of a Value Function*   Given a value function v, Jackson [349] defines its *monotonic cover* $\widehat{v}$ by

$$\widehat{v}(g) = \max_{g' \subset g} v(g').$$

Consider an efficient g and a component additive v. Define an allocation rule Y so that the allocation at g, v is described by

$$Y_i(g, v) = \sum_{S \subset N \setminus \{i\}} (\widehat{v}(g^{S \cup i}) - \widehat{v}(g^S)) \left(\frac{\#S!(n - \#S - 1)!}{n!} \right),$$

where g^S is the complete network on the nodes S (viewed as network on N). Show that this is the same allocation as the Shapley value of w_v, as defined in (12.4).

12.13 *The Core in Example 12.7* Determine the core networks and allocations (defined in Section 12.4.2) under the value functions in Example 12.7.

12.14 *The Networkolus and the Core* Consider a value function v for a three-player society such that $v(\{12\}) = v(\{23\}) = 1$, $v(\{12, 23\}) = w$, and $v(g) = 0$ for all other networks, where $w > 0$.

Find the Myerson value, the player-based flexible network allocation, the networkolus, and all of the core allocations.

CHAPTER 13

Observing and Measuring Social Interaction

To this point we have been focused on what is known about social and economic networks, as well as how to model and analyze social networks and their impact. I close with a chapter about measurement of social networks and inference from them. This discussion is not meant to be a primer on empirical work, but rather to provide background on some important issues that are particularly acute when doing empirical analyses involving social networks. Understanding these issues is important when interpreting and evaluating empirical and experimental studies of social and economic networks.

One challenge in working with social network data is that social structure is generally endogenous and related to multiple characteristics of the agents involved. People who have neighbors share many characteristics with them, as we saw in our discussion of homophily in Section 3.2.6, and they might be choosing their neighbors not just with those characteristics in mind but also using feedback from the outcomes or behaviors that we might be interested in studying. If we want to establish that a certain behavior is influenced by social network structure, then we have to properly account for numerous other related characteristics that may drive the behavior and for any feedback between the social structure, behavior, and the background characteristics. A model has to be properly specified so that different characteristics are "identifiable" and distinguishable given the data. In addition, there is also a related question of whether that social structure causes behavior or the behavior induces the structure observed. I discuss how these difficulties manifest themselves, as well as some techniques for overcoming them.

I also discuss how we can use the data to discover the latent social structures that underlie many social networks. This is not so much a challenge of social network analysis but rather an aspect of the analysis of social structure that involves methods specific to network settings. These methods include (stochastic) block modeling, identifying community structures, and latent space estimation. The idea is that the nodes of a network belong to groups or organizations that are not directly observed, or are part of an unobserved spatial structure. These unobserved social structures determine social networks and are interesting to understand in their own right. There are various techniques that have been developed for uncovering latent social structures based on social network data, and I provide an overview of some of the techniques and the ideas behind them.

13.1 ▪ Specification and Identification

A basic challenge arises in isolating the impact that social networks have on various behaviors and is related to a more general quest to identify peer effects. Social relationships come about for a variety of reasons, and social neighbors might display similar behaviors because they are influenced by common traits or experiences. It can be difficult to sort out whether individuals are behaving in a certain way because of the influence of their neighbors, the influences common to them and their neighbors, or other factors related to their network positions.

13.1.1 Specification and Omitted Variables

To clarify these issues, let us reconsider the study of Coleman, Katz, and Menzel [165] discussed in Section 3.2.10. Van den Bulte and Lilien [634] criticize the study, claiming that it did not properly account for the effects that marketing and advertising efforts by various pharmaceutical firms had on the diffusion process. To the extent that the direct effects of marketing or advertising, or other market forces, are correlated with the degree that Coleman, Katz, and Menzel measured, then these factors could be responsible for the diffusion, and the degree appears to drive diffusion because it is correlated with these other factors. Further studies (e.g., see Bhatia, Manchanda, and Nair [61]) have found evidence of network effects after controlling for marketing and other characteristics. Nevertheless, the point that all relevant factors must be included in the analysis is an important one.

As a simple example, suppose that degree in a network is correlated with age, so that older people have more connections. Now suppose that we hypothesize that people with higher degree behave differently from people with lower degree. So we estimate an equation of the form

$$Y_i = \alpha_d + \beta_d d_i + \varepsilon_i, \tag{13.1}$$

where Y_i is a measure of the behavior of interest, d_i is individual i's degree, and ε_i is an error term that captures unobserved idiosyncratic factors. However, suppose, for instance, that degree varies with age, so that

$$d_i = \delta + \gamma \, age_i + \eta_i, \tag{13.2}$$

where α_d is a base degree and η_i is an idiosyncratic term. If the true relationship is in fact of the form,

$$Y_i = \alpha_a + \beta_a \, age_i + v_i,$$

then when we fit (13.1), given (13.2), the estimate for the coefficient on degree, β_d, looks significant but is really just a proxy for the effect of age, which is filtered through degree.

Suppose instead that the true relationship is of the form

$$Y_i = \alpha_d + \beta_d d_i + \beta_a \, \text{age}_i + \varepsilon_i,$$

so that both degree and age affect the variable in question. For example, suppose that we are examining a model of social capital, and so Y_i is some measure of wealth or power accumulation, and we are testing for how this measure is influenced by a measure of social connectedness. If we then fit (13.1), including degree but omitting age, then given that degree and age are related according to (13.2), we will have a biased estimate of β_d, as it proxies for some of the effect of age. For instance, with positive relationships between each of the variables, by omitting age we would end up with an inflated estimate of how degree affects wealth.

While such problems of specification and omitted variables are not unique to the analysis of social networks, they are particularly acute in this field because of homophily. Recalling the discussion from Section 3.2.6, we know that people tend to associate with others who are similar to them in a wide variety of ways. So if the behavior of individuals tends to match patterns of the social network, we cannot attribute that behavior directly to network influence unless all other relevant factors have been taken into account.

Instrumental Variables It is generally difficult to identify all factors that might covary with both social relationships and behavior, and the lurking issue remains of omitted variables and other sorts of misspecifications when estimating the impact of social structure on behavior. The omission or misspecification can cause the variables whose effects are being estimated to be correlated with the error term, which may result in bias in the estimation (so that the estimator is inconsistent and even large data sets cannot overcome the problem).

A standard approach to dealing with such problems is to work with instrumental variables. To develop an intuitive understanding of instrumental variable methods, consider what goes wrong in the case of an omitted variable. Ideally, to estimate the effect of degree on wealth, we want a set of observations in which degree is varied and other variables are held constant to isolate the marginal effect of degree on wealth. A controlled experiment would be designed to yield such observations. The difficulty is that we do not have such control, and when the degree is high (low), the omitted variable of age also tends to be high (low). An instrumental variable should be correlated with degree but not with the error term—in this case, it is uncorrelated with the omitted variable of age. If we can find a variable that covaries with degree but is uncorrelated with the error term (and in this case with the omitted variables), then it can serve as a substitute for the controlled experiment that we would have liked to design. For example, we might find a factor that leads to changes in degree, such as a club that opens branches in some communities but not others (for exogenous reasons), but does so in a way that we expect to be uncorrelated with such factors as age that might lead to correlation with the error or omitted variables. This variable becomes the instrument and effectively causes variations in degree that are independent of other possible explanatory factors.

Estimates can then be derived based on the instruments using standard techniques developed for this purpose (or we can use a two-stage least squares estimation). That is, we first regress degree on the instrumental variable to develop estimated values of degree based on the instrument. Then in the place of degree in the original regression, we use these estimated values of degree, which are then independent of the error terms, as they are conditional on the instrument. In this way we obtain a consistent estimator.

Endogeneity   Another problem that arises in working with estimation based on social structure is that social structure often depends on the factors it affects. For example, if estimating the impact of social connectedness on wealth, it could also be that wealth affects connectedness (e.g., by providing additional access or opportunities).[1] While this problem is distinct from that of specification and omitted variables discussed above, it could also result in degree being correlated with the error term, causing biased and inconsistent estimations.

Instrumental variables can also be helpful in sorting out endogeneity issues. Again, we want an instrumental variable that is related to degree but not to the error terms. Thus it must be a variable that avoids being endogenous but is related to degree, so that variations in degree related to instrument are exogenous. For example, if a government randomly chooses some villages to receive subsidized communications equipment (in the form of, e.g., internet or telephone), which might enhance degree in some villages and not in others, then we could examine how these exogenous changes in degree influence wealth. More specifically, we could measure the induced degree differences attributed to the government program and then see the extent to which the program influences wealth, using two-stage least squares or other methods of working with instrumental variables.

While instrumental variables are a useful tool for dealing with the problems of omitted variable and endogeneity, they are not a panacea. First, finding good instruments—those related to the problematic variable(s) but are not endogenous or related to omitted variables and error terms—can be problematic. In fact, we can never really be sure that the instrument is ideal, as we never directly observe the true error terms. Second, even if we are reasonably confident in a chosen instrument, it may well not have a sufficiently strong relationship with the problematic variables to produce powerful estimates.

The example of a government program that randomly selects villages for improvements that could lead to changes in social capital is a form of *natural experiment*. That is, it sets up a control that provides just the sort of variation that is useful, so that we see what happens with and without an exogenously determined increase in social capital.[2] Often ideal instrumental variables are related to such exogenous variations. We saw examples of such natural experiments in Section 10.1 on the empirical analyses of how network connections shape labor market outcomes. That section discussed various exogenous variations in social structure, such as immigration due to drought in the home country, random assignment to military units, or assignment to a city by an agency for exogenous reasons.

---

1. See Durlauf [208] for more discussion of endogeneity issues associated with studying social capital.

2. For an interesting example of an analysis of social capital based on the presence of television (which depends on some exogenous programs and geography) in rural Indonesian villages, see Olken [513].

13.1.2 The Reflection Problem and Identification

Another problem in specifying a model of social influence on behavior has to do with making sure that the specified relationship is properly identified, meaning that the parameters of the model are uniquely determined by a data set. To caricature the identification problem in social influence: behavior is determined by behavior, and so there is a circularity of cause and effect, and it can be difficult to sort out the social influences of behavior on behavior. A well-known illustration of this circularity is due to Manski [448].

Manski [448], [449] describes an identification problem when examining social influences on individual behavior and provides a useful paradigm for understanding identification problems in such settings. To see the identification problem that Manski [448] refers to as the *reflection problem,* let us consider it in its starkest form. Consider a situation in which the behavior of an agent is a linear function of the average level of the behavior taken by other members of his or her cohort. To be specific, suppose that an agent i has some characteristics x_i and that the agent's cohort is other individuals who have the same attributes. The behavior of i is determined by what she expects her peers to do. Let Y_i denote i's behavior. Her expectation of what people with the same attributes x_i will do is simply $E[Y_i|x_i]$ and so the relationship is

$$Y_i = a + bE[Y_i|x_i] + \varepsilon_i, \tag{13.3}$$

where ε_i is a 0-mean random variable (conditional on x_i) that is a noise term capturing some unmeasured idiosyncracies. What happens if we try to estimate this relationship? Taking the expectation of both sides of (13.3) conditional on i's attributes x_i yields

$$E[Y_i|x_i] = a + bE[Y_i|x_i].$$

This equation has a unique solution (provided $E[Y_i|x_i]$ is nonzero), which is $a = 0$ and $b = 1$. Thus fitting (13.3) results in a tautological relationship:

$$Y_i = E[Y_i|x_i] + \varepsilon_i,$$

which simply says that Y_i is its expectation plus noise and tells us nothing about endogenous social interaction. This is what Manski [448] refers to as the reflection problem: i's behavior is a function of the expectation of peer behavior, which is just the expectation of i's behavior, which reflects i's behavior. We have not specified a system that effectively defines the relationship, and so it is not properly identified.[3]

Even if the specification is enriched to allow Y_i to also depend on i's attributes x_i directly, we still have an identification problem. That is, suppose we specify that

$$Y_i = a + b_1 E[Y_i|x_i] + b_2 x_i + \varepsilon_i. \tag{13.4}$$

---

3. The identification problem is different from a multiple equilibrium problem, which can also result in settings with interdependencies in behaviors. As we have seen in Section 9.3, graphical games with strategic complementarities can have multiple equilibria. The multiple equilibrium problem has its own challenges.

Again, taking expectations,

$$E[Y_i|x_i] = a + b_1 E[Y_i|x_i] + b_2 x_i.$$

If $b_1 \neq 1$ (to avoid having this parameter predetermined and tautological), then we can rewrite this equation as

$$E[Y_i|x_i] = \frac{a}{1-b_1} + \frac{b_2}{1-b_1} x_i.$$

Substituting this relation into (13.4) yields

$$Y_i = \frac{a}{1-b_1} + \frac{b_2}{1-b_1} x_i + \varepsilon_i. \tag{13.5}$$

Presuming that x_i is not constant, then we can estimate the two composite parameters $a/(1-b_1)$ and $b_2/(1-b_1)$. However, (13.5) does not uniquely identify the parameters a, b_1, and b_2, as there are many different values of these that lead to the same composites. We could set b_1 to any value (other than 1) and find values of a and b_2 that are consistent with the composite parameters. Adding in extra variables does not help: even though there are more variables, there are also more parameters to identify, as discussed in Exercise 13.1.

As Manski points out, one remedy is to use instrumental variables to sort out the identification, since part of the difficulty stems from the endogeneity of the behavior, which enters on both sides of the equations in the model. There are a variety of other ways around identification problems in social network settings. Let us examine several of them.

Social Structure and Identification   Note that the reflection problem stems from the fact that i's peers are not identified directly but are just assumed to be similar to i. We are ignoring any real social structure and information that might be available about i's neighbors. If we explicitly track i's neighbors, then that information can be used to identify a model.[4] To see how this approach works, let a possibly weighted and directed network matrix g govern interactions. To make the technique transparent, ignore constant terms as well as any node-specific characteristics, so that we are working directly with the interaction of behaviors, and let us stay with a linear relationship. In particular, suppose that

$$Y_i = b \sum_j g_{ij} Y_j + \varepsilon_i.$$

Thus each individual's behavior is a weighted average of his or her neighbors' behavior. If the g_{ij}s are not degenerate,[5] so that $(\mathbb{I} - bg)$ is invertible, then

$$Y = (\mathbb{I} - bg)^{-1} \varepsilon,$$

---

4. The techniques here are common to the spatial econometrics literature, as outlined in Anselin [18]. The specifics of the derivation that follows in this section are due to Marcel Fafchamps, who showed me how this approach adapts to network settings and pointed me to the related references.
5. This is relative to b, but generically in g the matrix will be invertible for any given b.

where Y and ε are the corresponding vectors, and $\mathbb{I}$ is the identity matrix. Letting ^{T} denote transpose, it follows that

$$E\left[YY^T\right] = (\mathbb{I} - bg)^{-1} E\left[\varepsilon\varepsilon^T\right] \left((\mathbb{I} - bg)^T\right)^{-1}. \qquad (13.6)$$

If we have knowledge of the social network matrix g and of the covariance matrix of the error terms $E\left[\varepsilon\varepsilon^T\right]$ (e.g., using a standard assumption that errors are independently and identically distributed with a finite variance), then b is determined by (13.6).

This technique relies on some knowledge of the error terms. This can be a problem, especially if omitted variables influence behavior; for instance, the technique could result in positive correlation in the errors. But this shortcoming does not preclude identification, as the above specification does not take advantage of other factors that influence behavior.

In particular, in the reflection problem, the behavior of i's peers was not identified other than through i's characteristics. If we have explicit information about i's peers, and they differ from i in characteristics, and the social relationships are not entirely symmetric, then using this information can overcome the reflection problem. As an extreme example suppose that there are different types of agents, say, young and old. Suppose that older agents do not pay attention to the younger agents, but the younger agents heed the older ones. In this case, a straightforward regression can estimate older agents' behavior, which can then be used to estimate the younger agents' behavioral relationship. While this example is extreme there are agents uninfluenced by social interaction, more generally some asymmetries on social influence are enough to provide identification. The example of the reflection problem was an extreme case of circularity, and so variation in background characteristics would not be helpful in identifying the relative behaviors and their mutual effects.

Nonlinearities in Social Interaction   The reflection problem can be overcome with richer information about interactions, but note that even without such information, it derives from the linear-in-means specification.[6] To see how linearity affects the problem, let us modify the specification in (13.3) so that instead of having an agent's behavior be proportional to the mean of his or her peers let i's behavior be influenced by the maximal action by his or her peers:

$$Y_i = a + b_1 E[\max_{j \in N_i} Y_j | x_i] + b_2 x_i + \varepsilon_i, \qquad (13.7)$$

where N_i are the neighbors of i, whose behavior might still be estimated based on x_i or observed more directly as a social network. Now, if we take the expectation of each side of (13.7), we no longer have a tautology but instead have

$$E[Y_i | x_i] = a + b_1 E[\max_{j \in N_i} Y_j | x_i] + b_2 x_i,$$

---

6. Manski [448] does consider a nonparametric version of the reflection problem, but it still requires a specific formulation to generate the circularity.

which is identified as long as $E[\max_{j \in N_i} Y_j | x_i]$ can be deduced in such a way that it and x_i are not constant or linearly dependent. Note that it is not essential that the specification be based on the maximum. There are a wide variety of specifications that avoid identification problems (see Brock and Durlauf [110] for more discussion and specifications). In fact, identification problems only arise for very particular specifications, such as the simple linear/average behavior specification.[7] Such problems emphasize the importance of building a model that properly captures the social interaction structure and incentives that are at the heart of the particular application. Chapters 7 and 9 provide a basis for modeling social interaction in ways that are generally nonlinear and identifiable.

Timing Another aspect of specification is timing. In the reflection problem, part of the difficulty stems from the contemporaneous interaction of the decisions. If decisions are repeatedly taken over time, then an agent's decision may depend on the observations of the *past* decisions of his or her peers. This dependence can help identify the relationship and sort out causation. One can see whether the current decisions of a given individual vary with the past decisions of his or her peers. For example, Conley and Udry [170] examine the timing of past neighbors' successes in experimenting with fertilizer in pineapple production in Ghana to test for social learning. But even taking advantage of the timing of such decisions may not solve the problem, as there can still be omitted variables that determine adoption. However, careful attention to the timing of behavior allows Conley and Udry to check whether increases in use by one agent are triggered by the past successes of that agent's neighbor(s), after properly sorting out other potential causes.

It is important to note that taking advantage of timing has been used not only to uncover whether an agent's behavior is influenced by his or her peers but also to examine the endogeneity of social structure. For example, does an agent act in a certain way because his or her friends do, does he or she select friends that act in similar ways? Such endogeneity questions are fundamental to social network analysis, especially when examining peer effects, and timing can be used to help sort out these effects. For example, Kandel [376] uses time series data on high school friendships—together with data on individual drug use, delinquency, political opinions, and educational aspirations—to examine whether friendships tend to form among agents with similar behaviors, or whether agents' behaviors are influenced by that of their friends.[8] By examining the formation of new friendships and the deterioration of old ones together with behavior over time, Kandel is able to identify whether either or both of these effects are present. The timing helps her show that both effects are present. Agents are relatively more likely to form new friendships with others whose past behavior matches their own, and to sever past friendships with others whose past behavior differs from theirs. In addition, when agents' behaviors change, it tends to be influenced by the past

---

7. One does need to be careful however. Nonlinear models can also face identification problems, as discussed in Exercise 13.2.
8. For alternative methods of studying the co-evolution of networks and behavior, see Snijders, Steglich, and Schweinberger [609].

behavior of the agents who remain their friends. Moreover, the magnitude of the effects differs across behaviors. Such an analysis is demanding in terms of the richness of the data that is necessary, especially if one also wants to control for other variables that might be adjusting over time and covarying with behavior and/or social structure. Nevertheless, longitudinal (time series) data are a powerful tool for sorting out both social influence and endogeneity.

13.1.3 Laboratory and Field Experiments

As we have seen in the previous sections, many of the challenges in empirical analysis of social structure and social interaction consist of separating out effects to determine which factors and behaviors are related and which might be causal. Various experiments on behavior, where the particular values of some variables are carefully controlled and varied, allow one to track which conditions have been altered and thus uncover their impact. Given the power of experiments, they are a rapidly emerging tool of network analysis.[9]

Experiments can be used in various ways. First, an experimenter can examine how social structure affects behavior, and how an agent's position within a network influences his or her behavior. For example, one can put agents in specific network contexts, such as in an exchange or bargaining network in the experiments of Cook and Emerson [172] and Charness, Corominas-Bosch, and Frechette [146]. Then comparing behavior across network position and tracking changes in behavior with alterations in overall network structure allow the experimenter to explicitly track the impact of social network structure on behavior. The only variable that is altered is the social structure, which helps show whether agents behave differently when embedded in different positions within a network or in different networks. One can also examine how well agents are able to communicate and learn in social contexts, and how that ability depends on the network structure (e.g., see Bavelas [47]; Leavitt [427]; and Choi, Gale, and Kariv [151]), or how well agents are able to coordinate their actions when embedded in a social network (e.g., see Kearns, Suri, and Montfort [386]). One can also investigate whether agents behave differently when the network is exogenously imposed or chosen by them (see Corbae and Duffy [176] and Riedl and Ule [562]). The control present in the experimental setting has proven to be a useful tool for identifying network effects.

Second, one can test alternative theories of behavior in network contexts. Here it is not necessary to use variations in treatments within an experiment; instead one can simply find a setting where the predictions of different theories lead to different outcomes. For example, one can test models of network formation, such as the myopic one based on pairwise stability, against a farsighted notion, as in Pantz and Ziegelmeyer [522]. The payoffs to network formation can be fixed but are chosen by the experimenter so that different networks should form, depending on whether agents are farsighted or myopic in their decisions to form links. Alternatively, the experimenter need not run a horse race of different theories against one another but can simply test whether a given theory's predictions hold. For example, one

---

9. For a survey of parts of the literature, see Kosfeld [410].

can examine variations on undirected or directed connections model and then see whether pairwise stable or Nash stable networks form (e.g., see Callander and Plott [122]; Falk and Kosfeld [241]; Deck and Johnson [187]; and Goeree, Riedl, and Ule [292]).

The first approach is useful because it allows one to adjust some aspect of the social setting and then isolate its effect on outcomes. The second approach requires only a single treatment, but one that distinguishes between different theories or hypothesized behaviors. The power of the experiment comes from being able to impose a certain structure and ensuring that the network interactions or relevant payoffs are observable. These characteristics can be difficult to observe from field data.

In addition to laboratory experiments, field experiments can be useful tools as well. One loses some of the control of a laboratory environment, but gains access to a social network in its natural setting. Milgram's [468] small-world experiments are examples of this technique. In such experiments the objective was to discover characteristics of a real social network (and how people navigate it), and the controlled aspect was to give subjects a particular task to perform in a way that elicits information about network structure, such as the distance between two agents. Such field experiments can be useful not only for discovering social network structure but also for seeing how it influences behavior. For instance, Goeree et al. [293] examine how players behave in a dictator game (where a player has a choice of the amount of money to keep and the amount to give to another player) as a function of the social network distance between the players. One can also sort through various theories of social capital, altruism, and reciprocity by examining specific behaviors as functions of the social network (e.g., see Leider et al. [429]). In such field experiments the control of a task, game, information seeding, and so forth, placed in the context of a (measured) real social network, helps uncover how behavior is related to social structure.

13.2 ▪ Community Structures, Block Models, and Latent Spaces

An aspect of empirical investigation that is special to social network analysis is uncovering the latent social structure of a network that led to the network's formation. Such structure is often not fully observable and so needs to be constructed from what is observed. Uncovering such social structure can be useful for a variety of reasons. For example, coupling such implicit structure with other attributes can help to determine whether there are specific biases in a society, such as in hiring or publishing. It can also help in classifying and categorizing political and other ideologies, as well as economic patterns of behavior.

The models for network formation that we have seen in earlier chapters, when fitted to network data, can provide some insight into underlying structure, especially in terms of fitting some of the strategic models to data. But there are also algorithmic and statistical methods specifically designed to discover underlying social structures or groupings. Let us examine these methods.

13.2.1   Communities and Blocks

There are various ways in which the structure underlying a social network can be modeled. One basic and standard way is to presume that the nodes of the network belong to different blocks or communities. The network that emerges depends on the underlying blocks or communities, which in turn can be recovered by examining the network. The early literature provided some definitions that can be used to capture underlying structures, and the subsequent literature has provided a number of algorithms for recovering them.

One notion of a block or community is that it consists of nodes that are somehow comparable or equivalent. An early concept of this sort is structural equivalence: two nodes are said to be structurally equivalent if their relationships to all other nodes are identical. This concept was first discussed by Lorrain and White [441] and further by White, Boorman, and Breiger [663]. In particular, two nodes i and j are *structurally equivalent* relative to a network g if $g_{ik} = g_{jk}$ for all $k \neq i, k \neq j$ (and the same for g_{ki} and g_{kj}). This equivalence notion defines equivalence classes of nodes, so that we can partition the set of nodes into sets of equivalent nodes.

It is clearly rare to find networks in which numerous pairs of nodes are structurally equivalent, as many factors affect network formation, resulting in substantial noise both in actual and observed relationships. In view of this problem, the literature has moved beyond such a strict definition to develop methods of grouping nodes into equivalence classes and defining how nodes relate to one another, and also to develop methods of uncovering basic blocks that underlie the society but are not directly observed.[10] To discuss such methods, I begin with a formal definition of such a grouping of nodes into equivalence classes, called a *community structure*.

A *community structure* is a partition of the set of nodes N, Π.[11] Thus it groups the set of nodes into separate communities.

For example, consider the nodes in the network pictured in Figure 13.1. One community structure from the network in Figure 13.1 is pictured in Figure 13.2, where the ovals group the nodes considered to be in the same community.

There are many different ways in which the nodes of any network might be partitioned, and which community structure seems most appropriate can depend on the context and on what we imagine a community to represent. Let us examine some of the more prominent methods for identifying community structures.

13.2.2   Methods for Identifying Community Structures

Just as we saw in Chapter 2, a proliferation of measurements of centrality and power, there are many different ways to conceptualize the notion that two nodes are equivalent or belong in the same equivalence class or community. I discuss a

---

10. For an overview of block modeling, see Doreian, Batagelj, and Ferligoj [203].
11. Recall that a partition Π is a collection of disjoint subsets of N whose union is N.

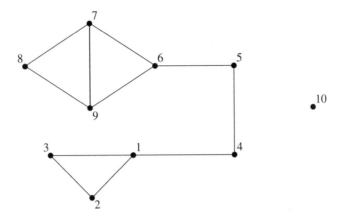

FIGURE 13.1   A sample network.

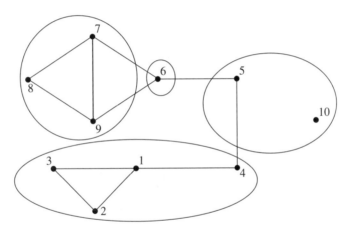

FIGURE 13.2   A community structure for the network in Figure 13.1.

few prominent approaches for classifying nodes into equivalence classes, chosen to give a feel for the spectrum of potential viewpoints.[12]

Let me begin with a criticism of some of the literature, which is important to keep in mind when examining the techniques presented below. To make sense of a community structure, we should have a well-specified notion of what a community represents. For example, is it based on common but unobserved traits of nodes? Is it defined by factors that influence nodes' behaviors? Is it defined by affinity that agents feel for one another? Is it meant to capture a natural complementarity in an association that is not directly observed but that favors link formation?

---

12. Wasserman and Faust [650], Snijders and Nowicki [608], and Newman [503], [506] provide additional background on various parts of this literature.

As we vary what a community represents, the optimal method for identifying communities will correspondingly change. In particular, it is important to have an idea of how community structure affects network formation. And if we do understand how community structure influences the observed network, then it is difficult to know how to recover the community structure from the network. Unfortunately, much of the literature takes a "I know a community when I see it" approach. That is, the researchers generally start with a simple algorithm for partitioning the nodes of a network, based on some heuristic, without a firm foundation in terms of defining what communities are, how they influence network formation, or why this algorithm is a natural way for uncovering them. Thus communities have tended to be defined as whatever the algorithms find rather than deriving the algorithms based on a well-defined notion of community. That is not to say that we could not determine the definition of the community corresponding to each technique, but rather that the literature has generally failed to be careful about this point. At the end of this chapter I discuss some approaches that are built from the ground up. Hopefully, such techniques will proliferate, and the many techniques for identifying community structures will be re-evaluated with more attention paid to what communities represent and how they relate to network structure.[13]

CONCOR   An early and widely used method for partitioning nodes into communities is CONCOR (for "convergence of iterated correlations"), as developed by Breiger, Boorman, and Arabie [107].[14] The idea is as follows. Start with an observed social network described by an adjacency matrix g, which can be either directed or undirected. Two nodes are thought of as being similar if they have a similar pattern of relationships with other nodes. One way to gauge how similar node i is to node j in terms of the network of relationships is to examine how similar row $(g_{i1}, \ldots, g_{in})$ is to row $(g_{j1}, \ldots, g_{jn})$.[15] A measure of how similar these rows are to each other is to examine the correlation between row i and row j.[16]

The CONCOR algorithm does not stop here. This first step leads to a correlation matrix C, where c_{ij} is the correlation between row g_i and row g_j of the adjacency matrix. Next, measure how similar row i of the correlation matrix C is to row j, using the same method. If nodes i and j are similar, then they should have similar rows of correlations with other nodes. So the algorithm then measures

---

13. To varying extents, the same criticism can be made of many measures in social network analysis. For example, there is still much to be learned about what different measures of centrality, prestige, and clustering really capture, beyond some of the heuristics discussed in Chapter 2.

14. Breiger, Boorman, and Arabie provide references to earlier versions of CONCOR that were used in some specific studies, and CONCOR was further elaborated upon by White, Boorman and Breiger [663] and Boorman and White [93].

15. In the case of directed networks, this measure misses some of the interaction, as it focuses on relationships that are directed outward, while in the undirected case it captures all of a node's relationships. One can also perform the described algorithm on columns. Using columns would not change the analysis in undirected networks, but could lead to a very different analysis in the case of a directed network, as it focuses on inward relationships, which can differ dramatically from outward ones.

16. That is, view the row g_i as a random variable that takes on value g_{ik} if state k is realized. Placing equal likelihood on the n states, the correlation is simply the correlation between g_i and g_j.

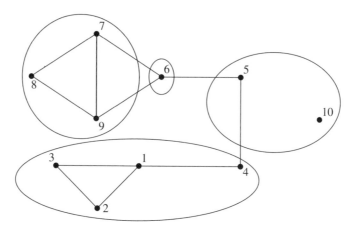

FIGURE 13.3 Community structure found by CONCOR on the network in Figure 13.1 when looking for four communities.

the correlation between rows c_i and c_j of the correlation matrix. In this way, we form a new correlation matrix $C^{(2)}$. Iterating, the algorithm generates a matrix of correlations of correlations and so on, denoted by $C^{(t)}$, after t iterations. Generally, this series of matrices converges as t grows, so that the entries of $C^{(t)}$ approach some limit. In fact, except in exceptional cases, this process converges to a matrix of (at most) two blocks in which the entries are 1 and -1.[17] That is, two blocks (which partition the nodes into two groups) emerge such that $C_{ij}^{(t)}$ converges to 1 when i and j are in the same block and $C_{ij}^{(t)}$ converges to -1 when i and j are in separate blocks.

As this process only produces two blocks, one can repeat the procedure on the network starting with each block separately to further subdivide the blocks. Choosing when to stop subdividing determines a community structure. To see how this works, let us prespecify that we wish to find a community structure with four communities. Then we operate CONCOR once to find two communities and operate it again on the resulting communities to find four communities. For example, in this way CONCOR finds the structure in Figure 13.3.

While CONCOR and its variations are included in many programs for network analysis, what one actually obtains through iterating on the correlation is not so obvious.[18] That is, what does it really mean for two nodes to be in the same block under this procedure? Why is this a reasonable way to group nodes? Another difficulty is that it is not obvious how many times to split the blocks.

Repeated Bisection CONCOR is a method that repeatedly bisects a set of nodes based on a specific technique for grouping nodes. There are other methods, developed out of the computer science literature, that also work by repeated

---

17. See Schwartz [582] for background on convergence properties of CONCOR and related methods of iterated covariance.

18. See Schwartz [582] for a criticism along these lines.

bisection. For example, a simple principle is to start by bisecting the set of nodes into two groups such that there is a minimal number of links between the two groups, with some rule based on what to do when more than one bisection leads to the same minimum. Then one can repeat the procedure on the emergent groups. Again, using some stopping rule, one ends up with a community structure.

There are many variations on this basic idea. Instead of minimizing the number of links between the two groups, one might try to maximize some measure of the number of links within each group minus the number of links across the groups (e.g., see Kernighan and Lin [389]). Or one might rely on more sophisticated methods that examine the eigenvectors of the adjacency matrix or an associated Laplacian matrix (e.g., see Fiedler [248] and Pothen, Simon, and Liou [545]).[19]

Beyond the question of determining how useful a bisection is, there is also a question of whether one searches over all possible bisections or examines only certain types of bisections. For example, one can prespecify the sizes of the two sets of nodes. One might only consider bisections into equally (or nearly equally) sized sets of nodes. This is not a minor technicality, as it can lead to a substantial difference in the number of potential bisections to be considered. Also, many of these methods tend to bisect the network very asymmetrically if no constraints are imposed, since, for instance, minimizing the links between groups might be found by identifying a single node with low degree and separating that from the rest of the network, and it is not clear that this approach is sensible. This observation relates back to the earlier criticism that we are not quite sure what we are looking for.

In addition to measuring the optimality of a bisection, and which bisections to consider, there is the stopping decision, which can be arbitrary. Finally, one can also move beyond bisections. For example, one can start with the problem of splitting a set of nodes into some number of (nearly) equally sized groups in a way that minimizes the links across groups and/or maximizes the number of links within groups.

Edge Removal   One method that avoids the issue of predetermining the sizes of the bisecting groups works by repeatedly removing edges of the network and keeping track of the component structure of the resulting graph to determine a community structure. Such a method, developed by Girvan and Newman [287], iteratively removes edges by calculating the betweenness of each link and then removing the link that has maximum betweenness. The logic is that if a link has a high betweenness score, then it is connecting (at least) two groups of nodes that are otherwise quite separate. These groups would then be natural candidates to be separate communities of nodes. This process can be done in various ways, depending on the notion of betweenness used.

Let us demonstrate this technique on the example from Figure 13.1, using the easy-to-calculate measure of the betweenness of a link that Girvan and Newman use (a variation on Freeman's notion discussed in Section 2.2.4), which is the number of shortest paths between pairs of nodes in the network that involve the link in question. That is, for each link we count the total number of geodesics in the network that include the link. The link with the highest count is the one we

19. See Newman [506] for an overview of some of those techniques.

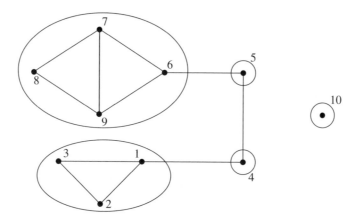

FIGURE 13.4 Community structure found by the Girvan-Newman algorithm on the network in Figure 13.1.

remove. (It could be that a pair of nodes has more than one shortest path that involves the same link. For example, in Figure 13.1, nodes 8 and 5 have two shortest paths between them that involve the link 56.) Once a link is removed, one repeats the process on the resulting subgraph, calculating new betweenness scores at each step. Figure 13.4 shows the resulting community structure under this algorithm if one stops when the betweenness of an edge reaches 2 (so that the highest betweenness measure of any link remaining in the network is 2; see Exercise 13.3 for the precise steps).

The calculations in the Girvan and Newman algorithm can become extensive as the number of nodes grows, because the betweenness measures involve finding all shortest paths through a given link. As a result, several alternative measures and algorithms have been developed (see Newman [506] for discussion and references).

This algorithm runs into the same issue that we saw with bisection methods: when to stop. Newman and Girvan [507] propose a method for determining when to stop this (or another) algorithm. For any community structure Π, one can calculate the following measure that Newman and Girvan call *modularity*. For two communities $\pi \in \Pi$ and $\pi' \in \Pi$, let $e_{\pi\pi'}(g)$ denote the fraction of all edges in the network that connect nodes in π to nodes in π'. Then the modularity of the community structure Π is

$$M(\Pi, g) = \sum_{\pi \in \Pi} e_{\pi\pi}(g) - \sum_{\pi \in \Pi, \pi' \in \Pi, \pi'' \in \Pi} e_{\pi\pi'}(g)e_{\pi'\pi''}(g).$$

Modularity measures the proportion of edges that lie within communities minus the expected value of the same quantity in a graph such that all nodes have the same degrees but links are generated uniformly at random (ignoring community structure). When this measure is 0, then the communities are not capturing much of anything. If the measure is positive, then the communities are capturing more of a fraction of links internally than one would expect at random. When the measure

450 Chapter 13 Observing and Measuring Social Interaction

is negative, then the community structure is cutting against the link pattern in that there are more links across communities and fewer within than one would see at random. So the idea is to maximize this modularity measure, and a rule for stopping an algorithm is that one stops if the modularity decreases by further edge removal. There can be local maximizers that are not global maximizers, and so one technique to avoid a premature stop is to continue the algorithm to exhaustion and then go back and pick the community structure that leads to the globally highest modularity measure.

Hierarchical Clustering   An approach that differs from CONCOR, repeated bisection techniques, and edge removal builds up communities by adding new nodes to groups successively by examining how similar pairs of nodes are, rather than starting from one large community and breaking it apart. This methodology underlies a whole class of algorithms called *hierarchical clustering* methods.

The ideas are as follows. The foundation is to have some measure of how similar two nodes are based on a given network. There are many such measures. One could use the correlation coefficients between the rows of the adjacency matrix, $C^{(1)}$, which was the first step of the CONCOR process. Instead, we could calculate the distance between the vectors g_i and g_j (using Euclidean distance; or city-block distance—the number of entries that differ across the two vectors). We could also use various measures built on path distances or other indications of how the roles of two nodes in the network compare. The important point is that there is some measure that the analyst believes captures the appropriate notion of similarity in the given application.

For the purpose of illustration, let us measure similarity between nodes i and j using a slight variation[20] on the city-block distance between rows. In particular, set the distance between nodes i and j to be

$$b_{ij}(g) = \#\{k | g_{ik} \neq g_{jk}, k \neq i, k \neq j\}.$$

Thus if $b_{ij}(g) = 0$, then i and j are structurally equivalent. More generally, $b_{ij}(g)$ indicates the number of other nodes k with whom i and j differ in their relationships. So it directly counts the differences between the neighborhoods of nodes i and j.

To use the distance information to construct a community structure, let us start with a threshold of 0. We form a graph (that is purely for algorithmic purposes and might look quite different from the original g), denoted by $g^{(0)}$, as follows. Link any two nodes together that have $b_{ij}(g) = 0$. Thus a link between two nodes indicates that we think they are similar in that they are a short distance from each other. If we stop here, then we end up with a community structure that is the partition induced by the components of this network, $g^{(0)}$.[21] As there may not be many pairs of nodes that are structurally equivalent, $g^{(0)}$ will tend to have very

---

20. The variation is that we do not examine the cases where $k = i$ or $k = j$, as those are not relevant for our purposes.

21. Note that structural equivalence is not a transitive relationship. That is, it is possible to have $b_{ij}(g) = 0 = b_{jk}(g)$, while $b_{ik}(g) = 1$. This happens when $g_{ij} \neq g_{kj}$. Thus it is possible to have components of $g^{(0)}$ that are not cliques.

few links, and so we usually have a sparse community structure with many small communities if we stop with a threshold of 0. This suggests raising the threshold for how distant linked (or "similar") nodes can be from each other. If we set a threshold t, then the resultant graph is

$$g^{(t)} = \{ij|b_{ij}(g) \le t\}.$$

As we raise the threshold, the components of the induced graph continue to grow as more edges are added. The community structure at some threshold t, denoted by $\Pi^{(t)}$, is the partition of nodes induced by the components of the network $g^{(t)}$ at some threshold t, so it is $\Pi(N, g^{(t)})$ (recalling the notation from Section 2.1.5). Eventually, as we raise t to $n - 1$, we end up with a completely connected graph and just one community.

To see how this process works, let us apply it to the network in Figure 13.2. First, note that nodes 2 and 3 are structurally equivalent, as are nodes 7 and 9, so that $b_{23}(g) = 0 = b_{79}$. As these are the only such pairs, if we set the threshold to be 0 then the resulting community structure is

$$\Pi^{(0)} = \{\{1\}, \{2, 3\}, \{4\}, \{5\}, \{6\}, \{7, 9\}, \{8\}, \{10\}\}.$$

If we raise the threshold to 1, then new pairs of nodes are defined as similar to each other: nodes 7 and 8 only differ by one relationship, and in fact $1 = b_{78}(g) = b_{89}(g) = b_{12}(g) = b_{13}(g)$. Then

$$\Pi^{(1)} = \{\{1, 2, 3\}, \{4\}, \{5\}, \{6\}, \{7, 8, 9\}, \{10\}\}.$$

Next, given that $b_{3,10}(g) = 2, b_{4,10}(g) = 2, b_{5,10}(g) = 2, b_{8,10}(g) = 2, b_{6,7}(g) = 2$, there is just a single component under $g^{(2)}$, and so the resulting community structure at a threshold of 2 is

$$\Pi^{(2)} = \{\{1, 2, 3, 4, 5, 6, 7, 8, 9, 10\}\}.$$

The term *hierarchical clustering* refers to the fact that as we raise the distance threshold (or lower the similarity threshold) for when we consider two nodes to be similar, groups of nodes continue to merge and a hierarchy is defined. The hierarchical tree, known as a *dendrogram,* generated for this example is pictured in Figure 13.5.

There are several challenges with such methods. First, we need to know when it is reasonable to stop the process. This can often be more of an art than a science, and so researchers often report the full hierarchy, indicating the thresholds at which different components coalesce. This report is often in the form of a dendrogram like the one in Figure 13.5. One can also use modularity, or some other method, to choose among potential community structures. Second, the way in which communities coalesce is a bit questionable. For example, once we set the threshold distance to be 2 in the above example, the similarities between node 10 and several other nodes is largely responsible for bringing all of the nodes together into one community. However, similarity measured in this way is not necessarily transitive. For example, nodes in the group {1, 2, 3} are now in the same community with the nodes in the group {7, 8, 9} because they are each sufficienty similar to

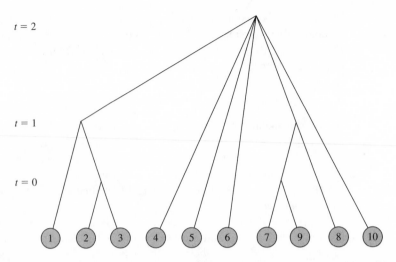

FIGURE 13.5 Dendrogram or hierarchical tree generated by running a hierarchical clustering algorithm based on a simple (dis)similarity measure on the example from Figure 13.1.

node 10, and yet any node from the first group is at a distance of at least 4 from any node in the second group, and in some cases they are at a distance of 6 (e.g., $b_{1,7}(g) = 6$). Thus a single node can cause many nodes to be grouped together that are not very similar to one another.

13.2.3 Stochastic Block Models and Communities

As mentioned at the outset of this section, a central difficulty behind all of the algorithmic methods discussed is that we are not sure what communities they are designed to uncover. If we iteratively remove edges from a network, what are we uncovering and why is that a sensible way to proceed? What sorts of communities emerge when we examine a similarity measure and identify communities by hierarchical clustering? The problem is that the methods described above are defined by the algorithms and not by having a theory or model of what a community structure is. Communities simply happen to be what we end up with. This is not to say that such methods cannot be built up from some foundation, but rather that such foundations do not yet exist.

A different approach is to start with an explicit idea of what a community is and how networks are generated as a function of the underlying community structure. Then we can work from the foundation: given the network structure, we can try to deduce which community structure is most likely to be present, as we know the likelihood with which different community structures lead to various networks. This is a standard statistical approach to the problem—one presumes a model of how data (here a network of relationships) are generated from underlying parameters (here a community structure) and then one examines the data to statistically infer the parameters of the model. When applied to social networks this approach is sometimes referred to as *a posteriori block modeling*.

A natural version of this approach is a variation of a model by Holland, Laskey, and Leinhardt [334], as analyzed by Snijders and Nowicki [608].[22] Each node belongs to a group, which can be thought of as a block or community. To adhere to the definitions above, let us call the group a community and work with community structures, so that each node belongs to exactly one community.

The model is that the probability of a link between two nodes depends on which communities the given nodes lie in. The probability, for instance, could be higher within a community than across communities. There might also be some pattern of link probabilities that depends on specific communities. In particular, the general form of the model is to have the probability of a link between a node in community π and a node in community π' be designated by a parameter $\eta_{\pi\pi'}$. The formation of each link is independent of the formation of every other link. The restriction of the model is that any two nodes in a given community are equivalent in the sense that the probability that either of them forms a link with any other node is the same. So one can think of this definition as a form of probabilistic structural equivalence. If the network is undirected, then $\eta_{\pi\pi'} = \eta_{\pi'\pi}$, while the probabilities can differ in a directed network.

A community structure with m communities, together with a list of the m^2 parameters $\eta_{\pi\pi'}$, leads to a well-defined probability that any given network will form. The probability that link ij forms is $\eta_{\pi_i\pi_j}$ where π_i is the community to which i belongs. So the probability (or likelihood) of a network g is

$$L(g|\pi, \eta) = \left(\prod_{ij \in g} \eta_{\pi_i\pi_j} \right) \left(\prod_{ij \notin g} (1 - \eta_{\pi_i\pi_j}) \right).$$

One of the difficulties with the model in its fullest generality is that it allows for many parameters and might not really be useful for making predicitons. For example, by setting the community structure to have each node in its own community and then setting link probabilities across communities to be 1 when a link exists and 0 otherwise, we would end up predicting that our observed network should have been the one that formed. There are so many free parameters that we can explain any possible network exactly. So for this approach to be meaningful, we need to place additional restrictions on the parameters of the model. These will generally be governed by the specific application and thus some additional information of what communities represent.

A special case of this model (studied by Copic, Jackson, and Kirman [175])[23] is when there is one probability for links within a community and another probability for links across communities. That is, there are just two probabilities, $1 \geq p_{in} > p_{out} \geq 0$, such that $\eta_{\pi\pi} = p_{in}$ for any $\pi \in \Pi$, and $\eta_{\pi\pi'} = p_{out}$ when $\pi \neq \pi'$. So communities are groups of nodes that are more likely to interact with one another, and interactions across communities are less likely.

---

22. This method is also related to other models, such as those of Holland and Leinhardt [333] and Fienberg and Wasserman [249]. It has experienced a recent resurgence (e.g., see Newman and Leicht [508]) in a portion of the literature unaware of its origin in the older literature.

23. They consider a more general variant, which is special in the ηs but also allows for multiple links and varying capacities across links.

To see how this model can be used to uncover a community structure, given network data, let us explore this two-probability case in more detail. As mentioned above, specifying Π, p_{in}, and p_{out} leads to a well-defined probability of observing any particular network g for the unweighted and undirected version of the model (for a weighted and directed version, see Copic, Jackson, and Kirman [175]).

Given a community structure Π, let $In(\Pi)$ be the set of all pairs of nodes that lie within the same community under Π and $Out(\Pi)$ is the set of all pairs of nodes that are in different communities under Π. That is,

$$In(\Pi) = \{ij \mid \text{there exists } \pi \in \Pi \text{ such that } \{i, j\} \subset \pi\},$$

and $Out(\Pi)$ is the set of all pairs of nodes that are not in $In(\Pi)$. Let

$$T_{in}(g, \Pi) = |g \cap In(\Pi)|$$

be the number of links that are in the network g and that lie within communities under Π. Similarly, let $T_{out}(g, \Pi) = |g \cap Out(\Pi)|$ be the number of links that are in the network g and that lie across communities under Π.

The probability of observing network g if Π is the community structure is described by the likelihood $L(g|\Pi, p_{in}, p_{out})$, where

$$L(g|\Pi, p_{in}, p_{out}) \tag{13.8}$$
$$= p_{in}^{T_{in}(g,\Pi)} \left(1 - p_{in}\right)^{|In(\Pi)| - T_{in}(g,\Pi)} p_{out}^{T_{out}(g,\Pi)} \left(1 - p_{out}\right)^{|Out(\Pi)| - T_{out}(g,\Pi)}.$$

Thus, given a community structure Π, (13.8) yields a well-defined probability of each possible network.

13.2.4 Maximum Likelihood Estimation of Communities

Given the probabilities of seeing each possible network as a function of the community structure, we can then invert the problem and ask which community structure leads to a highest likelihood of generating the network that we have actually observed. Maximizing this likelihood is equivalent to maximizing the log of the likelihood, which is often easier to manipulate. By taking the log of the likelihood function, $\ell(g|\Pi, p_{in}, p_{out}) = \log\left(L(g|\Pi, p_{in}, p_{out})\right)$, we define a very manageable expression for the relative likelihood of observing a particular network g if the community structure is Π:

$$\ell(g|\Pi, p_{in}, p_{out}) = \log(L(g|\Pi, p_{in}, p_{out})) = k_1 |In(\Pi)| + k_2 T_{in}(g, \Pi)$$
$$+ k_3 |Out(\Pi)| + k_4 T_{out}(g, \Pi), \tag{13.9}$$

where the ks depend only on p_{in} and p_{out}.[24]

---

24. In particular, $k_1 = \log\left(1 - p_{in}\right)$, $k_2 = \log\left(\frac{p_{in}}{1 - p_{in}}\right)$, $k_3 = \log\left(1 - p_{out}\right)$, and $k_4 = \log\left(\frac{p_{out}}{1 - p_{out}}\right)$.

Noting that

$$|Out(\Pi)| = \frac{n(n-1)}{2} - |In(\Pi)|$$

and

$$T_{out}(\Pi) = |g| - T_{in}(\Pi),$$

we can rewrite (13.9) as

$$\ell(g|\Pi, p_{in}, p_{out}) = (k_1 - k_3)\, |In(\Pi)| + (k_2 - k_4)\, T_{in}(g, \Pi) + r, \quad (13.10)$$

for r that depends only on $|g|$, p_{in}, and p_{out} and not on Π.

So the community structure that is identified by this method is the Π that maximizes $(k_1 - k_3)\, |In(\Pi)| + (k_2 - k_4)\, T_{in}(g, \Pi)$. We have reduced the problem of finding a community structure to a calculation that only involves examining a weighted difference of the number of pairs of nodes that lie within the same communities and the number of links that lie within communities. Noting that $k_1 - k_3 < 0$ and $k_2 - k_4 > 0$, as we group more nodes together we see two competing effects in terms of how the likelihood changes: the second term in (13.10) increases, while the first one decreases. The relative rates at which those terms change depend on the relative density of how many links are within a community compared to how many pairs of nodes lie within communities.

Algorithms for Maximum Likelihood Estimation   There are two challenges to implementing this method. One is that p_{in} and p_{out} need to be estimated along with the community structure, as they determine the k parameters. Dealing with this problem is relatively straightforward and involves an iterative procedure: one begins with an initial estimate of these parameters, which then leads to an initial estimate of a community structure. Next, based on the first estimation of the community structure, one can estimate p_{in} and p_{out} directly by examining the fraction of links within communities compared to the potential number, and similarly for across-community links. One can then iterate on this process.[25]

The second challenge is a bit more difficult. It is that the number of potential community structures grows exponentially with the number of nodes. For even moderate numbers of nodes, this growth makes it impossible to calculate the likelihoods of the given network for all possible partitions. For a small number of nodes, one can do the calculations directly, as for the network in Figure 13.1, which results in the community structure pictured in Figure 13.6.

However, for larger numbers of nodes, one has to employ an approximation technique. Various techniques can be found in Snijders and Nowicki [608]; Copic, Jackson, and Kirman [175]; and Newman and Leicht [508].

---

25. As Copic, Jackson, and Kirman [175] show, as the number of potential relationships is increased, there is a unique consistent pairing of a community structure Π with p_{in} and p_{out}, which leads to estimates of each other.

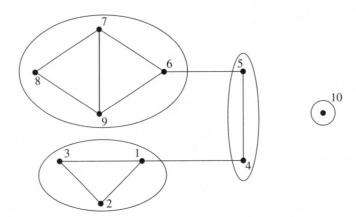

FIGURE 13.6 Community structure found by the maximum likelihood algorithm on the network in Figure 13.1.

13.2.5 Latent Space Estimation

The model of community structures in Section 13.2.3 is relatively simple in that it posits that link probabilities are completely governed by community membership. In some settings there are more complex relationships that underlie the links between nodes. There might be many attributes, including socioeconomic and geographic ones, and a variety of status attributes, such as profession, religion, gender, race, and membership in organizations, that influence the relationships. There might also be hierarchies of nodes according to various measures, and these might come together in different ways to influence the chance that two nodes are linked to each other. A community structure is an extreme model of this. More generally, one can model richer underlying structures that help determine relationships.

A straightforward generalization of the maximum likelihood approach outlined in Section 13.2.4 for community structures is to posit a model of structures, attributes, and how a given network is to emerge under various parameters of the model. Take some general structure S to be the primitive, which might include all sorts of information about nodes and layers of groupings. Each S then leads to a likelihood of observing a given network g, denoted $L(g|S)$. Given that we observe g, we can look at all potential Ss to find the one that maximizes the likelihood of having seen g—that is, the S that maximizes $L(g|S)$.[26]

An example of this model, which generalizes the basic community membership model, is known as *latent space estimation*. The idea is that nodes are located

---

26. One can alternatively do a Bayesian analysis, where one has a prior probability distribution, P, over possible Ss. Then given the likelihoods $L(g|S)$, one applies Bayes's rule to derive a posterior probability that S is actually in place. One chooses S to maximize

$$\frac{L(g|S)P(S)}{\sum_{S'} L(g|S')P(S')}.$$

in a space, and the probability that they are linked is dependent on their spatial locations (e.g., see Hoff, Raftery, and Handcock [331] and Hoff [330]), generally with the probability of a link increasing with decreasing distance between nodes.[27] The space can take many forms but provides an explicit model of how networks emerge and what we wish to uncover. The critical element in such an estimation is to ensure that the model has some limits in terms of the number of parameters to be estimated, so that one does not overfit the data.

13.3 • Exercises

13.1 *Identification with Contextual Effects* Consider the following generalization from Manski [448] of the model in (13.4).

$$Y_i = a + b_1 E[Y_i|x_i] + b_2 x_i + b_3 z_i + b_4 E[z_i|x_i] + \varepsilon_i.$$

Here z_i is a contextual effect, and i's expectation of z_i can matter too. Show that the parameters a, b_1, b_2, and b_4 are not identified.

13.2 *Identification Problems with Nonlinear Models*[*] Consider the following two alternative models of behavior. The first model is as follows. In each period a parent is replaced by his or her child, who then becomes a parent in the next period. The child makes a 0 or 1 decision (e.g., to attend university) dependent upon the parent's choice. If the parent took decision 1, then the child takes decision 1 with probability q_1, while if the parent took decision 0, then the child takes decision 1 with probability q_0, where $1 \geq q_1 \geq q_0 \geq 0$.

In the second model, there are two such families. In each period, one of the two families is selected by the toss of a fair coin. That family (and only that family) has its member ("the parent") die and be replaced by a child. In this model, the child then makes a decision. The child looks to the other family (its neighbor), and if that neighbor has taken decision 1, then the child takes decision 1 with probability p_1, while if the neighbor took decision 0, then the child takes decision 1 with probability p_0, where $1 \geq p_1 \geq p_0 \geq 0$.

Show the following result from Calvó-Armengol and Jackson [131]. For any such specification of $1 \geq q_1 \geq q_0 \geq 0$ in the first model, there is a specification of $1 \geq p_1 \geq p_0 \geq 0$ in the second model that leads to an observed probability of a child taking action 1 conditional on the parent's choice that is identical to that of the first model. Do this by calculating these conditional probabilities for any given p_1 and p_0, and show that the range is the set of (q_1, q_0) such that $1 \geq q_1 \geq q_0 \geq 0$.

13.3 *Applying the Girvan-Newman Algorithm to the Network in Figure 13.1* Which link has the highest betweenness score at the first step in the Girvan-Newman algorithm applied to the network in Figure 13.1, and what is that score? Indicate

---

This is equivalent to maximizing $L(g|S)P(S)$ and is the same as the maximum likelihood estimation if we set the prior, $P(S)$, to be equal across structures.

27. One can think of the community structure model as a special case in which the space is a hypercube with n vertices and all nodes in the same community are located at the same vertex.

the next two edges to be removed. Which edges would be removed next if we continued?

13.4 *Multiple Community Memberships* Consider the following variation on a model of community structures. There are a set of clubs that agents can be members of.[28] In the special case for which each agent must be in one and only one club, then this model reduces to a community structure (ignoring any empty clubs), but more generally we might allow agents to belong to more than one club, or even no clubs. An agent's likelihood to be linked to another agent depends on the number of clubs they are in together. The probability that two agents are linked is p_k, where k is the number of clubs that they are both members of, and p_k is increasing in k.

Describe a likelihood method for recovering club membership for a fixed number of clubs.

13.5 *Hierarchies and Communities* Consider augmenting a community structure by a hierarchy. That is, starting with a community structure Π, let h be a hierarchy function such that $h(\pi) \in \{1, 2, \ldots, K\}$ indicates which level of the hierarchy π lies in. Let the probability that a node links to another node in the same community be p_{in} and the probability that a node in a community in level k links to a node in a different community in level k' be $p_{kk'}$.

(a) Describe a likelihood method for recovering the community structure and hierarchy function, given that there are at most K levels to the hierarchy and fixing some starting estimates of $p_{kk'}$ values.

(b) Allowing for a directed network, describe a method for recovering the community structure, hierarchy function, and the $p_{kk'}$ values; under a constraint that $p_{kk'} > p_{k'k}$ when $k > k'$.

---

28. One can also think of these as activities that an agent can undertake, such as a sport, or as an attribute that an agent might have, such as ethnicity or profession.

Afterword

This book is intended to provide researchers with an overview of models of social and economic networks and the many techniques for analyzing them, while including an empirical perspective. By drawing from the multitude of disciplines that contribute to network analysis and the variety of perspectives underlying them, I have sacrificed being comprehensive with respect to any particular viewpoint. I have tried to provide a coherent view of the overall landscape, while still giving a good sense of the richness that emerges from the underlying mosaic.

There is much that is exciting about the study of social and economic networks. Researchers using the toolbox of powerful models and methods are poised for new discoveries. The emergence of increasingly large and extensive data sets, and the realization of researchers in more areas of study of the importance of social structure, have resulted in the growing reach of network analysis. Agent-based modeling and large-scale experiments are becoming more efficient and should also lead to new insights. Beyond the existing tools, network science is also exciting in that it calls out for further development of methods and models. We have seen many examples of this. We need more models that incorporate both the randomness that governs the opportunity for people to form relationships, and the incentives underlying the choices of which relationships appear and endure. We need more models that capture how networks influence behavior. There is still much to be learned about how the social network aspects of many economic and social interactions influence employment patterns, wage distributions, the pursuit of education, social mobility, conflict, learning, and various other behaviors. Moreover, we still know next to nothing about the co-evolution of networks and behavior, and clearly behavior and network structure influence each other. There are important open questions regarding how network structure affects the distribution of the benefits that accrue to different actors in a network, as well as how this allocation feeds back to network formation. There are also many fundamental methodological questions concerning what summary statistics of networks (clustering and centrality measures, community structures, and so forth) really capture. This partial list presents a formidable but enticing agenda for the future.

Bibliography

[1]   Achlioptas, D., A. Clauset, D. Kempe, and C. Moore (2005) "On the Bias of Traceroute Sampling," STOC, ACM 1581139608/ 05/0005, 1–10.

[2]   Adamic, L.A. (1999) "The Small World Web," in *Proceedings of the ECDL. Lecture Notes in Computer Science* 1696, Berlin: Springer-Verlag.

[3]   Adamic L.A., and E. Adar (2003) "Friends and Neighbors on the Web," *Social Networks* 25(3):211–230.

[4]   Adamic, L.A., and B.A. Huberman (2000) "The Nature of Markets in the World Wide Web," *Quarterly Journal of Electronic Commerce* 1:5–12.

[5]   Adamic, L.A., R.M. Lukose, and B.A. Huberman (2003) "Local Search in Unstructured Networks," in *Handbook of Graphs and Networks: From the Genome to the Internet,* S. Bornholdt and H.G. Schuster, eds., Berlin: Wiley VCH.

[6]   Adamic, L.A., R.M. Lukose, A.R. Puniyani, and B.A. Huberman (2001) "Search in Power-Law Networks," *Physical Review E* 64, 046135.

[7]   Aizer, A., and J. Currie (2004) "Networks or Neighborhoods? Correlations in the Use of Publicly-Funded Maternity Care in California," *Journal of Public Economics* 88:2573–2585.

[8]   Albert, R., and A.L. Barabási (2002) "Statistical Mechanics of Complex Networks," *Reviews of Modern Physics* 74:47–97.

[9]   Albert, R., H. Jeong, and A.L. Barabási (1999) "Diameter of the World Wide Web," *Nature* 401:130–131.

[10]   ——— (2000) "Attack and Error Tolerance of Complex Networks," *Nature* 406: 378–382.

[11]   Alderson, D. (2004) "Understanding Internet Robustness and Its Implications for Modeling Complex Networks," presentation, December 10, Northwestern University, Chicago.

[12]   Aliprantis, C.D., G. Camera, and D. Puzzello (2007) "A Random Matching Theory," *Games and Economic Behavior* 59:1–16.

[13]   Allen, F., and A. Babus (2007) "Networks in Finance," in *Network-Based Strategies and Competencies*, P. Kleindorfer and J. Wind, eds., Philadelphia: Wharton School Publishing, forthcoming.

[14]   Allport, W.G. (1954) *The Nature of Prejudice*, Cambridge, Mass.: Addison-Wesley.

[15]   Almeida, P., and B. Kogut (1999) "Localization of Knowledge and the Mobility of Engineers in Regional Networks," *Management Science* 45(7):905–917.

[16]   Anderson, C.J., S. Wasserman, and B. Crouch (1999) "A p^* Primer: Logit Models for Social Networks," *Social Networks* 21:37–66.

[17]   Anderson, R.M., and R.M. May (1988) Epidemiological Parameters of HIV Transmission," *Nature* 333:514–519.

[18]   Anselin, L. (1988) *Spatial Econometrics: Methods and Models*, Amsterdam: Kluwer Academic.

[19]   Appel, K., W. Haken, and J. Koch (1977) "Every Planar Map Is Four Colorable," *Journal of Mathematics* 21:439–567.

[20]   Arrow, K.J., and R. Borzekowski (2004) "Limited Network Connections and the Distribution of Wages," Federal Reserve Board Publication 41, Washington, D.C., mimeo.

[21]   Arthur, B.W. (1994) *Increasing Returns and Path-Dependence in the Economy*, Ann Arbor, Mich.: University of Michigan Press.

[22]   Auerbach, F. (1913) "Das Gesetz der Belvolkerungskoncentration," *Petermanns Geographische Mitteilungen* 59:74–76.

[23]   Aumann, R., and R. Myerson (1988) "Endogenous Formation of Links between Players and Coalitions: An Application of the Shapley Value," in *The Shapley Value*, A. Roth, ed., Cambridge: Cambridge University Press.

[24]   Axelrod, R. (1997) "Advancing the Art of Simulation in the Social Sciences," in *Simulating Social Phenomena,* R. Conte, R. Hegselmann, and P. Terna, eds., Berlin: Springer-Verlag.

[25]   —— (1997) *The Complexity of Cooperation: Agent-Based Models of Competition and Collaboration,* Princeton, N.J.: Princeton University Press.

[26]   Badasyan, N., and S. Chakrabarti (2003) "Private Peering among Internet Backbone Providers," Blacksburg Va., Virginia Tech, mimeo.

[27]   Baerveldt, C., M.A.J. Van Duijn, L. Vermeij, and D.A. Van Hemert (2004) "Ethnic Boundaries and Personal Choice: Assessing the Influence of Individual Inclinations to Choose Intra-Ethnic Relationships on Pupils' Networks," *Social Networks* 26:55–74.

[28]   Bailey, N.T.J. (1975) *The Mathematical Theory of Infectious Diseases*, London: Griffin.

[29]   Bala, V., and S. Goyal (1998) "Learning from Neighbors," *Review of Economic Studies* 65:595–621.

[30]   —— (2000) "A Strategic Analysis of Network Reliability," *Review of Economic Design* 5:205–228.

[31]   —— (2000) "A Non-Cooperative Model of Network Formation," *Econometrica* 68:1181–1230.

[32]   —— (2001) "Conformism and Diversity under Social Learning," *Economic Theory* 17:101–120.

[33]   Ballester, C., and A. Calvó-Armengol (2007) "Moderate Interactions in Games with Induced Complementarities," Barcelona, Universitat Autónoma Barcelona, mimeo.

[34]   Ballester, C., A. Calvó-Armengol, and Y. Zenou (2006) "Who's Who in Networks. Wanted: The Key Player," *Econometrica* 74:1403–1417.

[35]   Bandiera, O., and I. Rasul (2002) "Social Networks and Technology Adoption in Northern Mozambique," *Economic Journal* 116(514):869–902.

[36]   Banerjee, A.V. (1992) "A Simple Model of Herd Behavior," *Quarterly Journal of Economics* 107:797–817.

[37]   Banerjee, A., and B. Dutta (2004) "Local Network Externalities and Market Segmentation," Warwick Economics Research Paper Series 725, Coventry, England: University of Warwick.

[38]   Banerjee, A.V., and D. Fudenberg (2004) "Word-of-Mouth Learning," *Games and Economic Behavior* 46:1–22.

[39]   Banerjee, S. (1999) "Efficiency and Stability in Economic Networks," Boston, Boston University, mimeo.

[40]   Banerjee, S., H. Konishi, and T. Sönmez (2001) "Core in a Simple Coalition Formation Game," *Social Choice and Welfare* 18:135–154.

[41]   Barabási, A. (2002) *Linked*, Cambridge, Mass.: Perseus Publishing.

[42]   Barabási A., and R. Albert (1999) "Emergence of Scaling in Random Networks," *Science* 286:509–512.

[43]   Barabási, A., R. Albert, and H. Jeong (1999) "Mean-Field Theory for Scale-Free Random Networks," *Physica A* 272:173–187.

[44]   ———— (2000) "Scale-Free Characteristics of Random Networks: The Topology of the World-Wide Web," *Physica A* 281:69–77.

[45]   Barrat A., and M. Weigt (2000) "On the Properties of Small-World Network Models," *European Physical Journal B* 13:547–560.

[46]   Bass, F.M. (1969) "A New Product Growth Model for Consumer Durables," *Management Science* 15:215–227.

[47]   Bavelas, A. (1950) "Communication Patterns in Task-Oriented Groups," *Journal of the Acoustical Society of America* 22:725–730.

[48]   Bayer, P.J., S.L. Ross, and G. Topa (2005) "Place of Work and Place of Residence: Informal Hiring Networks and Labor Market Outcomes," Economic Growth Center Discussion Paper 927, New Haven, Conn., Yale University.

[49]   Béal, S., and N. Quérou (2007) "Bounded Rationality and Repeated Network Formation," *Mathematical Social Sciences* 54:71–89.

[50]   Beaman, L. (2007) "Refugee Resettlement: The Role of Social Networks and Job Information Flows in the Labor Market," New Haven: Conn., Yale University, manuscript.

[51]   Bearman, P., J. Moody, and K. Stovel (2004) "Chains of Affection: The Structure of Adolescent Romantic and Sexual Networks," Chicago, University of Chicago, manuscript.

[52]   Belleflamme, P., and F. Bloch (2004) "Market Sharing Agreements and Stable Collusive Networks," *International Economic Review* 45(2):387–411.

[53]   Benaïm, M., and J.W. Weibull (2003) "Deterministic Approximation of Stochastic Evolution in Games," *Econometrica* 71:873–903.

[54]   Bender, E.A., and E.R. Canfield (1978) "The Asymptotic Number of Labelled Graphs with Given Degree Sequences," *Journal of Combinatorial Theory A* 24:296–307.

[55]   Berger, N., C. Borgs, J. Chayes, and A. Saberi (2005) "The Epidemic Threshold in Scale-Free Graphs," in *Symposium on Discrete Algorithms Archive, Proceedings of the Sixteenth Annual ACM-SIAM Symposium on Discrete Algorithms, Society for Industrial and Applied Mathematics,* Philadelphia: ACM.

[56]   Berger, R.L. (1981) "A Necessary and Sufficient Condition for Reaching a Consensus Using De Groot's Method," *Journal of the American Statistical Association* 76:415–419.

[57]   Bernard, H.R. (2000) *Social Research Methods: Qualitative and Quantitative Approaches*, Thousand Oaks, Calif.: Sage Publications.

[58]   Bernard H.R., E.C. Johnsen, P.D. Killworth, and C. McCarty (1990) "Comparing Four Different Methods for Measuring Personal Social Networks," *Social Networks* 12:179–215.

[59]   Bertrand, M., E. Luttmer, and S. Mullainathan (2000) "Network Effects and Welfare Cultures," *Quarterly Journal of Economics* 115:1019–1055.

[60]   Besag, J.E. (1974) "Spatial Interaction and the Statistical Analysis of Lattice Systems (with Discussion)," *Journal of the Royal Statistical Society B* 36:196–236.

[61]   Bhatia, T., P. Manchanda, and H. Nair (2006) "Asymmetric Peer Effects in Physician Prescription Behavior: The Role of Opinion Leaders," Stanford University School of Business Research Paper 1970, Stanford, Calif., Stanford University.

[62]   Bianconi, G., and A.-L. Barabási (2001) "Competition and Multi-Scaling in Evolving Networks," *Europhysics Letters* 54:436–442.

[63] Bienenstock, E., and P. Bonacich (1993) "Game Theory Models for Social Exchange Networks: Experimental Results," *Sociological Perspectives* 36:117–136.

[64] ——— (1997) "Network Exchange as a Cooperative Game," *Rationality and Society* 9:37–65.

[65] Bikhchandani, B., D. Hirshleifer, and I. Welch (1992) "A Theory of Fads, Fashion, Custom, and Cultural Change as Informational Cascades," *Journal of Political Economy* 100(5):992–1026.

[66] Billand, P., and C. Bravard (2004) "Non-Cooperative Networks in Oligopolies," *International Journal of Industrial Organization* 22(5):593–609.

[67] ——— (2004) "A Note on the Characterization of Nash Networks," *Mathematical Social Sciences* 49(3):355–365.

[68] Billingsley, P. (1968) *Convergence of Probability Measures*, New York: Wiley.

[69] ——— (1979) *Probability and Measure*, New York: Wiley.

[70] Binmore, K.G. (1992) *Fun and Games*, Lexington, Mass.: D.C. Heath.

[71] Bisin, A., and T. Verdier (2000) "Beyond the Melting Pot: Cultural Transmission, Marriage, and the Evolution of Ethnic and Religious Traits," *Quarterly Journal of Economics* 115:955–988.

[72] Blalock Jr., H.M. (1982) *Race and Ethnic Relations*, Englewood Cliffs, N.J.: Prentice-Hall.

[73] Blau, P.M. (1964) *Exchange and Power in Social Life*, New York: Wiley.

[74] ——— (1977) *Inequality and Heterogeneity: A Primitive Theory of Social Structure*, New York: Free Press.

[75] Bloch, F. (2004) "Group and Network Formation in Industrial Organization," in *Group Formation in Economics; Networks, Clubs and Coalitions*, G. Demange and M. Wooders, eds., Cambridge: Cambridge University Press.

[76] Bloch, F., and B. Dutta (2005) "Communication Networks with Endogenous Link Strength," *Games and Economic Behavior,* forthcoming.

[77] ——— (2008) "Formation of Networks and Coalitions," in *Handbook of Social Economics,* J. Benhabib, A. Bisin, and M.O. Jackson, eds., New York: Elsevier, forthcoming.

[78] Bloch, F., and M.O. Jackson (2006) "Definitions of Equilibrium in Network Formation Games," *International Journal of Game Theory* 34(3):305–318.

[79] ——— (2007) "The Formation of Networks with Transfers among Players," *Journal of Economic Theory* 133(1):83–110.

[80] Bloch, F., G. Genicot, and D. Ray (2005) "Informal Insurance in Social Networks," in London, Washington D.C., and New York, GREQAM, Georgetown University, and New York University, mimeo.

[81] Blume, L. (1993) "The Statistical Mechanics of Strategic Interaction," *Games and Economic Behavior* 5:387–424.

[82] Blume, L., and S.N. Durlauf (2006) "Identifying Social Interactions: A Review," in *Methods in Social Epidemiology,* J.M. Oakes and J. Kaufman, eds., San Francisco: Jossey-Bass, forthcoming.

[83] Boguñá, M., R. Pastor-Satorras, and A. Vespignani (2003) "Epidemic Spreading in Complex Networks with Degree Correlations," *Lecture Notes in Physics* 625:127–147.

[84] Bollobás, B. (1980) "A Probabilistic Proof of an Asymptotic Formula for the Number of Labelled Regular Graphs," *European Journal of Combinatorics* 1:311–316.

[85] ——— (2000) *Modern Graph Theory,* New York: Springer-Verlag.

[86] ——— (2001) *Random Graphs,* second edition, Cambridge: Cambridge University Press.

[87] Bollobás, B., and O. Riordan (2002) "The Diameter of a Scale-Free Random Graph," *Combinatorica* 24(1):5–34.

[88] Bollobás, B., O. Riordan, J. Spencer, and G. Tusnady (2001) "The Degree Sequence of a Scale-Free Random Graph Process," *Random Structures and Algorithms* 18(3):279–290.

[89] Bonabeau, E. (2002) "Agent-Based Modeling: Methods and Techniques for Simulating Human Systems," *Proceedings of the National Academy of Sciences* 99(3):7280–7287.

[90] Bonacich, P. (1972) "Factoring and Weighting Approaches to Status Scores and Clique Identification," *Journal of Mathematical Sociology* 2:113–120.

[91] ——— (1987) "Power and Centrality: A Family of Measures," *American Journal of Sociology* 92:1170–1182.

[92] Boorman, S. (1975) "A Combinatorial Optimization Model for Transmission of Job Information through Contact Networks," *Bell Journal of Economics* 6:216–249.

[93] Boorman, S.A., and H.C. White (1976) "Social Structure from Multiple Networks. II. Role Structures," *American Journal of Sociology* 81(6):1384–1446.

[94] Borgatti, S.P. (2005) "Centrality and Network Flow," *Social Networks* 27:55–71.

[95] Borgatti, S.P., M.G. Everett, and L.C. Freeman (1999) *UCINET 5.0 Version 1.00.* Natick, Mass.: Analytic Technologies.

[96] ——— (1999) *UCINET 6.0 Version 1.00.* Natick, Mass.: Analytic Technologies.

[97] Borm, P., G. Owen, and S. Tijs (1992) "On the Position Value for Communication Situations," *SIAM Journal on Discrete Mathematics* 5:305–320.

[98] Boss, M., H. Elsinger, M. Summer, and S. Turner (2004) "The Network Topology of the Interbank Market," *Quantitative Finance* 4(6):677–684.

[99] Bourdieu, P. (1986) "Forms of Capital," in *Handbook of Theory and Research for the Sociology of Education,* J.G. Richardson, ed., Westport, Conn.: Greenwood Press.

[100] Bramoullé, Y. (2001) "Anti-Coordination and Social Interactions," Université de Laval, *Games and Economic Behavior* 58(1):30–49.

[101] ——— (2001) "Interdependent Utilities, Preference Indeterminacy and Social Networks," Nanterre, France: Theorie Economique, Modelisation et Applications (THEMA), Université de Paris X–Nanterre, mimeo.

[102] Bramoullé, Y., and R. Kranton (2005) "Risk-Sharing Networks," *Journal of Economic Behavior and Organization* 64(3–4):275–294.

[103] ——— (2007) "A Model of Public Goods: Experimentation and Social Learning," *Journal of Economic Theory* 135:478–494.

[104] Bramoullé, Y., D. Lopez-Pintado, S. Goyal, and F. Vega-Redondo (2004) "Network Formation and Anti-Coordination Games," *International Journal of Game Theory* 33:1–18.

[105] Brass, D.J. (1984) "Being in the Right Place: A Structural Analysis of Individual Influence in an Organization," *Administrative Science Quarterly* 29:518–539.

[106] Bratley, P., B. Fox, and L. Schrage (1987) *A Guide to Simulation,* second edition, New York: Springer-Verlag.

[107] Breiger, R.L., S.A. Boorman, and P. Arabie (1975) "An Algorithm for Clustering Relational Data with Applications to Social Network Analysis and Comparison with Multidimensional Scaling," *Journal of Mathematical Psychology* 12:328–383.

[108] Bridges, W.P., and W.J. Villemez (1986) "Informal Hiring and Income in the Labor Market," *American Sociological Review* 51(4):574–582.

[109] Brock, W., and S.N. Durlauf (2001) "Interactions-Based Models," in *Handbook of Econometrics, 5,* J. Heckman and E. Leamer, eds., Amsterdam: North-Holland.

[110] ——— (2001) "Discrete Choice with Social Interactions," *Review of Economic Studies* 68(2):235–260.

[111] Broder, A., R. Kumar, F. Maghoul, A. Raghavan, S. Rajagopalan, R. Stata, A. Tomkins, and J. Wiener (2000) "Graph Structure in the Web," *Computer Networks* 33:309–320.

[112] Brueckner, J.K. (2006) "Friendship Networks," *Journal of Regional Science* 46:847–865.

[113] Brueckner, J.K., and O. Smirnov (2007) "Workings of the Melting Pot: Social Networks and the Evolution of Population Attributes," *Journal of Regional Science* 47(2):209–228.

[114] Buechel, B., and V. Buskens (2008) "The Dynamics of Closeness and Betweenness," Bielefeld, Germany, University of Bielefeld, preprint.

[115] Burt, R.S. (1992) *Structural Holes: The Social Structure of Competition,* Cambridge, Mass.: Harvard University Press.

[116] —— (2001) "Structural Holes versus Network Closure as Social Capital," in *Social Capital: Theory and Research,* N. Lin, K.S. Cook, and R.S. Burt, eds., Edison, N.J.: Aldine Transaction.

[117] Burt, R.S., R.M. Hogarth, and C. Michaud (2000) "The Social Capital of French and American Managers," *Organization Science* 11(2):123–147.

[118] Burton, R. (1927) *The Anatomy of Melancholy,* New York: Farrar and Rinehart.

[119] Cabrales, A., A. Calvó-Armengol, and Y. Zenou (2005) "Building Socio-Economic Networks: How Many Conferences Should You Attend?" Barcelona, Universitat Pompeu Fabra, mimeo.

[120] Cahuc, P., and F. Fontaine (2002) "On the Efficiency of Job Search with Social Networks," Institute for the Study of Labor (IZA) Working Paper 583, Bonn.

[121] Caldarelli, G., A. Capocci, P. De Los Rios, and M.A. Muñoz (2002) "Scale-Free Networks from Varying Vertex Intrinsic Fitness," *Physical Review Letters* 89:258702.

[122] Callander, S., and C. Plott (2005) "Principles of Network Development and Evolution: An Experimental Study," *Study of Public Economics* 89:1469–1495.

[123] Callaway, D.S., M.E.J. Newman, S.H. Strogatz, and D.J. Watts (2000) "Network Robustness and Fragility: Percolation on Random Graphs," *Physical Review Letters* 85:5468–5471.

[124] Callaway, D.S., J.E. Hopcroft, J.M. Kleinberg, M.E.J. Newman, and S.H. Strogatz (2001) "Are Randomly Grown Graphs Really Random?" *Physical Review E.* 64:041902.

[125] Calvó, E., J. Lasaga, and A. van den Nouweland (1999) "Values of Games with Probabilistic Graphs," *Mathematical Social Sciences* 37:79–95.

[126] Calvó-Armengol, A. (2001) "Bargaining Power in Communication Networks," *Mathematical Social Sciences* 41:69–88.

[127] —— (2003) "Stable and Efficient Bargaining Networks," *Review of Economic Design* 7(4):411–428.

[128] —— (2004) "Job Contact Networks," *Journal of Economic Theory* 115:191–206.

[129] Calvó-Armengol, A., and R. Ilkilic (2004) "Pairwise Stability and Nash Equilibria in Network Formation," Barcelona, Universitat Autónoma de Barcelona, mimeo.

[130] Calvó-Armengol, A., and M.O. Jackson (2004) "The Effects of Social Networks on Employment and Inequality," *American Economic Review* 94(3):426–454.

[131] —— (2005) "Like Father, Like Son: Labor Market Networks and Social Mobility," *American Economic Journal: Microeconomics,* forthcoming.

[132] —— (2007) "Networks in Labor Markets: Wage and Employment Dynamics and Inequality," *Journal of Economic Theory* 132(1):27–46.

[133] Calvó-Armengol, A., and Y. Zenou (2004) "Social Networks and Crime Decisions: The Role of Social Structure in Facilitating Delinquent Behavior," *International Economic Review* 45:935–954.

[134] —— (2004) "Job Matching, Social Network and Word-of-Mouth Communication," *Journal of Urban Economics* 57:500–522.

[135] Calvó-Armengol, A., E. Patacchini, and Y. Zenou (2005) "Peer Effects and Social Networks in Education and Crime," Universitat Autónoma de Barcelona, mimeo.

[136] Capocci, A., G. Caldarelli, and P. De Los Rios (2003) "Quantitative Description and Modeling of Real Networks," *Physical Review E* 68:047101.

[137] Carayol, N., and P. Roux (2003) "Knowledge Flow and the Geography of Networks: A Strategic Model of Small Worlds Formation," *Journal of Economic Behavior and Organization,* forthcoming.

[138] ——— (2005) "'Collective Innovation' in a Model of Network Formation with Preferential Meeting," in *Nonlinear Dynamics and Heterogeneous Interacting Agents,* T. Lux, S. Reitz, and E. Samanidon, eds., Heidelberg: Springer-Verlag.

[139] Carlson, J., and J. Doyle (1999) "Highly Optimized Tolerance: A Mechanism for Power Laws in Designed Systems, *Physical Review E* 60(2):1412–1427.

[140] Casella, A., and J. Rauch (2002) "Anonymous Market and Group Ties in International Trade," *Journal of International Economics* 58(1):19–47.

[141] Cassar, A. (2007) "Coordination and Cooperation in Local, Random and Small World Networks: Experimental Evidence," *Games and Economic Behavior* 58:209–230.

[142] Castells, J. (2000) *The Rise of the Network Society,* New York: Blackwell.

[143] Caulier, J.F., A. Mauleon, and V. Vannetelbosch (2007) "Contractually Stable Networks," Louvain, Belgium, CORE, mimeo.

[144] Celen, B., S. Kariv, and A. Schotter (2004) "Learning in Networks: An Experimental Study," New York and Berkeley, Columbia University, University of California at Berkeley, and New York University, mimeo.

[145] Charness, G., and M.O. Jackson (2007) "Group Play in Games and the Role of Consent in Network Formation," *Journal of Economic Theory* 136(1):417–445.

[146] Charness, G., M. Corominas-Bosch, and G.R. Frechette (2007) "Bargaining and Network Structure: An Experiment," *Journal of Economic Theory* 136(1):28–65.

[147] Chatterjee, K., B. Dutta, D. Ray, and S. Sengupta (1993) "A Noncooperative Theory of Coalitional Bargaining," *Review of Economic Studies* 60(1):463–477.

[148] Chatterjee, S., and E. Seneta (1977) "Towards Consensus: Some Convergence Theorems on Repeated Averaging," *Journal of Applied Probability* 14:159–164.

[149] Cheeger, J. (1970) "A Lower Bound for the Smallest Eigenvalue of the Laplacian," in *Problems in Analysis (Papers Dedicated to Salomon Bochner, 1969),* Princeton, N.J.: Princeton University Press.

[150] Chiappori, P.-A., S. Levitt, and T. Groseclose (2002) "Testing Mixed-Strategy Equilibria When Players Are Heterogeneous: The Case of Penalty Kicks in Soccer," *American Economic Review* 92(4):1138–1151.

[151] Choi, S., D. Gale, and S. Kariv (2005) "Behavioral Aspects of Learning in Social Networks: An Experimental Study," in *Advances in Applied Microeconomics*, Vol. 13, Behavioral and Experimental Economics, J. Morgan, ed., Berkeley, Calif.: BE Press.

[152] ——— (2007) "Social Learning in Networks: A Quantal Response Equilibrium Analysis of Experimental Data," Berkeley, University of California at Berkeley, preprint.

[153] Christakis, N.A., and J.H. Fowler (2007) "The Spread of Obesity in a Large Social Network over 32 Years," *New England Journal of Medicine* 357:370–379.

[154] Chun, B.-G., R. Fonseca, I. Stoica, and J. Kubiatowicz (2004) "Characterizing Selfishly Constructed Overlay Routing Networks," *INFOCOM 2004, Twenty-Third Annual Joint Conference of the IEEE Computer and Communication Societies* 2:1329–1339.

[155] Chung, F., and L. Lu (2002) "The Average Distances in Random Graphs with Given Expected Degrees," *Proceedings of the National Academy of Sciences* 99:15879–15882.

[156] —— (2002) "Connected Components in Random Graphs with Given Degree Sequences," *Annals of Combinatorics* 6:125–145.

[157] Chvátal, V. (1972) "On Hamilton's Ideals," *Journal of Combinatorial Mathematics* 12:163–168.

[158] Chwe, M.S.-Y. (1994) "Farsighted Coalitional Stability," *Journal of Economic Theory* 63:299–325.

[159] —— (2000) "Communication and Coordination in Social Networks," *Review of Economic Studies* 67:1–16.

[160] Cingano, F., and A. Rosolia (2004) "People I Know: Workplace Networks and Job Search Outcomes," Economic Reserch Paper 600, Rome Bank of Italy.

[161] Cohen, R., K. Erez, D. ben-Avraham, and S. Havlin, S. (2000) "Resilience of the Internet to Random Breakdowns," *Physical Review Letters* 85:4626–4628.

[162] —— (2001) "Breakdown of the Internet under Intentional Attack," *Physical Review Letters* 86:3682–3685.

[163] Coleman, J.S. (1988) "Social Capital in the Creation of Human Capital," *American Journal of Sociology* 94(Supplement: Organizations and Institutions: Sociological and Economic Approaches to the Analysis of Social Structure):S95–S120.

[164] —— (1990) *Foundations of Social Theory*, Cambridge Mass.: Harvard University Press.

[165] Coleman, J.S., E. Katz, and H. Menzel (1966) *Medical Innovation: A Diffusion Study,* Indianapolis, Ind.: Bobbs-Merrill.

[166] Comola, M. (2007) "The Network Structure of Informal Arrangements: Evidence from Rural Tanzania," Barcelona, Universitat Pompeu Fabra, mimeo.

[167] Conley, T.G., and G. Topa (2001) "Socio-Economic Distance and Spatial Patterns in Unemployment," *Journal of Applied Economics* 17(4):303–327.

[168] Conley, T.G., and C.R. Udry (2001) "Social Learning through Networks: The Adoption of New Agricultural Technologies in Ghana," *American Journal of Agricultural Economics* 83(3):668–673.

[169] —— (2004) "Social Networks in Ghana," New Haven, Conn., Economic Growth Center, Yale University, Discussion Paper 888.

[170] —— (2004) "Learning About a New Technology: Pineapple in Ghana," Economic Growth Center Working Paper 817, New Haven, Conn., Yale University.

[171] —— (2004) "The Adoption of New Agricultural Technologies in Ghana," *American Journal of Agricultural Economics* 83(3):668–673.

[172] Cook, K.S., and R.M. Emerson (1978) "Power, Equity and Commitment in Exchange Networks," *American Sociological Review* 43:721–739.

[173] Cook, K.S., and J.M. Whitmeyer (1992) "Two Approaches to Social Structure: Exchange Theory and Network Analysis," *Annual Review of Sociology* 18:109–127.

[174] Cooper, C., and A. Frieze (2003) "A General Model of Web Graphs," *Random Structures and Algorithms* 22(3):311–335.

[175] Copic, J., M.O. Jackson, and A. Kirman (2005) "Identifying Community Structures from Network Data." Available at http://www.stanford.edu/~jacksonm/netcommunity.pdf.

[176] Corbae, D., and J. Duffy (2007) "Experiments with Network Economies," *Games and Economic Behavior,* forthcoming.

[177] Corcoran, M., L. Datcher, and G. Duncan (1980) "Information and Influence Networks in Labor Markets," in *Five Thousand American Families,* volume 8, G. Duncan and J. Morgan, eds., Ann Arbor: University of Michigan.

[178] Corominas-Bosch, M. (2004) "On Two-Sided Network Markets," *Journal of Economic Theory* 115:35–77.

[179] Cournot, A.A. (1838) "Recherches sur les principes mathematiques de la théorie des richesses," translated as: "Researches into the Mathematical Principles of the Theory of Wealth," New York: Macmillan (1897).

[180] Currarini, S. (2004) "Network Design in Games with Spillovers," *Review of Economic Design* 10(4):305–326.

[181] Currarini, S., and M. Morelli (2000) "Network Formation with Sequential Demands," *Review of Economic Design* 5:229–250.

[182] Currarini, S., M.O. Jackson, and P. Pin (2007) "An Economic Model of Friendship: Homophily, Minorities and Segregation," *Econometrica,* forthcoming.

[183] Daskalakis, C., and C.H. Papadimitriou (2005) "The Complexity of Games on Highly Regular Graphs," in *Lecture Notes in Computer Science* 3669, Berlin: Springer-Verlag.

[184] Daskalakis, C., A.G. Dimakis, and E. Mossel (2007) "Connectivity and Equilibrium in Random Games," arXiv:math/0703902.

[185] Daskalakis, C., P.W. Goldberg, and C.H. Papadimitriou (2006) "The Complexity of Computing a Nash Equilibrium," in *Proceedings of the Thirty-Eighth Annual ACM Symposium on Theory of Computing,* Seattle, Wash.

[186] de Castro, R., and J.W. Grossman (1999) "Famous Trails to Paul Erdos," *Mathematical Intelligencer* 21:51–63.

[187] Deck, C., and C. Johnson (2002) "Link Bidding in a Laboratory Network," *Review of Economic Design* 8(4):359–372.

[188] Deffuant G., D. Neau, F. Amblard, and G. Weisbuch (2000) "Mixing Beliefs among Interacting Agents," *Advances in Complex Systems* 3:87–98.

[189] DeGroot, M.H. (1974) "Reaching a Consensus," *Journal of the American Statistical Association* 69:118–121.

[190] DeGroot, M.H., and M.J. Schervish (2002) *Probability and Statistics,* Boston: Addison Wesley.

[191] Deijfen, M., and W. Kets (2007) "Random Intersection Graphs with Tunable Degree Distribution and Clustering," Center for Economic Research Working Paper 2007–08, Tilburg, The Netherlands, Tilburg University.

[192] Demange, G. (2004) "On Group Stability in Hierarchies and Networks," *Journal of Political Economy* 112(4):754–778.

[193] DeMarzo, P., D. Vayanos, and J. Zwiebel (2003) "Persuasion Bias, Social Influence, and Unidimensional Opinions," *Quarterly Journal of Economics* 118:909–968.

[194] Deroïan, F. (2003) "Farsighted Strategies in the Formation of a Communication Network," *Economics Letters* 80:343–349.

[195] ——— (2008) "Dissemination of Spillovers in Cost-Reducing Alliances," *Research in Economics* 62:34–44.

[196] Deroïan, F., and F. Gannon (2006) "Quality-Improving Alliances in Differentiated Oligopoly," *International Journal of Industrial Organization* 24(3):629–637.

[197] De Weerdt, J. (2004) "Risk-Sharing and Endogenous Network Formation," in *Insurance Against Poverty,* S. Dercon, ed., Oxford: Oxford University Press.

[198] De Weerdt, J., and S. Dercon (2006) "Risk-Sharing Networks and Insurance against Illness," *Journal of Development Economics* 81:337–356.

[199] Diamond, P. (1981) "Mobility Costs, Frictional Unemployment, and Efficiency," *Journal of Political Economy* 89:798–812.

[200] Diestel, R. (2000) *Graph Theory,* Heidelberg: Springer-Verlag.

[201] Dirac, G.A. (1952) "Some Theorems on Abstract Graphs," *Proceedings of the London Mathematical Society* 2(3):69–81.

[202] Dodds, P.S., R. Muhamad, and D.J. Watts (2003) "An Experimental Study of Search in Global Social Networks," *Science* 301:327–329.

[203] Doreian, P., V. Batagelj, and A. Ferligoj (2004) *Generalized Blockmodeling*, Cambridge: Cambridge University Press.

[204] Dorogovtsev, S.N., and A.V. Goltsev (2007) "Critical Phenomena in Complex Networks," *Condensed Matter Statistical Mechanics,* arXiv:0705.0010v2.

[205] Dorogovtsev, S.N., and J.F.F. Mendes (2001) "Scaling Properties of Scale-Free Evolving Networks: Continuous Approach," *Physical Review Letters* 63:056125.

[206] Droste, E., R.P. Gilles, and C. Johnson (2000) "Evolution of Conventions in Endogenous Social Networks," in *Econometric Society World Congress 2000 Contributed Papers* 0594, Seattle: Econometric Society.

[207] Durieu J., H. Haller, and P. Solal (2005) "Contagion and Dominating Sets," in *Cognitive Economics: New Trends,* R. Topol and B. Walliser, eds., Amsterdam: Elsevier.

[208] Durlauf, S.N. (2002) "On the Empirics of Social Capital," *Economic Journal* 112:459–479.

[209] Durlauf, S.N., and M. Fafchamps (2006) "Social Capital," in *Handbook of Economic Growth*, P. Aghion and S. Durlauf, eds., Amsterdam: North-Holland.

[210] Dutta, B., and M.O. Jackson (2000) "The Stability and Efficiency of Directed Communication Networks," *Review of Economic Design* 5:251–272.

[211] ——— (2003) "On the Formation of Networks and Groups," in *Networks and Groups: Models of Strategic Formation*, B. Dutta and M.O. Jackson, eds., Heidelberg: Springer-Verlag.

[212] ——— (2003) *Networks and Groups: Models of Strategic Formation*, Heidelberg: Springer-Verlag.

[213] Dutta, B., and S. Mutuswami (1997) "Stable Networks," *Journal of Economic Theory* 76:322–344.

[214] Dutta, B., S. Ghosal, and D. Ray (2005) "Farsighted Network Formation," *Journal of Economic Theory* 122:143–164.

[215] Dutta, B., A. van den Nouweland, and S. Tijs (1998) "Link Formation in Cooperative Situations," *International Journal of Game Theory* 27:245–256.

[216] Economides, N. (1996) "The Economics of Networks," *International Journal of Industrial Organization* 16(4):673–699.

[217] Eeckhout, J. (2004) "Gibrat's Law for (All) Cities," *American Economic Review* 94(5):1429–1451.

[218] Ehrhardt, G., M. Marsili, and F. Vega-Redondo (2006) "Diffusion and Growth in an Evolving Network," *International Journal of Game Theory* 34:383–397.

[219] Eiron, N., and K.S. McCurley (2003) "Locality, Hierarchy, and Bidirectionality in the Web," extended abstract, Second Workshop on Algorithms and Models for the Web-Graph (WAW 2003).

[220] Elbittar, A., R. Harrison, and R. Munoz (2007) "Network Structure in a Link-Formation Game: An Experimental Study," Working Paper 331, Santiago, Chile, Instituo de Economia, Pontifica Universidad Catholica de Chile.

[221] Ellison, G. (1993) "Learning, Local Interaction, and Coordination," *Econometrica* 61:1047–1071.

[222] Ellison, G., and D. Fudenberg (1995) "Word-of-Mouth Communication and Social Learning," *Quarterly Journal of Economics* 110:93–126.

[223] Emerson, R.M. (1962) "Power-Dependence Relations," *American Sociological Review* 27(3):1–40.

[224] ——— (1967) "Exchange Theory, Part I: A Psychological Basis for Social Exchange," in *Sociological Theories in Progress*, J. Berger, M. Zelditch, Jr., and B. Anderson, eds., Boston: Houghton-Mifflin.

[225] ——— (1967) "Exchange Theory, Part II: Exchange Relations and Networks," in *Sociological Theories in Progress,* J. Berger. M. Zelditch, Jr., and B. Anderson, eds., Boston: Houghton-Mifflin.

[226] Epstein, J.M., and R. Axtell (1996) *Growing Artificial Societies: Social Science from the Bottom Up,* Cambridge, Mass.: MIT Press.

[227] Erdös, P., and A. Rényi (1959) "On Random Graphs," *Publicationes Mathematicae Debrecen* 6:290–297.

[228] ——— (1960) "On the Evolution of Random Graphs," *Publication of the Mathematical Institute of the Hungarian Academy of Sciences* 5:17–61.

[229] ——— (1961) "On the Strength of Connectedness of a Random Graph," *Acta Mathamatica Academy of Sciences of Hungarica* 12:261–267.

[230] Esary, J.D., F. Proschan, and D. W. Walkup (1967) "Association of Random Variables, with Applications," *Annals of Mathematical Statistics* 38:1466–1474.

[231] Estoup, J.B. (1916) *Gammes Stenographiques,* Paris: Institut Stenographique de France.

[232] Fabrikant, A., E. Koutsoupias, and C. Papadimitriou (2002) "Heuristically Optimized Tradeoffs: A New Paradigm for Power Laws in the Internet," in *Lecture Notes in Computer Science* 2380, Berlin: Springer-Verlag.

[233] Fabrikant, A., A. Luthra, E. Maneva, C. Papadimitriou, and S. Shenker (2003) "On a Network Creation Game," in *Proceedings of ACM Symposium on Principles of Distributed Systems,* New York: ACM.

[234] Fafchamps, M. (2003) "Ethnicity and Networks in African Trade," in *Contributions to Economic Analysis and Policy,* Berkeley Electronic Press: www.bepress.com 2(1):article 14.

[235] ——— (2004) "Market Institutions in Sub-Saharan Africa" Cambridge, Mass.: MIT Press.

[236] Fafchamps, M., and F. Gubert (2007) "The Formation of Risk-Sharing Networks," *Journal of Development Economics* 83(2):326–350.

[237] Fafchamps, M., and S. Lund (2003) "Risk-Sharing Networks in Rural Philippines," *Journal of Development Economics* 71:261–287.

[238] Fafchamps, M., and B. Minten (2001) "Social Capital and Agricultural Trade," *American Journal of Agricultural Economics* 83(3):680–685.

[239] ——— (2002) "Returns to Social Network Capital among Traders," *Oxford Economic Papers* 54:73–206.

[240] Fafchamps, M., S. Goyal, and M. van der Leij (2005) "Scientific Networks and Coauthorship," Economics Series Working Papers 256, Oxford, University of Oxford.

[241] Falk, A., and M. Kosfeld (2003) "It's All About Connections: Evidence on Network Formation," Institute for the Study of Labor (IZA) Discussion Paper 777, Zurich IEER Working Paper 146, Zurich, Switzerland.

[242] Faloutsos, M., P. Faloutsos, and C. Faloutsos (2004) "On Power-Law Relationships of the Internet Topology," Riverside University of California at Riverside, preprint, Zurich IEER Working Paper 146, Zurich, Switzerland.

[243] Feri, F. (2005) "Network Formation with Endogenous Decay," Working Paper 2005.35, Venice, Università Ca Foscari, mimeo.

[244] ——— (2007) "Stochastic Stability in Networks with Decay," *Journal of Economic Theory* 135(1):442–457.

[245] Fernandez, R.M., and N. Weinberg (1997) "Sifting and Sorting: Personal Contacts and Hiring in a Retail Bank," *American Sociological Review* 62(6):883–902.

[246] Fernandez, R.M., E.J. Castilla, and P. Moore (2000) "Social Capital at Work: Networks and Employment at a Phone Center," *American Journal of Sociology* 105:1288–1356.

[247] Festinger, L. (1957) *A Theory of Cognitive Dissonance,* Stanford, Calif.: Stanford University Press.

[248] Fiedler, M. (1973) "Algebraic Connectivity of Graphs," *Czechoslovak Mathematical Journal* 23:298–305.

[249] Fienberg, S.E., and S.S. Wasserman (1981) "Categorical Data Analysis of Single Sociometric Relations," *Sociological Methodology* 12:156–192.

[250] Fisher, F.M. (1989) *Disequilibrium Foundations of Equilibrium Economics,* Cambridge: Cambridge University Press.

[251] Flinn, C., and J.J. Heckman (1982) "New Methods for Analyzing Structural Models of Labor Force Dynamics," *Journal of Econometrics* 18:115–168.

[252] Fong, E., amd W.W. Isajiw (2000) "Determinants of Friendship Choices in Multiethnic Society," *Sociological Forum* 15(2):249–271.

[253] Fontaine, F. (2004) "Why Are Similar Workers Paid Differently? The Role of Social Networks," Institute for the Study of Labor (IZA) Discussion Paper 1786, Bonn, EUREQua–CNRS.

[254] Frank, O., and D. Strauss (1986) "Markov Graphs," *Journal of the American Statistical Association* 81:832–842.

[255] Freeman, L.C. (1977) "A Set of Measures of Centrality Based on Betweenness," *Sociometry* 40:35–41.

[256] ——— (2004) *The Development of Social Network Analysis: A Study in the Sociology of Science,* Vancouver: Empirical Press.

[257] Freidlin, M.I., and A.D. Wentzell (1984) *Random Perturbations of Dynamical Systems,* New York: Springer-Verlag.

[258] Friedkin, N.E. (1980) "A Test of Structural Features of Granovetter's Strength of Weak Ties Theory," *Social Networks* 2:411–422.

[259] Friedkin, N.E., and E.C. Johnsen (1990) "Social Influence and Opinions," *Journal of Mathematical Sociology* 15:193–206.

[260] ——— (1997) "Social Positions in Influence Networks," *Social Networks* 19:209–222.

[261] Fronczak, A., P. Fronczak, and J.A. Holyst (2003) "Mean-Field Theory for Clustering Coefficients in Barabási-Albert Networks," arXiv:cond-math/0306255v1.

[262] Fryer, R. (2007) "Guess Who's Been Coming to Dinner? Trends in Interracial Marriage over the 20th Century," *Journal of Economic Perspectives* 21(2):71–90.

[263] Fudenberg, D., and J. Tirole (1993) *Game Theory,* Cambridge, Mass.: MIT Press.

[264] Furusawa, T., and H. Konishi (2005) "Free Trade Networks," *Japanese Economic Review* 56:144–164.

[265] Gabaix, X. (1999) "Zipf's Law for Cities: An Explanation," *Quarterly Journal of Economics* 114(3):739–767.

[266] Gale, D., and S. Kariv (2003) "Bayesian Learning in Social Networks," *Games and Economic Behavior* 45(2):329–346.

[267] Gale, D., and L.S. Shapley (1962) "College Admissions and the Stability of Marriage," *American Mathematical Monthly* 69:9–15.

[268] Galeotti, A. (2005) "Consumers' Networks and Search Equilibria," Tinbergen Institute Discussion Paper 04-075-1, Essex, University of Essex.

[269] ——— (2006) "One-Way Flow Networks: The Role of Heterogeneous Players," *Economic Theory* 29:163–179.

[270] Galeotti, A., and M.A. Meléndez-Jiménez (2004) "Exploitation and Cooperation in Networks," Tingbergen Discussion Paper 04-076-1, Essex, University of Essex.

[271] Galeotti A., and G. Mueller (2005) "Friendship Relations in the School Class and Adult Economic Attainment," Institute for the Study of Labor (IZA) Discussion Paper 1682, Bonn.

[272] Galeotti, A., and F. Vega-Redondo (2005) "Strategic Analysis in Complex Networks with Local Externalities," California Institute of Technology Working Paper 1224, Pasadena, Calif.

[273] Galeotti, A., S. Goyal, and J. Kamphorst (2006) "Network Formation with Heterogeneous Players," *Games and Economic Behavior* 54(2):353–372.

[274] Galeotti, A., S. Goyal, M.O. Jackson, F. Vega-Redondo, and L. Yariv (2005) "Network Games," available at http://www.stanford.edu/~jacksonm/netgames.pdf.

[275] Gantmacher, F.R. (1959) *The Theory of Matrices*, Providence, R.I.: American Mathematical Society.

[276] Garfield, E. (1979) "It Is a Small World After All," *Current Contents* 43:5–10.

[277] Geertz, C. (1979) "Suq: The Bazaar Economy in Sefrou," in *Meaning and Order in Moroccan Society*, C. Geertz, H. Geertz, and L. Rosen, eds., Cambridge: Cambridge University Press.

[278] Gell-Mann, M. (1994) *The Quark and the Jaguar*, New York: Freeman.

[279] Gilbert, N., and K. Troitzsch (2005) *Simulation for the Social Scientist*, New York: McGraw-Hill International.

[280] Giles, M.W., and A. Evans (1986) "The Power Approach to Intergroup Hostility," *Journal of Conflict Resolution* 30(3):469–486.

[281] Gilles, R.P., and S. Sarangi (2003) "The Role of Trust in Costly Network Formation," Blackburg, Va., and Baton Rouge, Virginia Tech and Louisiana State University, mimeo.

[282] —— (2003) "Rationalizing Trust in Network Formation," Blackburg, Va., and Baton Rouge, Virginia Tech and Louisiana State University, mimeo.

[283] —— (2005) "Stable Networks and Convex Payoffs," *Review of Economic Design*, forthcoming.

[284] —— (2005) "Building Social Networks," Blackburg, Va., and Baton Rouge, Virginia Tech and Louisiana State University, mimeo.

[285] —— (2007) "Network Potentials," *Review of Economic Design* 11:13–52.

[286] Gilles, R.P., S. Chakrabarti, S. Sarangi, and N. Badasyan (2004) "Critical Agents in Networks," *Mathematical Social Sciences,* forthcoming.

[287] Girvan, M., and M.E.J. Newman (2002) "Community Structure in Social and Biological Networks," *Proceedings of the National Academy of Sciences* 99:7821–7826.

[288] Gladwell, M. (2000) *The Tipping Point: How Little Things Can Make a Big Difference,* New York: Little, Brown, and Company.

[289] Glaeser, E.L., and J. Scheinkman (2003) "Non-Market Interactions," in *Advances in Economics and Econometrics: Theory and Applications, Eight World Congress,* M. Dewatripont, L.P. Hansen, and S. Turnovsky, eds., Cambridge: Cambridge University Press.

[290] Glaeser, E., B. Sacerdote, and J. Scheinkman (1996) "Crime and Social Interactions," *Quarterly Journal of Economics* 111:507–548.

[291] Gleiser, P.M. (2007) "How to Become a Superhero," *Journal of Statistical Mechanics: Theory and Experiments* P09020.

[292] Goeree, J.K., A. Riedl, and A. Ule (2003) "In Search of Stars: Network Formation among Heterogeneous Agents," Institute for the Study of Labor (IZA) Discussion Paper 1754, Bonn.

[293] Goeree, J.K., M.A. McConnell, T. Mitchell, T. Tromp, and L. Yariv (2007) "Linking and Giving Among Teenage Girls," Pasadena, California Institute of Technology, mimeo.

[294] Golub, B., and M.O. Jackson (2007) "Naive Learning in Social Networks: Convergence, Influence, and the Wisdom of Crowds," available at http://www.stanford.edu/~jacksonm/naivelearning.pdf.

[295] Gomez, D., E. Gonzalez-Aranguena, C. Manuel, G. Owen, M. Del Pozo, and J. Tejada (2003) "Centrality and Power in Social Networks: A Game Theoretic Approach," *Mathematical Social Sciences* 46(1):27–54.

[296] Goyal, S. (1993) "Sustainable Communication Networks," Discussion Paper TI 93-250, Amsterdam-Rotterdam, Tinbergen Institute.

[297] ——— (2005) "Learning in Networks," in *Group Formation in Economics: Networks, Clubs, and Coalitions*, G. Demange and M. Wooders, eds., Cambridge: Cambridge University Press.

[298] Goyal, S., and S. Joshi (2003) "Networks of Collaboration in Oligopoly," *Games and Economic Behavior* 43:57–85.

[299] ——— (2006) "Bilateralism and Free Trade," *International Economic Review* 47(3):749–778.

[300] ——— (2006) "Unequal Connections," *International Journal of Game Theory* 34(3):319–349.

[301] Goyal, S., and J.-L. Moraga-González (2001) "R and D Networks," *Rand Journal of Economics* 32:686–707.

[302] Goyal, S., and F. Vega-Redondo (2005) "Learning, Network Formation and Coordination," *Games and Economic Behavior* 50:178–207.

[303] Goyal, S., A. Konovalov, and J.-L. Moraga (2007) "Hybrid R and D," *Journal of the European Economic Association,* forthcoming.

[304] Goyal, S., M. van der Leij, and J.-L. Moraga-González (2006) "Economics: An Emerging Small World," *Journal of Political Economy* 114(2):403–412.

[305] Granovetter, M. (1973) "The Strength of Weak Ties," *American Journal of Sociology* 78:1360–1380.

[306] ——— (1978) "Threshold Models of Collective Behavior," *American Journal of Sociology* 83(6):1420–1443.

[307] ——— (1985) "Economic Action and Social Structure: The Problem of Embeddedness," *American Journal of Sociology* 91(3):481–510.

[308] ——— (1995) *Getting a Job: A Study of Contacts and Careers*, second edition, Chicago: University of Chicago Press.

[309] Grassberger, P. (1983) "On the Critical Behavior of the General Epidemic Process and Dynamical Percolation," *Mathematical Biosciences* 63:157–172.

[310] Greenberg, J., and E. Postalci (2004) "Characterization of Stable Nash Networks with Production," Montreal, McGill University, mimeo.

[311] Griliches, Z. (1957) "Hybrid Corn: An Exploration in the Economics of Technological Change," *Econometrica* 25(4):501–522.

[312] Grimm, V., and S.F. Railsback (2005) *Individual-Based Modeling and Ecology,* Princeton, N.J.: Princeton University Press.

[313] Grimmett, G. (1999) *Percolation*, Heidelberg: Springer-Verlag.

[314] Grossman, J.W. (2002) "The Evolution of the Mathematical Research Collaboration Graph," in *Proceedings of the 33rd Southeastern Conference on Combinatorics* (Congressus Numerantium, Vol. 158).

[315] Grossman, J.W., and P.D.F. Ion (1995) "On a Portion of the Well-Known Collaboration Graph," in *Proceedings of the 26th Southeastern Conference on Combinatorics* (Congressus Numerantium, Vol. 108).

[316] Haag, M., and R. Lagunoff (2006) "Norms, Local Interaction, and Neighborhood Planning," *International Economic Review* 47(1):265–296.

[317] Hagerstrand, T. (1967) *Innovation Diffusion as a Spatial Process,* Chicago: University of Chicago Press.

[318] Haller, H. (1990) "Large Random Graphs in Pseudo-Metric Spaces," *Mathematical Social Sciences* 20:147–164.

[319] Haller, H., and S. Sarangi (2005) "Nash Networks with Heterogeneous Links," *Mathematical Social Sciences* 50:181–201.

[320] Haller, H., J. Kamphorst, and S. Sarangi (2007) "(Non-)Existence and Scope of Nash Networks," *Economic Theory* 31(3):597–604.

[321] Handcock, M.S., and M. Morris (2006) "A Simple Model for Complex Networks," in *Statistical Network Analysis: Models, Issues, and New Directions*, E. Airoldi, D. Blei, and S.E. Fienberg, eds., New York: Springer-Verlag.

[322] Harrison, R., and R. Munoz (2007) "Stability and Equilibrium Selection in a Link Formation Game," *Economic Theory*, forthcoming.

[323] Hartfiel, D.J., and C.D. Meyer (1998) "On the Structure of Stochastic Matrices with a Subdominant Eigenvalue Near 1," *Linear Algebra and Its Applications* 272(1):193–203.

[324] Heckman, J.J., and G. Borjas (1980) "Does Unemployment Cause Future Unemployment? Definitions, Questions and Answers from a Continuous Time Model of Heterogeneity and State Dependence," *Economica* 47:247–283.

[325] Hegselmann, R., and U. Krause (2002) "Opinion Dynamics and Bounded Confidence Models, Analysis, and Simulations," *Journal of Artifical Societies and Social Simulation* 5(3):1–33.

[326] Hendricks, K., M. Piccione, and C. Tan (1995) "The Economics of Hubs: The Case of Monopoly," *Review of Economic Studies* 62:83–100.

[327] Herings, P.J.J., A. Mauleon, and V. Vannetelbosch (2004) "Rationalizability for Social Environments," *Games and Economic Behavior* 49:135–156.

[328] ——— (2006) "Farsightedly Stable Networks," CORE Discussion Paper 2006-92, Louvain, Belgium.

[329] Hirshleifer, J. (1983) "From Weakest-Link to Best-Shot: The Voluntary Provision of Public Goods," *Public Choice* 41(3):371–386.

[330] Hoff, P.D. (2006) "Multiplicative Latent Factor Models for Description and Prediction of Social Networks," Seattle, University of Washington, mimeo.

[331] Hoff, P.D., A.E. Raftery, and M.S. Handcock (2002) "Latent Space Approaches to Social Network Analysis," *Journal of the American Statistical Association* 97:1090–1098.

[332] Hojman, D., and A. Szeidl (2006) "Endogenous Networks, Social Games and Evolution," *Games and Economic Behavior* 55(1):112–130.

[333] Holland, P.W., and S. Leinhardt (1977) "A Dynamic Model for Social Networks," *Journal of Mathematical Sociology* 5:5–20.

[334] Holland, P.W., K.B. Laskey, and S. Leinhardt (1983) "Stochastic Blockmodels: First Steps," *Social Networks* 5:109–137.

[335] Holme, P., and B.J. Kim (2002) "Vertex Overload Breakdown in Evolving Networks," *Physical Review E* 65:066109.

[336] Homans, C.G. (1958) "Social Behavior as Exchange," *American Journal of Sociology* 62:597–606.

[337] ——— (1961) *Social Behavior: Its Elementary Forms*, New York: Harcourt, Brace and World.

[338] Hoory, S., N. Linial, and A. Widgerson (2006) "Expander Graphs and Their Applications," *Bulletin of the American Mathematical Society* 43(4):439–561.

[339] Horst, U., and J.A. Scheinkman (2006) "Equilibria in Systems of Social Interacion," *Journal of Economic Theory* 130:44–77.

[340] Huckfeldt, R., and J. Sprague (1987) "Networks in Context: The Social Flow of Political Information," *American Political Science Review* 81(4):1197–1216.

[341] Ilkilic, R. (2004) "Pairwise Stability: Externalities and Existence," Barcelona, Universitat Autonoma de Barcelona, mimeo.

[342] Ioannides, Y.M. (1997) "Evolution of Trading Structures," in *The Economy as an Evolving Complex System II*, B. Arthur, S. Durlauf, and D. Lane, eds., Reading, Mass.: Addison-Wesley.

[343] ——— (2004) "Random Graphs and Social Networks: An Economics Perspective," Boston, Tufts University, mimeo.

[344] ——— (2006) "Topologies of Social Interactions," *Economic Theory* 28(3):559–584.

[345] Ioannides, Y.M., and L. Datcher Loury (2004) "Job Information Networks, Neighborhood Effects and Inequality," *Journal of Economic Literature* 42(4):1056–1093.

[346] Ioannides, Y.M., and A. Soetevent (2005) "Social Networking and Individual Outcomes: Beyond the Mean Field Case," *Journal of Economic Behavior and Organization* 64(3–4):369–390.

[347] Jackson, M.O. (2003). "The Stability and Efficiency of Economic and Social Networks," in *Advances in Economic Design*, S. Koray and M. Sertel, eds., Heidelberg: Springer-Verlag. Reprinted [2003] in *Networks and Groups: Models of Strategic Formation*, B. Dutta and M.O. Jackson, eds., Heidelberg: Springer-Verlag.

[348] ——— (2005) "A Survey of Models of Network Formation: Stability and Efficiency," in *Group Formation in Economics; Networks, Clubs and Coalitions*, G. Demange and M. Wooders, eds., Cambridge: Cambridge University Press.

[349] ——— (2005) "Allocation Rules for Network Games," *Games and Economic Behavior* 51:128–154.

[350] ——— (2006) "The Economics of Social Networks," in *Advances in Economics and Econometrics, Theory and Applications: Ninth World Congress of the Econometric Society*, Vol. 3, R. Blundell, W. Newey, and T. Persson, eds., Cambridge: Cambridge University Press.

[351] ——— (2007) "The Study of Social Networks in Economics," in *The Missing Links: Formation and Decay of Economic Networks,* J. Podolny and J.E. Rauch, eds., New York: Russell Sage Foundation.

[352] ——— (2008) "Social Networks in Economics," in *Handbook of Social Economics,* J. Benhabib, A. Bisin, and M.O. Jackson, eds., New York: Elsevier.

[353] Jackson, M.O., and B.W. Rogers (2005) "The Economics of Small Worlds," *Journal of the European Economic Association (Papers and Proceedings)* 3(2–3):617–627.

[354] ——— (2007) "Relating Network Structure to Diffusion Properties through Stochastic Dominance," *B.E. Press Journal of Theoretical Economics* 7(1):1–13.

[355] ——— (2007) "Meeting Strangers and Friends of Friends: How Random Are Social Networks?" *American Economic Review* 97(3):890–915.

[356] Jackson, M.O., and A. van den Nouweland (2005) "Strongly Stable Networks," *Games and Economic Behavior* 51:420–444.

[357] Jackson, M.O., and A. Watts (2001) "The Existence of Pairwise Stable Networks," *Seoul Journal of Economics* 14(3):299–321.

[358] ——— (2002) "The Evolution of Social and Economic Networks," *Journal of Economic Theory* 106(2):265–295.

[359] ——— (2002) "On the Formation of Interaction Networks in Social Coordination Games," *Games and Economic Behavior* 41(2):265–291.

[360] ——— (2005) "Social Games: Matching and the Play of Finitely Repeated Games," *Games and Economic Behavior,* forthcoming.

[361] Jackson, M.O., and J. Wolinsky (1996) "A Strategic Model of Social and Economic Networks," *Journal of Economic Theory* 71(1):44–74.

[362] Jackson, M.O., and L. Yariv (2005) "Diffusion on Social Networks," *Économie Publique* 16(1):3–16.

[363] ——— (2006) "Social Networks and the Diffusion of Behavior," *Yale Economic Review* 3(2):42–47.

[364] ——— (2007) "The Diffusion of Behavior and Equilibrium Structure on Social Networks," *American Economic Review* (papers and proceedings) 97(2):92–98.

[365] ——— (2008) "Diffusion, Strategic Interaction, and Social Structure," in *Handbook of Social Economics,* J. Benhabib, A. Bisin, and M.O. Jackson, eds., New York: Elsevier.

[366] Johari, R., S. Mannor, and J.N. Tsitsiklis (2006) "A Contract-Based Model for Directed Network Formation," *Games and Economic Behavior* 56(2):201–224.

[367] Johnson, C., and R.P. Gilles (2000) "Spatial Social Networks," *Review of Economic Design* 5:273–300.

[368] Johnson, D.S., C.H. Papadimitriou, and M. Yannakakis (1988) "On Generating All Maximal Independent Sets," *Information Processing Letters* 27(3):119–123.

[369] Judd, J.S., and M. Kearns (2008) "Behavioral Experiments in Networked Trade," Philadelphia, Wharton School of Business, University of Pennsylvania, preprint.

[370] Jun, T., K.-Y. Kim, B.J. Kim, and M.Y. Choi (2006) "Consumer Referral in a Small World Network," *Social Networks* 28(3):232–246.

[371] ——— (2006) "Network Marketing on Small World Network," *Physica A: Statistical Mechanics and Its Applications* 360(2):493–504.

[372] Kakade, S.M., M. Kearns, and L.E. Ortiz (2004) *Graphical Economics.* Lecture Notes in Computer Science 3120, Berlin: Springer-Verlag.

[373] Kakade, S.M., M. Kearns, J. Langford, and L.E. Ortiz (2003) "Correlated Equilibria in Graphical Games," in *Proceedings of the 4th ACM Conference on Electronic Commerce,* New York: ACM.

[374] Kakade, S.M., M. Kearns, L.E. Ortiz, R. Pemantle, and S. Suri (2004) "Economic Properties of Social Networks," *Proceedings of Neural and Information Processing Systems (NIPS),* Cambridge, Mass.: MIT Press.

[375] Kalai, E. (2004) "Large Robust Games," *Econometrica* 72:1631–1666.

[376] Kandel, D.B. (1978) "Homophily, Selection, and Socialization in Adolescent Friendships," *American Journal of Sociology* 14:427–436.

[377] Kandori, M., G. Mailath, and R. Rob (1993) "Learning, Mutation, and Long-Run Equilibria in Games," *Econometrica* 61:29–56.

[378] Kannan, R., L. Ray, and S. Sarangi (2005) "The Structure of Information Networks," *Economic Theory*, forthcoming.

[379] Karlin, S., and H. Taylor (1975) *A First Course in Stochastic Processes,* San Diego: Academic Press.

[380] Katz, E., and P.F. Lazarsfeld (1955) *Personal Influence: The Part Played by People in the Flow of Mass Communication,* New York: Free Press.

[381] Katz, L. (1953) "A New Status Index Derived from Sociometric Analysis," *Psychometrica* 18:39–43.

[382] Katz, M., and C. Shapiro (1994) "Systems Competition and Networks Effects," *Journal of Economic Perspectives* 8:93–115.

[383] Kawamata, K., and Y. Tamada (2004) "Work Partnering and Network Formation," Waseda Economics Papers, Tokyo, Waseda University.

[384] Kearns, M.J., and S. Suri (2005) "Networks Preserving Evolutionary Stability and the Power of Randomization," Philadelphia, University of Pennsylvania, preprint.

[385] Kearns, M.J., M. Littman, and S. Singh (2001) "Graphical Models for Game Theory," in *Proceedings of the 17th Conference on Uncertainty in Artificial Intelligence,* J.S. Breese and D. Koller, eds., San Francisco: Morgan Kaufmann.

[386] Kearns, M.J., S. Suri, and N. Montfort (2006) "An Experimental Study of the Coloring Problem on Human Subject Networks," *Science* 313:824–827.

[387] Kent, D. (1978) *The Rise of the Medici: Faction in Florence 1426–1434,* Oxford: Oxford University Press.

[388] Kermack, W.O., and A.G. McKendrick (1927) "A Contribution to the Mathematical Theory of Epidemics," *Proceedings of the Royal Society of London* Series A, *Containing Papers of a Mathematical and Physical Character* 115:700–721.

[389] Kernighan, B.W., and S. Lin (1970) "An Efficient Heuristic Procedure for Partitioning Graphs," *Bell System Technical Journal* 49:291–307.

[390] Kesten, H. (1973) "Random Difference Equations and Renewal Theory for Products of Random Matrices," *Acta Mathematica* 131:207–248.

[391] Kets, W. (2008) "Networks and Learning in Game Theory," dissertation, Tilburg University, Tilburg, The Netherlands.

[392] Khwaja, A.I., A. Mian, and A. Qamar (2005) "The Value of Business Networks," Chicago, University of Chicago, mimeo.

[393] Killworth, B., and H. Bernard (1976) "Informant Accuracy in Social Network Data," *Human Organization* 35:269–286.

[394] ——— (1978/1979) "The Reversal Small-World Experiment," *Social Networks* 1:159–192.

[395] ——— (1979) "A Pseudomodel of the Small World Problem," *Social Forces* 58:477–505.

[396] Kirman, A.P. (1983) "Communication in Markets: A Suggested Approach," *Economics Letters* 12:1–5.

[397] ——— (1993) "Ants, Rationality, and Recruitment," *Quarterly Journal of Economics* 108(1):137–156.

[398] ——— (1997) "The Economy as an Evolving Network," *Journal of Evolutionary Economics* 7:339–353.

[399] ——— (2001) "Market Organization and Individual Behavior: Evidence from Fish Markets," in *Networks and Markets,* J. Rauch and A. Cassella, eds., New York: Russell Sage Foundation.

[400] Kirman, A.P., C. Oddou, and S. Weber (1986) "Stochastic Communication and Coalition Formation," *Econometrica* 54:129–138.

[401] Kleinberg, J.M. (2000) "Navigation in a Small World," *Nature* 406:845.

[402] ——— (2000) "The Small-World Phenomenon: An Algorithmic Perspective," *Annual ACM Symposium on Theory of Computing, Proceedings of the Thirty-Second Annual ACM Symposium on Theory of Computing,* New York: ACM.

[403] Kleinberg, J.M., S.R. Kumar, P. Raghavan, S. Rajagopalan, and A. Tomkins (1999) "The Web as a Graph: Measurements, Models and Methods," in *Proceedings of the International Conference on Combinatorics and Computing, Lecture Notes in Computer Science* 1627, Berlin: Springer-Verlag.

[404] Klemm, K., and V.M. Eguíluz (2002) "Growing Scale-Free Networks with Small World Behavior," *Physical Review E* 65:036123.

[405] ——— (2002) "Highly Clustered Scale-Free Networks," *Physical Review E* 65:057102.

[406] Kochen, M. (1989) *The Small World*, Norwood, N.J.: Albex.

[407] Kogut, B. (2000) "The Network as Knowledge: Generative Rules and the Emergence of Structure," *Strategic Management Journal* 21(3):405–425.

[408] Koller, D., and B. Milch (2003) "Multi-agent Influence Diagrams for Representing and Solving Games," *Games and Economic Behavior* 45:181–221.

[409] Konishi, H., and M. Utku Ünver (2006) "Credible Group Stability in Many-to-Many Matching Problems," *Journal of Economic Theory* 129:57–80.

[410] Kosfeld, M. (2003) "Network Experiments," Zuirch IEER Working Paper 152, University of Zurich.

[411] Krackhardt, D. (1987) "Cognitive Social Structures," *Social Networks* 9:109–134.

[412] Kranton, R., and D. Minehart (2000) "Competition in Buyer-Seller Networks," *Review of Economic Design* 5(3):301–331.

[413] ———— (2000) "Competition for Goods in Buyer-Seller Networks," *Review of Economic Design* 5:301–332.

[414] ———— (2001) "A Theory of Buyer-Seller Networks," *American Economic Review* 91(3):485–508.

[415] Krapivsky, P.L., and S. Redner (2001) "Organization of Growing Random Networks," *Physical Review E* 63:066123.

[416] ———— (2002) "A Statistical Physics Perspective on Web Growth," *Computer Networks* 39(3):261–276.

[417] Krause, U. (2000) "A Discrete Nonlinear and Nonautonomous Model of Consensus Formation," in *Communications in Difference Equations*, S. Elaydi, G. Ladas, J. Popenda, and J. Rakowski, eds., Amsterdam: Gordon and Breach.

[418] Kretschmar, M., and M. Morris (1996) "Measures of Concurrency in Networks and the Spread of Infectious Disease," *Mathematical Bioscience* 133:165–195.

[419] Krishnan, P., and E. Sciubba (2005) "Links and Architecture in Village Networks," Working Papers in Economics and Science 0614, Cambridge and London, University of Cambridge and Birkbeck College.

[420] Kumar, R., P. Raghavan, S. Rajagopalan, D. Sivakumar, A. Tomkins, and E. Upfal (2000) "Stochastic Models for the Web Graph," *Foundations of Computer Science 2000,* available at www.csa.com.

[421] La Mura, P. (2000) "Game Networks," in *Proceedings of the 16th Conference on Uncertainty in Artificial Intelligence,* San Francisco: Morgan Kaufmann.

[422] Landers, D., and L. Rogge (1987) "Laws of Large Numbers for Pairwise Independent Uniformly Integrable Random Variables," *Mathematische Nachrichten* 130:189–192.

[423] Laschever, R. (2007) "The Doughboys Network: Social Interactions and Labor Market Outcomes of World War I Veterans," University of Illinois Working Paper, Champaign-Urbana.

[424] Lavezzi, A.M., and N. Meccheri (2005) "Social Networks in Labor Markets: The Effects of Symmetry, Randomness, and Exclusion on Output and Inequality," Pisa, University of Pisa, mimeo.

[425] Lazarsfeld, P.F., and R.K. Merton (1954) "Friendship as a Social Process: A Substantive and Methodological Analysis," in *Freedom and Control in Modern Society*, M. Berger, ed., New York: Van Nostrand.

[426] Lazarsfeld, P.F., B. Berelson, and H. Gaudet (1944) *The People's Choice: How the Voter Makes Up His Mind in a Presidential Campaign*, New York: Columbia University Press.

[427] Leavitt, H.J. (1951) "Some Effects of Certain Communication Patterns on Group Performance," *Journal of Abnormal and Social Psychology* 46:38–50.

[428] Lehrer, K. (1975) "Social Consensus and Rational Agnoiology," *Synthese* 31:141–160.

[429] Leider, S., M.M. Mobius, T. Rosenblat, and Q.-A. Do (2007) "How Much Is a Friend Worth? Directed Altruism and Enforced Reciprocity in Social Networks," revision of NBER Working Paper 13135, Cambridge, Mass., National Bureau of Economics Research.

[430] Leskovec, J., L.A. Adamic, and B.A. Huberman (2007) "The Dynamics of Viral Marketing," *ACM Transactions on the Web (TWEB) archive* 1:1: article 5.

[431] Levene, M., T. Fenner, G. Loizou, and R. Wheeldon (2002) "A Stochastic Model for the Evolution of the Web," *Computer Networks* 39:277–287.

[432] Li, L., D. Alderson, W. Willinger, J. Doyle, R. Tanaka, and S. Low (2004) "A First Principles Approach to Understanding the Internet's Router Technology," in *Applications, Technologies, Architectures and Protocols for Computer Communication, Procedings of the 2004 Conference,* New York: ACM.

[433] Liebowitz, S., and S. Margolis (1994) "Network Externality: An Uncommon Tragedy," *Journal of Economic Perspectives* 8:133–150.

[434] Lin, N., and M. Dumin (1986) "Access to Occupations through Social Ties," *Social Networks* 8:365–385.

[435] Lin, N., W.M. Ensel, and J.C. Vaughn (1981) "Social Resources and Strength of Ties: Structural Factors in Occupational Status Attainment," *American Sociological Review* 44(4):393–405.

[436] Lippert, S., and G. Spagnolo (2003) "Networks of Relations," Discussion Paper 28, Mannheim, Germany, University of Mannheim.

[437] Littman, M.L., M.J. Kearns, and S.P. Singh (2001) "An Efficient, Exact Algorithm for Solving Tree-Structured Graphical Games," in *Advances in Neural Information Processing Systems*, Cambridge, Mass.: MIT Press.

[438] Lopez-Pintado, D. (2008) "Diffusion in Complex Social Networks," *Games and Economic Behavior* 62(2):573–590.

[439] Lopez-Pintado, D., and D. Watts (2006) "Social Influence, Binary Decisions and Collective Dynamics," New York, Department of Sociology, Columbia University, mimeo.

[440] Lorenz, J. (2005) "A Stabilization Theorem for Dynamics of Continuous Opinions," *Physica A* 355:217–223.

[441] Lorrain, F., and H.C. White (1971) "Structural Equivalence of Individuals in Social Networks," *Journal of Mathematical Sociology* 1:49–80.

[442] Lotka, A.J. (1926) "The Frequency Distribution of Scientific Productivity," *Journal of the Washington Academy of Sciences* 16:317–323.

[443] Luce, R.D., and A.D. Perry (1949) "A Method of Matrix Analysis of Group Structure," *Journal Psychometrika* 14:95–116.

[444] Luce, R.D., and H. Raiffa (1957) *Games and Decisions,* New York: Wiley.

[445] Lynch, L.M. (1989) "The Youth Labor Market in the Eighties: Determinants of Reemployment Probabilities for Young Men and Women," *Review of Economics and Statistics* 71:37–54.

[446] MacRae, J. (1960) "Direct Factor Analysis of Sociometric Data," *Sociometry* 23:360–371.

[447] Mairesse, J., and L. Turner (2006) "Measurement and Explanation of the Intensity of Co-Publication in Scientific Research: An Analysis at the Laboratory Level," in *New Frontiers in the Economics of Innovation and New Technology: Essays in Honor of Paul A. David,* C. Antonelli, D. Foray, B. Hall, et al., eds., Northhampton, Mass.: Elgar Publishing.

[448] Manski, C.F. (1993) "Identification of Endogenous Social Effects: The Reflection Problem," *Review of Economic Studies* 60:419–431.

[449] ——— (2000) "Economic Analysis of Social Interactions," *Journal of Economic Perspectives* 14(3):115–136.

[450] Marsali, M., F. Vega-Redondo, and F. Slanina (2007) "The Rise and Fall of a Networked Society: A Formal Model," *Proceedings of the National Academy of Sciences of the U.S.A.* 101:1439–1442.

[451] Marsden, P.V. (1987) "Core Discussion Networks of Americans," *American Sociological Review* 52:122–131.

[452] ——— (1988) "Homogeneity in Confiding Relations," *Social Networks* 10:57–76.

[453] ——— (1990) "Network Data and Measurement," *Annual Review of Sociology* 16:435–463.

[454] Marsden, P.V., and E. Gorman (2001) "Social Networks, Job Changes, and Recruitment," in *The Sourcebook of Labor Markets: Evolving Structure and Processes* I. Berg and A.L. Kalleberg, eds., New York: Kluwer.

[455] Marsden, P.V., and J. Podolny (1990) "Dynamic Analysis of Network Diffusion Processes," in *Social Networks through Time*, J. Weesie and H. Flap, eds., Utrecht: ISOR.

[456] Matsubayashi, N., and S. Yamakawa (2004) "A Network Formation Game with an Endogenous Cost Allocation Rule," Tokyo, Tokyo University of Science, mimeo.

[457] Mauleon A., and V.J. Vannetelbosch (2004) "Farsightedness and Cautiousness in Coalition Formation," *Theory and Decision* 56:291–324.

[458] Mauleon A., J. Sempere-Monerris, and V.J. Vannetelbosch (2005) "Networks of Manufacturers and Retailers," Louvain, Belgium, CORE, mimeo.

[459] —— (2005) "R and D Networks among Unionized Firms," Louvain, Belgium, CORE, mimeo.

[460] McBride, M. (2002) "Position-Specific Information in Social Networks: Are You Connected?" University of California at Irvine, mimeo.

[461] —— (2006) "Imperfect Monitoring in Communication Networks," *Journal of Economic Theory* 126:97–119.

[462] McPherson, M., L. Smith-Lovin, and J.M. Cook (2001) "Birds of a Feather: Homophily in Social Networks," *Annual Review of Sociololgy* 27:415–444

[463] Meessen, R. (1988) "Communication Games," master's thesis, University of Nijmegen, Nijmegen, The Netherlands.

[464] Meléndez-Jiménez, M.A. (2008) "A Bargaining Approach to Coordination in Networks," *International Journal of Game Theory,* forthcoming.

[465] Merton, R. (1973) *The Sociology of Science: Theoretical and Empirical Investigations,* Chicago: University of Chicago Press.

[466] Meyer, C.D. (2000) *Matrix Analysis and Applied Linear Algebra*, Philadelphia: SIAM.

[467] Milchtaich, I. (1996) "Congestion Games with Player–Specific Payoff Functions," *Games and Economic Behavior* 13:111–124.

[468] Milgram, S. (1967) "The Small-World Problem," *Psychology Today* 2:60–67.

[469] Milgrom, P.R., and D.J. Roberts (1990) "Rationalizability, Learning and Equilibrium in Games with Strategic Complementarities," *Econometrica* 58(6):1255–1277.

[470] Mitzenmacher, M. (2004) "A Brief History of Generative Models for Power Law and Lognormal Distributions," manuscript available at http://www.eecs.harvard.edu/~michaelm/ListByYear.html.

[471] Mobius, M.M., and T.S. Rosenblatt (2003) "Experimental Evidence on Trading Favors in Networks," Cambridge, Mass., and Middletown, Conn., Harvard University and Wesleyan University, mimeo.

[472] Mobius, M.M., and A. Szeidl (2006) "Trust and Social Collateral," Cambridge, Mass., and Berkeley, Harvard University and the University of California at Berkeley.

[473] Molloy, M., and B. Reed (1995) "A Critical Point for Random Graphs with a Given Degree Sequence," *Random Structures and Algorithms* 6:161–179.

[474] Monasson, R. (1999) "Diffusion, Localization and Dispersion Relations on 'Small-World' Lattices," *European Physics Journal B* 12:555–567.

[475] Monderer, D., and L.S. Shapley (1996) "Potential Games," *Games and Economic Behavior* 14:124–143.

[476] Monsuur, H. (2002) "Centrality and Network Dynamics," Den Helder, Royal Netherlands Naval Academy, mimeo.

[477] Monsuur, H., and T. Storcken (2004) "Centrality Orderings in Social Networks," in *Chapters in Game Theory: In Honor of Stef Tijs*, Theory and Decision Library Series C, Vol. 31, Game Theory, Mathematical Programming and Operations Research, Berlin: Springer-Verlag.

[478] ——— (2004) "Centers in Connected Undirected Graphs: An Axiomatic Approach," *Operations Research* 52(1):54–64.

[479] Montgomery, J. (1991) "Social Networks and Labor Market Outcomes," *American Economic Review* 81:1408–1418.

[480] ——— (1992) "Job Search and Network Composition: Implications of the Strength-of-Weak-Ties Hypothesis," *American Sociological Review* 57:586–596.

[481] ——— (1994) "Weak Ties, Employment, and Inequality: An Equilibrium Analysis," *American Journal of Sociology* 99:1212–1236.

[482] Moody, J. (2001) "Race, School Integration, and Friendship Segregation in America," *American Journal of Sociology* 107(3):679–716.

[483] Moreno, Y., J.B. Gómez, and A.F. Pacheco (2002) "Instability of Scale-Free Networks under Node-Breaking Avalanches," *Europhysics Letters* 58:630–636.

[484] Moreno, Y., R. Pastor-Satorras, and A. Vespignani (2002) "Epidemic Outbreaks in Complex Heterogeneous Networks," *European Physics Journal B* 26:521–529.

[485] Moreno, Y., R. Pastor-Satorras, A. V'azquez, and A. Vespignani (2003) "Critical Load and Congestion Instabilities in Scale-Free Networks," *Europhysics Letters* 62:292–298.

[486] Morris, M. (1993) "Epidemiology and Social Networks: Modeling Structured Diffusion," *Sociological Methods and Research* 22(1):99–126.

[487] Morris, S. (2000) "Contagion," *Review of Economic Studies* 67:57–78.

[488] Morris, S., and H. Shin (2002) "Heterogeneity and Uniqueness in Interaction Games," Cowles Foundation Discussion Paper 1275R, New Haven, Conn., Yale University.

[489] ——— (2003) "Global Games: Theory and Applications," in *Advances in Economics and Econometrics. Proceedings of the Eighth World Congress of the Econometric Society,* M. Dewatripont, L. Hansen, and S. Turnovsky, eds., Cambridge: Cambridge University Press.

[490] Mortensen, D.T., and C.A. Pissarides (1994) "Job Creation and Job Destruction in the Theory of Unemployment," *Review of Economic Studies* 61(3):397–415.

[491] Motter, A.E., and Y-C. Lai (2002) "Cascade-Based Attacks on Complex Networks," *Physical Review E* 66:065102.

[492] Moulin, H. (1988) "Axioms of Cooperative Decision-Making," Cambridge: Cambridge University Press.

[493] Munshi, K. (2003) "Networks in the Modern Economy: Mexican Migrants in the U.S. Labor Market," *Quarterly Journal of Economics* 118(2):549–597.

[494] Mutuswami, S., and E. Winter (2002) "Subscription Mechanisms for Network Formation," *Journal of Economic Theory* 106:242–264.

[495] Myers C.A., and G.P. Shultz (1951) *The Dynamics of a Labor Market*, New York: Prentice-Hall.

[496] Myerson, R. (1977) "Graphs and Cooperation in Games," *Mathematics of Operations Research* 2:225–229.

[497] ——— (1991) *Game Theory: Analysis of Conflict*, Cambridge, Mass.: Harvard University Press.

[498] Nash, J.F. (1951) "Non-Cooperative Games," *Annals of Mathematics* 54:286–295.

[499] Navarro, N. (2007) "Fair Allocation in Networks with Externalities," *Games and Economic Behavior* 58(2):354–364.

[500] ——— (2007) "A Sensitive Flexible-Network Approach," Málaga Spain, Universidad de Málaga, mimeo.

[501] Navarro, N., and A. Perea (2001) "Bargaining in Networks and the Myerson Value," Madrid, Universidad Carlos III de Madrid, mimeo.

[502] Newman, M.E.J. (2002) "The Spread of Epidemic Disease on Networks," *Physical Review E* 66:016128.

[503] ——— (2003) "The Structure and Function of Complex Networks," *SIAM Review* 45:167–256.

[504] ——— (2003) "Properties of Highly Clustered Networks," *Physical Review E* 68:026121.

[505] ——— (2004) "Coauthorship Networks and Patterns of Scientific Collaboration," *Proceedings of the National Academy of Sciences* 101:5200–5205.

[506] ——— (2004) "Detecting Community Structure in Networks," *Physical Review E* 69:066133.

[507] Newman, M.E.J., and M. Girvan (2004) "Finding and Evaluating Community Structure in Networks," *Physical Review E* 69:026113.

[508] Newman, M.E.J., and E.A. Leicht (2007) "Models and Exploratory Analysis in Networks," *Proceedings of the National Academy of Sciences, U.S.A.* 104:9564–9569.

[509] Newman, M.E.J., and D.J. Watts (1999) "Renormalization Group Analysis of the Small-World Network Model," *Physical Review E* 263:341–346.

[510] Newman, M.E.J., S.H. Strogatz, and D.J. Watts (2001) "Random Graphs with Arbitrary Degree Distributions and Their Applications," *Physical Review E* 64:026118.

[511] Nieva, R. (2002) "An Extension of the Aumann-Myerson Solution for Reasonable Empty-Core Games," Rochester, N.Y., Rochester Institute of Technology, mimeo.

[512] Ochs, J., and I.-U. Park (2004) "Overcoming the Coordination Problem: Dynamic Formation of Networks," Pittsburgh, Penn., University of Pittsburgh, mimeo.

[513] Olken, B. (2006) "Do Television and Radio Destroy Social Capital? Evidence from Indonesian Villages," NBER Working Paper 12561, Cambridge, Mass., National Bureau of Economic Research.

[514] Osborne, M., and A. Rubinstein (1994) *A Course in Game Theory*, Cambridge, Mass.: MIT Press.

[515] Ozsoylev, H.N. (2003) "Knowing Thy Neighbor: Rational Expectations and Social Interaction in Financial Markets," Minneapolis, University of Minnesota, mimeo.

[516] Padgett, J.F., and C.K. Ansell (1993) "Robust Action and the Rise of the Medici, 1400–1434," *American Journal of Sociology* 98:1259–1319.

[517] Page, F., and S. Kamat (2004) "Farsighted Stability in Network formation," in *Group Formation in Economics: Networks, Clubs, and Coalitions*, G. Demange and M. Wooders, eds., Cambridge: Cambridge University Press.

[518] Page, F., and M.H. Wooders (2005) "Strategic Basins of Attraction, the Farsighted Core, and Network Formation Games," *Games and Economic Behavior,* forthcoming.

[519] Page, F., M.H. Wooders, and S. Kamat (2005) "Networks and Farsighted Stability," *Journal of Economic Theory* 120(2):257–269.

[520] Palacios-Huerta, I., and O. Volij (2004) "The Measurement of Intellectual Influence," *Econometrica* 72(3):963–977.

[521] ——— (2006) "Professionals Play Minimax in Laboratory Experiments," Providence, R.I., and Ames, Brown University and Iowa State University, mimeo.

[522] Pantz, K., and A. Ziegelmeyer (2003) "An Experimental Study of Network Formation," Garching, Germany, Max Planck Institute, mimeo.

[523] Papadimitriou, C.H. (1995) *Computational Complexity*, San Diego, Calif.: Addison Wesley.

[524] ——— (2001) "Algorithms, Games, and the Internet," in *Proceedings of the 33rd Annual ACM Symposium on the Theory of Computing*, New York: ACM.

[525] Papadimitriou, C.H., and T. Roughgarden (2005) "Computing Equilibria in Multi-Player Games," in *Proceedings of the Sixteenth Annual ACM-SIAM Symposium on Discrete Algorithms*, New York: ACM.

[526] Pareto, V. (1896) *Cours d'Economie Politique,* Geneva: Droz.

[527] Park, J., and M.E.J. Newman (2004) "Exact Solution for the Properties of a Clustered Network," arXiv:cond-mat/0412579.

[528] Pasini, G., P. Pin, and S. Weidenholzer (2007) "A Network Model of Price Dispersion: Theory and Evidence," Venice University of Venice, mimeo.

[529] Pastor-Satorras, R., and A. Vespignani (2000) "Epidemic Spreading in Scale-Free Networks," *Physical Review Letters* 86:3200–3203.

[530] ——— (2001) "Epidemic Dynamics and Endemic States in Complex Networks," *Physical Review E* 63:066117.

[531] ——— (2002) "Immunization of Complex Networks" *Physical Review E* 65: 036104.

[532] Pattanaik, P.K., and Y. Xu (2007) "On Measuring Personal Connections and the Extent of Social Networks," *Analyse und Kritik* 26:290–310.

[533] Pattison, P.E. (1993) *Algebraic Models for Social Networks,* Cambridge: Cambridge University Press.

[534] Pattison, P.E., and S. Wasserman (1999) "Logit Models and Logistic Regressions for Social Networks: II. Multivariate Relations," *British Journal of Mathematical and Statistical Psychology* 52:169–193.

[535] Pellizzari, M. (2004) "Do Friends and Relatives Really Help in Getting a Job?" Centre of Economic Research Discussion Paper 623, London, London School of Economics.

[536] Pemantle, R., and B. Skyrms (2004) "Network Formation by Reinforcement Learning: The Long and Medium Run," *Mathematical Social Sciences* 48:315–327.

[537] ——— (2004) "Time to Absorption in Discounted Reinforcement Models," *Stochastic Processes and Their Applications* 109:1–12.

[538] Pennock, D.M., G.W. Flake, S. Lawrence, E.J. Glover, and C.L. Giles (2002) "Winners Don't Take All: Characterizing the Competition for Links on the Web," *Proceedings of the National Academy of Sciences, U.S.A.* 99(8):5207–5211.

[539] Podolny, J.M. (1994) "Market Uncertainty and the Social Character of Economic Exchange," *Administrative Science Quarterly* 39(3):458–483.

[540] Podolny, J.M., and J.N. Baron (1997) "Resources and Relationships: Social Networks and Mobility in the Workplace," *American Sociological Review* 62(5):673–693.

[541] Polanski, A. (2007) "Bilateral Bargaining in Networks," *Journal of Economic Theory* 134:557–565.

[542] Polanyi, K. (1944) *The Great Transformation: The Political and Economic Origins of Our Time,* New York: Holt Rinehart.

[543] Pool I.S., and M. Kochen (1978/1979) "Contacts and Influence," *Social Networks* 1(5):5–51.

[544] Porter, M.A., P.J. Mucha, M.E.J. Newman, and C.M. Warmbrand (2005) "A Network Analysis of Committees in the U.S. House of Representatives," *Proceedings of the National Academy of Sciences, U.S.A.* 102:7057–7062.

[545] Pothen, A., H. Simon, and K.-P. Liou (1990) "Partitioning Sparse Matrices with Eigenvectors of Graphs," *SIAM Journal on Matrix Analysis and Applications* 11:430–452.

[546] Price, D.J.S. (1965) "Networks of Scientific Papers," *Science* 149:510–515.

[547] ——— (1976) "A General Theory of Bibliometric and Other Cumulative Advantage Processes," *Journal American Society of Information Science* 27:292–306.

[548] Putnam, R. (1995) "Bowling Alone: America's Declining Social Capital," *Journal of Democracy* 6:65–78.

[549] ——— (2000) *Bowling Alone: The Collapse and Revival of American Community,* New York: Simon and Schuster.

[550] Qin, C.-Z. (1996) "Endogenous Formation of Cooperation Structures," *Journal of Economic Theory* 69:218–226.

[551] Rapoport, A. (1953) "Spread of Information through a Population with Socio-Structural Bias: I. Assumption of Transitivity," *Bulletin of Mathematical Biophysics* 15:523–533.

[552] ——— (1953) "Spread of Information through a Population with Socio-Structural Bias: II. Various Models with Partial Transitivity," *Bulletin of Mathematical Biophysics* 15:535–546.

[553] ——— (1957) "A Contribution to the Theory of Random and Biased Nets," *Bulletin of Mathematical Biophysics* 19:257–271.

[554] ——— (1963) "Mathematical Models of Social Interaction," in *Handbook of Mathematical Psychology*, R.D. Luce, R.R. Bush, and E. Galanter, eds., New York: Wiley.

[555] Rauch, J., and A. Cassella, eds. (2001) *Networks and Markets*, New York: Russell Sage Foundation.

[556] Reed, B. (2003) "The Height of a Random Binary Search Tree," *Journal of the ACM* 50(3):306–332.

[557] Rees, A. (1966) "Information Networks in Labor Markets," *American Economic Review* 56:559–566.

[558] Rees, A., and G.P. Shultz (1970) *Workers in an Urban Labor Market*, Chicago: University of Chicago Press.

[559] Reiss, A.J. (1980) "Understanding Changes in Crime Rates," in *Indicators of Crime and Criminal Justice: Quantitative Studies*, Washington, D.C.: Bureau of Justice Statistics.

[560] ——— (1988) "Co-Offending and Criminal Careers," in *Crime and Justice: A Review of Research,* Vol. 10, M. Tonry, ed., Chicago: University of Chicago Press.

[561] Renault, J., and T. Tomala (1998) "Repeated Proximity Games," *International Journal of Game Theory* 27:539–559.

[562] Riedl, A., and A. Ule (2007) "Segregation and Cooperation in Network Formation Experiments," Maastricht, Maastricht University, mimeo.

[563] Rogers, B.W. (2006) "Learning and Status in Social Networks," Ph.D. dissertation, California Institute of Technology, Pasadena.

[564] Rogers, E.M. (1995) *Diffusion of Innovations,* fourth edition, New York: Simon and Schuster.

[565] Rogers, E.M., and E. Rogers (2003) *Diffusion of Innovations,* fifth edition, New York: Free Press.

[566] Romer, P.M. (1986) "Returns and Long-Run Growth," *Journal of Political Economy* 94:1002–1037.

[567] Rosenberg, D., E. Solan, and N. Vieille (2007) "Informational Externalities and Emergence of Consensus," Paris, HEC School of Management, mimeo.

[568] Roth, A., and M. Sotomayor (1989) *Two-Sided Matching,* Econometric Society Monograph 18, Cambridge: Cambridge University Press.

[569] Rothschild, M., and J. Stiglitz (1970) "Increasing Risk: I. A Definition, "*Journal of Economic Theory* 2:225–243.

[570] Roughgarden, T. (2008) "Computing Equilibria: A Computational Complexity Perspective," Stanford, Calif., Stanford University, mimeo.

[571] Roughgarden, T., and É. Tardos (2002) "How Bad Is Selfish Routing?" *Journal of the ACM* 49(2):236–259.

[572] Rozenfeld, A.F., R. Cohen, D. ben Avraham, and S. Havlin (2002) "Scale-Free Networks on Lattices," *Physical Review Letters* 89:218701.

[573] Ryan, B., and N.C. Gross (1943) "The Diffusion of Hybrid Seed Corn in Two Iowa Communities," *Rural Sociology* 8:15–24.

[574] Sander, L.M., C.P. Warren, I.M. Sokolov, C. Simon, and J. Koopman (2002) "Percolation on Heterogeneous Networks as a Model for Epidemics," *Mathematical Biosciences* 180:293–305.

[575] Santamaria-Garcia, J. (2003) "Gathering Information through Social Contacts: An Empirical Analysis of Labor Market Outcomes," Alicante, Spain, University of Alicante, mimeo.

[576] Sarangi, S., R. Kannan, and L. Ray (2003) "The Structure of Information Networks," *Economic Theory,* forthcoming.

[577] Sattenspiel, L., and C.P. Simon (1988) "The Spread and Persistence of Infectious Diseases in Structured Populations," *Mathematical Biosciences* 90:341–366.

[578] Schelling, T.C. (1969) "Models of Segregation," *American Economic Review (Papers and Proceedings)* 59(2):488–493.

[579] ——— (1971) "Dynamic Models of Segregation," *Journal of Mathematical Sociology* 1:143–186.

[580] ——— (1978) *Micromotives and Macrobehavior,* New York: W.W. Norton.

[581] Schmeidler, D. (1969) "The Nucleolus of a Characteristic Function Game," *SIAM Journal of Applied Math* 17:1163–1170.

[582] Schwartz, J.E. (1977) "An Examination of CONCOR and Related Methods for Blocking Sociometric Data," *Sociological Methodology* 8:255–282.

[583] Schweitzer, S.O., and R.E. Smith (1974) "The Persistence of the Discouraged Worker Effect," *Industrial and Labor Relations Review* 27:249–260.

[584] Scott, J. (1988) "Social Network Analysis," *Sociology* 22(1):109–127.

[585] Seeley, J.R. (1949) "The Net Reciprocal Influence: A Problem in Treating Sociometric Data," *Canadian Journal of Psychology* 3:234–240.

[586] Selten, R. (1975) "Reexamination of the Perfectness Concept for Equilibrium Points in Extensive Games," *International Journal of Game Theory* 4:25–55.

[587] Seneta, E. (1973) *Non-Negative Matrices,* New York: John Wiley and Sons.

[588] Serrano, M.A., and M. Boguñá (2003) "Topology of the World Trade Web," *Physical Review E* 68:01510(R).

[589] ——— (2006) "Percolation and Epidemic Thresholds in Clustered Networks," *Physical Review Letters* 97:088701.

[590] Shachter, R.D. (1986) "Evaluating Influence Diagrams," *Operations Research* 34:871–882.

[591] Shapley, L.S. (1953) "A Value for n-Person Games," in *Contributions to the Theory of Games,* Vol. 2, H.W. Kuhn and A.W. Tucker, eds., Annals of Mathematical Studies 28, Princeton, N.J.: Princeton University Press.

[592] Shrum, W., N.H. Cheek, Jr., and S.M. Hunter (1988) "Friendship in School: Gender and Racial Homophily," *Sociology of Education* 61:227–239.

[593] Silverman, G. (2001) *The Secrets of Word-of-Mouth Marketing: How to Trigger Exponential Sales through Runaway Word of Mouth,* New York: Amacom Books.

[594] Simmel, G. (1908) *Sociology: Investigations on the Forms of Sociation,* Berlin: Duncker and Humblot.

[595] Simon, H. (1955) "On a Class of Skew Distribution Functions," *Biometrika* 42(3–4):425–440.

[596] Skyrms, B., and R. Pemantle (2000) "A Dynamic Model of Social Network Formation," *Proceedings of the National Academy of Sciences, U.S.A.* 97:9340–9346.

[597] ——— (2004) "Learning to Network," in *Probability in Science,* E. Eells and J. Fetzer, eds., New York: Open Court.

[598] Slikker, M. (2000) *Decision Making and Cooperation Structures,* CentER Dissertation Series, Tilburg, The Netherlands: Tilburg University.

[599] ——— (2005) "Endogenously Arising Network Allocation Rules," Eindhoven, The Netherlands, Universiteit Eindhoven, mimeo.

[600] Slikker, M., and A. van den Nouweland (2000) "Network Formation Models with Costs for Establishing Links," *Review of Economic Design* 5:333–362.

[601] ——— (2001) *Social and Economic Networks in Cooperative Game Theory*, Amsterdam: Kluwer.

[602] ——— (2001) "A One-Stage Model of Link Formation and Payoff Division," *Games and Economic Behavior* 34:153–175.

[603] Slikker, M., B. Dutta, A. van den Nouweland, and S. Tijs (2000) "Potential Maximizers and Network Formation," *Mathematical Social Sciences* 39:55–70.

[604] Slikker, M., R.P. Gilles, H. Norde, and S. Tijs (2005) "Directed Networks, Allocation Properties and Hierarchy Formation," *Mathematical Social Sciences* 49:55–80.

[605] Smith, A. (1776) *An Inquiry into the Nature and Causes of the Wealth of Nations*, London: Strahan and Cadell Publishers.

[606] Snijders, T.A.B. (2001) "The Statistical Evaluation of Social Network Dynamics," in *Sociological Methodology*, M.E. Sobel and M.P. Becker, eds., Boston and London: Basil Blackwell.

[607] ——— (2002) "Markov Chain Monte Carlo Estimation of Exponential Random Graph Models," *Journal of Social Structure* 3(2):2–40.

[608] Snijders, T.A.B., and K. Nowicki (1997) "Estimation and Prediction for Stochastic Block Models for Graphs with Latent Block Structure," *Journal of Classification* 14:75–100.

[609] Snijders, T.A.B., C.E.G. Steglich, and M. Schweinberger (2007) "Modeling the Coevolution of Networks and Behavior," in *Longitudinal Models in the Behavioral and Related Sciences*, K. van Montfort, H. Oud, and A. Satorra, eds., New York: Routledge.

[610] Sobel, J. (2002) "Can We Trust Social Capital?" *Journal of Economic Literature* 40: 139–154.

[611] Solomonoff, R., and A. Rapoport (1951) "Connectivity of Random Nets," *Bulletin of Mathematical Biophysics* 13:107–117.

[612] Song, H., and V. Vannetelbosch (2005) "International R and D Collaboration Networks," Louvain, Belgium, Université catholique de Louvain, mimeo.

[613] Starr, R.M., and M.B. Stinchcombe (1992) "Efficient Transportation Routing and Natural Monopoly in the Airline Industry: An Economic Analysis of Hub-Spoke and Related Systems," Economics Department Working Paper 92-25, San Diego, University of California at San Diego.

[614] ——— (1999) "Exchange in a Network of Trading Posts," in *Markets, Information and Uncertainty*, G. Chichilnisky, ed., Cambridge: Cambridge University Press.

[615] Stauffer, D., and A. Aharony (1994) *Introduction to Percolation Theory*, second edition, CRC Press (online).

[616] Stole, L., and J. Zwiebel (1996) "Intra-Firm Bargaining under Non-Binding Constraints," *Review of Economic Studies* 63:375–410.

[617] Strang, D., and S.A. Soule (1998) "Diffusion in Organizations and Social Movements: From Hybrid Corn to Poison Pills," *Annual Review of Sociology* 24:265–290.

[618] Sundararajan, A. (2007) "Local Network Effects and Network Structure," *BE Journal of Theoretical Economics* 71(1):article 46.

[619] Tarde, G. (1903) *Les Lois de l'Imitation: Étude Sociologique*, Boston: Adamant Media Corporation.

[620] Tardos, É., and T. Wexler (2007) "Network Formation Games and the Potential Function Method," in *Algorithmic Game Theory*, N. Nisan, T. Roughgarden, É. Tardos, and V. Vazirani, eds., Cambridge: Cambridge University Press.

[621] Tassier, T. (2005) "A Markov Model of Referral Based Hiring and Workplace Segregation," *Journal of Mathematical Sociology* 29:233–262.

[622] Tassier, T., and F. Menczer (2007) "Social Network Structure, Equality and Segregation in a Labor Market with Referral Hiring," *Journal of Economic Behavior and Organization,* forthcoming.

[623] Tercieux, O., and V. Vannetelbosch (2006) "A Characterization of Stochastically Stable Networks," *International Journal of Game Theory* 34(3):351–369.

[624] Tesfatsion, L. (1997) "A Trade Network Game with Endogenous Partner Selection," in *Computational Approaches to Economic Problems*, H. Amman and B. Rustem, eds., Amsterdam: Kluwer Academic.

[625] ——— (2002) "Agent-Based Computational Economics: Growing Economies from the Bottom Up," *Artificial Life* 8(1):55–82.

[626] Tesfatsion, L., and K.L. Judd (2006) *Handbook of Agent-Based Computational Modeling*, Amsterdam: Elsevier/North Holland.

[627] Thibaut, J., and H.H. Kelley (1959) *The Social Psychology of Groups*, New York: Wiley.

[628] Thomson, W. (2003) "Axiomatic and Game-Theoretic Analysis of Bankruptcy and Taxation Problems: A Survey," *Mathematical Social Sciences* 45:249–297.

[629] Topa, G. (2001) "Social Interactions, Local Spillovers and Unemployment," *Review of Economic Studies* 68:261–296.

[630] Topkis, D.M. (1998) *Supermodularity and Complementarity*, Princeton N.J.: Princeton University Press.

[631] Ule, A., and A. Riedl (2004) "Cooperation in Network Formation Games," Amsterdam, Center for Research in Experimental Economics and Political Decision-Making (CREED), mimeo.

[632] Uzzi, B. (1996) "The Sources and Consequences of Embeddedness for the Economic Performance of Organizations: The Network Effect," *American Sociological Review* 61:674–698.

[633] van den Brink, R., and R.P. Gilles (2000) "Measuring Domination in Directed Networks," *Social Networks* 22:141–157.

[634] van den Bulte, C., and G.L. Lilien (2001) "Medical Innovation Revisited: Social Contagion versus Marketing Effort," *American Journal of Sociology* 106(5):1409–1435.

[635] van den Nouweland, A. (2004) "Static Networks and Coalition Formation," in *Group Formation in Economics: Networks, Clubs, and Coalitions*, G. Demange and M. Wooders, eds., Cambridge: Cambridge University Press.

[636] van Zandt, T., and X. Vives (2005) "Monotone Equilibria in Bayesian Games of Strategic Complementarities," *Journal of Economic Theory* 134(1):339–360.

[637] Vázquez, A. (2003) "Growing Network with Local Rules: Preferential Attachment, Clustering Hierarchy, and Degree Correlations," *Physical Review E* 67(5):056104.

[638] Vázquez, A., and M. Weigt (2003) "Computational Complexity Arising from Degree Correlations in Networks," *Physical Review E* 67:027101.

[639] Vega-Redondo, F. (2006) "Building Up Social Capital in a Changing World," *Journal of Economic Dynamics and Control* 30(11):2305–2338.

[640] ——— (2007) *Complex Social Networks,* Cambridge: Cambridge University Press.

[641] Vergara Caffarelli, F. (2004) "Non-Cooperative Network Formation with Network Maintenance Costs," European University Institute Working Paper 2004/18, San Domenico di Fiesole, Italy.

[642] Vives, X. (1990) "Nash Equilibrium with Strategic Complementarities," *Journal of Mathematical Economics* 19:305–321.

[643] Vivier-Lirimont, S. (2004) "Interbanking Networks: Towards a Small Financial World? Bonn, Institute for the Study of Labor (IZA), mimeo.

[644] Wahba, J., and Y. Zenou (2005) "Density, Social Networks and Job Search Methods: Theory and Application to Egypt," *Journal of Development Economics* 78:443–473.

[645] Walker, G., B. Kogut, and W. Shan (1997) "Social Capital, Structural Holes and the Formation of an Industry Network," *Organization Science* 8(2):109–125.

[646] Walker, M., and J. Wooders (2001) "Minimax Play at Wimbledon," *American Economic Review* 91(5):1521–1538.

[647] Wang, P., and A. Watts (2002) "Formation of Buyer-Seller Trade Networks in a Quality-Differentiated Product Market," State College, Penn., and Carbondale, Ill., Penn State University and Southern Illinois University, mimeo.

[648] Wang, P., and Q. Wen (1998) "Network Bargaining," State College, Penn., Penn State University, mimeo.

[649] Warren, C.P., L.M. Sander, and I. Sokolov (2002) "Geography in a Scale-Free Network Model," *Physical Review E* 66:056105.

[650] Wasserman, S., and K. Faust (1994) *Social Network Analysis: Methods and Applications,* Cambridge: Cambridge University Press.

[651] Wasserman, S., and P. Pattison (1996) "Logit Models and Logistic Regressions for Social Networks: I. An Introduction to Markov Graphs and $P*$," *Psychometrika* 61:401–425.

[652] Watts, A. (2001) "A Dynamic Model of Network Formation," *Games and Economic Behavior* 34:331–341.

[653] ———— (2002) "Non-Myopic Formation of Circle Networks," *Economics Letters* 74:277–282.

[654] ———— (2007) "Formation of Segregated and Integrated Groups," *International Journal of Game Theory* 35:505–519.

[655] Watts, D.J. (1999) *Small Worlds: The Dynamics of Networks between Order and Randomness,* Princton, N.J.: Princeton University Press.

[656] ———— (2002) "A Simple Model of Global Cascades on Random Networks," *Proceedings of the National Academy of Sciences, U.S.A.* 99:5766–5771.

[657] ———— (2004) "The New Science of Networks," *Annual Sociological Review* 30:243–270.

[658] Watts, D.J., and S. Strogatz (1998) "Collective Dynamics of 'Small-World' Networks," *Nature* 393:440–442.

[659] Watts, D.J., P.S. Dodds, and M.E.J. Newman (2002) "Identity and Search in Social Networks," *Science* 296:1302–1305.

[660] Watts, D.J., R. Muhamad, D.C. Medina, and P.S. Dodds (2005) "Multiscale, Resurgent Epidemics in a Hierarchical Metapopulation Model," *Proceedings of the National Academy of Sciences, U.S.A.* 102:11157–11162.

[661] Weisbuch, G., A. Kirman, and D. Herreiner (2000) "Market Organization," *Economica* 110:411–436.

[662] Wellman, B., and S.D. Berkowitz, eds. (1988) *Social Structures: A Network Approach,* New York: Cambridge University Press.

[663] White, H.C., S.A. Boorman, and R.L. Breiger (1976) "Social Structure from Multiple Networks. I. Blockmodels of Roles and Positions," *American Journal of Sociology* 81(4):730–780.

[664] Williamson, O. (1983) *Markets and Hierarchies: Analysis and Antitrust Implications*, New York: Free Press.

[665] Willis, J.C., and G.U. Yule (1922) "Statistics of Evolution and Geographical Distribution in Plants and Animals, and Their Significance," *Nature* 109:177–179.

[666] Woodward, C.J., and D.C. Parkes (2004) "Strategy-Proof Mechanisms for Ad Hoc Network Formation," Cambridge, Mass., Harvard University, mimeo.

[667] Wormald, N.C. (1984) "Generating Random Regular Graphs," *Journal of Algorithms* 5:247–280.

[668] Young, H.P. (1993) "The Evolution of Conventions," *Econometrica* 61:57–84.

[669]   —— (1998) *Individual Strategy and Social Structure,* Princeton, N.J.: Princeton University Press.

[670]   —— (2003) "The Diffusion of Innovations on Social Networks," in *The Economy as a Complex Evolving System,* Vol. 3, L.E. Blume and S. Durlauf, eds., Oxford: Oxford University Press.

[671]   Young, P. (2007) "Innovation Diffusion in Heterogeneous Populations: Contagion, Social Influence, and Social Learning," Oxford, Nuffield College, mimeo.

[672]   Yule, G. (1925) "A Mathematical Theory of Evolution Based on the Conclusions of Dr. J.C. Willis," *F.R.S. Philosophical Transactions of the Royal Society of London B* 213:21–87.

[673]   Zachary, W.W. (1977) "An Information Flow Model for Conflict and Fission in Small Groups," *Journal of Anthropological Research* 33(4):452–473.

[674]   Zheng, T., M.J. Salganik, and A. Gelman (2006) "'How Many People Do You Know in Prison?' Using Overdispersion in Count Data to Estimate Social Structure in Networks," *Journal of the American Statistical Association* 101:409–423.

[675]   Zipf, G. (1949) *Human Behavior and the Principle of Least Effort,* Cambridge, Mass.: Addison-Wesley.

Index

Page numbers for entries occurring in figures are followed by an *f;* those for entries occurring in notes, by an *n;* and those for entries occurring in tables, by a *t.*